COLOR ATLAS and MANUAL of

MICROSCOPY for CRIMINALISTS, CHEMISTS, and CONSERVATORS

COLOR ATLAS and MANUAL of

MICROSCOPY for CRIMINALISTS, CHEMISTS, and CONSERVATORS

NICHOLAS PETRACO
THOMAS KUBIC

Boca Raton London New York Washington, D.C.

Library of Congress Cataloging-in-Publication Data

Petraco, Nicholas.

Color atlas and manual of microscopy for criminalists, chemists, and conservators / Nicholas Petraco, Thomas Kubic.

p. cm.
Includes bibliographical references and index.
ISBN 0-8493-1245-0
1. Microscopy—Atlases. 2. Microscopy—Handbooks, manuals, etc. 3. Forensic sciences—Atlases. 4. Forensic sciences—Handbooks, manuals, etc. I. Kubic, Thomas. II. Title.

QH205.2.P486 2003
363.25′6—dc22 2003061051

Direct all inquiries to CRC Press LLC, 2000 N.W. Corporate Blvd., Boca Raton, Florida 33431.

International Standard Book Number 0-8493-1245-0
Library of Congress Card Number 2003061051
Printed in the United States of America 1 2 3 4 5 6 7 8 9 0
Printed on acid-free paper

Dedication

We dedicate this work to the memory of Dr. Walter McCrone, whose tireless energy, availability, research, and teaching of the uses of polarized light microscopy served to convert, stimulate, and inspire these two disciples.

Acknowledgments

The authors would like to thank the many people who made this work possible. First, we want to acknowledge and thank Criminalist III Mary Eng who supervises the Hair and Fiber Unit and Criminalist III Lisa Faber for their diligent editing of this manuscript and invaluable suggestions and contributions. We also thank members of the New York City Forensic Investigation Division, in particular, Criminalist IV Judy O'Connor, Criminalist III Valerie Wade Alisan, Melanie McMillian, and Melvin Shaw for their contributions.

We also want to convey our thanks to the faculty at John Jay College of Criminal Justice, especially President Gerald Lynch, Provost Basil Wilson, and Assistant Provost Lawrence Kobilinsky for their kind support. In addition, we would like to acknowledge and thank all the members of John Jay's science department, especially Dr. Selman Burger, chairperson of the science department; Dr. Charles Kingston, professor emeritus and former science chairperson; Dr. Peter R. Deforest, professor of criminalistics; and Dr. Alexander Josephs, who created the forensic science program at John Jay College. Finally, we especially thank Dr. John A. Reffner for endlessly sharing his enormous knowledge and experience with polarized light and analytical microscopy techniques.

We thank our parents for their love and nurturing without which we would not be here. Most of all, we thank MaryBeth and Lynn, and our children (Dr. Nick, John, M.J., Alex, Georgann, and Ana) for their patience, love, and unwavering support in good times and bad times.

Authors

Nicholas Petraco earned a B.S. in chemistry and an M.S. in forensic science from John Jay College of the City University of New York. He served as a detective and criminalist at New York City's Police Laboratory from 1968 to 1990 and held the position of senior forensic microscopist of the laboratory's trace section between 1982 and 1990, when he became a private forensic consultant. In addition to participating in the professional activities listed below, Petraco became the first contracted consultant in criminalistics and trace evidence engaged by the New York City Police Department in 1999. He continues to provide consultant services to the department.

Throughout his career, he also served as an adjunct lecturer at John Jay College and an associate professor at St. John's University, educated hundreds of forensic scientists and art conservators in light microscopy, worked on more than 4000 death investigations on behalf of prosecution and defense attorneys, and testified as an expert in more than 400 trials conducted in local, state, and federal courts.

Petraco is a fellow of the New York Microscopical Society (FNYMS), fellow of the American Academy of Forensic Scientists (FAAFS), diplomate of the American Board of Criminalistics (DABC), and a member of the Committee on Forensic Hair Comparisons. He served as chairperson of the SWGMAT forensic hair committee from 1997 to 2001.

He has published extensively in the forensic literature, and has authored and co-authored six book chapters and several books and CDs about forensic science and microscopy.

Thomas A. Kubic is currently an instructor in forensic chemical instrumentation, scientific and expert testimony, electron microscopy, and advanced trace-evidence analysis at John Jay College of Criminal Justice of the City University of New York. He is also a director of TAKA, a New York not-for-profit educational institution concentrating on training in microscopy.

Kubic earned an M.S. in chemistry from Long Island University and a law degree from St. John's University. In addition, he earned his Ph.D. in forensic science at City University of New York. He received the Paul L. Kirk award (1997) from the Criminalistics Section of the American Academy of Forensic Sciences for his contributions to criminalistics and forensic science. The Graduate School of the City University of New York awarded him the Arthur Neiderhoffer Scholarship for significant contributions to the field of criminal justice.

Kubic retired from a municipal crime laboratory where he spent more than 20 years as a forensic microscopist. He also served as director of a New York State accredited environmental laboratory and as an instructor of a number of microscopy and industrial-hygiene courses offered through Rutgers University's continuing education program. He maintains a successful consulting company specializing in the identification and characterization of particulate materials by light and electron microscopy in cases of forensic import. He is the author of 20 scientific and technical articles and a number of textbook chapters and has made more than 50 presentations at technical conferences.

Kubic is a member of the FBI-sponsored SWGMAT committee on forensic glass analysis, and is a technical expert in NIST's laboratory accreditation program (NVLAP). His research interests include the application of light and electron microscopy to the analysis of particulates and microscopic forensic evidence and assisting graduate students in research at John Jay. He is a fellow of the American Academy of Forensic Sciences, a certified criminalist, and holds memberships in the New York Microscopical Society, the Microscopy Society of America, the Microbeam Analysis Society, the American Chemical Society, and other professional organizations.

Table of Contents

1 Basic Light Microscopy

INTRODUCTION

The ability to identify and characterize a wide variety of materials rapidly and accurately has long been a desirable goal in the disciplines of forensic science, chemistry, art history, and art, architectural, and archeological conservation. Practitioners of these disciplines often must identify a diverse range of the materials used in everyday life in the past and in modern times.

Forensic scientists frequently encounter many different types of materials during criminal investigations. The identification and comparison of miniscule traces of a wide range of diverse materials such as synthetic and natural fibers, wood fragments, paint chips, minerals, feathers, pollen grains, hairs, and so on can be crucial elements in the reconstruction of an event.[1–15] Chemists are often asked to identify tiny samples of unknown substances.[16–18] Art historians must know from what material works of art are made before they can render opinions about their true origins. Art conservators are asked to repair, restore, and preserve rare paintings, ancient tapestries, old textiles, statuary, and historic buildings.[19–44] Before they can do so, they must first ascertain the composition of such *objects d'art* in order to properly refurbish and conserve them. Often the people working at these disciplines are not analytical chemists. Consequently, swift, simple, and irrefutable analytical tests and procedures that can be used on minute specimens have been in great demand for centuries.

When researchers peruse the available literature, the vast number of proven and unproven analytical and microscopic techniques they encounter appear overwhelming. They soon learn that although methods and/or techniques that will help solve their analytical problems exist, the equipment and required test materials are often difficult, if not downright impossible, to obtain. They may find that the methods and procedures are spread around the literatures of many different disciplines, and descriptions of tests or articles may be difficult to obtain or unavailable. Procedures may be written in a specific technical jargon that makes them, problematical for the average nonchemist to interpret and understand. In addition, the availability of certain required reagents and equipment is often questionable or unknown. Finally, they soon learn that few reference texts present clear, concise, and simple rationales they can use to solve their analytical problems. Consequently, they learn that they must devise their own procedures by drawing from the vast jumble of literature to solve a particular analytical problem.

During the authors' tenure as forensic microscopists, much of their work involved the study of all types of microscopic specimens. The data obtained from their microanalyses were often used by them to associate people, places, and things involved in crimes; deduce the occupations of principals involved in crimes; and reconstruct the crime scene and/or the crime.[45–46] As private consultants, the authors have had numerous opportunities to assist analytical chemists, art historians, and art conservators in their analyses of specimens. Table 1.1 is a partial listing of the materials the authors have been asked to characterize, identify, and compare.

Polarized light microscopy (PLM) has been the method of choice for microscopists concerned with identifying many of the substances depicted in Table 1.1. For nearly two centuries, professionals from various disciplines have used PLM to solve their microanalytical quandaries. Geologists, chemists, and material scientists solidly established PLM as the initial method of choice for characterizing, studying, and identifying a broad range of microscopic-sized bits of matter. In recent years, art historians, art conservators, archeologists, architects, forensic chemists, environmental chemists, and gemologists, to name only a few, have joined the ranks of professions that use PLM methodologies in their work. This phenomenon spawned the need for a text short on theory and long on basic fundamental techniques. The primary goal of this text is to fill the gap in the current literature by providing a working manual that demonstrates simple, concise rationales that can be used by members of all these disciplines to characterize and identify broad ranges of materials, thereby enabling them to solve many of their analytical problems.

The two fundamental types of compound microscopes (CMs) used in identification microscopy are the polarized light microscope (PLM) and the stereomicroscope (SM). The CM acquired its name because the image it produces is formed in two stages. The primary image of the specimen formed by the objective is magnified by the ocular to generate the final image. Much has been written about the theories of light, lenses, geometric optics, and image formation in the CM. Thus, only a nomenclature and brief explanations of theory will be given in this chapter. Stereomicroscopy will be discussed briefly in Chapter 2. When appropriate, additional concepts will be introduced and explained throughout this text. Where appropriate, interested readers will be referred to further literature.

TABLE 1.1
Materials Encountered in Casework

Hairs	Fibers	Inorganic Compounds	Organic Compounds	Paints/Pigments
Human	Synthetic fibers	Salts	Explosives	Resins
Head	Acetate	Sodium chloride, nitrate, hydroxide, carbonate, and bicarbonate	RDX	Acrylic
Pubic	Acrylic	Potassium nitrate, nitrate, chloride, perchlorate	TNT	Alkyd
Limb	Aramid	Ammonium nitrate	PETN	Urethane
Body	Modacrylic	Calcium carbonate	EGDN	Linseed
Facial	Nylon	Silver sulfide	NG	Vegetable oils
	Olefin		Smokeless powder	Latexes
	Polyester			Casein
Animal	Rayon	Minerals and mineral-like materials	Drugs	Albumin
Cat	Spandex	Quartz	Cocaine	Animal glues
Wools	Foil	Feldspar	Crack	Damar
Llama	Fiberglass	Dolomite	Heroin	Copal
Vicuña	Rock wool	Calcite	Opiates	Waxes
Alpaca	Slag wool	Gypsum	Marijuana, hashish	Gum arabic
Bear	Carbon	Muscovite	Procaine, lidocaine	Lacquers
Beaver		Biotite	Amphetamines	Starch paste
Bison	Natural fibers	Phlogophite	PCP	
Camel	Cotton	Garnet	Barbiturates	Extenders
Cow	Hemp	Diatoms	LSD	Mica, barium sulfate
Coyote	Jute	Tourmaline	Starches	
Deer	Flax	Hornblende	Aspirin	Pigments
Dog	Linen	Chlorites	Vitamin C	White: chalk, whiting, lead white, zinc white, titanium white
Fox	Abaca	Olivine	Sugars	Yellow: chrome yellow, ochre, Naples yellow, cadmium yellow, litharge, lemon yellow, orpiment
Hare	Sisal	Limonite	Caffeine	Red: realgar, ochre, cadmium red, rose madder, carmine, cinnabar (vermilion)
Hog	Ramie	Limestone		Orange: cadmium orange, chrome orange
Horse	Feathers	Magnetite	Biological materials	Green: chrome green, viridian, terre verte, verdigris, chrome oxide
Mink	Paper	Iron oxides	Epithelial cells	Blue: indigo, Prussian blue, Cerulean blue, cobalt blue, lapis lazuli (ultramarine), small
Mouse	Wood	Obsidian	Blood, blood cells	Brown: raw and burnt umber, raw and burnt Sienna, Van Dyke brown
Muskrat	Webs	Glasses	Seeds, twigs, leaves	Black: Charcoal, carbon black, Mars black, ivory black
Rat		Sands	Plant parts	
Sable		Slate	Spices and herbs	
Seal		Talc	Algae, fungi	
Squirrel	Minerals	Zircon	Mold spores	
	Asbestos	Perlite	Pollens	
	Quartz wool	Pumice	Trichomes	
	Talc	Cinders, coal, bricks	Tobacco dust	
		Concretes and mortars	Sawdust	
		Asphalt	Insects and insect parts	
		Metal fragments	Shrimp, clams, mussels	
		Iron, copper, lead	Bone fragments	
		Silver, gold, aluminum		
		Zinc		

BASIC PARTS OF COMPOUND MICROSCOPE WITH EMPHASIS ON PLM

A CM is composed of certain basic elements: an illuminator providing light that illuminates the specimen (Figure 1.1); a substage condenser controlling the contrast in the image and brightness of illumination (Figure 1.2); a stage (S) that holds the specimen (Figure 1.3); an objective lens (OL) that forms the primary image (PI) of the specimen (Figure 1.4); and an ocular or eyepiece (OC) that acts as a simple magnifying lens and magnifies the PI formed by the OL (Figure 1.5).

The essential difference between the common brightfield microscope (BFM) and PLM — two common forms of CM — is that the PLM has a condenser with a removable auxiliary top lens, rotatable and centerable circular stage, compensation slot (CPS), Bertrand lens (BL), and two polarizing light filters (one in the condenser [C], called the polarizer, and one between the OL and OC, called the analyzer) as shown in Figure 1.6.

FUNDAMENTAL COMPOUND MICROSCOPE THEORY WITH EMPHASIS ON PLM

Image formation in a CM involves the interaction of light and matter. Polychromatic (having many colors) or white light vibrating in all directions passes through the condenser lens (CL) and field diaphragm (FD) found in the illuminator (I). The FD controls the area of the specimen illuminated. The light from the I passes into the substage condenser (C). The C-aperture diaphragm (AD) controls the angle of the light interacting with the specimen. The undeviated light and the light diffracted and refracted by the specimen are collected by the objective lens (OL). See Figure 1.7.

In the PLM, white light from the illuminator is passed through a light-polarizing filter and AD found in the condenser. The illumination that interacts with the specimen is now plane-polarized, vibrating in an east ↔ west privileged direction (PD). After the specimen is focused using a 10× objective, the PPL and image of the FD (located in the illuminator) are focused by the condenser lens system onto the specimen plane (see Figure 1.7). Figure 1.8 depicts a typical PLM illumination system configuration.

The front lens of an objective collects light coming from the specimen. The quantity of image-forming light collected by the objective is related to the focal length of the objective, the refractive index (n) of the medium between the specimen and the front lens of the objective, and the numerical aperture (NA) of the objective. The larger the NA of an objective, the greater its ability to collect light. The numerical aperture is defined mathematically as:

$$\text{NA} = n \sin \tfrac{1}{2}\text{AA}$$

The angle aperture (AA) varies with the focal length of the OL system. The light-gathering capacity of an objective and its ability to resolve fine detail in a specimen are directly related to the AA of an objective. The resolving power of a compound microscope is also dependent on the condenser NA. Consequently, the approximate resolving power (R) of a microscope can be defined mathematically as:

$$R = \frac{\lambda}{\text{NA objective} + \text{NA condenser}}$$

where R is the smallest distance between two side-by-side structural details in a specimen that can be observed as distinct structures and λ is the wavelength of the interfering light. The higher the useful NA of a system, the greater its resolving power. Under ideal conditions, the diffraction pattern that forms the primary image of the specimen can be seen at the back focal plane of the objective as seen in Figure 1.9. Clear, detailed explanations of basic CM and PLM theory can be found in several texts.[47–52]

The PI of the specimen formed by the OL is projected to the intermediate image plane (IMP) of the OC. At this position along the optic axis, the image of the specimen is both real (can be projected) and inverted (reversed). When the illumination is set according to Köhler's instructions, the image of the FD and specimen are in focus at the intermediate image plane (IMP) of the OC, and the image of the AD and filament are in focus at the back focal plane of the objective (BFO). An excellent explanation and series of illustrations of how to set up Köhler illumination is presented by Delly.[53] When the PI of the specimen magnified by the ocular is viewed, the observer can perceive an enlarged, virtual, and inverted image of the specimen that appears to be below the condenser, approximately 10 inches from the observer, as shown in Figure 1.10. The figure also depicts the positions of the AD&SP and FD&SP conjugate foci along the OA. Note that each image of AD&F is alternated with an image of FD&SP.

FUNDAMENTAL POLARIZED LIGHT MICROSCOPE THEORY

A PLM is essentially a CM with two polarizers placed along its optic axis. The value of a PLM rests in its ability to supply additional data about a specimen's nature not obtainable with other forms of CM. In addition to the physical properties (color, shape, size) visible with most forms of CM, PLM enables an analyst to acquire data concerning optical properties. Is a material isotropic or anisotropic with respect to its optical properties? Is a colored specimen pleochroic? What is a substance's birefringence? These are only a few of the questions a PLM can help an analyst answer in the quest to characterize and identify many types of materials.

Figure 1.11 illustrates the ease with which PLM can provide additional information and identify a specimen. Before continuing with PLM theory, a brief explanation of the behavior of light traveling through transparent materials is in order.

White light contains all the colors of the rainbow (red, orange, yellow, green, blue, indigo, violet or *roy-g-biv*). When white light traveling from a more optically dense transparent medium (TM) at 90° to its surface travels into a less optically dense TM, it changes speed while traveling straight through the material. It does not separate into the component colors of white light. When white light is incident on the surface of a more optically dense TM at any angle other than 90°, it bends toward the normal to the surface of the TM. The phenomenon of light bending as it passes from one transparent medium into another is known as refraction.

Figure 1.12 depicts white light entering a glass prism while traveling through the air. The angle of incidence (I) is approximately 42°; the refracted light rays bend toward the normal. The white light disperses (separates) into its monochromatic (one color) component wavelengths because each color of monochromatic light actually travels at a slightly different speed and bends at a different angle. Red light travels at the slowest rate of speed and is refracted (bent) least, while violet light travels at the fastest speed and is refracted most. The degree to which light refracts (refractive index or RI) can be defined mathematically as the ratio of the speed of light in a vacuum to the ratio of the speed of light in another medium:

$$\text{Refractive Index} = \text{RI or } n_{\text{medium}} = \frac{\text{Speed of Light in a Vacuum}}{\text{Speed of Light in a Medium}}$$

The larger the RI of a medium, the more light bends as it passes through that medium. Since we now know how light behaves as it passes through a transparent medium, we can continue our brief study of PLM theory. Again, interested readers are referred to the literature for detailed explanations of the behavior of light when it interacts with matter.[54–56]

When a transparent specimen possessing only one refractive index is viewed between crossed polarizers (PDs of the analyzer and polarizer are perpendicular), the specimen appears black (extinct) against a black (extinct) background. Specimen appearance remains unchanged when the microscope stage is rotated 360°. A transparent specimen that behaves in this manner is said to be isotropic (having one RI) with respect to its optical properties. Amorphous materials such as glass and some crystalline materials are known to have only one RI. Figure 1.13 (left) demonstrates what occurs when plane-polarized light (PPL) from the condenser (E ↔ W) interacts with an isotropic specimen lying parallel to the PDs. The PPL interferes with the specimen and continues vibrating in the E ↔ W direction after exiting the specimen. The PPL from the specimen is collected by the OL, which forms a PI of the specimen. Next, the light exiting the OL enters the analyzer. The analyzer (having an N ↔ S privileged direction) stops all the PPL coming from the OL vibrating in the E ↔ W direction. Consequently, the specimen and the background both appear black or extinct. When an isotropic specimen is rotated between crossed polarizers, the specimen and background remain extinct in all orientations as seen in Figure 1.13 (right).

Light behaves differently when it passes through a doubly refractive (anisotropic) material. When light normal to a specimen's plane enters a doubly refractive specimen, it is split into two primary rays. The two rays vibrate perpendicularly. The first or fast ray (f) travels through the thickness of the material at a faster rate of speed than the second ray (s), known as the slow ray. When the f and s rays emerge from the top surface of the material, the real, measurable, linear distance between them is known as the retardation distance (R). See Figure 1.14.

A transparent specimen having more than one primary refractive index acts differently from an isotropic substance when viewed between crossed polarizers. If the specimen lies parallel to the PD of the polarizer or analyzer, it behaves as if it were an isotropic material. If, on the other hand, the specimen is rotated into a position other than parallel to one of the polarizers, it will display polarization colors against a black background. As the specimen is rotated a full 360°, it will appear to go black four times. A transparent specimen that behaves in this manner is said to be anisotropic with respect to its optical properties. When a specimen is positioned at any angle other than 90° to the PD of one of the polarizers, it will display polarization (interference) colors (ICs). The distance between the f ray and the s ray (retardation or R) is measured in nanometers (nm). The quantity of retardation depends on the thickness (t) of the specimen measured in micrometers (µm) and the difference between the indices of the fast and slow rays — a nondimensional number, known as birefringence (Bi). When the two rays enter the analyzer, they form a retardation vector (RV) that can be described mathematically:

$$\text{RV} = t \times (s - f) \times 1000 = t\ \mu\text{m} \times (\text{Bi}) \times 1000\ \text{nm}$$

where RV is the retardation vector, t is the specimen thickness, s is the RI of the slow ray, f is the RI of the fast ray, and Bi (birefringence) is the difference between the RI values of the s and f rays. A transparent specimen that behaves in this manner is said to be anisotropic with respect to its optical properties.

Figure 1.15 shows the passage of PPL through a doubly refractive transparent material when a specimen is viewed between crossed polarizers (CPs). At left, normal PPL from the polarizer interacts with the specimen lying parallel to the PD of the polarizer. The light leaving the specimen is still plane-polarized in the E ↔ W PD. It travels up the OA where it is collected by the front lens of the objective. The OL forms a primary image of the specimen that travels up the OA to the analyzer. The light exiting the OL is still plane-polarized, vibrating in the E ↔ W PD. Since the PD of the analyzer is perpendicular (at 90°) to the PD of the polarizer, the analyzer absorbs all the light coming from the OL. Consequently, light from the OL cannot pass through the analyzer. Therefore, the value of the RV formed by the analyzer is zero, and the specimen appears extinct (black) against a black background when observed through the oculars.

The right side of Figure 1.15 shows how PPL from the polarizer, normal to the specimen surface, enters the specimen and is split into two primary rays of light. Both rays are plane-polarized, vibrating at 90° to each other in opposite planes. The slow (s) ray is traveling behind the fast (f) ray; therefore the two rays cannot interfere until they enter the analyzer — a device that analyzes light coming from the OL and forms a retardation vector (RV). Reference to an interference chart (see Figures 1.16 and 1.18) can allow an observer to determine the amount of retardation in nanometers from the interference colors manifested in the specimen image.

When an anisotropic transparent substance is rotated between CPs, it alternates between displaying interference colors and proceeding to extinction every 90°. The brightest colors of anisotropic materials are exhibited when they are viewed 45° off extinction as shown in Figure 1.17.

The birefringence of an anisotropic specimen can be accurately estimated by using an interference chart. As demonstrated in Figure 1.15 (right), a 20 μm thick (t) specimen displays a red/orange interference color (IC) when viewed between crossed polars 45° off extinction. By comparing the displayed IC to an interference chart, the specimen retardation is estimated to be 400 nm. Figure 1.18 contains three lime-colored lines. Line t on the left is drawn along the line representing 20 μm of thickness; on the bottom, line r is drawn straight up the section of the chart displaying 400 nm of retardation; at X (the intersection of lines t and r), a diagonal line Bi is drawn from the origin, through the intersection, until it meets the line representing birefringence. The specimen birefringence is then determined at the point of intersection to be 0.020.

Sometimes it is difficult to interpret interference colors or determine where the specimen IC falls on the Michel–Levy chart, and a device known as a compensator is often employed. The two primary types are fixed compensators and variable compensators. Only fixed compensators will be discussed in this chapter. Variable compensators will be briefly discussed in Chapter 7. A fixed compensator is made from an anisotropic material cut a into disc of exacting thickness so that it can deliver a precise and predictable amount of retardation to the optical system. The quantity of retardation is noted directly on the device. In addition, the disc is usually set into a mount with its slow (Z) ray oriented at 90° to the compensator length. The direction of the slow (s) ray of the compensator is marked directly on the compensator. Several kinds of fixed compensators are available for most PLM systems. The three most popular are the 1/4 wave plate that has a value of approximately 130 nm, the full wave plate that has a value of 525 nm, and the Sénarmont compensator, often used for determining precise values of retardation for fibers exhibiting low-order interference colors. The Sénarmont compensator will be discussed in Chapter 7. Compensators are made to fit snugly into a microscope compensator slot (CPS).

Figure 1.19 depicts the use of a fixed compensator (FCp). In the center, an FCp is added to the optic axis of a PLM. The PPL from the polarizer (P) is not refracted as it passes through the anisotropic specimen. The specimen behaves like an isotropic substance because the incident PPL is traveling normal to the specimen's plane and no retardation occurs. Light from the specimen is collected by objective, which forms a primary image. Light from the objective then passes into the compensator (Cp).

Because the Cp is composed of an anisotropic material, light passing through it is resolved into fast (F) and slow (S) rays that have mutually perpendicular privileged directions and, in this case, a retardation (R) of 525 nm between the F and S rays. Upon entering the analyzer (A), all the PPL vibrating perpendicular to the PD of the A is blocked. Only the light vibrating at a 45° angle to the PD of the A is allowed to pass into the A. At this point, the A analyzes the light entering it and forms a retardation vector (RV) having a value equal to that of the Cp (in this case, 525 nm). The specimen and background display an interference color representing the quantity of retardation due to the utilized compensator (magenta in this case). The left side shows anisotropic substance viewed between CPs 45° off extinction. The specimen exhibits its maximum interference color (IC) of 125 nm. On the right, the specimen is viewed 45° off extinction. Here again, the specimen shows its maximum IC. When retardation due to the specimen and FCp enter the analyzer, RV of 650 nm is formed (125 nm + 525 nm = 650 nm) and the specimen appears blue against a magenta background. The addition of interference colors is demonstrated in Figure 1.20.

Another important concern is whether the compensation is added or subtracted from the resulting retardation vector. The general rule is that when the slow (S) rays of the specimen and the compensator are parallel to each other, additive compensation occurs; when the S rays are perpendicular, subtractive compensation takes place.

Figure 1.21 demonstrates this phenomenon. On the right, the slow (S) rays of the specimen and the compensator are parallel. Additive compensation occurs and a final RV displaying higher order interference colors (IC) is formed (525 nm + 125 nm = 650 nm). On the left, the specimen S ray and the Cp S ray are oriented perpendicularly. Subtractive compensation takes place, and a final RV exhibiting lower order IC is formed (525 nm – 125 nm = 400 nm).

The optical property known as sign of elongation (SE) is another important parameter to determine when characterizing and identifying various anisotropic materials with PLM. The SE is concerned with the orientation of the slow (larger) refractive index in a multiply refractive transparent substance. By convention, if the slow ray (S) or largest refractive index of an anisotropic is oriented along the geometric length of a fiber or crystal, the specimen is said to possess a positive SE (+SE). Conversely, if the S ray or largest refractive index of an anisotropic is aligned with the width of a fiber or crystal, the specimen is said to have negative SE (–SE). Fixed compensators (FCps) are normally utilized in determining the SE of a birefringent sample. The procedure is shown in Figure 1.22.

An elongated anisotropic specimen is viewed between CPs (bottom). The specimen exhibits a low-order white IC of approximately 125 nm (center). Upon viewing the resulting IC, the analyst is left wondering about the orientation of the F and S rays of the material. When a full wave of compensation is added along the PLM's optic axis, the specimen demonstrates subtractive retardation (left). To achieve subtractive retardation, the S rays of the specimen and the FCp must be perpendicular. Therefore, the S ray of the specimen must be oriented along its width and its SE must be negative. Conversely, on the right, another specimen demonstrates additive retardation when a full wave of compensation is added along the PLM's optic axis. Again, to achieve additive retardation, the S rays of the specimen and FCp S must be parallel to each other. This specimen's S ray must be oriented along its length and its SE must be positive.

Finally, definitions for most microscopy terms can be found in a glossary prepared by members of the New York Microscopical Society — oldest microscope society in the U.S. The glossary is an excellent resource that should be consulted whenever the meaning of a microscopical term is in question.[57]

REFERENCES

1. Gross, H., *Criminal Investigation*, adapted from Adams, J.C., *System der Kriminalistik*, London, Sweet & Maxwell, 1924, p. 144.
2. Locard, E., L' analyse des poussieres en criminalistique, *Rev. Int. Criminalistique*, 1, 176, 1929.
3. Locard, E., The analysis of dust traces, *Am. J. Police Sci.*, Parts I–III, 1, 276, 1930.
4. Smith, S. and Glaister, Jr., *Recent Advances in Forensic Medicine*, 2nd ed., Philadelphia, Blakiston's, 1939, p. 118.
5. Kirk, P.L., Microscopic evidence: its use in the investigation of crime, *J. Crim. Law Criminol. Police Sci.*, 40, 362, 1949.
6. Burd, B. and Kirk, P.L., Clothing fibers as evidence, *J. Crim. Law Criminol. Police Sci.*, 32, 353, 1941.
7. Longhetti, A. and Roche, G., Microscopic identification of man-made fibers from the criminalistics point of view, *J. Forens. Sci.*, 3, 303, 1958.
8. Nickolls, L.C., The identification of stains of nonbiological origin, in *Methods of Forensic Science*, Vol. 1, Lundquist, F., Ed., New York, Interscience, 1962, p. 335.
9. Frei-Sulzer, M., Colored fibres in criminal investigation with special reference to natural fibres, in *Methods of Forensic Science*, Vol. 4, Curry, A.S., Ed., New York, Interscience, 1965, p. 141.
10. McCrone, W.C., Particle analysis in the crime laboratory, in *The Particle Atlas*, Vol. 5, McCrone, W.C., Delly, J.G., and Palenik, S.J., Eds., Ann Arbor, MI, Ann Arbor Science Publishers, 1979, p. 1379.
11. Grieve, M.C., The role of fibers in forensic science examinations, *J. Forens. Sci.*, 28, 877, 1983.
12. Fong, W., Fiber evidence: laboratory methods and observation from casework, *J. Forens. Sci.*, 29, 55, 1984.
13. Deadman, H., Fiber evidence and the Wayne Williams trial, *FBI Law Enf. Bull.*, March 1984, p. 12; May 1984, p. 10.
14. Gaudette, B.D., Fibre evidence, *R.C.M.P. Gazette*, 47, 18, 1985.
15. Petraco, N., The occurrence of trace evidence in one examiner's casework, *J. Forens. Sci.*, 30, 486, 1985.
16. Schaeffer, H.F., *Microscopy for Chemists*, New York, Van Nostrand, 1953.
17. Feigl, F., *Qualitative Analysis by Spot Tests*, 3rd ed., New York, Elsevier, 1946.
18. Jungreis, E., *Spot Test Analysis*, 2nd ed., New York, Wiley Interscience, 1997.
19. Cennini, C., *Il Libro dell Arte (The Craftsman's Handbook)* written in the 15th century, translated by D.V. Thompson, Jr. for Yale University Press, 1933, republished, New York, Dover, 1954.
20. Cellini, B., *The Treatises of Benvenuto Cellini on Goldsmithing and Sculpture*, translated by C.R. Ashbee, London, Edward Arnold, 1898.
21. Toft, A., *Modelling and Sculpture*, Philadelphia, Lippincott, 1924.
22. Lent, F.A., *Trade Names and Descriptions of Stones*, New York, New Stone Publishing, 1925.
23. Fink, C.G. and Eldridge, C.H., *The Restoration of Bronzes and Other Alloys*, New York, Metropolitan Museum of Art, 1925.
24. Beaufort, T.R., *Pictures and How To Clean Them*, New York, Frederick A. Stokes, 1926.
25. DeWild, M.A., *The Scientific Examination of Pictures* (translated from Dutch), London, G. Bell, 1929.
26. Toch, M., *Paint, Paintings, and Restoration*, New York, Van Nostrand, 1931.

27. Lucas, A., *Antiques, Their Restoration and Preservation*, rev. ed., London, Edward Arnold, 1932.
28. Fogg Museum, *Technical Studies in the Field of the Fine Arts*, Cambridge, Harvard University, 1932–1942.
29. Obermeyer, H., *Stop That Smoke!* New York, Harpers, 1933.
30. Plenderleith, H.J., *The Conservation of Prints and Drawings*, London, Museums Association, 1937.
31. Gettens, R.J. and Stout, G.L., *Painting Materials: A Short Encyclopedia*, New York, Van Nostrand; reprinted, New York, Dover, 1966.
32. Stout, G.L., *The Card of Paintings*, New York, Columbia University Press, 1948.
33. Bradley, M.C., *The Treatment of Pictures*, Cambridge, Art Technology, 1950.
34. *Studies in Conservation*, Vol. 1, London, National Gallery, 1952.
35. Keck, C.K., *How To Take Care of Your Pictures*, New York, Museum of Modern Art; Brooklyn Museum, 1954.
36. Gettens, R.J. and Upsilon, B.M., *Abstracts of Technical Studies in Art and Archaeology 1943–1952*, Washington, D.C., Freer Gallery of Art, 1955.
37. *Art and Archaeology Technical Abstracts*, New York University Center for Conservation of Historic and Artistic Works, 1955–1965.
38. Keck, C.K., *Handbook on the Care of Paintings*, New York, Watson Guptill Publications, 1965.
39. Ruhemann, H., *The Cleaning of Paintings: Problems and Potentialities*, London, Faber & Faber, 1968.
40. Plenderleith, H.J. and Werner, A.E.A., *Conservation of Antiquities and Works of Art: Treatment, Repair, and Restoration*, 2nd ed., New York, Oxford University Press, 1971.
41. Clapp, A.F., *Curatorial Care of Works on Paper*, Oberlin, OH, Intermuseum Laboratory.
42. Bromelle and Smith, Eds., *Conservation and Restoration of Pictorial Art*, London, Butterworths, 1976.
43. Edlin, H.L., *What Wood Is That?* New York, Viking Press, 1977.
44. Mayer, R., *The Artist's Handbook of Materials and Techniques*, 5th ed., New York, Viking Press, 1991.
45. Petraco, N., Trace evidence: the invisible witness, *JFSCA*, 31, 321. 1986.
46. Petraco, N. and DeForest, P.R., Trajectory reconstructions. I. Trace evidence in flight, *JFSCA*, 35, 1284, 1990.
47. Zieler, H.W., *The Optical Performance of the Light Microscope*, Parts 1 and 2, Chicago, Microscope Publications, 1972.
48. McCrone, W.C., McCrone, L.C., and Delly, J.G., *Polarized Light Microscopy*, Ann Arbor, MI, Ann Arbor Science Publishers, 1978.
49. Delly, J.G., *Photography through the Microscope*, 8th ed., New York, Eastman Kodak Co., 1980.
50. DeForest, P.R., Foundations of forensic microscopy, in *Forensic Science Handbook*, Vol. 1, Saferstein, R., Ed., Englewood Cliffs, NJ, Regents/Prentice-Hall, 1982, p. 417.
51. Abramowitz, M., *Microscope Basics and Beyond*, Vol.1, Melville, NY, Olympus America, 1988.
52. Bloss, F.D., *Optical Crystallography*, Monograph Series 5, Washington, D.C., Mineralogical Society of America, 1999.
53. Delly, J.G., *Photography through the Microscope*, 8th ed., New York, Eastman Kodak Co., 1980, p. 24.
54. Wagner, A.F., *Experimental Optics*, New York, John Wiley & Sons, 1929.
55. Zieler, H.W., *The Optical Performance of the Light Microscope*, Part 1, Chicago, Microscope Publications, 1972.
56. Zieler, H.W., *The Optical Performance of the Light Microscope*, Part 2, Chicago, Microscope Publications, 1972.
57. Aschoff, W.W., Kobilinsky, L., Loveland, R.P., McCrone, W.C., and Rochow, T.G., *Glossary of Microscopical Terms and Definitions*, Chicago, McCrone Research Institute, 1989.

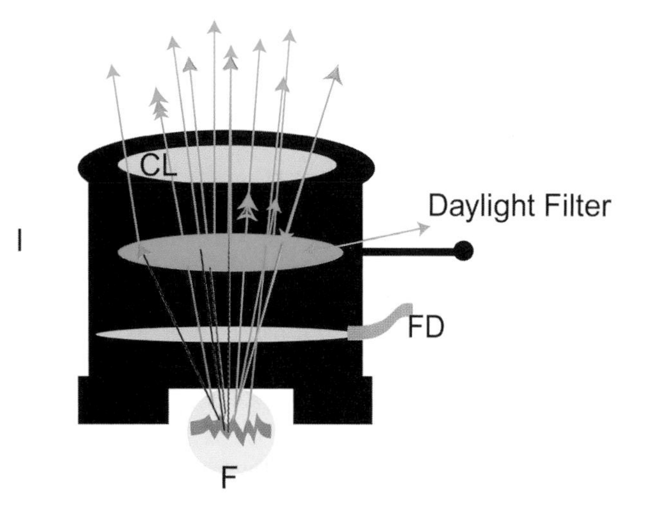

FIGURE 1.1 An illuminator (I) contains a light source (F), a field diaphragm (FD), a condenser lens (CL), and a mount for filters (e.g., a daylight filter that replicates sunlight).

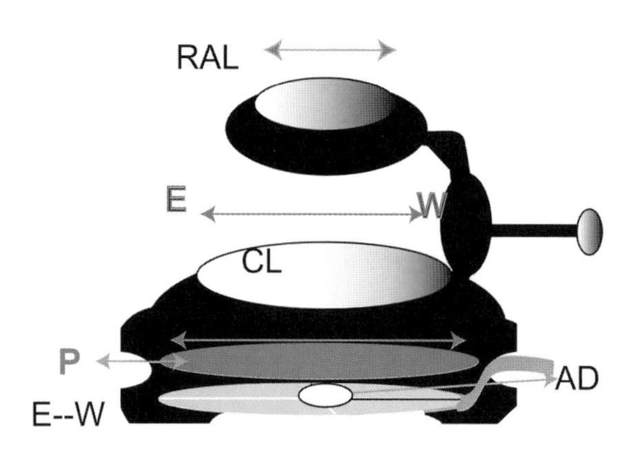

FIGURE 1.2 A typical PLM condenser (C) contains an aperture diaphragm (AD), a polarizer with an E↔W privileged direction, a condenser lens (CL), and a removable auxiliary lens (RAL).

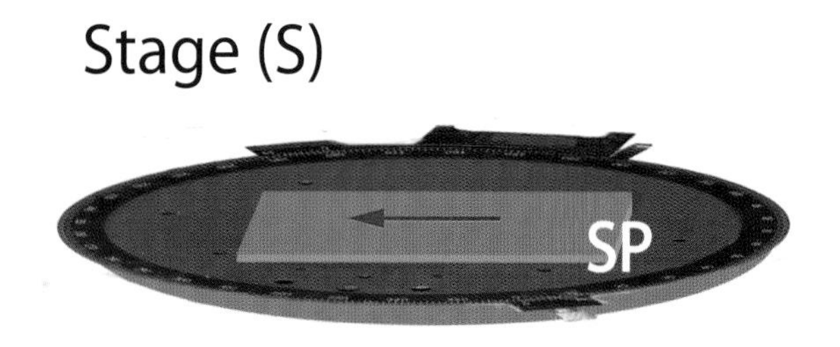

FIGURE 1.3 Typical PLM rotatable circular-shaped stage (S) that holds a specimen slide at the specimen plane (SP).

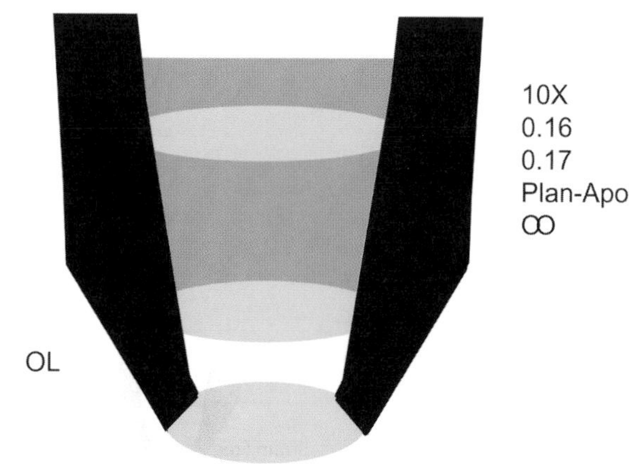

FIGURE 1.4 The objective lens (OL) forms the primary image of a specimen; it is the most important part of a CM. Information such as magnification (10×), focal length (0.16 mm), cover glass thickness (0.17 mm), objective type (Plan–Apo), and tube length may be marked on the barrel.

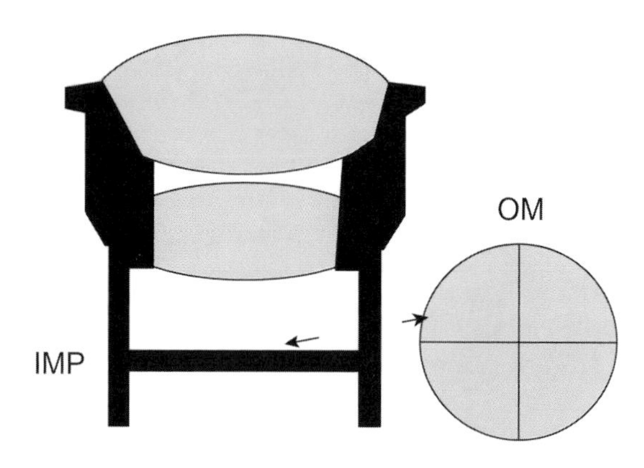

FIGURE 1.5 The ocular or eyepiece (OC) acts as a simple magnifying lens and magnifies the primary image formed by the OL. In a PLM, the OC contains a cross-hair ocular micrometer (OM) placed at the intermediate image plane (IMP) of the OC.

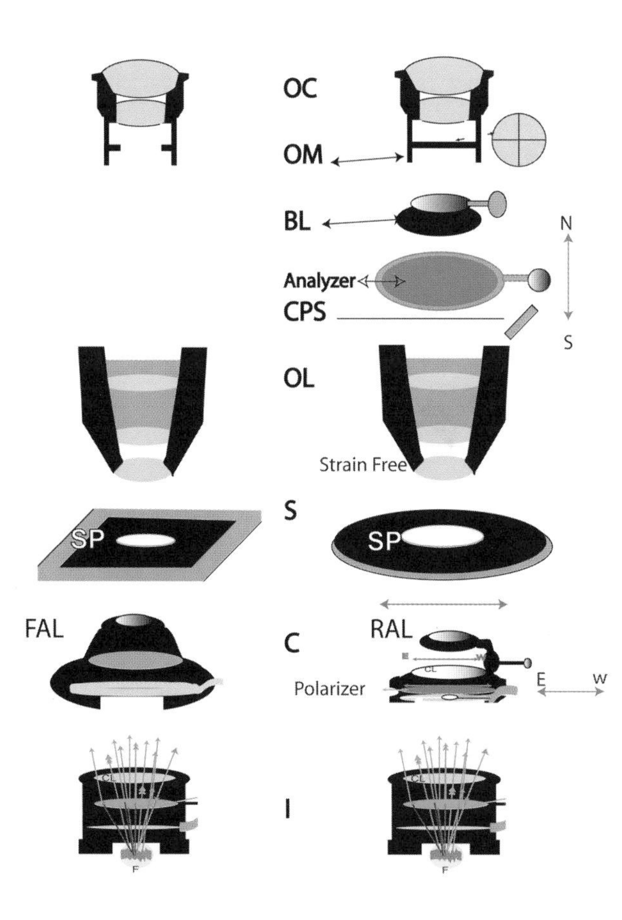

FIGURE 1.6 Left: basic components of a typical bright-field microscope (BFM). Right: typical polarized light microscope (PLM). The illuminators are essentially the same in both forms of CM. The C of a BFM has a fixed auxiliary lens (FAL), whereas the C of a PLM has a removable auxiliary lens (RAL). The PLM C also contains a polarizer that converts light entering the C from vibrating in all planes at 360° (diffused light) to plane-polarized light (PPL) vibrating in the E ↔ W privileged direction (PD). The stage (S) in a BFM is square or rectangular and normally in a fixed position. The S in a PLM is circular and can freely rotate a full 360°. The objective lens (OL) of a BFM may allow strain; in PLM, the OL must be strain-free. The space between the OL and OC of a BFM is normally empty, but may contain auxiliary contrast enhancement devices. The space between the OL and OC of a PLM contains a compensation slot (CSP) cut at 45° to the microscope's optic axis (OA); an analyzer, positioned above the CPS; it is removable from the light path and can be rotated 90°; and a Bertrand lens. The PD of the analyzer is N ↔ S 90° or perpendicular to the direction of the polarizer PD. When the Bertrand lens above the analyzer is placed in the light path, it allows easy viewing of the objective's back focal plane. Finally, the OC in a PLM normally has some type of cross-hair ocular micrometer (OM) at the IMP.

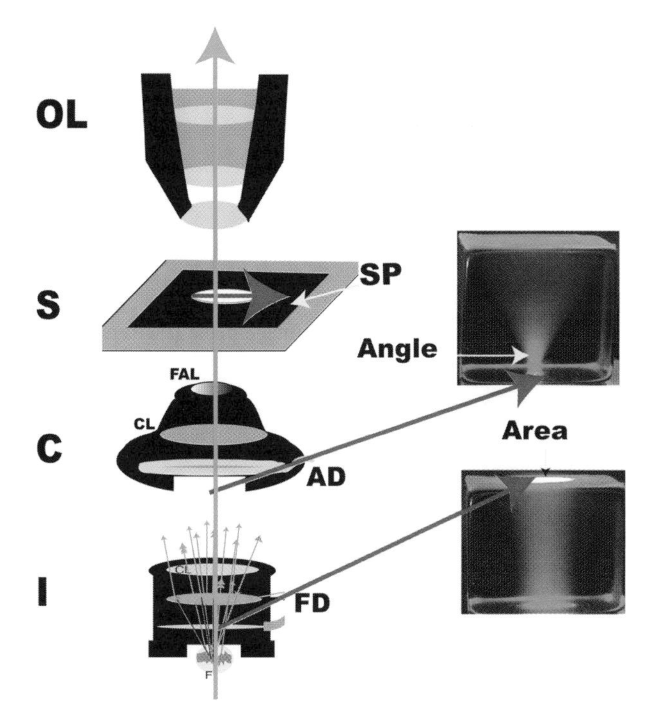

FIGURE 1.7 Illumination system located beneath the stage (S) consisting of an illuminator (I) and substage condenser (C). The illuminator contains the light source (F) and field diaphragm (FD). The FD controls the area of the specimen to be illuminated and the field of observation. A typical BFM substage condenser contains a fixed auxiliary top lens (FAL), a condenser lens (CL), and an aperture diaphragm (AD). The condenser AD controls the size and angle of the cone of light illuminating the specimen, thereby controlling the contrast and brightness of the PI formed by the OL.

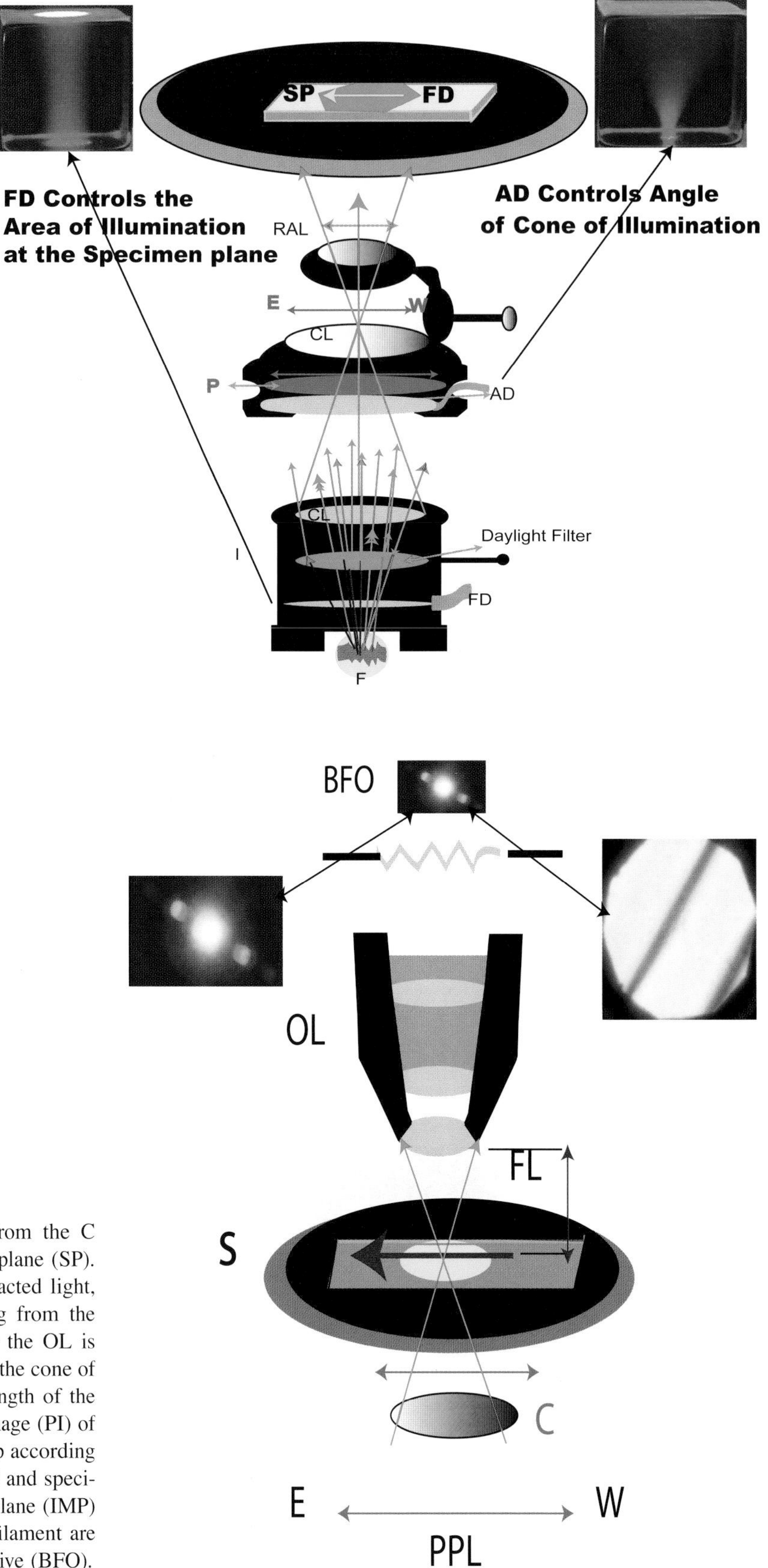

FIGURE 1.8 Typical PLM substage condenser (C) beneath the circular stage (S); it contains a removable auxiliary top lens (RAL), condenser lens (CL), polarizing filter (polarizer), and aperture diaphragm (AD). The polarizing filter transmits plane-polarized light (PPL) vibrating in an E ↔ W privileged direction. The AD controls the angle of the cone of light illuminating the specimen, thereby controlling the contrast and brightness of the PI formed by the OL. Top left: the auxiliary lens (AL) is employed when using higher numerical aperture (NA) objective lenses. Top right: the AL is removed from the optic axis (light path) to obtain axial illumination.

FIGURE 1.9 Plane-polarized light (PPL) from the C interacts with the specimen at the specimen plane (SP). The front lens of the objective collects diffracted light, refracted light, and undeviated light coming from the specimen. The amount of light gathered by the OL is directly related to the angle aperture (AA) of the cone of light captured by the front lens and focal length of the objective. The OL then forms the primary image (PI) of the specimen. When the illumination is set up according to Köhler's instructions, the image of the FD and specimen are in focus at the intermediate image plane (IMP) of the OC, while the image of the AD and filament are in focus at the back focal plane of the objective (BFO).

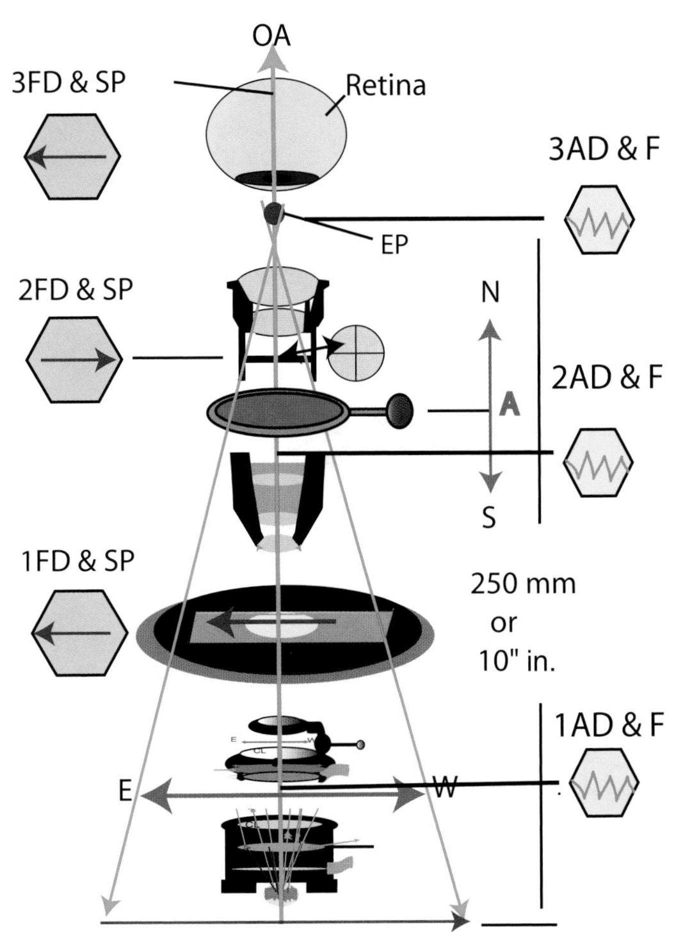

FIGURE 1.10 Image of FD & SP (1) formed by the OL is projected to the IMP (2) of the OC. The image of the FD & SP (3) magnified by the OC is projected and formed on the retina. The image of the AD & F (1) formed by the OL is projected to the BFO of the OL (2). The image of the AD & F (3) magnified by the OC is projected to the eye point (EP). The image of the FD & SP (3) formed on the retina is real and erected. The image of the FD & SP (4) perceived below the illuminator is virtual and inverted and appears to be about 10 inches away from the viewer.

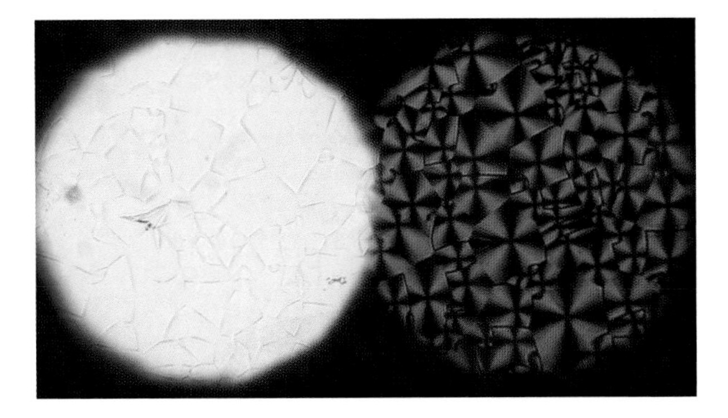

FIGURE 1.11 Left: specimen of cholesterol acetate viewed with BFM; it tells the analyst little about the specimen. Right: same specimen viewed with PLM between crossed polars; it will help a trained analyst identify the specimen.

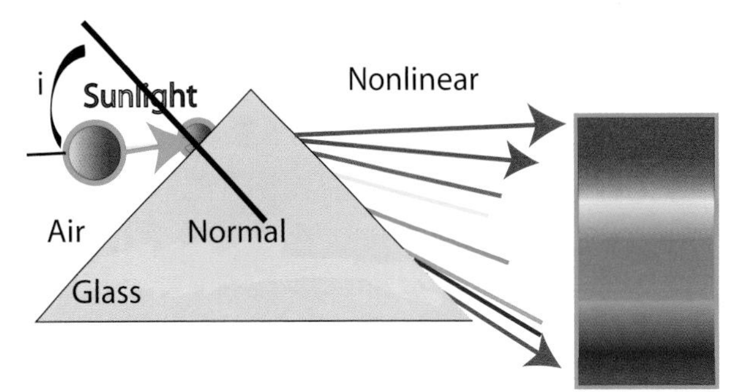

FIGURE 1.12 Polychromatic light slows up or bends, dispersing into its monochromatic colors when it travels through air (RI 1.00029) into a glass prism (RI 1.520). This phenomenon is known as refraction; the degree of bend is known as the refractive index (RI). Different colors of light have slightly different RI values when traveling through the same material. Consequently, because white light is composed of different colors of light, it spreads out (disperses) into its component colors (roy-g-biv) when it passes through a glass prism into air.

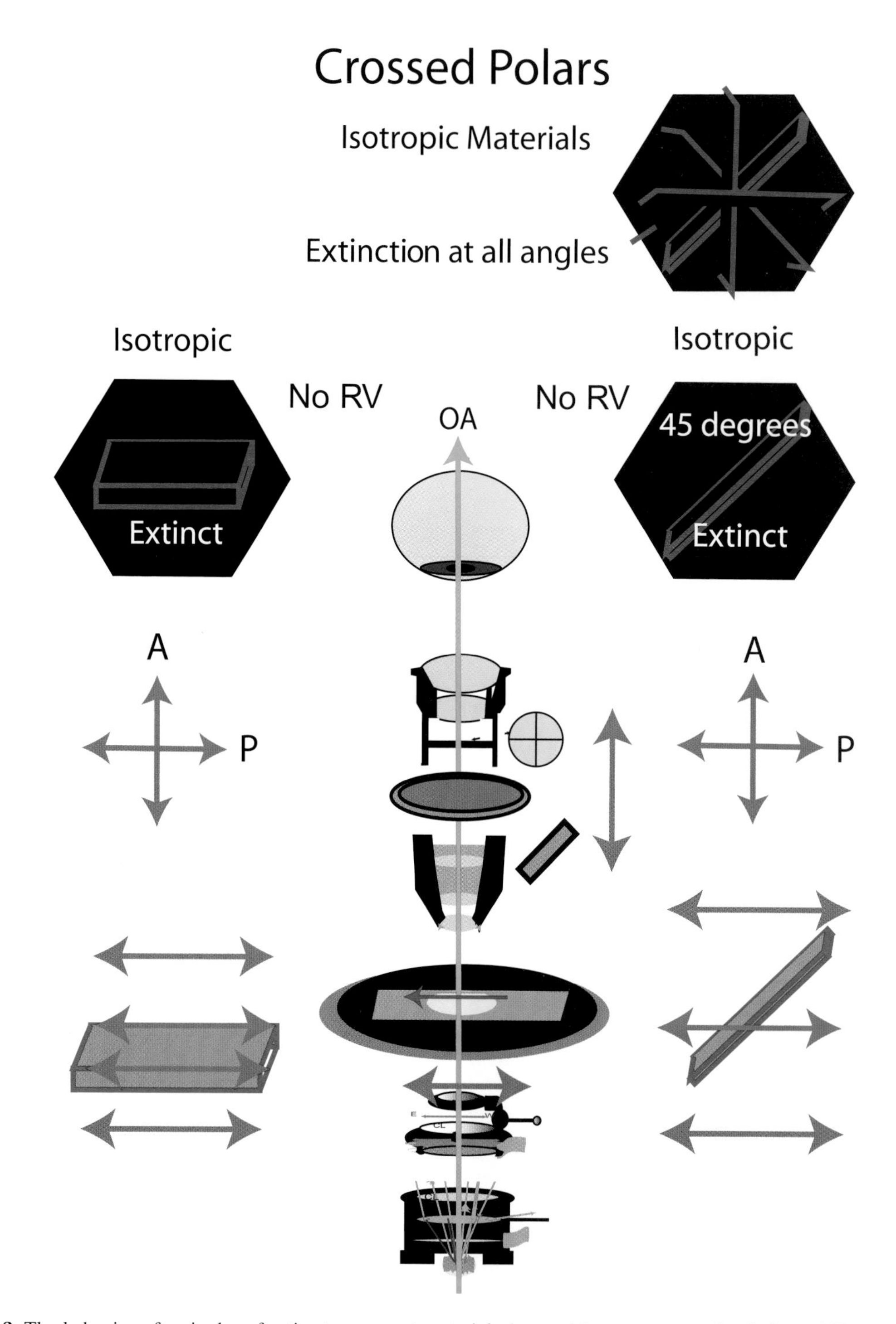

FIGURE 1.13 The behavior of a singly refractive transparent material observed between crossed polarizers (CPs). Plane-polarized light from the polarizer interferes with the specimen. Light leaving the specimen is still polarized in the E ↔ W PD. It travels up the microscope OA where it is collected by the front lens of the objective. The OL forms a primary image of the specimen that then travels up the OA to the analyzer. The light exiting the OL is still plane-polarized, vibrating in the E ↔ W PD. Since the analyzer's privileged direction is perpendicular to that of the polarizer, the analyzer stops all light coming from the OL; no light can pass through the analyzer. The value of the retardation vector (RV) formed by the analyzer is zero and the specimen appears extinct (black) against a black background. A material behaving in this manner is said to have isotropic optical properties. *Note:* The specimens are depicted roughly the way they would appear if viewed through the ocular.

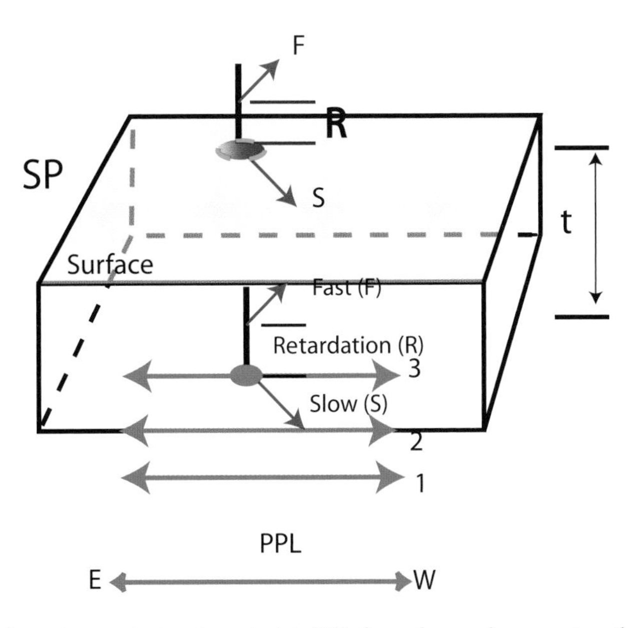

FIGURE 1.14 Passage of PPL through an anisotropic material. PPL from the condenser enters the specimen and is split into two primary rays that vibrate perpendicular to each other. The fast ray (F) travels through the material thickness at a higher speed than the second primary ray. Consequently, the second or slow ray (S) travels through the material thickness at a slower speed. Therefore, the S ray is located some distance behind the F ray. When both rays emerge from the material, the linear distance between them is known as the retardation distance (R). The two plane-polarized rays vibrate at 90°, the S ray travels behind the F ray, and they cannot interfere until they enter the analyzer.

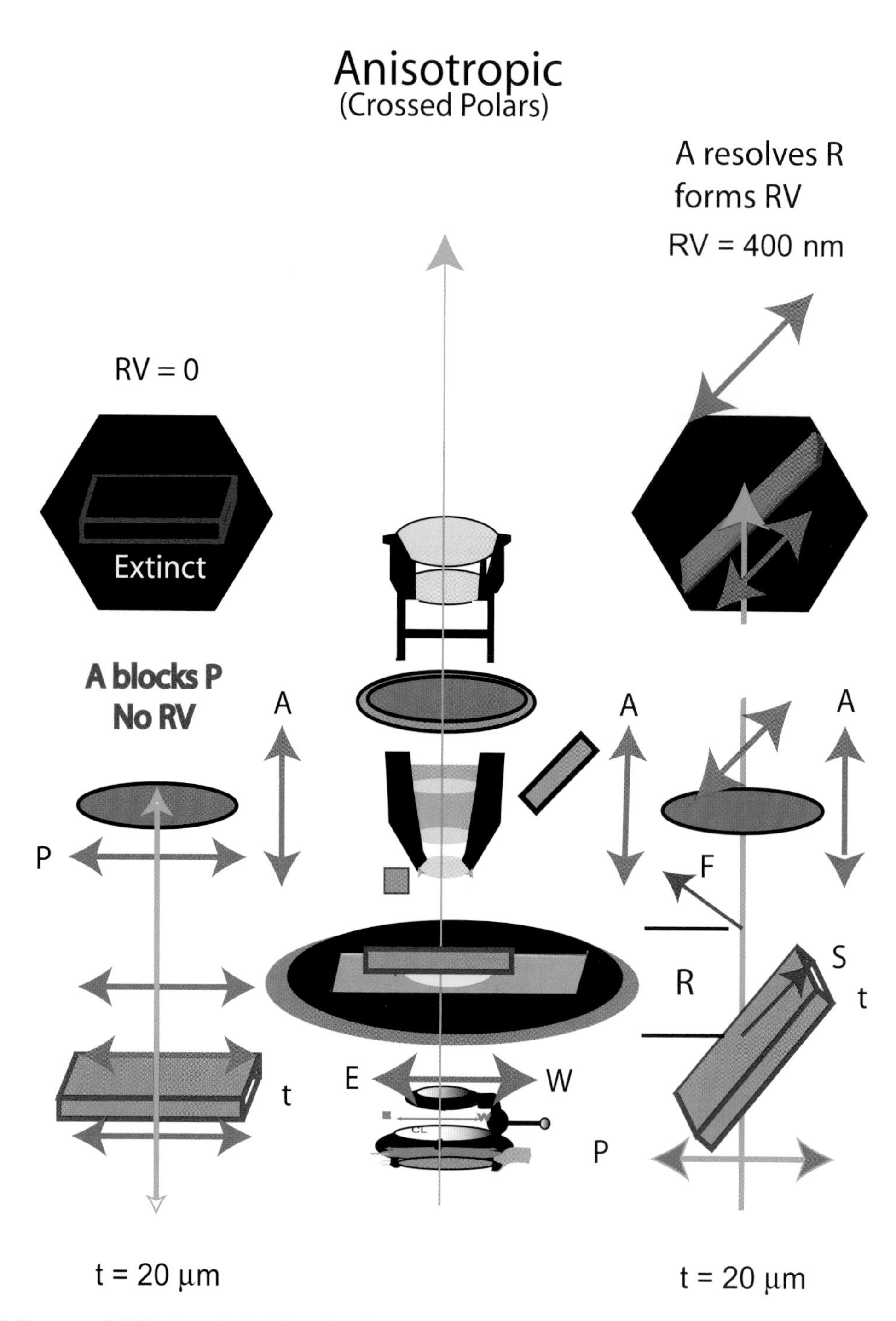

FIGURE 1.15 Passage of PPL through doubly refractive transparent material observed between crossed polarizers (CPs). Left: specimen behaves as an isotropic material; RV = 0. Right: specimen behaves as an anisotropic material; RV = 400 nm. *Note:* The specimens are depicted as they would roughly appear if viewed through the ocular.

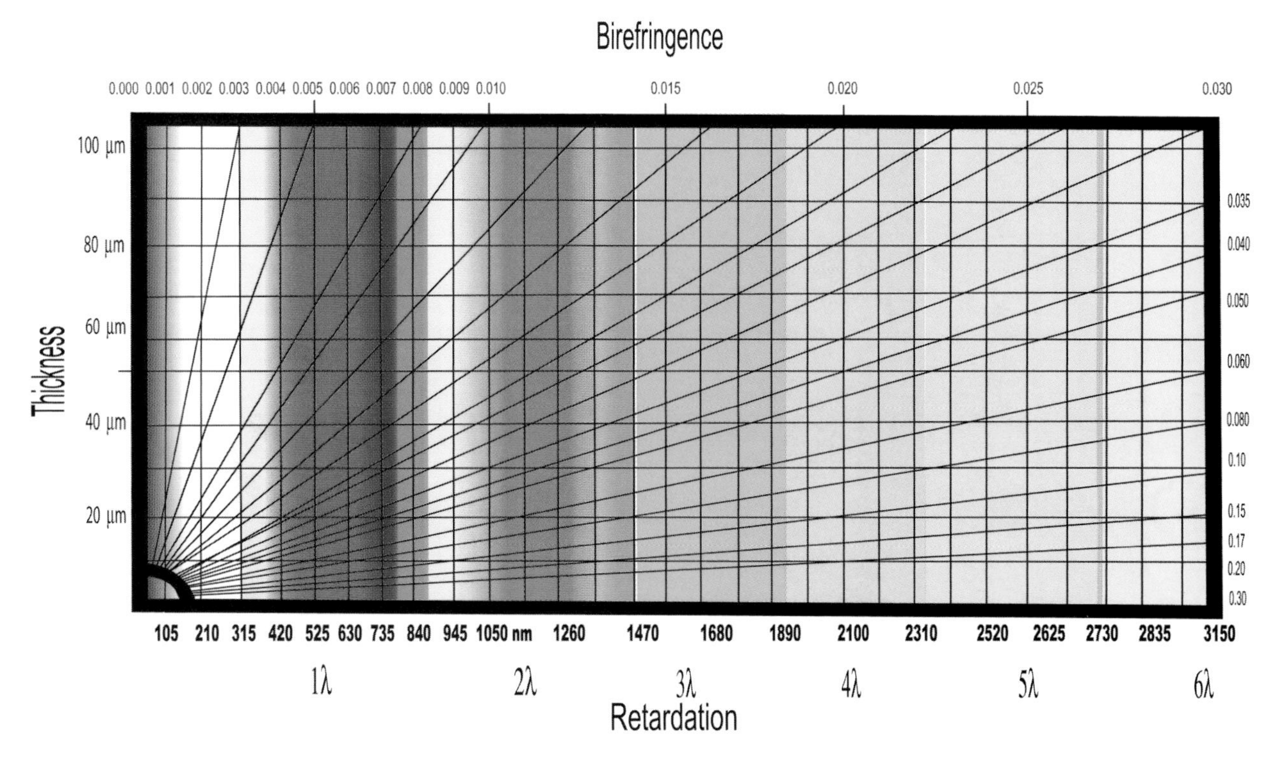

FIGURE 1.16 An interference color chart showing the relationship of specimen thickness, displayed retardation colors, and birefringence. If two of the quantities are known, the third can be determined from this chart.

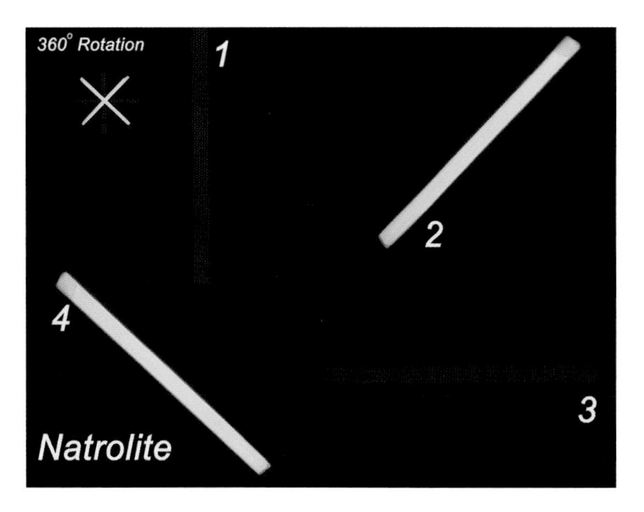

FIGURE 1.17 When an anisotropic substance is rotated between CPs, it alternates between displaying polarization colors and becoming extinct every 90°. The polarization colors are also called interference colors (ICs). It is important to note that an anisotropic material exhibits its brightest IC at 45° off extinction.

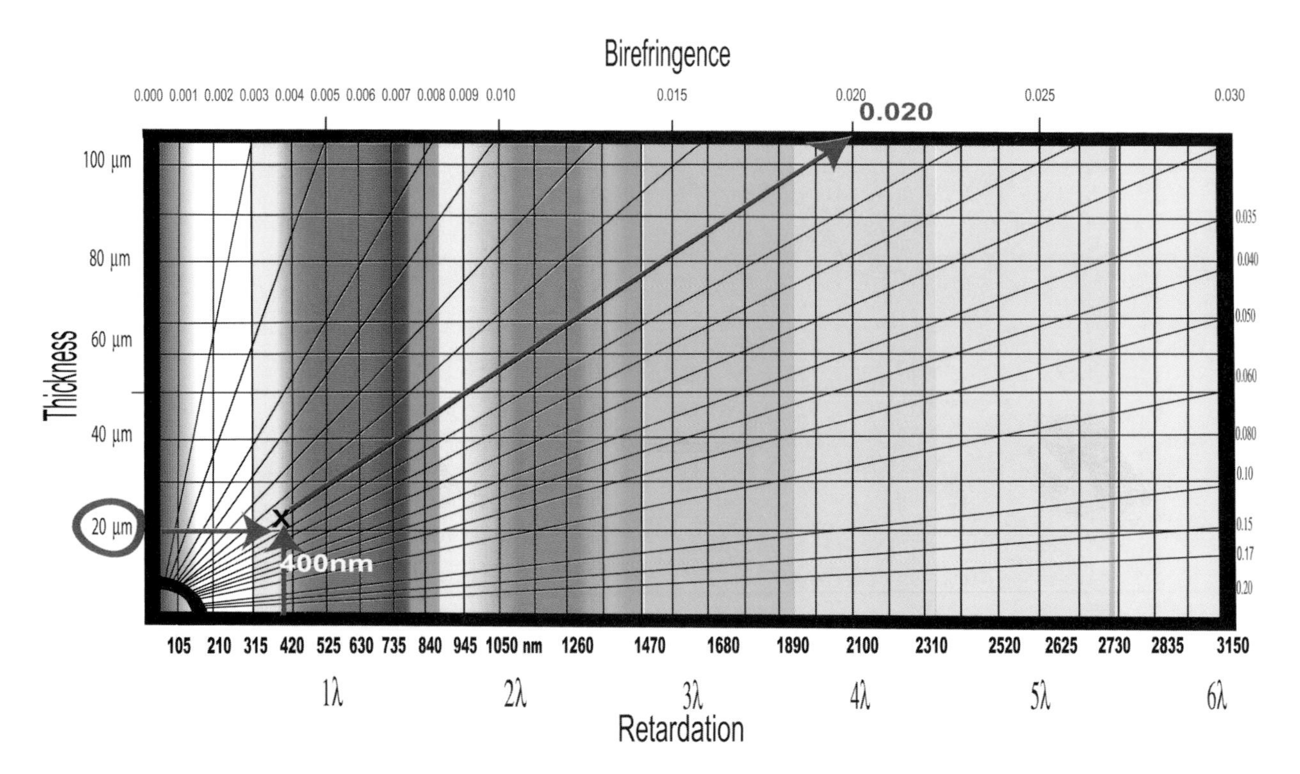

FIGURE 1.18 The birefringence of an anisotropic specimen is determined by using an interference chart. A 20-μm thick (t) specimen displays a red/orange interference color (IC) when viewed between crossed polars. By comparing the specimen IC to an interference chart, the retardation is estimated to be 400 nm. These values for thickness and retardation mean the specimen birefringence is estimated to be 0.020.

Fixed Compensation

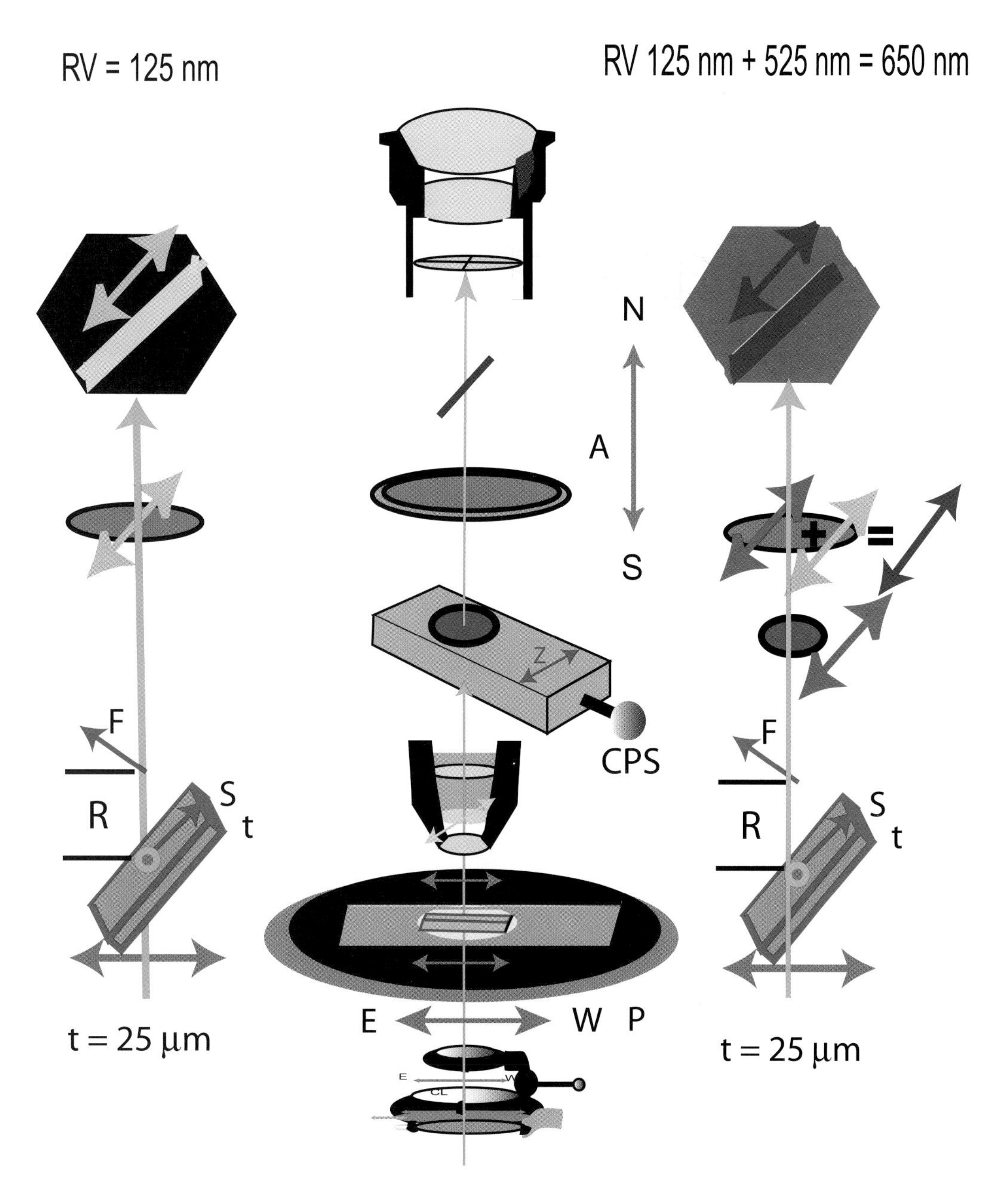

FIGURE 1.19 Center: addition of a fixed compensator (FCp) to the optic axis of a PLM. Left: anisotropic substance viewed between crossed polars (CPs) at 45° off extinction; specimen exhibits its maximum interference color (IC) of 125 nm and gray/white against black background. Right: specimen is viewed at 45° off extinction with the addition of a full-wave FCp. When retardation due to the specimen and FCp enter the analyzer, RV of 650 nm is formed and the specimen appears blue against a magenta background. *Note:* The specimens are depicted as they would roughly appear if viewed through the ocular.

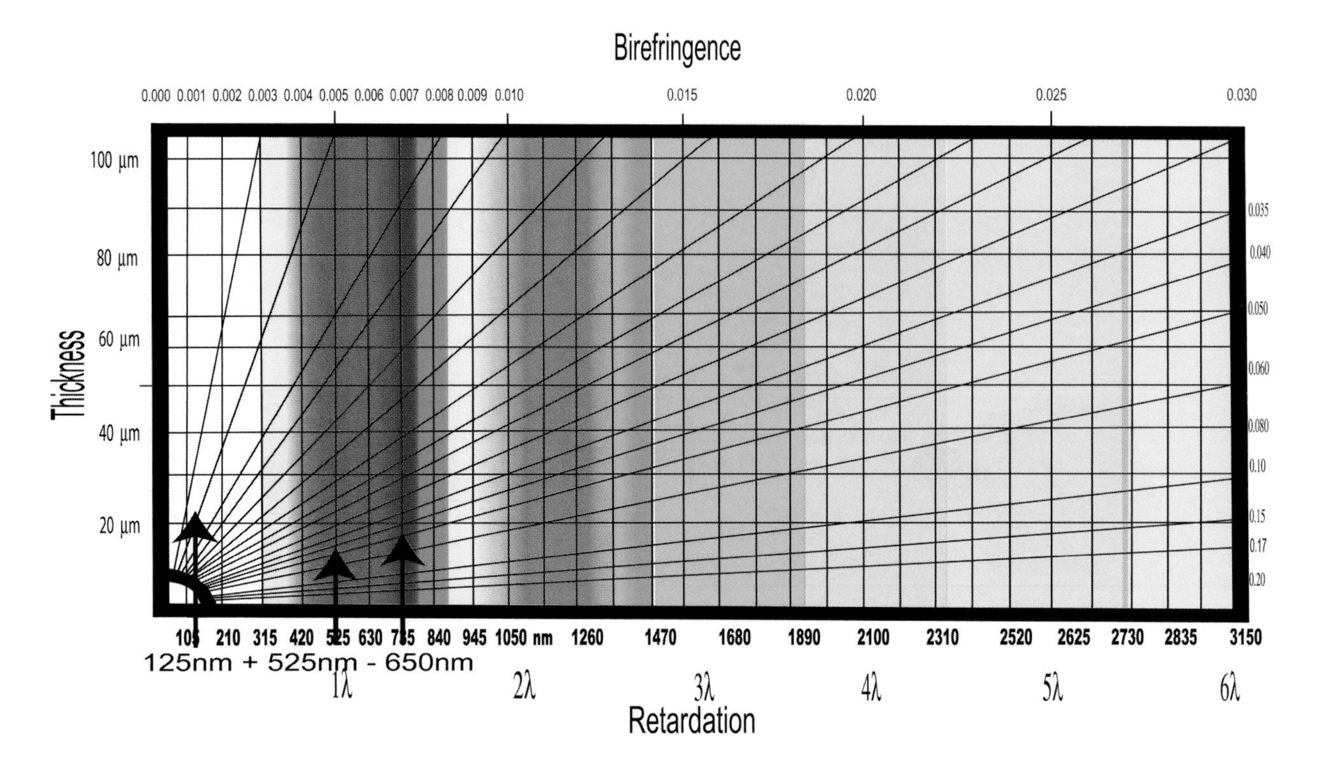

FIGURE 1.20 The retardation vector (RV) is formed by the addition of the interference colors as shown above. RV = 125 nm + 525 nm = 650 nm. When 525 nm of retardation are added to the specimen's 125 nm of retardation, the specimen's interference color changes from low-order gray to first-order blue (see yellow dots).

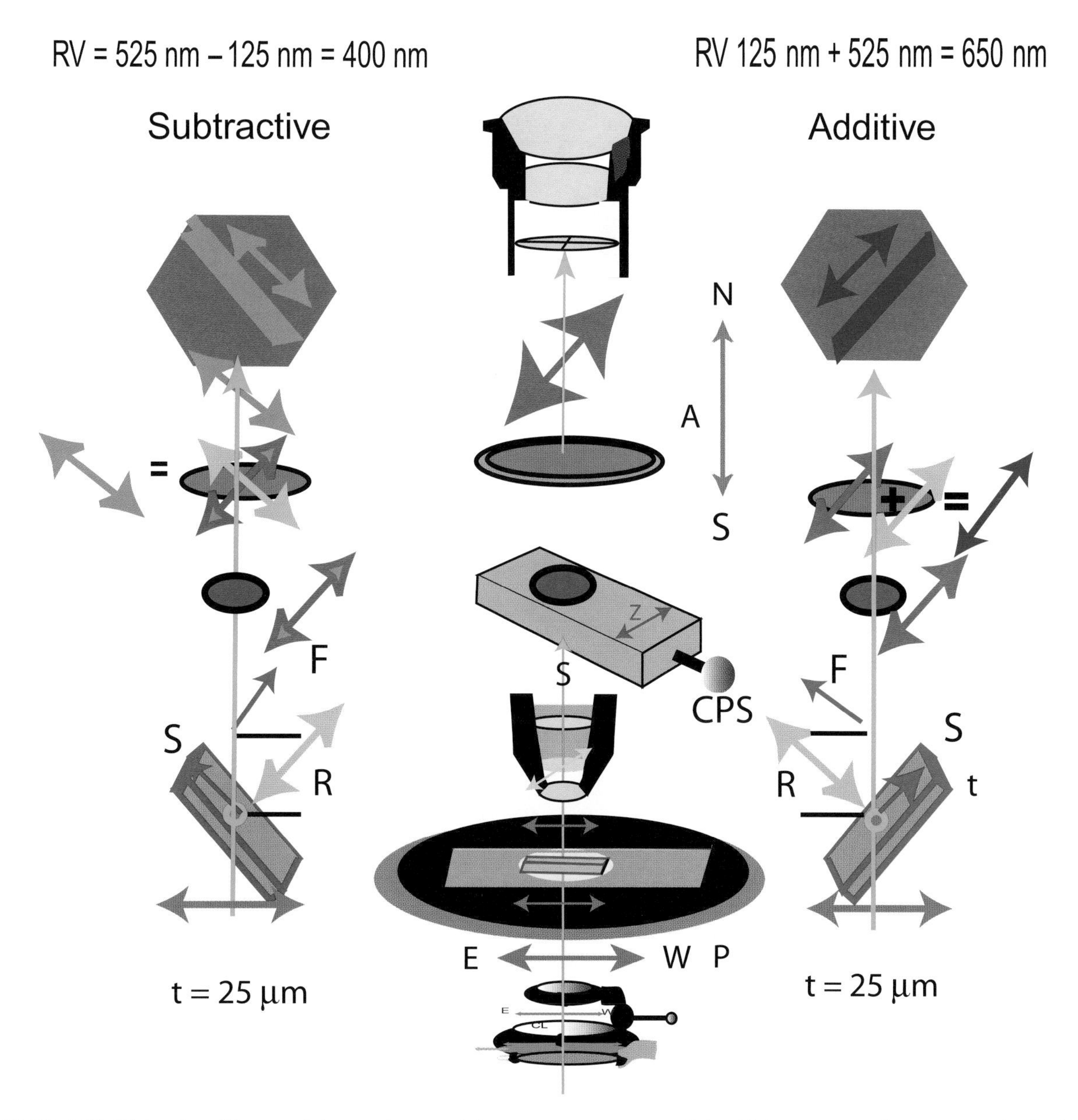

FIGURE 1.21 Right: slow (S) rays of the specimen and the compensator are parallel; additive compensation occurs and a final RV of 650 nm displaying higher-order interference colors (IC) is formed. Left: S ray and Cp S ray are oriented perpendicularly; subtractive compensation takes place, and a final RV of 400 nm exhibiting lower-order IC is achieved. *Note:* The specimens are depicted as they would roughly appear if viewed through the ocular.

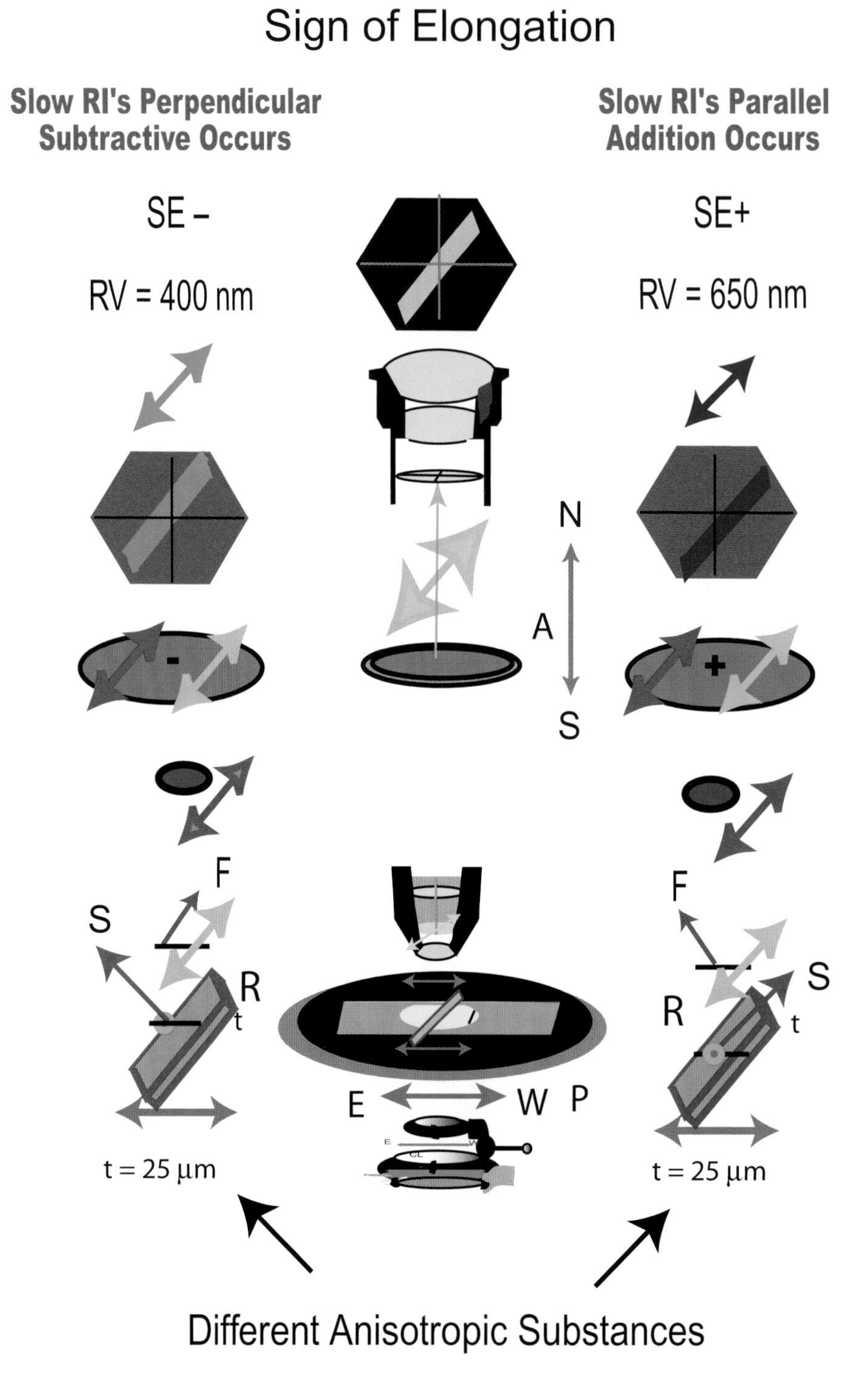

FIGURE 1.22 Use of FCp to determine sign of elongation. Center: anisotropic material is viewed between CPs. The specimen exhibits a low-order white IC of approximately 125 nm. Upon viewing the resulting IC, the analyst is left wondering about the orientation of the F and S rays of the material. Left: when full wave of compensation is added along a PLM's optic axis, the specimen demonstrates subtractive retardation. To achieve subtractive retardation, the S ray must be oriented along the specimen width. Right: another specimen demonstrates additive retardation when a full wave of compensation is added along the optic axis. To achieve additive retardation, the S ray must be oriented along the specimen's length. *Note:* The specimens are depicted as they would roughly appear if viewed through the ocular.

2 Preliminary Examination, Stereomicroscopy, and Basic Sample Preparation

PRELIMINARY EXAMINATION AND STEREOMICROSCOPY

The purpose of any examination should be established prior to the start of a preliminary assessment. Is the specimen a questioned sample or is it a source of known material or information? Is the sample to be identified or otherwise classified? Is it to be compared to another sample in an attempt to establish common origin or is it to be cleaned or restored? Where should the questioned or known specimen be removed from the object? How should the specimen be removed?

These are only a few questions that must be asked and answered before work is commenced. It is important to note that all the disciplines targeted in this work have well-established protocols and procedures. Practitioners of these fields should be fully familiar with the established methodologies of their disciplines. Finally, individuals conducting the examinations should have basic knowledge of the analytical methods they will employ.

The most essential item of scientific equipment (besides eyes) needed for a preliminary examination is a quality stereomicroscope (SM) capable of various forms of illumination, i.e., transmitted light, dark field, and axial and oblique epi-illumination. A camera system is also essential for documentation purposes. The two primary types of stereomicroscopes are the Greenough and the common main objective (CMO). For speedy documentation, a Polaroid® or modern digital camera can be employed.

The Greenough SM is composed of two separate objective lenses each paired to one eyepiece or ocular. This configuration provides a compound microscope (CM) for each eye, and as a result produces two different images of the specimen. The CMO SM has one main objective lens and two oculars. This configuration also provides two different images of the same specimen. Both types of stereomicroscopes produce 3-dimensional images because they provide two different angles of view of the same specimen. The different images provided for both eyes result in stereoscopic vision. Figure 2.1 is a schematic of a Greenough SM — an earlier form of a stereomicroscope. Figure 2.2 shows the newer CMO form of SM.

Although stereomicroscopes are designed to magnify specimens up to 150 diameters, SM is best employed at low power — 2 to 50×. Stereomicroscopy was developed to allow users to observe objects in 3 dimensions, similar to the way they normally see with unaided eyes. As an observer increases the magnification of a specimen, the depth of field (3-dimensional appearance) of the specimen tends to decrease, thereby causing the final image to appear as a 2-dimensional image, thus nullifying the purpose of the design: making specimen manipulation and macroscopic observation of physical structure and properties easier by providing the viewer with a 3-dimensional image. Figures 2.3a and b depict CMO SMs; Figure 2.4 shows a Greenough SM.

Stereomicroscopy is also used for myriad tasks in industry and the arts. Various stands have been developed to facilitate its use in many endeavors. Figures 2.5a and 2.5b illustrate two configurations of SMs and stands used in the scientific disciplines covered in this text.

In addition, various auxiliary sample manipulation devices such as spatulas, scalpels, needles, tweezers, and many other tools have proven indispensable for conducting preliminary examinations (Figure 2.6).

Each individual specimen should be visually evaluated with a stereomicroscope. If a specimen appears homogeneous, i.e., a single fiber type or class of animal hair, a representative sample should be isolated and examined in detail. A representative sample of a heterogeneous dust specimen can be mounted in the same manner. The substances composing a heterogeneous specimen should be sorted as to type of material and isolated as shown in Figure 2.7. A data sheet such as Figure 2.8 should be prepared for each sample. The data collected can be used for the initial classification of materials contained in a specimen and can guide an examiner to the appropriate identification procedure or scheme.

MICROSCOPE SLIDES AND PREPARATION

A large array of microscope slides (Ms) and cover glasses (Cgs) are available to the analyst. The authors have found two standard glass-slide sizes (25 × 75 mm and 1.0 mm thick and 50 × 75 mm and 1.0 mm thick) the most practical in their work. In addition, the No. 1½ square (18 × 18 mm and 22 × 22 mm) and the No. 1½ rectangular Cgs (22 × 40, 24 × 50, and 45 × 70 mm) are most useful.

Although Ms and Cgs are labeled precleaned when they arrive from manufacturers, the authors have found it necessary to clean them before use. All glass Ms and Cgs are placed in grooved Coplin jars filled with Windex®. The slides are allowed to soak at least 15 minutes. Upon removal from the cleaning solution, the slides should be dried with lint-free paper. Figure 2.9 shows two sizes of Coplin jars. The large size jar is for Ms and the small size jar is for Cg.

SPECIMEN PREPARATION

The preparations of various types of mounts used in the microscopy studies of particulates and fibrous materials are well documented in the literature.[1–22] The authors have found many published methods of enormous value in conducting microscopical studies and have developed certain methods for use in their work that will be discussed below. This section covers two basic categories of specimens: fibrous materials; and particulates. Specialized procedures will be presented in appropriate chapters.

Human Hair, Animal Hair, and Natural and Synthetic Fibers

Hair and fiber specimens should be mounted longitudinally in a mounting medium with a refractive index range between 1.510 and 1.560 for the N_D line at 25°C. Figure 2.10 shows the preparation of a temporary mount in Cargille® refractive index oil. Such oil mounts allow easy manipulation of specimens and painless remounting (Figure 2.11).[23]

Permanent mounts in either Permount® (Figure 2.12) or Melt Mount® 1.539 media (Figure 2.13) can be kept for many years.[24–26] Both media are very stable and chemically inert to a wide variety of materials. If prepared properly, materials mounted in them do not change and remain as they were when originally mounted. The authors have standard collections of hairs, fibers, pigments, minerals, and other substances prepared in these mounting media that have remain unchanged for more than 20 years.

Preparation of dry or air mounts is also recommended (Figure 2.14). Dry mounts allow examination of a specimen's surface structure, i.e., scale patterns and other physical properties, and possible surface debris, i.e., blood or paint.

Scale casts of hairs should be prepared for animal guard hair identification. They can also be used in the study of human hair cuticles. It is recommended that both temporary and permanent casts be prepared and studied. A temporary cast can be prepared in (1) most types of clear, quick-drying lacquer such as nail polish; (2) Polaroid® black-and-white photographic coating solution[27]; or (3) Melt Mount® 1.539 resin (Figure 2.15).[28,29] When preparing scale casts, hair specimens should be cast from basal (B) to tip (T) ends and from T to B to make certain that the entire guard hair scale pattern configuration is documented.

A permanent scale cast can be prepared on Rinzl® clear plastic microscope slides (Figure 2.16).

Cross-sectional (X-S) mounts of hairs and fibers should be prepared when necessary for identification and comparison purposes. For X-S observation, the authors prepare temporary mounts with Rinzl clear plastic slides (Figure 2.17)[30] and permanent mounts with a Hardy hand microtome (Figure 2.18).[31,32] Other types of microtomes and methods are available for the preparation of X-S mounts. Their utilization depends on the equipment and materials available to the examiner.

The first method requires a Rinzl® clear plastic microscope slide (3 × 1 × 0.5 mm) as shown in Figure 2.17, a length of heavy-duty thread, a size 5/10 sewing needle, and a new Teflon®-coated single-edge razor blade. Double-thread the needle and use it to puncture a 1-mm hole in the center of the slide (1b). Draw the sewing needle completely through the plastic slide, leaving a loop of thread on the opposite side and remove the needle (1c). Place the bundle of hairs or fibers to be cross-sectioned into the loop of thread; pull the thread downward until the bundle is drawn halfway through the slide as shown in the figure (1d and e). An equal amount of the bundle should remain on each side of the slide (1e). Use a new Teflon-coated single-edge razor blade to cut the hair or fiber bundle. Hold the cutting edge of the blade at approximately a 30° angle to the surface of the slide and draw it across the slide surface (2 in the figure). The tuft of fibrous material should be cut flush to the slide on both sides (3a). The cross-sections can be viewed directly or after mounting in an appropriate mounting medium; 3b is a photomicrograph of rayon fibers prepared for X-S viewing.

The steps in the operation of the Hardy microtome for cutting X-S of hairs and fibers are straightforward and quick. Separate the two sections of the Hardy microtome as shown in Figure 2.18 (top) Insert the hairs or fibers into the specimen slot and loosely join both sections. The bundle can now be treated with a resin (colloidin) that will hold the materials together. After resin application, firmly push the microtome sections together and cut the fiber bundle flush to the slide surfaces with a Teflon-coated single-edge razor blade. Turn the plunger micrometer to push the fiber bundle above the plate surface, and the specimen bundle is sectioned, and mounted for observation.[33]

Particles, Glass, and Paint Chips

Longitudinal mounts of particles prepared in a mounting medium with a refractive index liquid that allows specimen morphology to be seen clearly may prove valuable. Figure 2.19 shows preparation of a temporary longitudinal mount in Cargille® refractive index oil. Oil mounts such as the one shown in the figure allow clear observation of particles with various refractive indices, easy manipulation and separation of particles, and problem-free transfers of particles from one mount to another.

Preparation of permanent and dry mounts of particles can be carried out in the manner described for hairs and fibers (Figures 2.12 through 2.14). Finally, cross-sections of paint chips and other thick particulate materials may be prepared by following the procedure shown in Figure 2.20. Resin, pigment, contaminates, and other materials can be isolated easily from cross-sections produced in this manner.

After preliminary examination of a multilayered paint chip, a double-sticky press glue tab is applied to the bottom of one of the small compartments of a polyethylene ice cube tray. The cross-section of the chip is positioned in the center of the compartment containing the resin and attached to the bottom of the tray. A 2-mL aliquot of catalyzed polyester casting resin is poured into the compartment and allowed to harden under a fume hood. After the cured resin cube is removed from the tray, the stratigraphy of paint layers is exposed by sanding and polishing the bottom surface of the cube with increasingly fine grades of micron-graded (30, 20, 15, 9, 3, 2, and 1 μm) finishing paper available from 3M (St. Paul, MN). The layer structure can be viewed with a low-magnification stereomicroscope or a microscope equipped with reflected light illumination (see Figure 2.20).[34]

Finally, when deemed necessary, details on using specialized procedures to view minerals, gemstones, and other materials will be presented in appropriate chapters.

REFERENCES

1. Glaister, J., *A Study of Hairs and Wools Belonging to the Mammalian Group of Animals, Including a Special Study of Human, Considered from the Medicolegal Aspect*, Publication 2, Cairo, MISR Press, 1931.
2. Hardy, J.I. and Plitt, T.M., An Improved Method for Revealing the Surface Structure of Fur Fibers, Wildlife Circular 7, Washington, D.C., U.S. Department of the Interior, 1940, p. 10.
3. Kirk, P.L. and Gamble, L.H., Human hair studies. II. Scale counts, *J. Crim. Law Criminol. Pol. Sci.*, 31, 627, 1941.
4. Kirk, P.L. and Gamble, L.H., Further investigation of scale count of human hair, *J. Crim. Law Criminol. Pol. Sci.*, 33, 1942.
5. Kirk, P.L., Magagnose, S., and Salisbury, D., Casting of hairs: its technique and application to species and personal identification, *J. Crim. Law Criminol. Pol. Sci.*, 236, 40, 1949.
6. Kirk, P.L., *Crime Investigation*, New York, Interscience, 1953, p. 152.
7. Kirk, P.L. and Cooper, R.M., An improved technique for sectioning hairs, *J. Crim. Law Criminol. Pol. Sci.*, 44, 1953.
8. Wildman, A.B., *The Microscopy of Animal Textile Fibres*, Leeds, U.K., Wool Industries Research Association, 1954, p. 26.
9. Heyn, A.N.J., *Fiber Microscopy. A Textbook and Laboratory Manual*, New York, Interscience, 1954.
10. Wildman, A.B., The identification of animal fibres, *J. Forens. Sci. Soc.*, 1, 1961.
11. Adorjan, A.S. and Kolenosky, G.B., *A Manual for the Identification of Hairs of Selected Ontario Mammals*, Wildlife Research Report 90, Ontario, Department of Lands and Forests, 1969.
12. McCrone, W.C. and Delly, J.G., Eds., *The Particle Atlas*, 2nd ed., Ann Arbor, MI, Ann Arbor Science Publishers (5 volumes), 1979.
13. Ogle, R.R. and Mitosinka, G.T., A rapid technique for preparing hair cuticle scale casts, *J. Forens. Sci.*, 18, 82, 1973.
14. Moore, T.D., Spence, L.E., Dugnolle, C.E., and Hepworth, W.G., *Identification of the Dorsal Guard Hairs of Some Mammals of Wyoming*, Bulletin 14, Cheyenne, Wyoming Fish and Game Department, 1974.
15. *Identification of Textile Materials*, 7th ed., Manchester, U.K., Textile Institute, 1975.
16. Metropolitan Police Forensic Science Laboratory, *Biology Methods Manual*, London, Commissioner of the Metropolis, 1978.
17. Appleyard, H.M., *Guide to the Identification of Animal Fibres*, 2nd ed., Leeds, U.K., Wira, 1978.
18. Bisbing, R.E., The forensic identification and association of human hair, in *Forensic Science Handbook*, Vol. 1, Saferstein, R., Ed., Englewood Cliffs, NJ, Prentice-Hall, 1982, p. 184.
19. DeForest, P.R., Foundations of forensic microscopy, in *Forensic Science Handbook*, Vol. 1, Saferstein, R., Ed., Englewood Cliffs, NJ, Prentice-Hall, 1982, p. 416.
20. Palenik, S., Microscopy and microchemistry of physical evidence, in *Forensic Science Handbook*, Vol. 2, Saferstein, R., Ed., Englewood Cliffs, NJ, Prentice-Hall, 1982, p. 161.
21. Gaudette, B.D., The forensic aspects of textile fiber examination, in *Forensic Science Handbook*, Vol. 2, Saferstein, R., Ed., Englewood Cliffs, NJ, Prentice-Hall, 1988, p. 209.
22. Petraco, N. and DeForest, P.R., A guide to the analysis of forensic dust specimens, in *Forensic Science Handbook*, Vol. 3, Saferstein, R., Ed., Englewood Cliffs, NJ, Prentice-Hall, 1993, p. 24.
23. Allen, R.M., *Practical Refractometry by Means of the Microscope*, 2nd ed., Cedar Grove, NJ, Cargille Laboratories, 1962.
24. Bisbing, R.E., The forensic identification and association of human hair, in *Forensic Science Handbook*, Vol. 1, Saferstein, R., Ed., Englewood Cliffs, NJ, Prentice-Hall, 1982, p. 209.
25. DeForest, P.R., Shankles, B., Sacher, R.L., and Petraco, N., Melt Mount 1.539 as a mounting medium for hair, *Microscope*, 35, 249, 1987.
26. DeForest, P.R., Ryan, S., and Petraco, N., Melt Mount stick mounting medium, *Microscope*, 35, 261, 1987.

27. Ogle, R.R. and Mitosinka, G.T., A rapid technique for preparing hair cuticle scale casts, *J. Forens. Sci.*, 18, 82, 1973.
28. Petraco, N., The replication of hair cuticle scale patterns in Melt Mount, *Microscope*, 34, 341, 1986.
29. Petraco, N., A microscopical method to aid in the identification of animal hair, *Microscope*, 35, 83, 1987.
30. Petraco, N., A modified technique for the cross-sectioning of hairs and fibers, *J. Pol. Sci. Admin.*, 9, 448, 1981.
31. Wildman, A.B., *The Microscopy of Animal Textile Fibres*, Leeds, U.K., Wool Industries Research Association, 1954, p. 26.
32. Heyn, A.N.J., *Fiber Microscopy: A Textbook and Laboratory Manual*, New York, Interscience, 1954, p. 142.
33. Wildman, A.B., *The Microscopy of Animal Textile Fibres*, Leeds, U.K., Wool Industries Research Association, 1954, p. 27.
34. Petraco, N. and Gale, F., A rapid method for cross-sectioning of multilayered paint chips, *J. Forens. Sci.*, 29, 597, 1984.

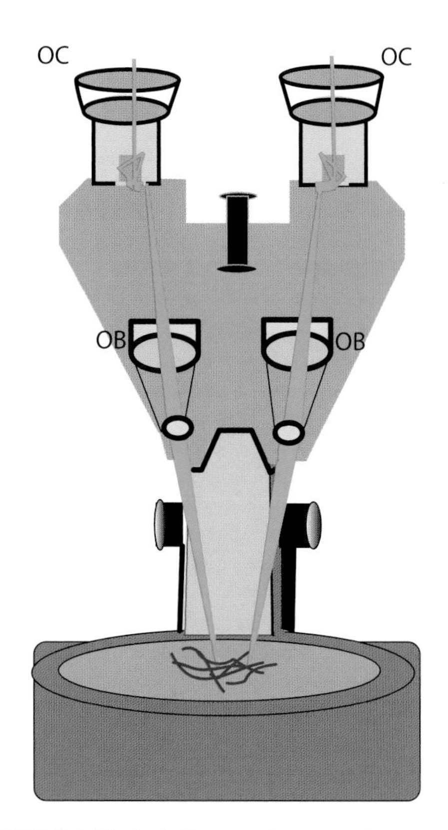

FIGURE 2.1 Typical Greenough stereomicroscope. Note the positions of the two objective lenses (OB) and two oculars (OC).

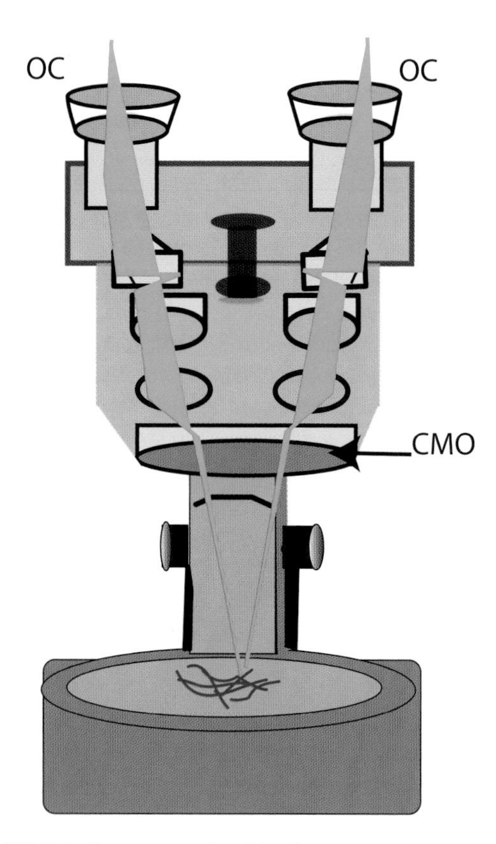

FIGURE 2.2 Common main objective stereomicroscope. Note the position of the main objective lens (CMO) and the two oculars (OC).

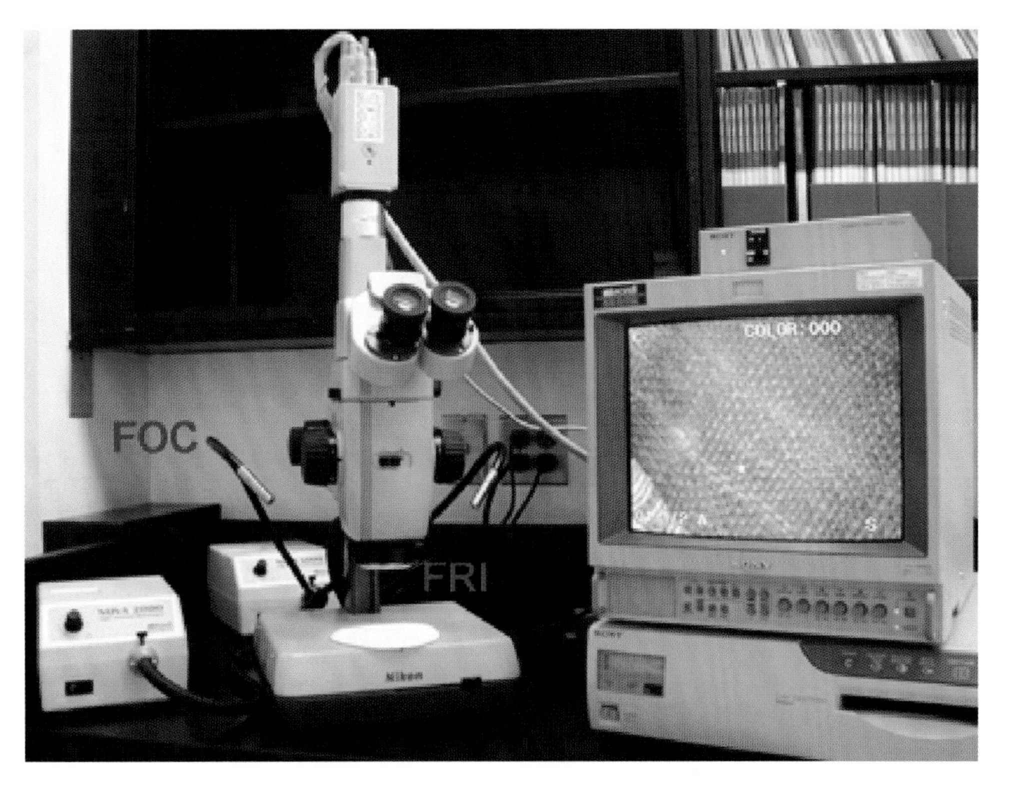

FIGURE 2.3a CMO stereomicroscope shown with a reflected light, fiber optic ring illuminator (FRI) providing axial illumination, auxiliary bifurcated fiber optic illumination system (FOC) providing oblique illumination, and digital video system. This SM is ideal for examining textiles, threads, cordage, papers, and flat materials.

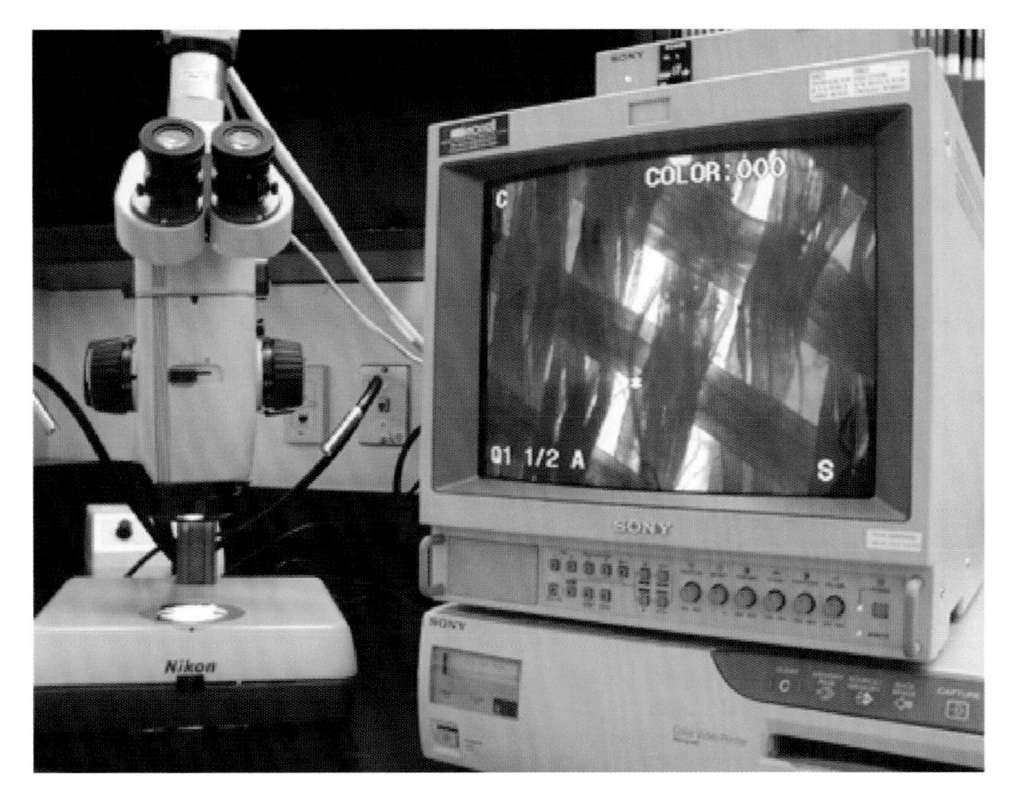

FIGURE 2.3b CMO stereomicroscope with transmitted illumination (also capable of dark-field illumination) and a digital video/photographic system. This SM is ideal for examining thin transparent or translucent substances such as thin wood sections, paper, and fine tapestries.

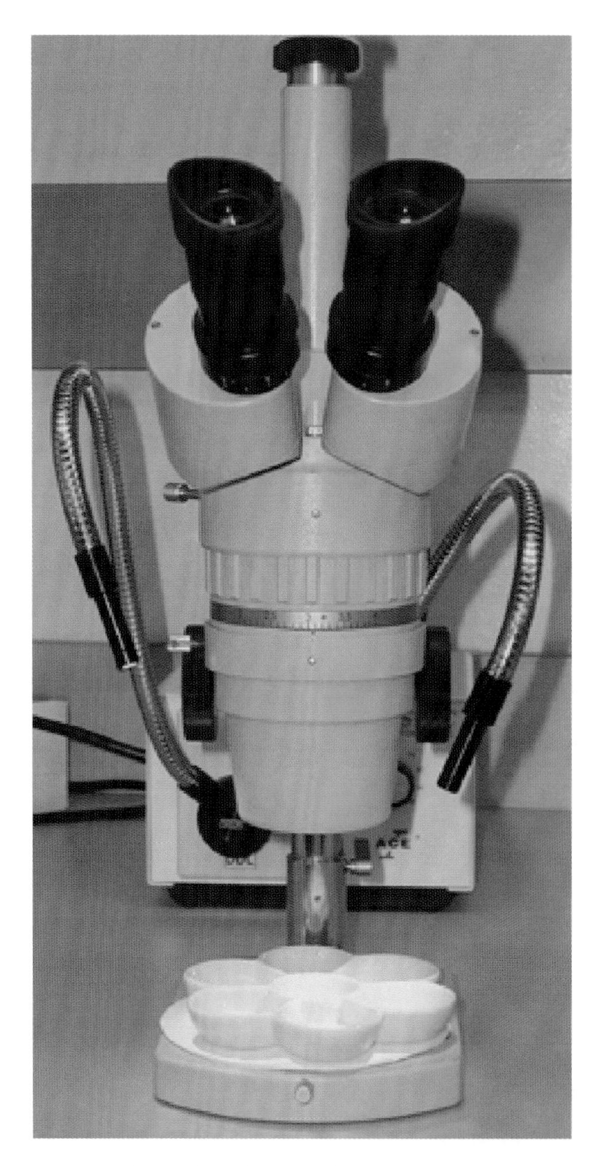

FIGURE 2.4 Greenough stereomicroscope with a reflected light-bifurcated fiber optic oblique illumination system and porcelain watercolor palette used as a specimen-sorting dish. This form of SM is useful for examining three-dimensional objects, manipulating samples, isolating small fibers or particulates, collecting known specimens for artifacts, and in chemical microscopy.

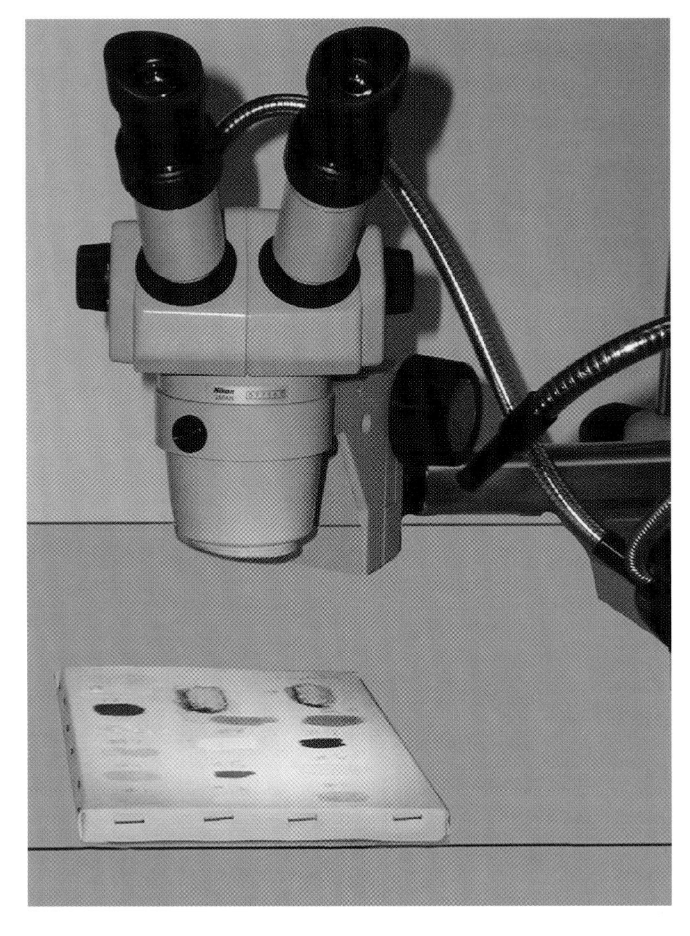

FIGURE 2.5a Boom stand is integrated into this tabletop stereomicroscope. It is useful for examining objects such as small paintings, uniforms and clothing, statues, firearms, and other small- to medium-sized objects.

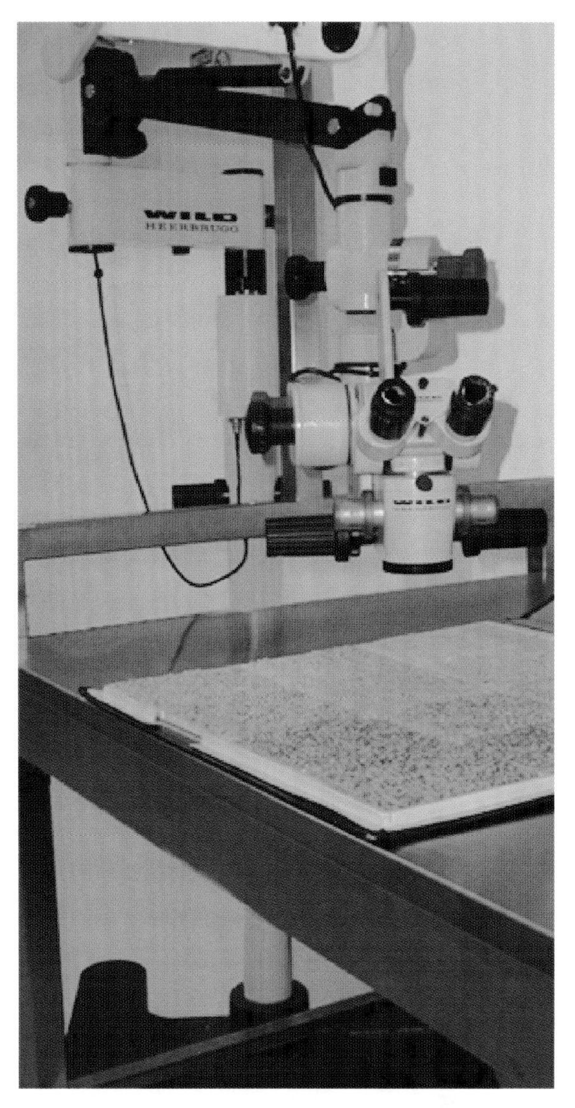

FIGURE 2.5b Floor stand stereomicroscope. This model is useful for examining and restoring large objects such as engine blocks, large paintings, wall frescoes, tapestries, costumes, furniture items, bronze and marble statues, and other large of items.

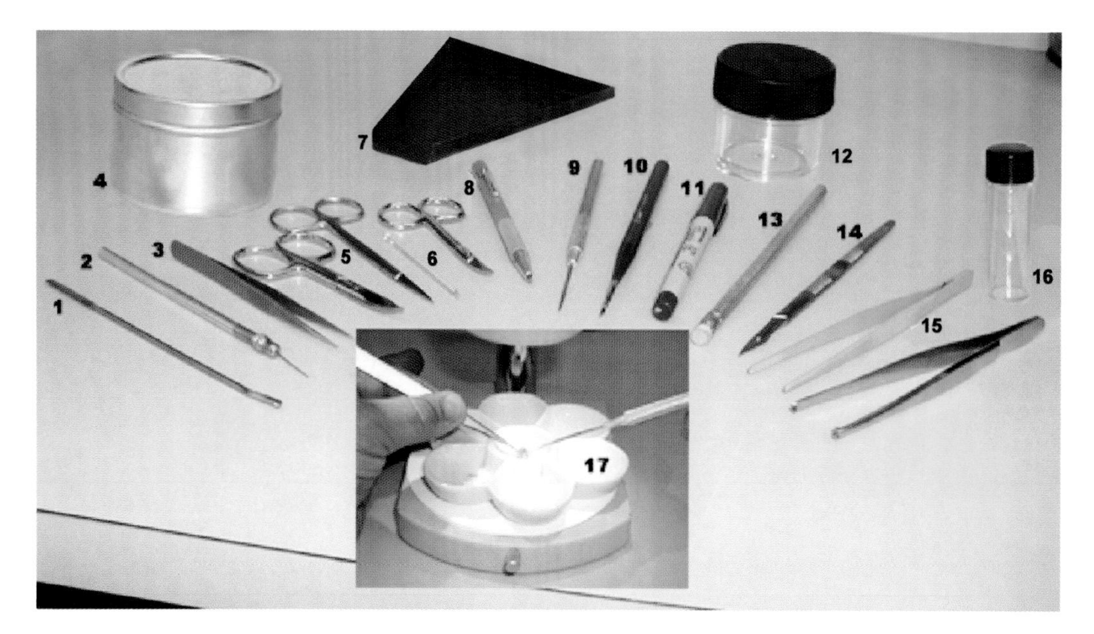

FIGURE 2.6 Assortment of implements useful for preliminary examination: (1) microspatula, (2) fine needle, (3) fine tweezers, (4) aluminum canister, (5) fine scissors, (6) glass rod, (7) colored tray, (8) diamond scribe with magnet, (9) needle probe, (10) fine brush, (11) permanent marker, (12) heavy-duty plastic jar, (13) pencil with eraser, (14) scalpel, (15) heavy-duty Teflon-coated and stainless steel tweezers, (16) glass vial, and (17) porcelain sorting dish.

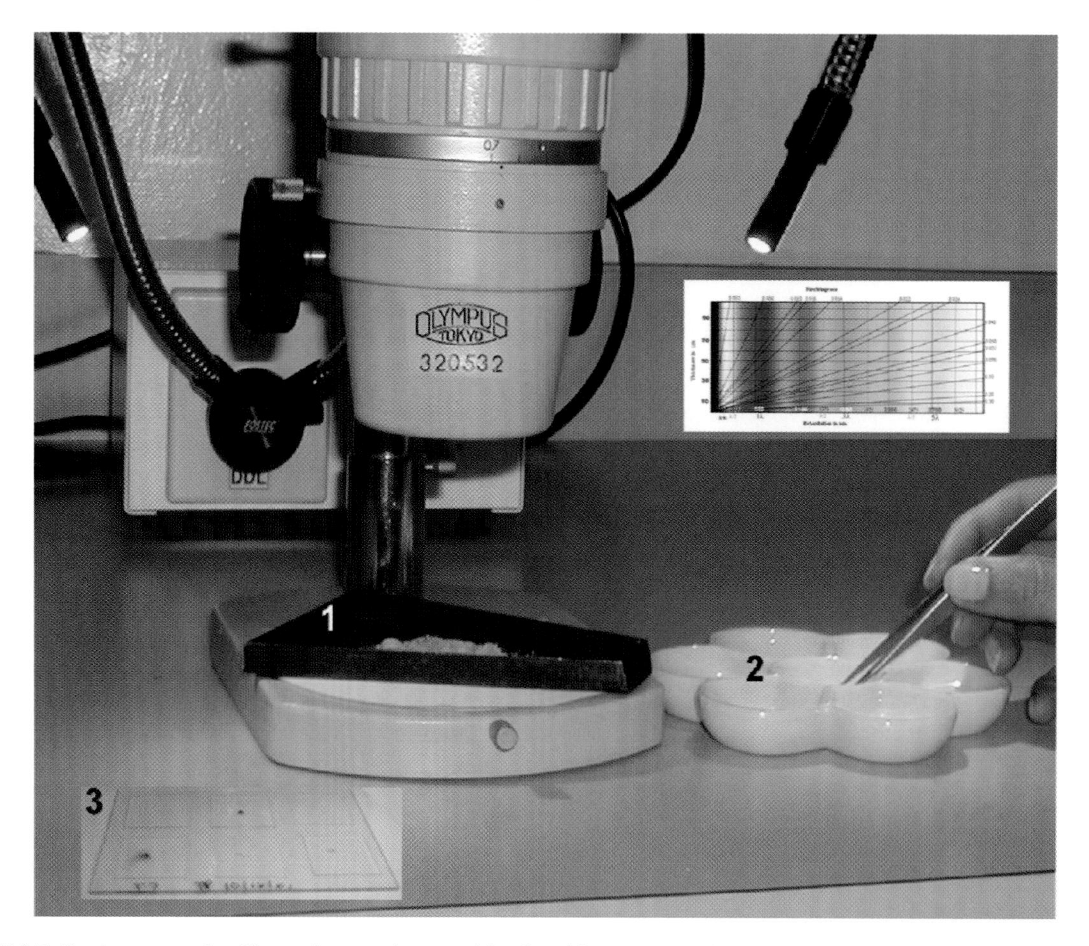

FIGURE 2.7 Sorting a sample: (1) specimen to be sorted is placed in an appropriate colored tray; (2) each type of substance is removed from the specimen and placed into a separate section of a porcelain sorting dish; (3) materials are mounted onto microscope slides for further investigation. Each type of material collected should be mounted individually or stored in an appropriate container.

PRELIMINARY DATA SHEET

I - Morphology (Shape)

Fibrous___ Particulate____ Both_________

II - Homogeneity

Homogeneous:	Yes____ No____		
Fiber:	Separate__	Cluster__	Both___
Particle:	Separate__	Cluster__	Both___
Heterogeneous:	Yes___	No ___	
Aggregate of Both	Primary Forms	Yes___	No___

No. of possible Fiber types_________

No. of possible Particle types________

III - Initial Classification

Shape:________ Sketch:

If Fiber:

Hair: Human ______________ Animal_____

Synthetic Fiber ______________________

Vegetable Fiber ______________________

Feather______________________________

Other Fiber ___________________________

Other Fiber ___________________________

If Particles:

Mineral Grain(s)_________ Glass Chip(s)_______ Paint Chips________

Other Particulate ___________________________

FIGURE 2.8 Preliminary examination data sheet.

FIGURE 2.9 Grooved Coplin jars filled with glass cleaner; examples of precleaned-from-package and cleaned microscope slides.

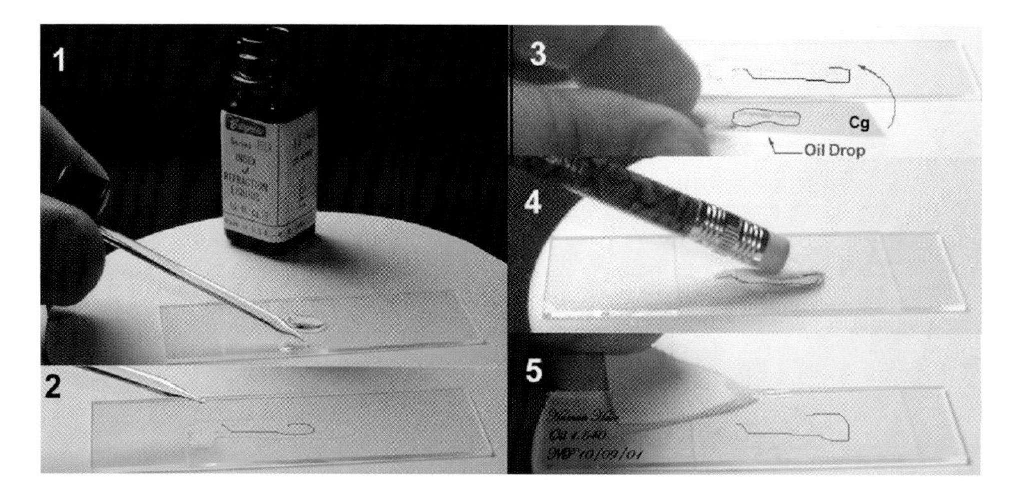

FIGURE 2.10 Temporary mount in Cargille oil: (1) place one or two drops of the oil on the slide (the glass rod has been modified for easier application of oil); (2) place specimen in oil; (3) place cover glass with oil on top of specimen; (4) use a pencil eraser to gently press the cover glass to remove air; and (5) drain off excess oil with a small piece of bibulous paper and label slide.

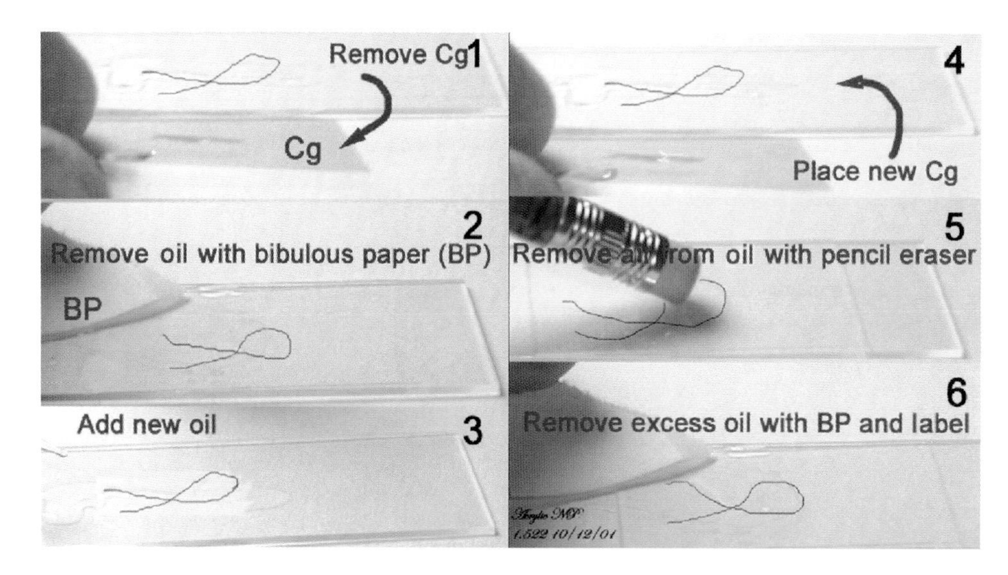

FIGURE 2.11 Remounting in Cargille oil: (1) remove cover glass; (2) remove old oil with bibulous paper; (3) add one or two drops of new oil to specimen with modified glass rod; (4) place a new cover glass with oil onto specimen; (5) use a pencil eraser to gently press the cover glass to remove air; and (6) drain off excess oil with a small piece of bibulous paper and label slide.

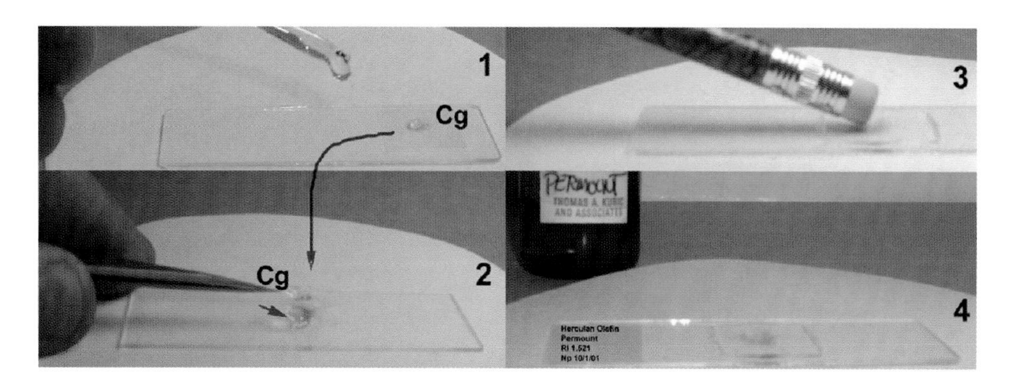

FIGURE 2.12 Permanent mount in Permount®: (1) place one or two drops of Permount on a slide and cover glass (the glass rod has been modified for easier application of medium); (2) place specimen in Permount and then place the cover glass on top of specimen; (3) Use a pencil eraser to gently press the cover glass to remove air; and (4) remove excess medium with a small piece of bibulous paper and label slide.

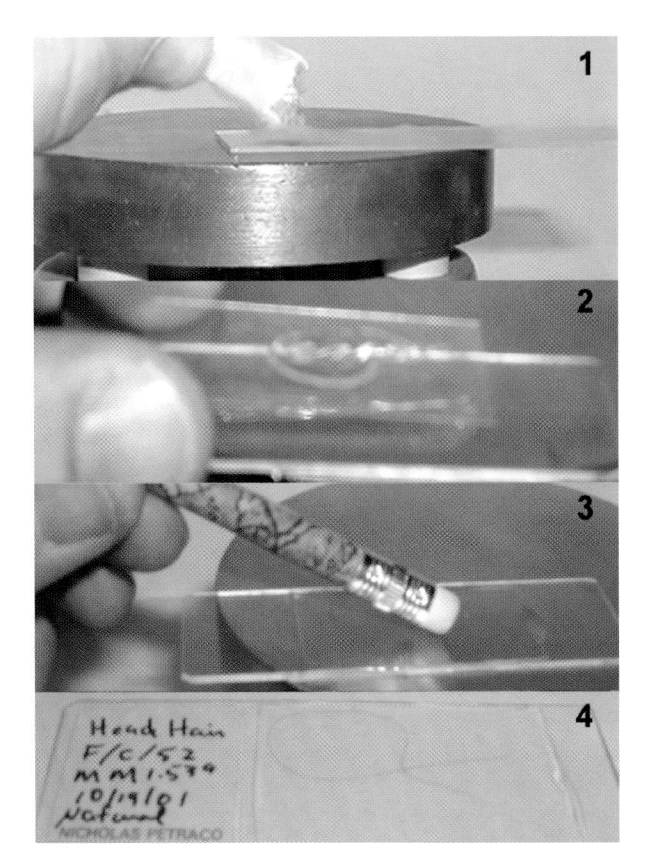

FIGURE 2.13 Permanent preservation in Melt Mount 1.539: (1) place a microscope slide onto a hotplate (70 to 80°C, drop Melt Mount onto the slide, and spread evenly with a cover glass; (2) place the specimen on the area of the slide containing Melt Mount, place a dab of Melt Mount on a cover glass, then place the cover glass face down onto the warmed preparation on the slide; (3) use a pencil eraser to press the cover glass to remove air; and (4) allow the slide to cool and label the preparation.

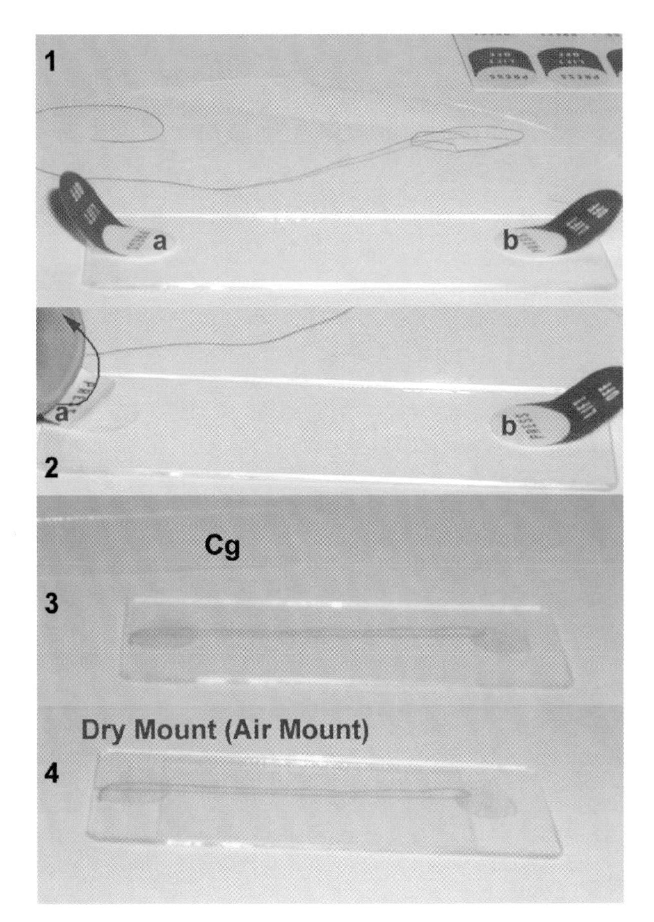

FIGURE 2.14 Dry or air mount: (1) place double sticky tabs (a and b) onto a microscope slide; (2) press tabs a and b down and peel the lift-off portions away from the slide; (3) press the specimen ends into the adhesive left by the tabs; and (4) place a cover glass onto the resin deposits and press down lightly. The cover glass and specimen will adhere to the slide. Double-sticky tape can also be utilized to secure the specimen and cover glass. Dry mounts can be viewed with stereomicroscopes and light microscopes.

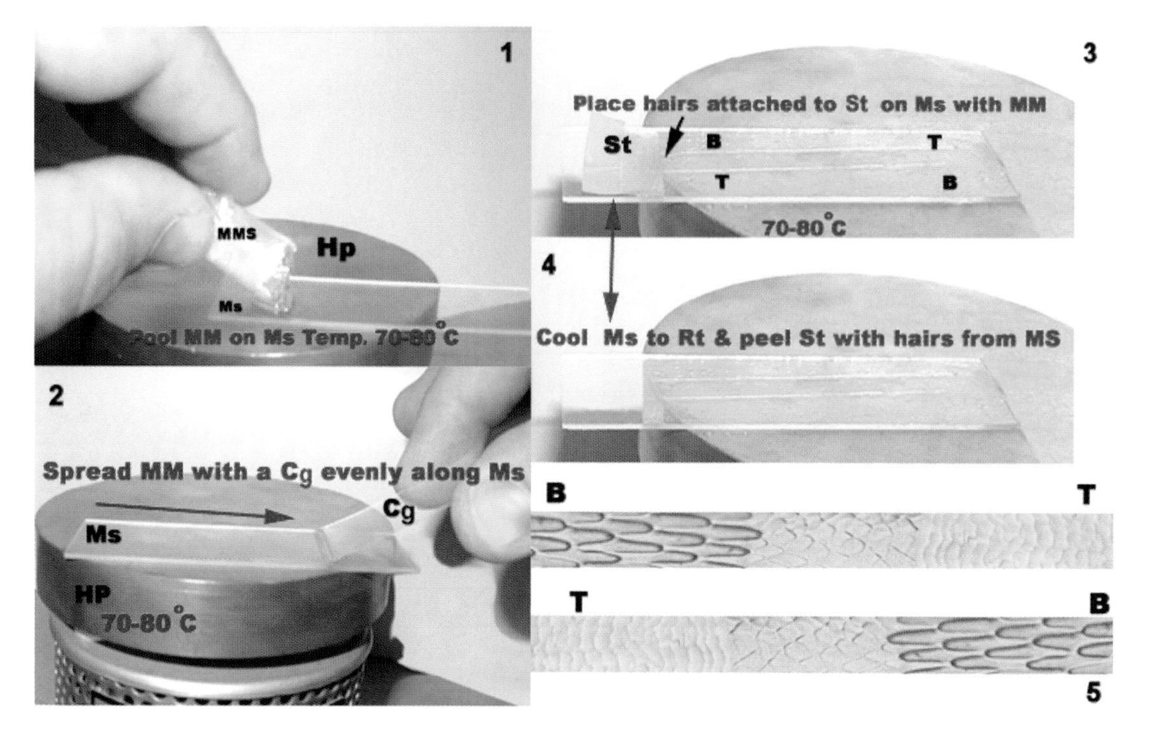

FIGURE 2.15 Preparation of temporary scale cast in Melt Mount 1.539: (1) place a microscope slide onto a hotplate (HP) set at 70 to 80°C and melt a stick of Melt Mount onto the slide; (2) spread the Melt Mount evenly with a cover glass; (3) place hair specimens attached to Scotch™ tape (St) onto the portion of the slide containing the soft Melt Mount; (4) cool the slide to room temperature (Rt) and carefully peel off the tape holding the hairs; and (5) view scale patterns with microscope set for 100 to 200× magnification.

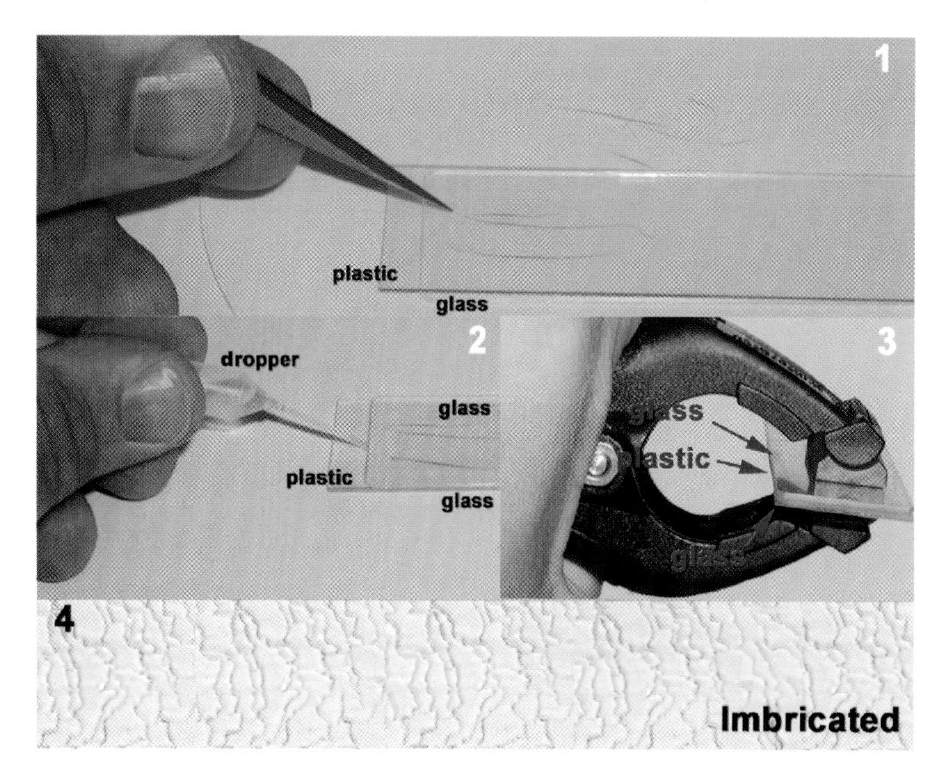

FIGURE 2.16 Preparation of permanent scale cast with Rinzl® clear plastic microscope slide: (1) place a plastic slide on top of a glass slide and arrange the hair specimens on the plastic slide; (2) under a hood, add a few drops of methylene chloride to the hair specimens and place a second glass slide on top of the plastic slide; (3) apply pressure to the preparation with a hand clamp for 1 min; and (4) after removing the clamp and top glass slide, peel the hairs from the plastic slide with tweezers and view the scale patterns as dry mounts with a compound microscope set for 100 to 200× magnification.

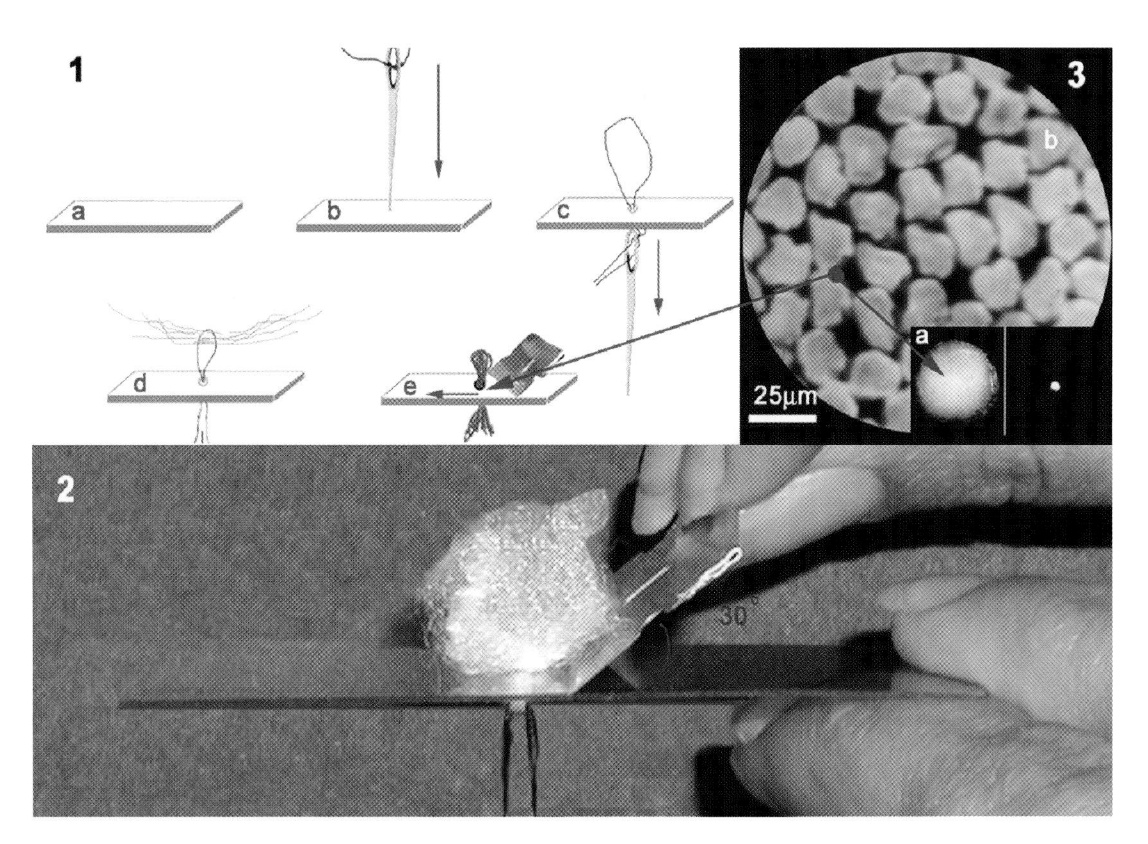

FIGURE 2.17 Preparation of hair and fiber cross-section on plastic slides: (1) flow chart showing steps for preparing samples on Rinzl clear plastic slides; (2) cutting a tuft of fibers with a single-edge, Teflon®-coated razor blade; and (3) light microscopy view of a cross-section of rayon fibers (200×).

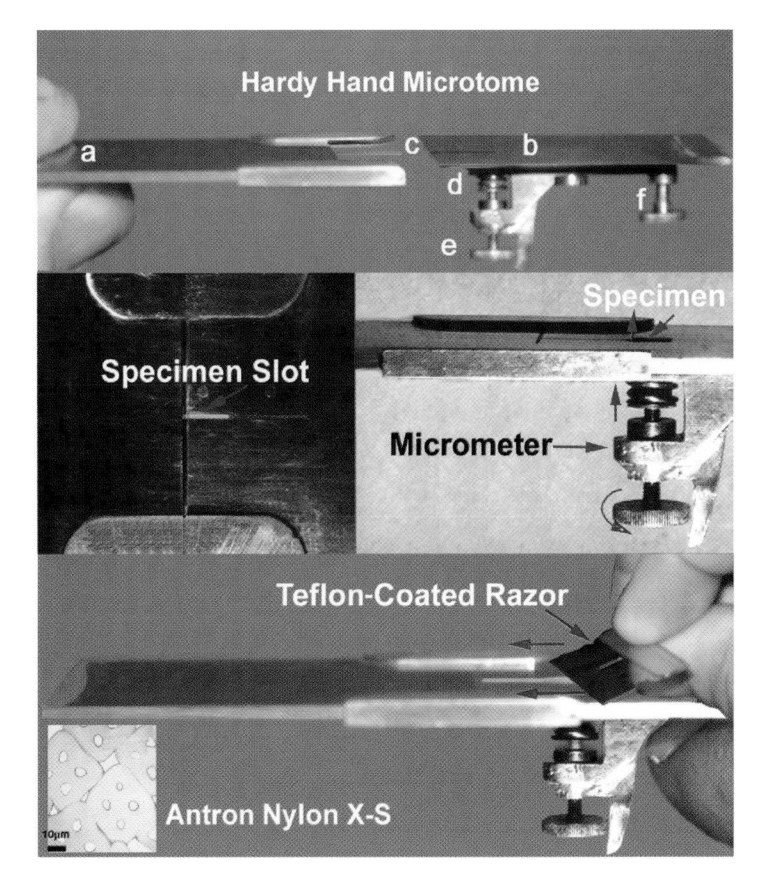

FIGURE 2.18 Preparation of hair and fiber cross-sections with a hand microtome. Top: (a) left slide section; (b) right slide section; (c) specimen slot; (d) specimen plunger; (e) micrometer dial; and (f) micrometer lock. Middle: fibers are placed into the specimen slot and the left and right portions of the slide are fitted and gently squeezed together (left); as the micrometer is rotated, the plunger pushes the fiber upward (right). Bottom: the fiber specimen is cut with a single-edge, Teflon-coated razor blade, mounted on a slide and viewed with a light microscope at 200×.

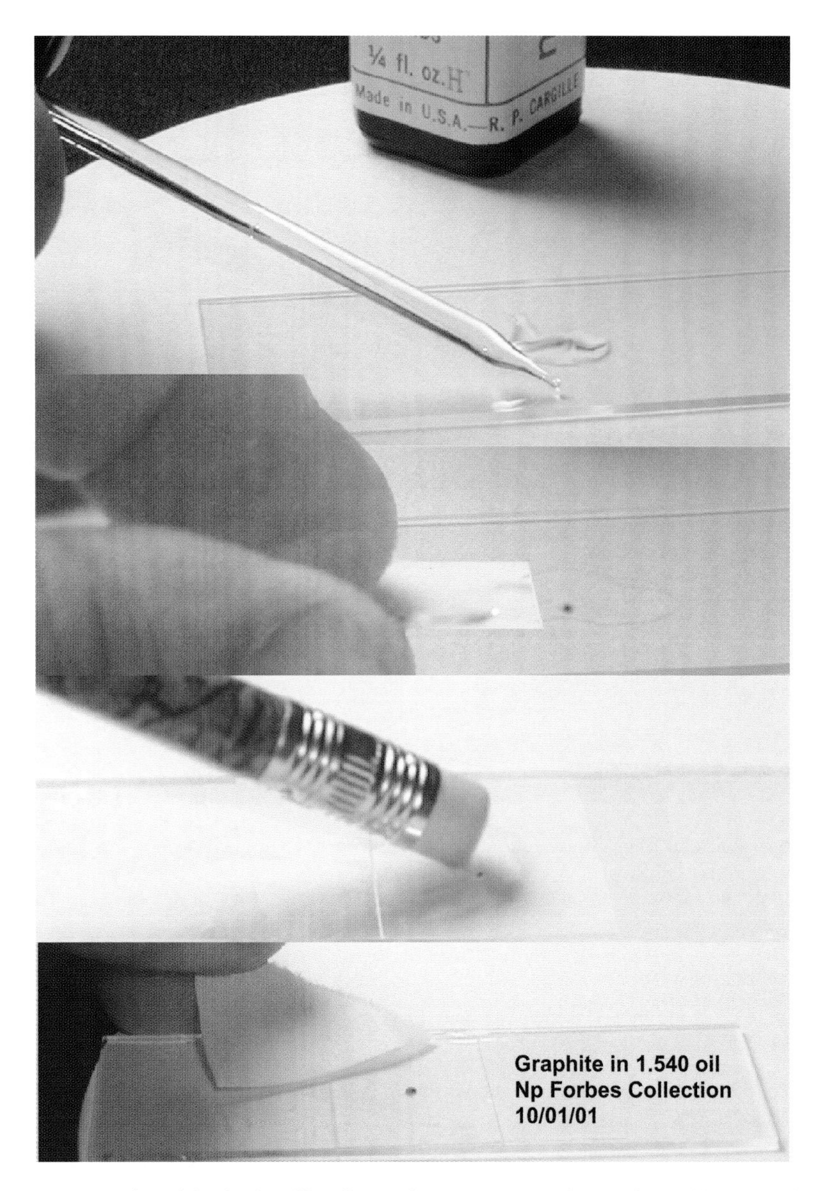

FIGURE 2.19 Temporary mount of particles in Cargille oil: (1) place one or two drops of the oil onto a microscope slide (the glass rod has been modified for easier application of oil); (2) place specimen in oil; (3) place cover glass with oil on top of specimen; (4) gently press on cover glass with a pencil to remove air; and (5) drain off excess oil with a small piece of bibulous paper and label slide.

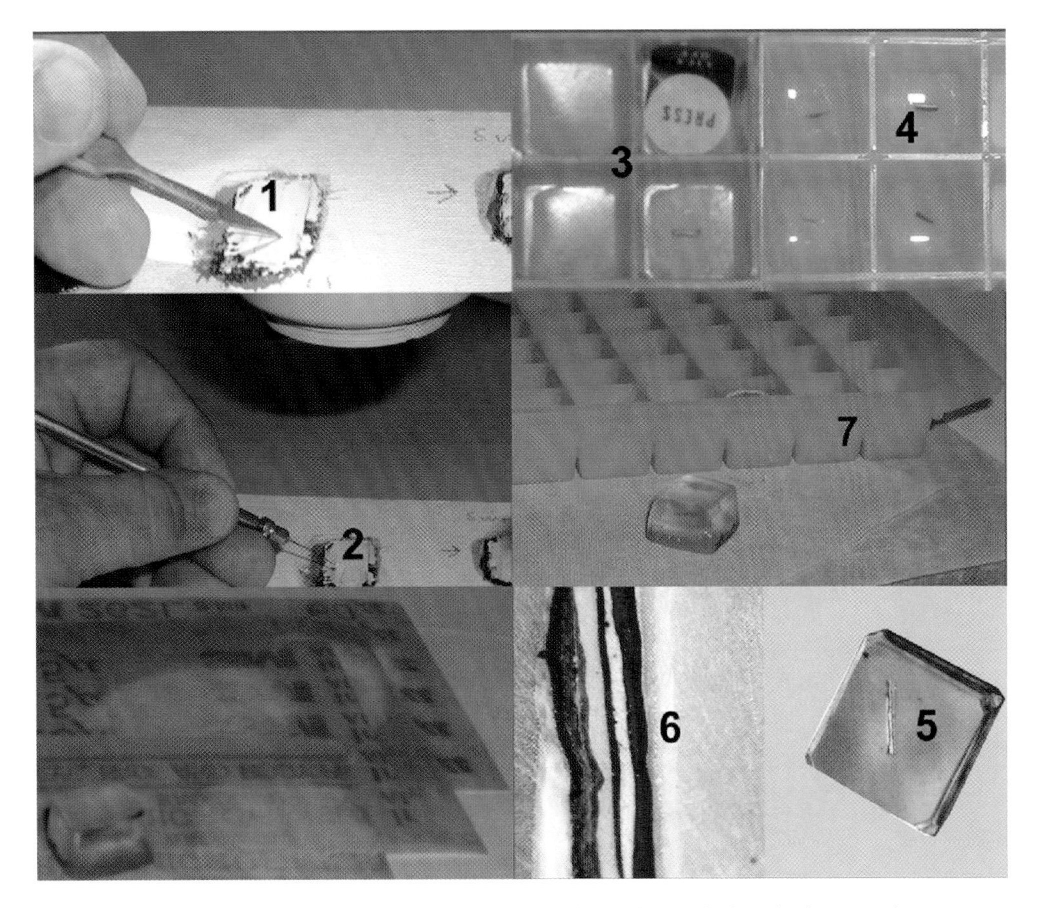

FIGURE 2.20 After removal (1 and 2), the specimen to be cross-sectioned is attached to the bottom of a compartment of an ice cube tray into which a tab of double-sticky press glue has been applied (3 and 4). The cross-section of the specimen chip is positioned in the center of the compartment (4). A 2-mL aliquot of catalyzed polyester casting resin is poured into the compartment and allowed to harden under a fume hood (4). After the cured resin cube is removed from the tray, the stratigraphy of paint layers is exposed by sanding and polishing the bottom surface of the cube with increasingly fine grades of micron-graded (30, 20, 15, 9, 3, 2, and 1 μm) finishing paper available from 3M (5). The layer structure can then be viewed with a low magnification stereomicroscope or compound microscope equipped with reflected light illumination (6 and 7).

3 Basic Observations and Measurements with Polarized Light Microscopy

REFRACTIVE INDEX AND RELIEF

To collect fundamental information about the appearance or physical structure of a specimen, an observer must be able to see the specimen clearly. The magnitude to which a colorless to lightly colored transparent particle or fiber can be seen when immersed in a colorless or nearly colorless mounting medium (MTM) is known as relief. If the specimen and MTM have the same refractive index (RI), the specimen will not be visible with a PLM under plane-polarized light (PPL). If the RI of the MTM is near that of the specimen, low relief results.

If RI of an MTM is somewhat different from the RI of the specimen, moderate relief results and the specimen will be fairly visible in the MTM. If the difference between the RIs of the MTM and specimen is large, high relief will result and the specimen will appear to stand out from the MTM. Figure 3.1 depicts all three relief conditions. Whether the particle or the MTM has a higher or lower RI does not matter. The degree of relief will appear the same in either case as long as the RI difference between the specimen and MTM is of the same magnitude.

The RI of a liquid or solid is one of its most significant properties that can be used for identification. Various methods have been developed to determine the RIs of unknown liquids and solids. Most involve the immersion of a small portion of an unknown substance into a liquid with known RI, and then observing the appearance and behavior of the substance in the known liquid. Collectively, they are known as immersion methods.

Methods for determining RI have been published. Whether a solid or MTM has a higher or lower RI can be determined by Becke line (BL) movement. The BL is a bright halo observed along the edges of a particle or fiber when the focus of a 10× objective is raised after the specimen is in sharp focus. The BL always moves toward the substance with the higher RI when the focus of a sharply focused specimen is raised (working distance increases; see Figure 3.2).[1–14]

Two common methods involve the immersion of an unknown solid into a known liquid and observing the BL that develops around the edges of the solid when the focus is raised. Polychromatic illumination is used for one method. The second utilizes monochromatic illumination.

When light passes through a small piece of transparent solid immersed in a liquid having a lower RI, it bends toward the solid. As light passes through a small piece of transparent solid immersed in a liquid having a higher RI, it bends toward the liquid (Figure 3.3). If white or polychromatic light is used to illuminate a solid, a BL in the shape of a bright white halo forms when the focus is raised. The movement of the BL is then used to determine whether the solid or MTM has the higher RI. Remember, when the focus of a 10× objective is raised after being sharply focused on a specimen, the BL always moves toward the medium of higher RI. Figures 3.3 and 3.4 illustrate BL development and movement. The BL is shown as a bright white halo moving toward the particle (Figure 3.4, left). The BL is a bright white halo moving toward the MTM (Figure 3.4, right). The solid in the center section is in sharp focus and exhibits no BL.

Figure 3.5 shows BL movement when the specimen is viewed with plane-polarized light and a PLM. When the focus is raised, the white BL may move inward (left), indicating that the RI of the glass is higher than that of the liquid (HTL); the white BL may move outward (right), indicating that the RI of the glass is lower than that of the liquid (LTL). A colored BL is formed (center) when the focus is raised and the particle and liquid RI have nearly the same RI values.

When determining the RI of a material, it is crucial to remember that refractive index is wavelength dependent. For most materials, refractive indices are higher for smaller wavelengths (violet to blue) and lower for larger wavelengths (green to red). Due to this phenomenon, a standard wavelength of monochromatic light had to be established in order to achieve consistent and accurate RI readings. Over the years, the light emitted when sodium burns has been accepted by convention as the reference wavelength. The values of standard RI liquids and solids are normally reported for the sodium D line (589 nm), yellow light at 25°C. Other wavelengths such as the hydrogen C line (656 nm), red light, and F line (486 nm), blue light, are used to report dispersion values of liquids and solids.

Figure 3.6 depicts the dispersion of white light when a solid particle is observed under polychromatic and monochromatic lights. On the right, the RI of the solid is near that of the MTM and the white light is dispersed into its component colors. RI is normally greater for shorter wavelengths than for longer wavelengths. Thus, when the focus is raised, the blue wavelengths are dispersed outward toward the liquid while the red wavelengths are dispersed inward toward the solid. The yellow or matching wavelengths (λ_m) stay around the edges of the solid. When monochromatic sodium D line (589 nm) is used to illuminate the specimen (Figure 3.5, left), the yellow or matching wavelength λ_{589nm} stays around the edges when the focus is raised.

Many microscopists use another immersion method known as dispersion staining (DS). One form of this method involves placing a black central stop (CS) at the back focal plane of a 10× achromatic objective. A common practice is to place a round dot (2 to 3 mm in diameter) of black ink or flat black paint at the center of an 18-mm round cover glass. The treated cover glass is placed at the back focal plane of the objective. The particles or fibers to be examined are mounted in an RI oil standard and viewed with a PLM using plane-polarized light (PPL). If the RI values of the isotropic particle and the oil it is mounted in match, the polychromatic light (~6000 K) illuminating the specimen is dispersed into its component colors. When focused, the particle's edges will appear as bluish/violet against a black background. When the focus is raised the blue wavelengths will refract toward the oil and the red wavelengths will refract toward the particle. The blue wavelengths get past the CS, while most of the remaining wavelengths are blocked by the CS. Thus, each particle will have a bluish halo around its edges, while the background appears black. It is important to note that the resulting colored halos may vary in hue depending on several factors: the dispersion properties of the mounting oil and specimen involved, the color temperature of the light source, the illumination intensity, and the degree to which the beam of illumination is axial. As this phenomenon is somewhat complex the interested reader is encouraged to review the literature regarding the theory and practice of the various dispersion staining methods and techniques. Figure 3.6 illustrates a simple example employing the CS dispersion staining technique.

PHYSICAL AND MORPHOLOGICAL PROPERTIES

Now that we know how to obtain a clear image of a specimen, data concerning its physical structure can be easily and accurately collected. Features such as color, size, thickness, width, length, shape, surface texture, surface appearance, opacity, optical density, crystal system, and interfacial angles can all help reveal the identity of a sample and they must be meticulously studied and recorded. Particular attention must be given to specimen shape. A shape that is elongated or short, flat or cubic, round or square, has sharp or rounded edges, appears irregular or has structure indicates much about the identity of an unknown. A PLM can be used to collect all of these data and more.

Although crystallography is a complex subject that will only be touched upon in this work, it is important to recognize basic crystal shapes. Interested readers are referred to the cited literature for additional study. Every microanalyst should understand fundamental crystallography. Many materials encountered by microscopists fall into one of the six basic crystal systems. Thus, it is important to be able to determine whether a specimen has a crystalline or amorphous internal structure. This is achieved easily by viewing a specimen between crossed polars. Amorphous substances and crystals of the cubic (isometric) system will remain black (extinct) on stage rotation, while fibers, grains of minerals, and inorganic or organic salts belonging to one of the other five systems will display the most intense interference colors every 90°. Figure 3.7 depicts the six crystal systems.

Pure crystalline substances normally grow into common shapes known as habits. However, it is not unusual for crystals to distort as they grow. Phenomena such as polymorphism (many crystalline forms of the same compound or element), isomorphism (similar crystalline forms and habits for different compounds), and twinning (intergrowth of two or more crystals) can create many appearance variations. The effects of rain, running water, wind, ice, snow, and temperature changes can also help determine crystal shape. Consequently, crystals can present many manifestations of their basic shapes (Figure 3.8).

MICROMETRY

Measurement of small linear distances, angles, and areas with a microscope is known as micrometry. Quantitative measurement with a microscope involves the use of various types of ocular scales, some of which are calibrated with a stage micrometer. The normal unit of measurement for length, width, and thickness is the micrometer (μm). One micrometer equals a millionth of a meter or 10^6 m.

Two distinct types of micrometer devices are used with microscopes. The first is the ocular micrometer (OCM), usually an arbitrary scale produced on a round disc of glass placed at the primary focal plane of the ocular. The ocular scale can take on many configurations. It can be a cross-hair, a small ruler made from a vertical or horizontal line with 100 equal divisions, a cross-hair made by combining vertical and horizontal rulers, a grid divided into 100 equal squares, or other form.

The second type is the stage micrometer (SM), typically a scale of known length, normally 1 mm, with 100 equal divisions, each having a value equal to 10 μm (K). To determine the value for each ocular micrometer division (OCMD), align the image of the SM with the image of the OCM. Both micrometers are aligned as a vernier, and the numbers of OCMDs and SM divisions (SMDs) are counted and recorded. The value in micrometers for one OCMD in Figure 3.9 is computed as follows:

$$1 \text{ OCMD} = \frac{\text{No. of SMD} \times \text{K}}{\text{No. of OCMD}} \quad (\text{where K} = 10\ \mu\text{m})$$

$$1 \text{ OCMD} = \frac{6 \times 10\ \mu m}{60} = \frac{60\ \mu m}{60} = 1\ \mu m$$

This procedure must be followed for each objective lens. The value for each objective should be recorded and taped on the base of the microscope for quick reference. The values will remain the same as long as the objectives, oculars, and body tube or head are not changed or altered. Figure 3.10 illustrates the alignment of stage and ocular micrometers. Figure 3.11 shows actual images observed when a 20× objective was calibrated. It depicts the measurement of a human head hair cross-section with a calibrated OCM.

The design of the PLM enables it to make accurate angular and linear measurements. The cross-hair micrometer can be used to make quick angular estimates, i.e., parallel, oblique, and symmetrical extinction angles. The circular stage divided into 360° can be used to measure oblique extinction angles, interfacial angles, and many others. Accurate angular measurements are made in the following manner:

1. The angle to be measured is aligned parallel to the cross-hair micrometer (CHM), and a vernier reading is taken and noted.
2. The stage is then rotated until the plane of angle to be measured is parallel to the CHM. A second vernier reading is taken.
3. The value of the measured angle is obtained by subtracting the first reading from the second (Figure 3.12).

Finally, specimen thickness can be measured by employing the fine focus micrometer of a microscope. The top surface of the specimen to be measured is sharply focused. A micrometer reading is then taken from the fine adjustment knob and noted. The microscope focus is lowered until features at the bottom of the specimen or minute structures adjacent to the specimen are in sharp focus. A final micrometer reading is then made. Specimen thickness is computed by subtracting the final reading from the first reading and multiplying the remainder by the value for one micrometer division. When making thickness measurements, several readings should be taken and averaged. Figure 3.13 depicts the thickness-measurement procedure.

DETERMINING OPTICAL PROPERTIES

Now that we can clearly see and make accurate measurements, we can use a PLM to collect data concerning physical properties. We can observe features such as color, size, thickness, width, length, shape, surface texture, surface appearance, opacity, optical density, crystal system, and interfacial angles for identification and comparison purposes.

The specimen is first mounted in an MTM with the appropriate RI and covered with a No. $1^1/_2$ cover glass. Information about specimen morphology is collected first. The specimen is then observed between crossed polars to determine whether it is isotropic (one RI) or anisotropic (more than one RI). If the specimen is anisotropic, its thickness is measured and the degree of retardation it exhibits is estimated (Figure 3.14). Next, birefringence (difference between the specimen's smallest and largest refractive indices) is estimated and calculated using the collected data and an interference chart (Figure 3.15).

Based on particle morphology and estimated birefringence, the authors believe the specimen in Figure 3.15 is a fragment of quartz. Quartz has a published birefringence value of 0.009. The value estimated by the authors is 0.0089 — remarkably accurate for a quick estimate made without determining RI values.

Another important optical property that can help identify a mineral or characterize and compare a dyed fiber is pleochroism — the ability of a mineral or dyed fiber to differentially absorb various wavelengths of transmitted light when rotated in PPL and thus display different colors in different orientations. Figure 3.16 is a photomicrograph of the hornblende mineral when rotated in PPL.

Finally, all the information collected about a specimen's physical and optical properties should be recorded on a data sheet such as the sample that appears at the end of this chapter. Chemical properties will be discussed in the next chapter.

REFERENCES

1. Allen, R.M., *The Microscope*, New York, Van Nostrand, 1940.
2. Gibb, T.R.P., Jr., *Optical Methods of Chemical Analysis*, New York, McGraw-Hill, 1942.
3. Shillaber, C.P., *Photomicrography in Theory and Practice*, New York, John Wiley & Sons, 1944.
4. Hartshorne, N.H. and Stuart, A., *Crystals and the Polarizing Microscope*, 2nd ed., London, Edward Arnold, 1950.
5. Kirk, P.L., *Density and Refractive Index*, Springfield, IL, Charles C. Thomas, 1951.
6. Schaefer, H.F., *Microscopy for Chemists*, New York, Van Nostrand, 1953.
7. Kirk, P.L., *Crime Investigation*, New York, Interscience, 1953.
8. Bloss, F.D., *An Introduction to the Methods of Optical Crystallography*, New York, Holt, Rinehart & Winston, 1961.
9. Allen, R.M., *Practical Refractometry by Means of the Microscope*, 2nd ed., Cedar Grove, NJ, Cargille Laboratories, 1962.
10. McCrone, W.C., McCrone, L.C., and Delly, J.G., *Polarized Light Microscopy*, Ann Arbor, MI, Ann Arbor Science Publishers, 1978.

11. Miller, E.T., Forensic glass comparisons, in *Forensic Science Handbook*, Saferstein, R., Ed., Englewood Cliffs, NJ, Prentice-Hall, 1982, chap. 4.
12. DeForest, P.R., Foundations of forensic microscopy, in *Forensic Science Handbook*, Saferstein, R., Ed., Englewood Cliffs, NJ, Prentice-Hall, 1982, chap. 9.
13. Stoiber, R.E. and Morse, S.A., *Crystal Identification with the Polarizing Microscope*, New York, Chapman & Hall, 1994.
14. Bloss, F.D., *Optical Crystallography*, Washington, D.C., Mineralogical Society of America, 1999.

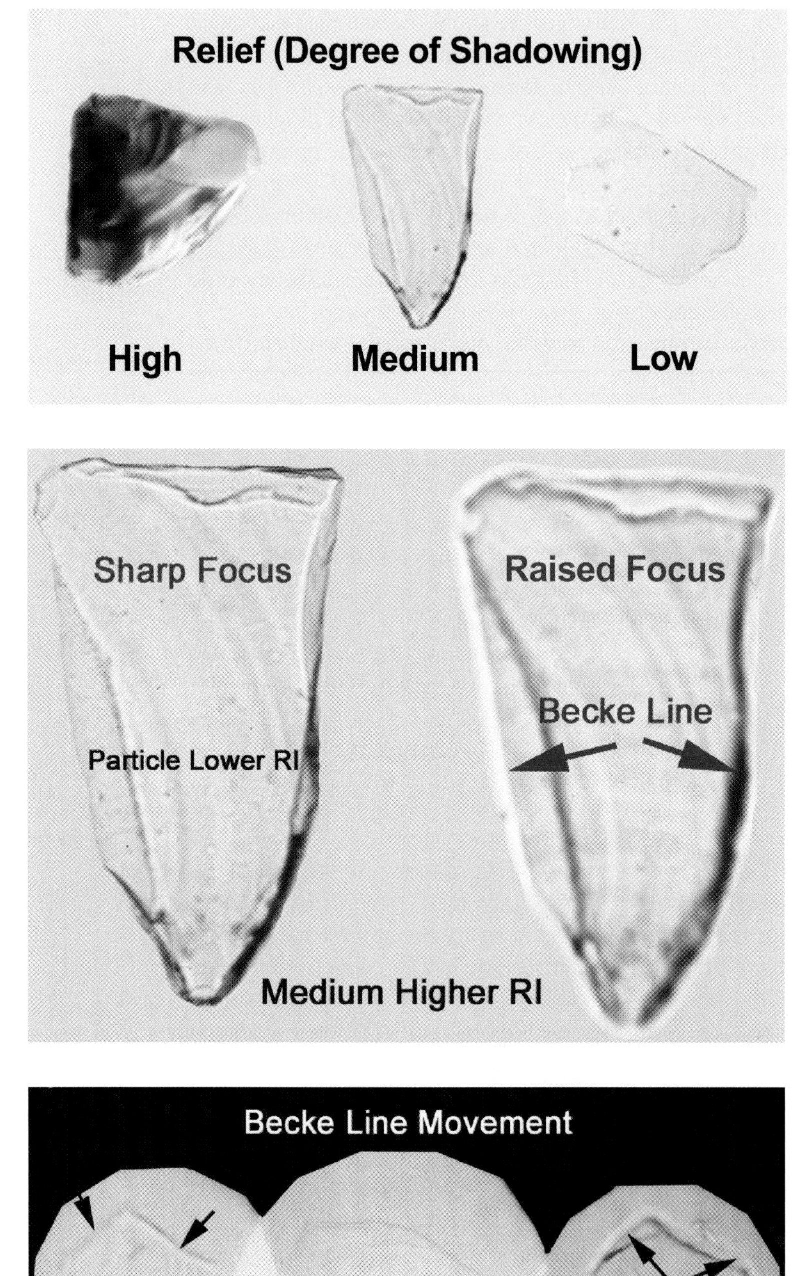

FIGURE 3.1 High, medium, and low degrees of relief.

FIGURE 3.2 Left: a particle of glass (isotropic; having one RI) is in sharp focus with a 10× objective. Right: a bright halo (BL) is formed when the focus is raised. The BL appears white, if white light is the source of illumination. In this situation, the movement of the BL toward the MTM indicates that the MTM has the higher RI.

FIGURE 3.3 Center: specimen is sharply focused and has no BL. Left: when the focus is raised, a white BL forms and moves inward toward the solid, indicating that the RI of the glass is higher than that of the liquid. Right: when the focus is raised, a white BL forms and moves outward toward the liquid, indicating that the RI of the glass is lower than that of the liquid.

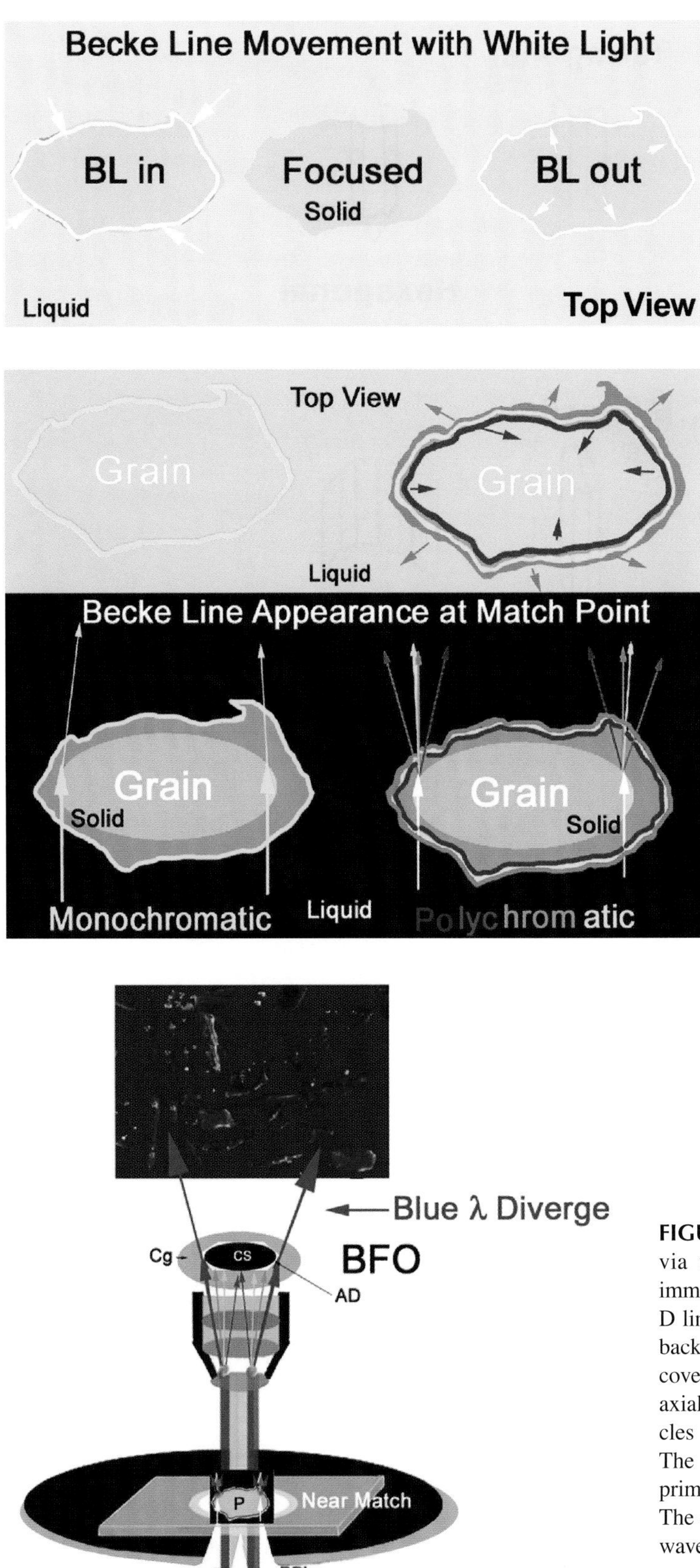

FIGURE 3.4 Three micrographs that demonstrate BL behavior when illuminated with white light. Left: when the focus is raised, the white BL moves inward, indicating that the RI of the glass is higher than that of the liquid (HTL). Right: when the focus is raised, the white BL moves outward, indicating that the RI of the glass is lower than that of the liquid (LTL). Center: when the focus is raised, a colored BL is formed, indicating that the particle and liquid RI have nearly the same RI values.

FIGURE 3.5 Appearance of the BL formed when the focus is raised. Right: monochromatic yellow light forms a yellow BL that remains around the particle edges. Left: polychromatic light forms a colored BL that disperses into its component colors when the focus is raised. The blue light moves toward the liquid, the red light moves toward the solid, and the yellow light remains around the edges.

FIGURE 3.6 Photomicrograph of glass partricles examined via the CS dispersion staining method. Glass particles immersed in oil have nearly the same RI values for the sodium D line observed with a 10× achromatic objective fitted at its back focal plane with a central stop. The CS is made from a cover glass and black ink. The PPL, polychromatic (white), axial illumination from the condenser interferes with the particles and is dispersed into its monochromatic wavelengths. The light is then collected by the objective and forms a primary image that is magnified by the eyepiece (not shown). The edges of the glass particles appear blue because the blue wavelengths (BIV) are dispersed toward the oil, while the red wavelengths (ROY) are dispersed inward toward the particle. The green (G) or matching wavelengths (λ_m) travel straight through the specimen. The condenser stops ROY and G wavelengths; BIV wavelengths are not stopped. Thus the particle edges appear blue against a dark background.

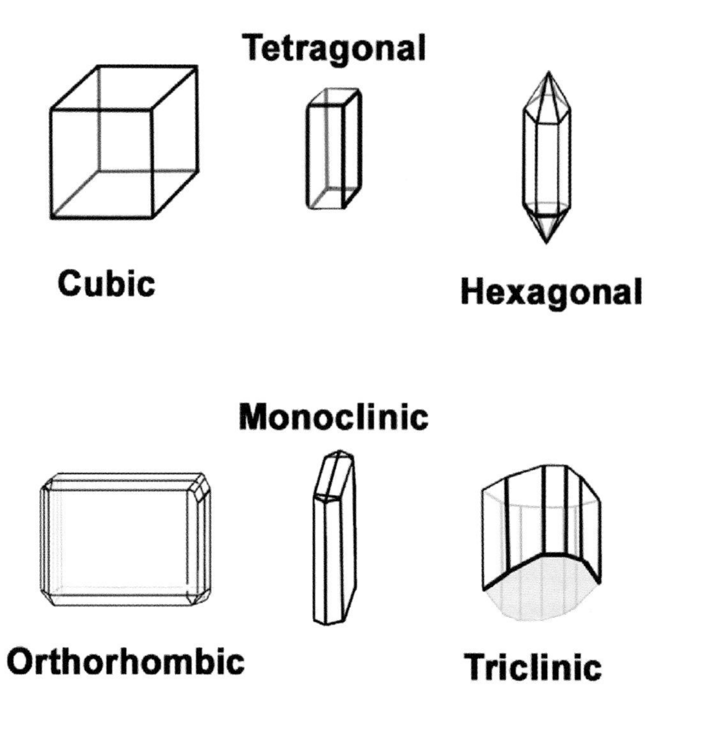

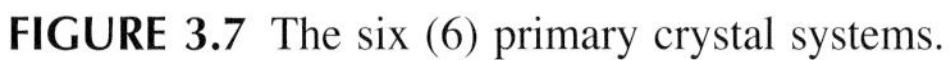

FIGURE 3.7 The six (6) primary crystal systems.

FIGURE 3.8 Microscopic appearances of common minerals. Top, right to left: chrysotile, hornblende, mica, barite, and epidote. Middle, right to left: quartz, zircon, calcite, and orthoclase. Bottom, right to left: obsidian, opal, garnet, halite, and fluorite.

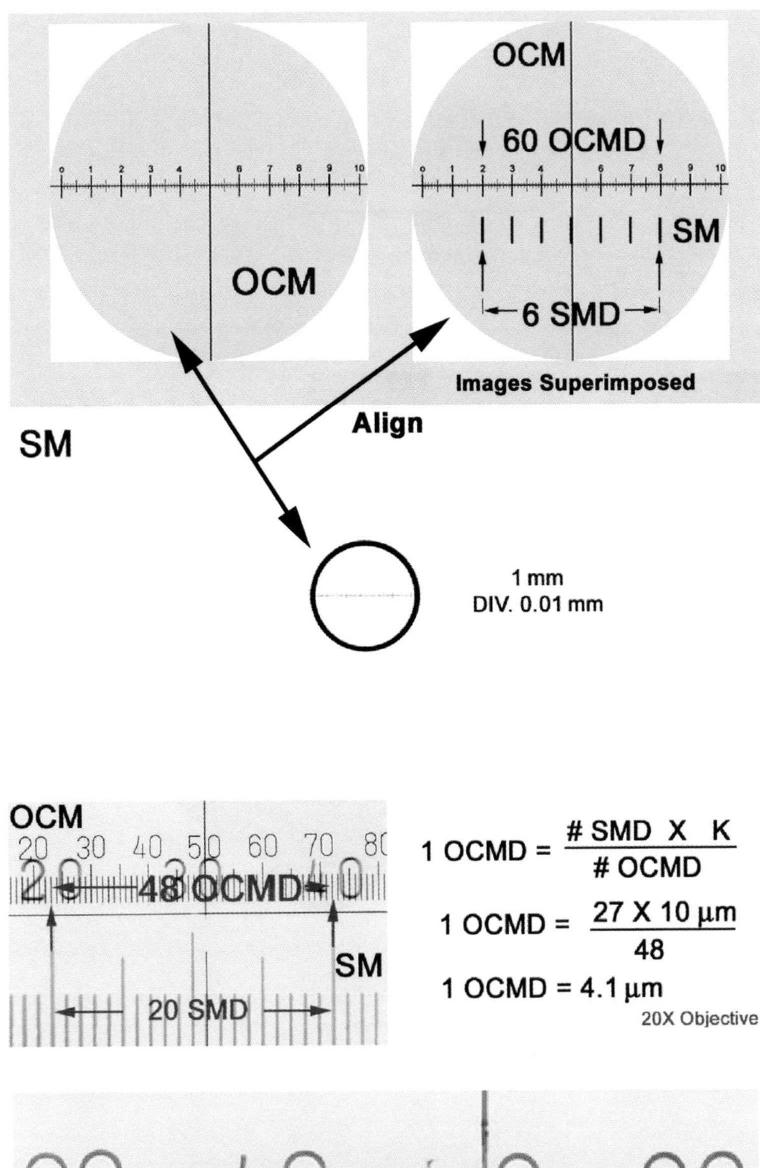

FIGURE 3.9 Top left: typical ocular micrometer (OCM). Bottom: stage micrometer (SM). Top right: calibration of an ocular micrometer (OCM). To determine the value for each ocular micrometer division (OCMD), align the images of the SM and the OCM. Both micrometers are aligned as a vernier, and the number of OCMD and stage micrometer divisions (SMDs) is counted and recorded. The value in micrometers for one OCMD equals the number of SMDs times the value for each SMD divided by the number of OCMDs.

FIGURE 3.10 To determine the value for each OCMD for a 20× objective, the objective must be focused on a SM. Next, the images of the OCM and SM are aligned as one would align a vernier scale. The number of OCMDs and SMDs that align with each other are counted and recorded. The value in micrometers for one OCMD is equal to the number of SMDs (20) times the value for each SMD (10 μm) divided by the number of OCMDs (48); 1 OCMD = 4.1 μm.

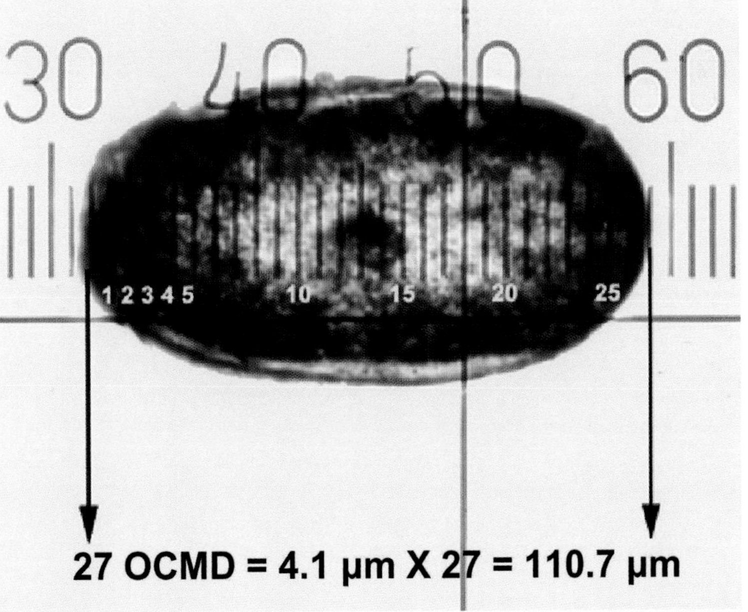

FIGURE 3.11 Measurement of a human head hair cross-section (X-S) with the calibrated OCM and 20× objective combination shown in Figure 3.10. To determine width of the hair, the image of the hair is aligned with the OCM divisions. The number of OCMDs that align with the hair cross-section are counted and recorded. The value in micrometers for the hair cross-section is equal to the number of OCMDs (27) times the value for each division (4.1 μm); the X-S equals 27 × 4.1 μm or 110.7 μm.

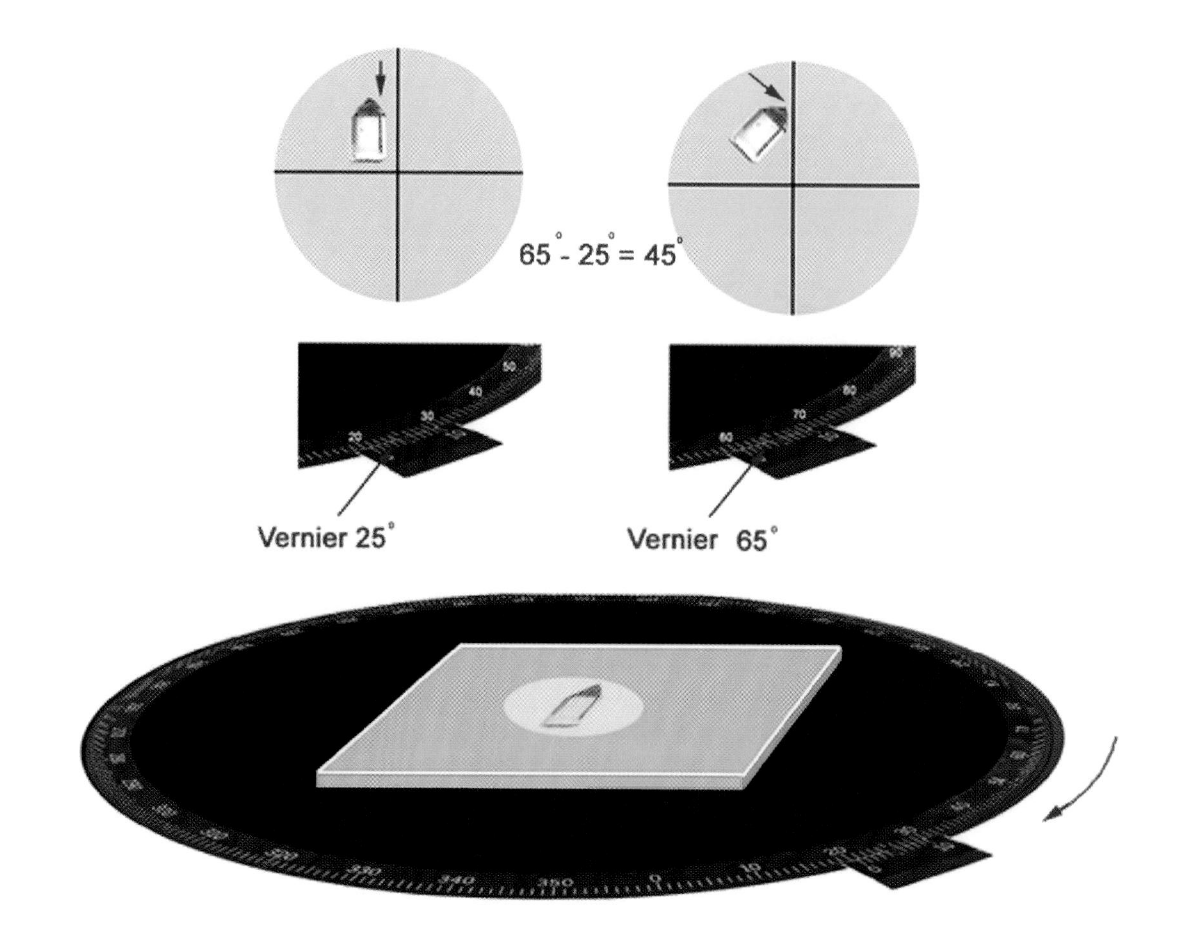

FIGURE 3.12 Top left: the image of the specimen is aligned with the cross-hair micrometer (CHM). A vernier reading is taken and the stage is rotated until the surface being measured is parallel to the CHM (right). A second reading is taken and the questioned angle is determined by subtracting the first reading from the second.

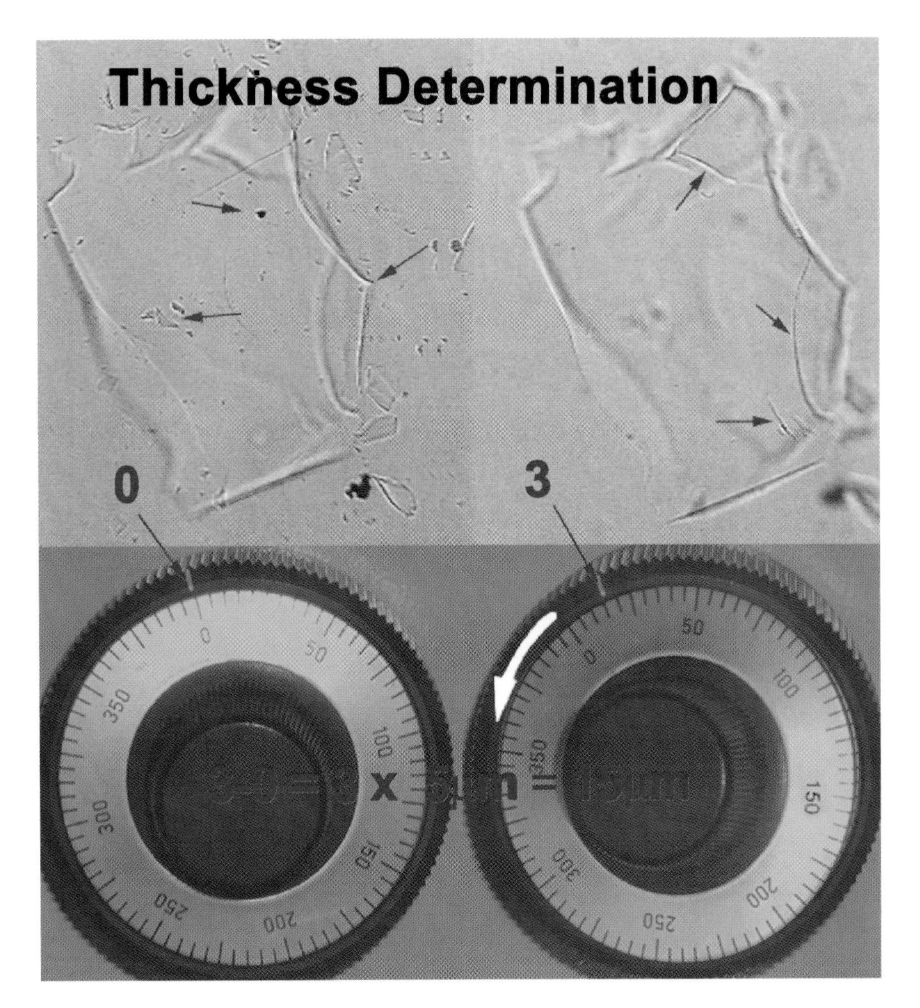

FIGURE 3.13 Left: the top surface of the specimen is sharply focused. The red arrows indicate sharply focused features on the specimen surface. A micrometer reading is then taken and noted (0). Right: the focus is lowered until fine structures at the bottom of the specimen or minute particles adjacent to the specimen are sharply focused (red arrows). A second micrometer reading is taken and noted. The first micrometer reading is subtracted from the final reading (3 – 0 = 3). To determine specimen thickness, the remainder is multiplied by the value for a single micrometer division (3 × 5 μm = 15 μm). Several readings should be made and averaged to obtain an accurate thickness value. The fine adjustment micrometer division value varies among microscope brands. The value for each micrometer division can usually be found in the manual supplied with the microscope.

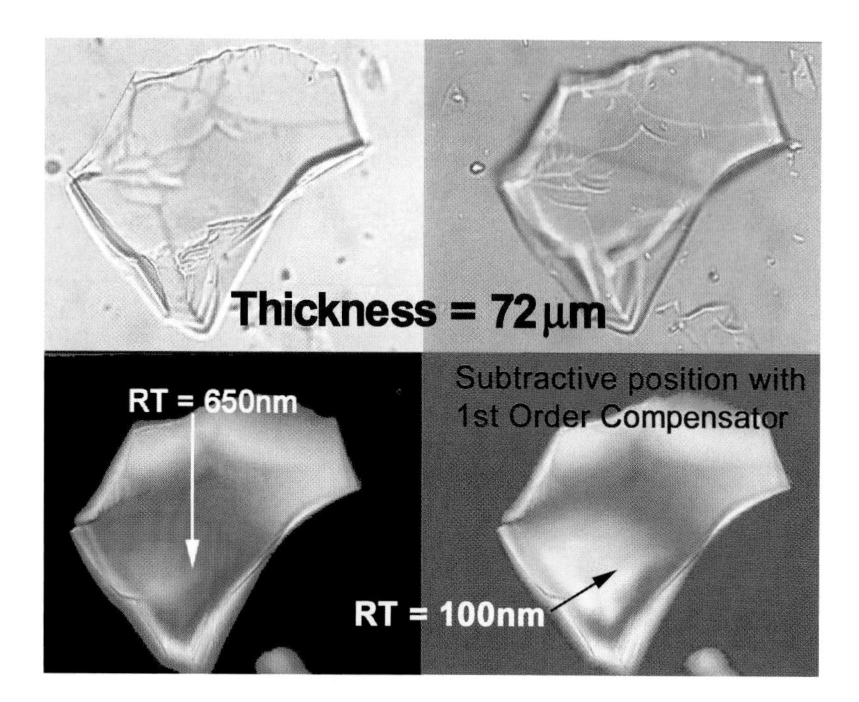

FIGURE 3.14 Top: specimen thickness is measured with a fine adjustment micrometer. Bottom left: degree of retardation due to the thickness is estimated by comparing its interference color (IC) to an interference chart. A fixed compensator can be used to achieve a more precise reading of the specimen IC.

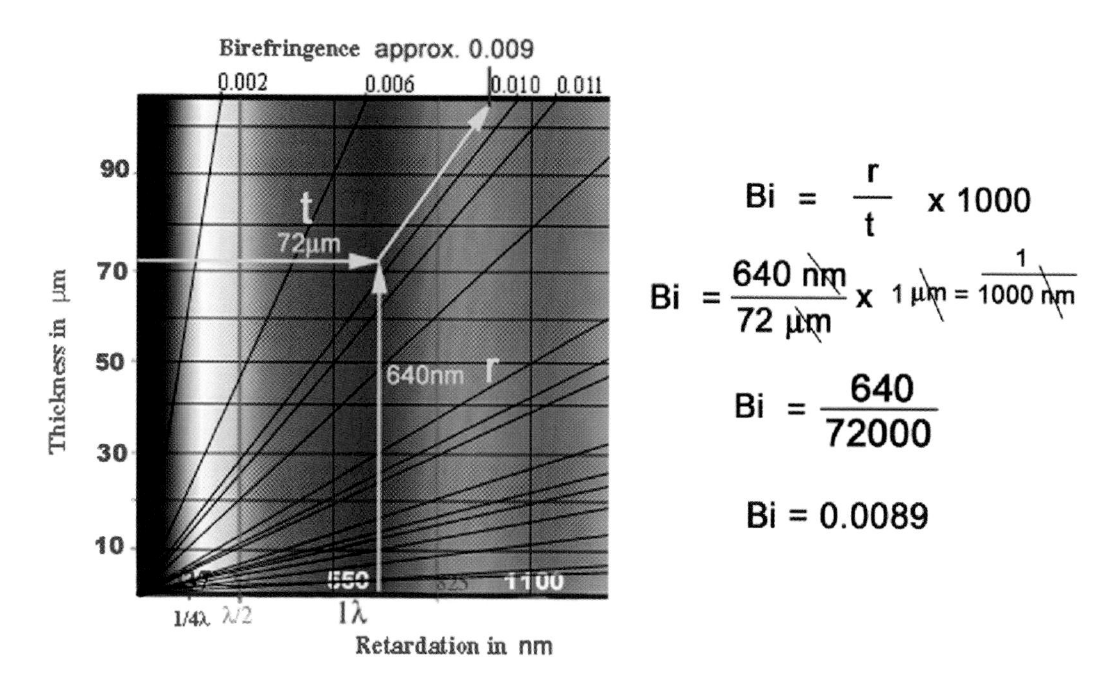

FIGURE 3.15 Interference color chart showing relationship of specimen thickness, displayed retardation colors, and birefringence (Bi). If two of the quantities are known, the third can be determined from this chart. For this specimen, thickness (t) and retardation (r) are plotted on the interference chart and its Bi is estimated and/or calculated using the cited formula.

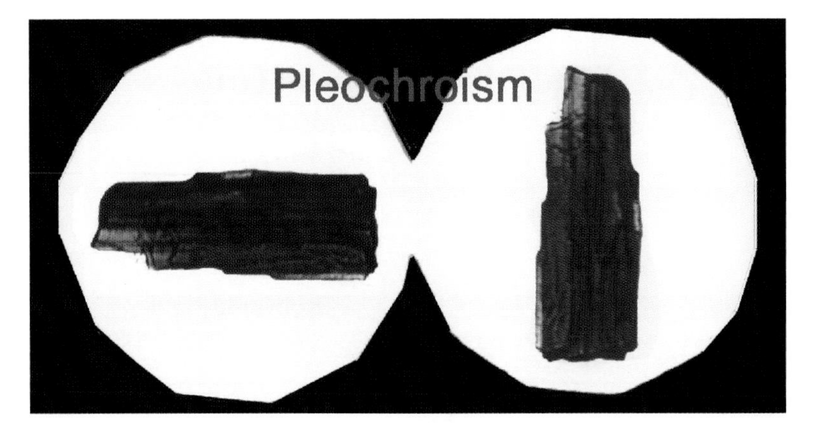

FIGURE 3.16 Fragment of hornblende, a pleochroic mineral, is rotated 90 degrees in PPL. Note change in displayed color.

SPECIMEN DATA SHEET

File/Case # ____________________

Specimen source ____________________ Sample #____________

Color: Transmitted light ____________ Reflected light ____________

Transparency: Transparent ________ Opalescent ________ Opaque ________

RI of mounting medium (MTM) used ____________________

Degree of relief displayed by specimen in MTM ____________________

Becke line appearance and movement ____________________

Specimen color: Transmitted light ____________ Reflected light ____________

Shape: Fiber ____________ Particle ____________

Sketch:

Crystallinity: ________ System ________ Amorphous ____________

Surface texture or appearance ____________________

Diameter in µm ____________________

Thickness in µm ____________________

Dimensions in µm: Length ____________ Width ____________

Pleochroism ____________________

Crossed polars: Isotropic ____________ Anisotropic ____________

Extinction: Parallel ______ Symmetrical ______ Oblique ______ Undulose ______

From observed interference colors, estimated retardation in nm ____________

Estimated birefringence ____________________

Other Observations ____________________

____

4 Chemical Microscopy and Microtechnique

The utilization and value of chemical microscopy, microchemical tests, and microtechniques in characterizing, identifying, and comparing all types of materials are well established.[1–18] For nearly two centuries, chemical microscopists have used the polarized light microscope (PLM) as their primary tool to identify and quantify all manner of inorganic and organic substances. Polarized light microscopy has been used to study the physical, optical, and chemical properties of pure materials, heterogeneous mixtures, and amalgamations of substances. Microchemical methods have several advantages:

1. They are well established, easily obtainable, and accepted by the general scientific community.
2. The chemistry of most of the reactions is understood and well documented.
3. The tests are very sensitive and often highly specific.
4. The methods are easy to learn.
5. The tests require only very small quantities of test reagents.
6. Minute test sample sizes are needed for examination.
7. The tests can be used on pure elements, inorganic compounds, and organic molecules.
8. The cost of each test is minuscule.
9. The chemical data collected can be useful in the identification and comparison of unknowns.
10. Except for the PLM, the necessary equipment is inexpensive and readily available.
11. A microkit such as the one depicted in Figure 4.1 is economical and easy to assemble.

BASIC METHODS AND TECHNIQUES OF CHEMICAL MICROSCOPY

Chamot and Mason first published the standard methods of chemical microscopy in their classic *Handbook of Chemical Microscopy*. This text is still the "bible" for practitioners of chemical microscopy. The most common Chamot and Mason methods are illustrated and discussed in this chapter. For a more comprehensive discussion of these and other methods, interested readers are referred to the literature.

Solubility Testing

A small drop of solvent is placed on a clean microscope slide (Ms). A tiny fragment of the test substance (TS) is placed about 1 mm away from the solvent drop. During observation with a stereomicroscope, the fine point of a needle, drawn glass rod, or the end of a new wooden toothpick is used to gently push the TS into the solvent. The results are noted and recorded as soluble (S), insoluble (I), or slightly soluble (SS). See Figure 4.2.[19]

Evaporation

The solvent containing the TS is lightly heated on a warmed hot plate and allowed to evaporate; the solute (TS) remains as a residue on the Ms, as shown in Figure 4.3.[20]

Decantation

The solvent is decanted from a preparation containing a precipitate by carefully tilting the Ms containing the preparation, while drawing off the liquid with a glass rod. The channel of liquid that develops is cut with a piece of filter paper (FP). The excess liquid is taken up with a piece of FP. The process is depicted in Figure 4.4.[21]

Sublimation

The phenomenon occurring when an element, molecule, or compound changes from the solid phase into the gaseous phase without first passing through the liquid phase is known as sublimation. Sublimation on a microscopic level is performed as follows:

1. A microscope slide (Ms) is placed on a hot plate (Hp).
2. A glass sublimation cup containing the TS is covered with a round cover glass (Cg) and placed on top of the Ms.
3. The preparation is gently warmed on a Hp.
4. The crystals that form on the undersurface of the Cg are viewed at 100× (Figure 4.5).[22]

Fusion

Fusion is achieved on a microscopic level by placing a mixture of a flux and TS onto the looped end of a length of platinum (Pt) wire. The looped end is heated with a microburner. After fusion occurs, the Pt wire containing the hot glassy bead is dissolved in a small drop of acid (Figure 4.6).[23]

Chamot and Mason Reagent Application Methods[24]

Method I — A drop of reagent (RD) is drawn into a test drop (TD), thus creating a channel of RD and TD as shown in Figure 4.7. The microcrystals formed are normally viewed with a PLM at 100 to 400× magnification.

Method IA — The RD is placed directly into the TD (Figure 4.8). The precipitate or microcrystals that form are normally viewed with a PLM at 100 to 400× magnification.

Method IB — Small quantities of reagent and test solution are collected into a glass tube by capillary action and allowed to react (Figure 4.9). The resulting reaction is viewed with a PLM at 100× magnification.

Method II— A small fragment of the solid reagent is placed directly into the TD. The resulting reaction is viewed with a PLM (Figure 4.10).

Method III — The RD is drawn directly into a dried residue of TD. The resulting reaction is viewed with a PLM (Figure 4.11). Note the barrel-shaped zinc squarate crystals that form in the reagent solution.

The illustrated methods are utilized most often by chemical microscopists. Chamot and Mason described step-by-step details of those and many other reagent application methods. Interested readers are referred to the literature for further study.

The last two tests to be discussed in this work are simple and very easy to perform. The first determines pH value. Is a TD acid or basic? The pH of a solution can be checked by simply applying a tiny drop of the solution to be tested to pH paper. A red to orange color reaction indicates an acid; a green to blue reaction is indicative of a base.

Finally, whether a substance is magnetic or nonmagnetic can often be an important clue to its identity. To determine whether a material is magnetic, place a magnet on one side of a preparation on the stage of a PLM set for 100× magnification. If the specimen is magnetic, the particles will point or be drawn to the side of the preparation nearest the magnet.

REFERENCES

1. Chamot, E.M. and Mason, C.W., *Handbook of Chemical Microscopy*, Vols. 1 and 2, New York, John Wiley & Sons, 1930 and 1931.
2. *The Merck Index*, Rahway, NJ, Merck & Co., 1940.
3. Vesce, V.C., *Classification and Microscopic Identification of Organic Pigments*, Mattiello, J.J., Ed., New York, John Wiley & Sons, 1942.
4. Fiegl, F., *Qualitative Analysis by Spot Test: Inorganic and Organic Applications*, 3rd ed., New York, Elsevier, 1946.
5. Kirk, P.L., *Crime Investigation*, New York, Interscience, 1953.
6. Schaefer, H.F., *Microscopy for Chemists*, New York, Van Nostrand, 1953.
7. McCrone, W.C., *Fusion Methods in Chemical Microscopy*, New York, Interscience, 1957.
8. Bloss, F.D., *An Introduction to the Methods of Optical Crystallography*, New York, Holt, Rinehart & Winston, 1961.
9. Schneider, F.L., *Qualitative Organic Microanalysis*, New York, Academic Press, 1964.
10. Fulton, C.C., *Modern Microcrystal Tests for Drugs*, New York, Interscience, 1969.
11. Fiegl, F., *Spot Tests in Inorganic Analysis*, 6th ed., Amsterdam, Elsevier, 1972.
12. Stevens, R.E., Squaric acid: a novel reagent in chemical microscopy, *Microscope*, 22, 163, 1974.
13. Wills, W.F., Jr. and Whitman, V.L., Extended use of squaric acid as a reagent in chemical microscopy, *Microscope*, 25, 113, 1977.
14. Teetsov, A.S., Techniques of small particle manipulation, *Microscope*, 25, 103, 1977.
15. McCrone, W.C., Delly, J.G., and Palenik, S.J., *The Particle Atlas*, 2nd ed., Vol. 5, Ann Arbor, MI, Ann Arbor Science Publishers, 1979, p. 117.
16. Delly, J.G., Microchemical tests for selected cations, *Microscope*, 37, 139, 1989.
17. Wills, W.F., Jr., Squaric acid revisited, *Microscope*, 38, 169, 1990.
18. Jungreis, E., *Spot Test Analysis: Clinical, Environmental, Forensic, and Geochemical Applications*, 2nd ed., New York, John Wiley & Sons, 1997.
19. Chamot, E.M. and Mason, C.W., *Handbook of Chemical Microscopy*, Vol. 2, New York, John Wiley & Sons, 1931, p. 4.
20. Chamot, E.M. and Mason, C.W., *Handbook of Chemical Microscopy*, Vol. 2, New York, John Wiley & Sons, 1931, p. 7.
21. Chamot, E.M. and Mason, C.W., *Handbook of Chemical Microscopy*, Vol. 2, New York, John Wiley & Sons, 1931, p. 8.
22. Chamot, E.M. and Mason, C.W., *Handbook of Chemical Microscopy*, Vol. 2, New York, John Wiley & Sons, 1931, p. 15.
23. Chamot, E.M. and Mason, C.W., *Handbook of Chemical Microscopy*, Vol. 2, New York, John Wiley & Sons, 1931, p. 189.
24. Chamot, E.M. and Mason, C.W., *Handbook of Chemical Microscopy*, Vol. 2, New York, John Wiley & Sons, 1931, p. 30.

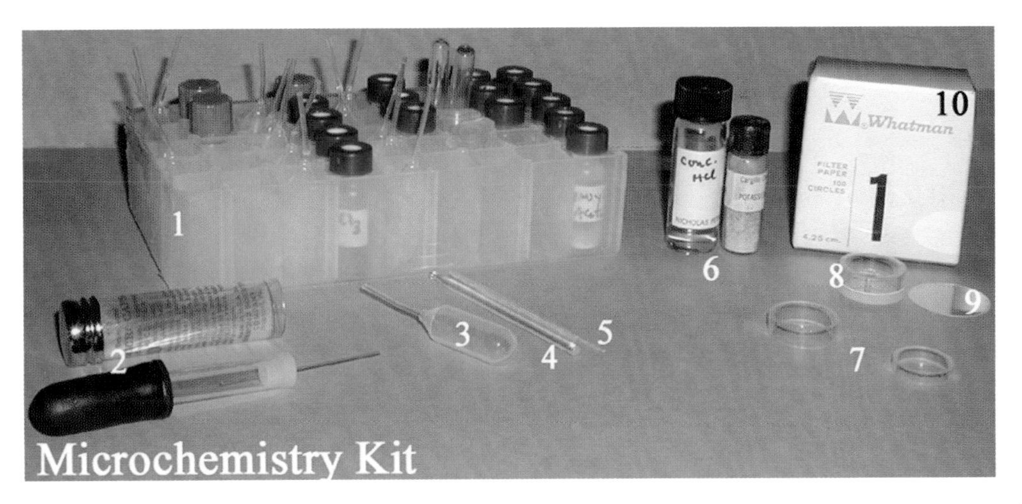

FIGURE 4.1 Microchemistry kit components: (1) polypropylene tray; (2) micropipettes; (3) microdropper; (4) small glass rod; (5) microspatula (flat wooden toothpick); (6) reagents; (7) glass ring; (8) sublimation cup; (9) round cover slip; and (10) filter paper. Additional items not shown are pH paper and a magnet.

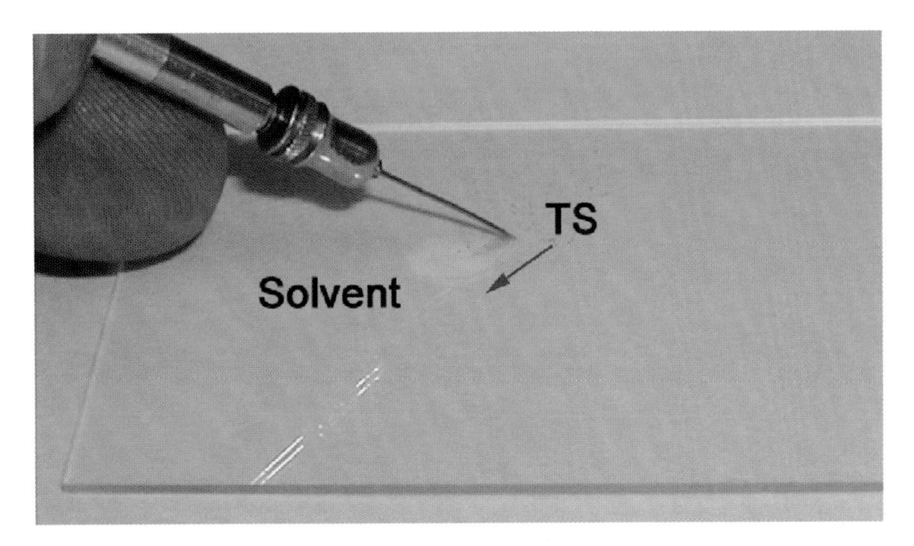

FIGURE 4.2 The test for solubility of a test substance (TS). A needle is used to push a small fragment of TS into solvent; results are recorded.

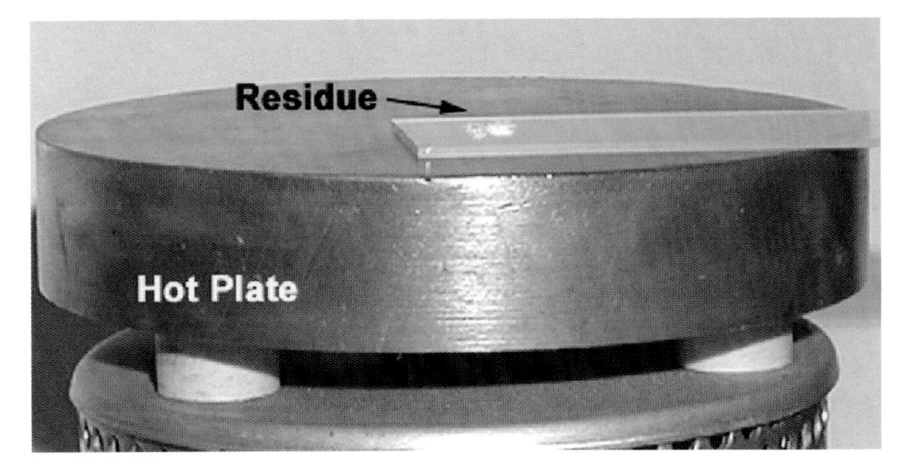

FIGURE 4.3 Specimen is allowed to go to dryness.

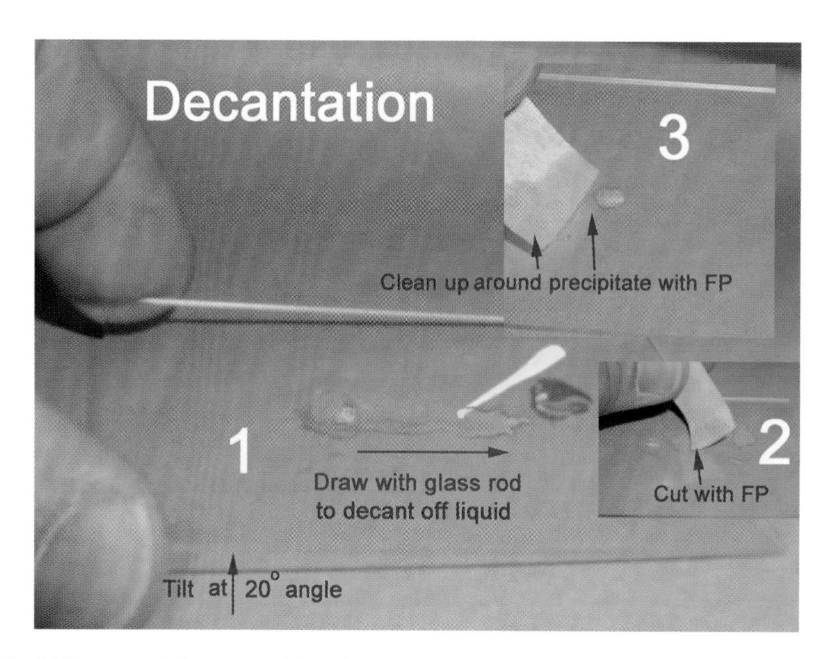

FIGURE 4.4 Excess liquid is removed from a precipitate by decantation: (1) the slide carefully tilted 20° while the liquid is drawn off with a glass rod; (2) the channel of liquid that forms is cut with a piece of filter paper (FP); and (3) the remaining liquid is removed with another piece of FP.

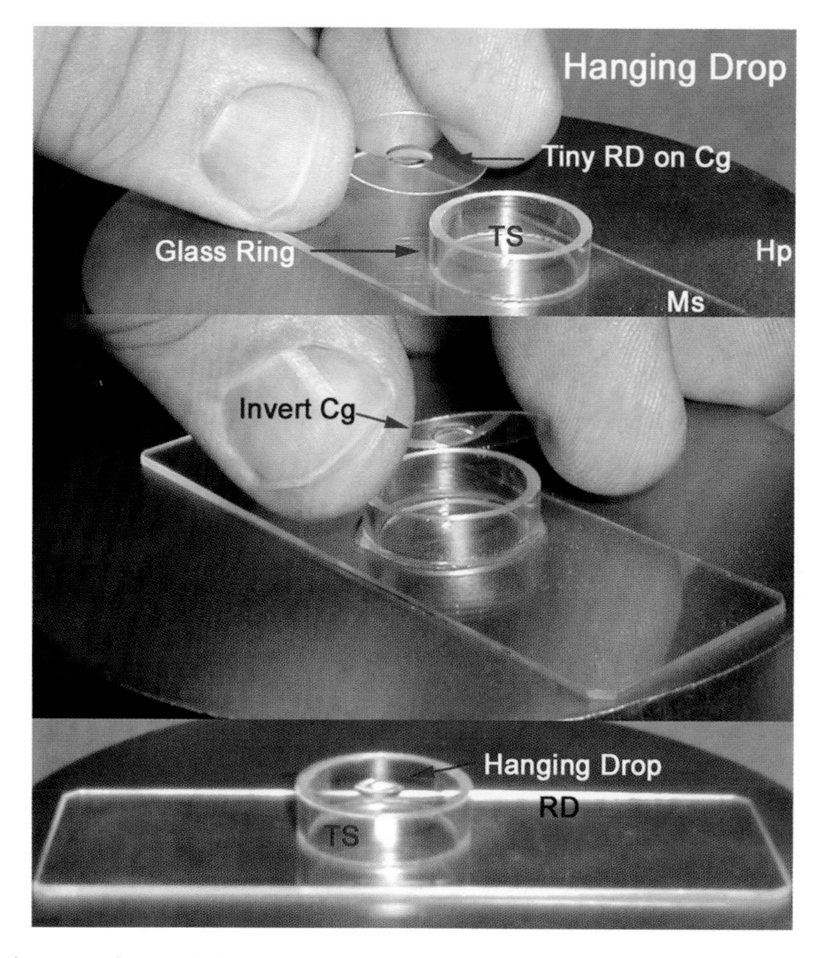

FIGURE 4.5 Sublimation on a microscopic level: (1) a microscope slide is placed on a hot plate (Hp); (2) a glass sublimation cup containing the TS is covered with a round cover glass and placed on top of the slide; (3) the preparation is gently warmed on the Hp; and (4) the sublimed crystals form on the undersurface of the cover glass.

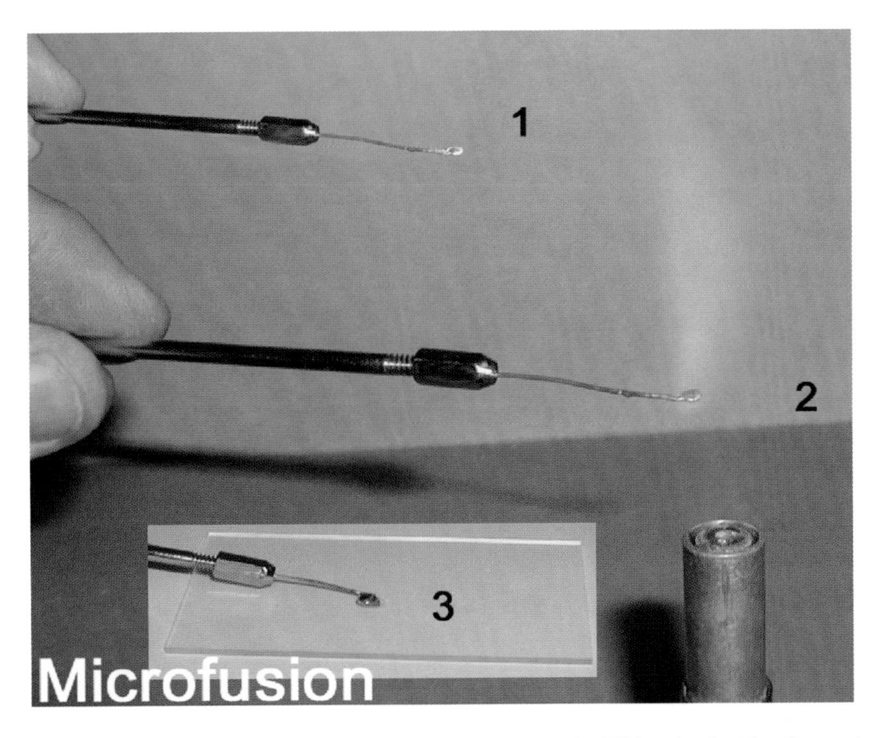

FIGURE 4.6 The fusion of substances that are insoluble in most reagents: (1) the TS is mixed with a flux such as borax and placed on the loop end of a platinum wire; (2) the preparation is heated with a microburner; and (3) the hot preparation is placed into a small drop of concentrated mineral acid.

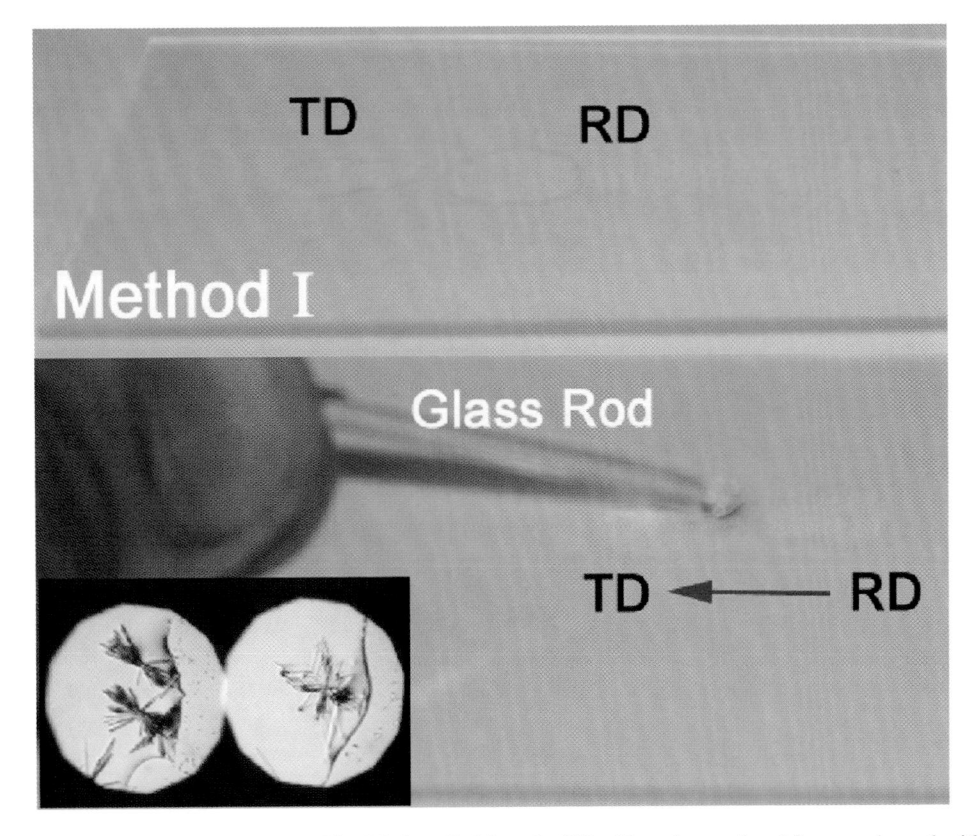

FIGURE 4.7 Method I of Chamot and Mason: The RD is pulled into the TD with a glass rod and forms a channel with a gradient of test substance (TS) and reagent. Inset: crystals of calcium II ion formed with 5% H_2SO_4. The crystals form around the gradient's edges.

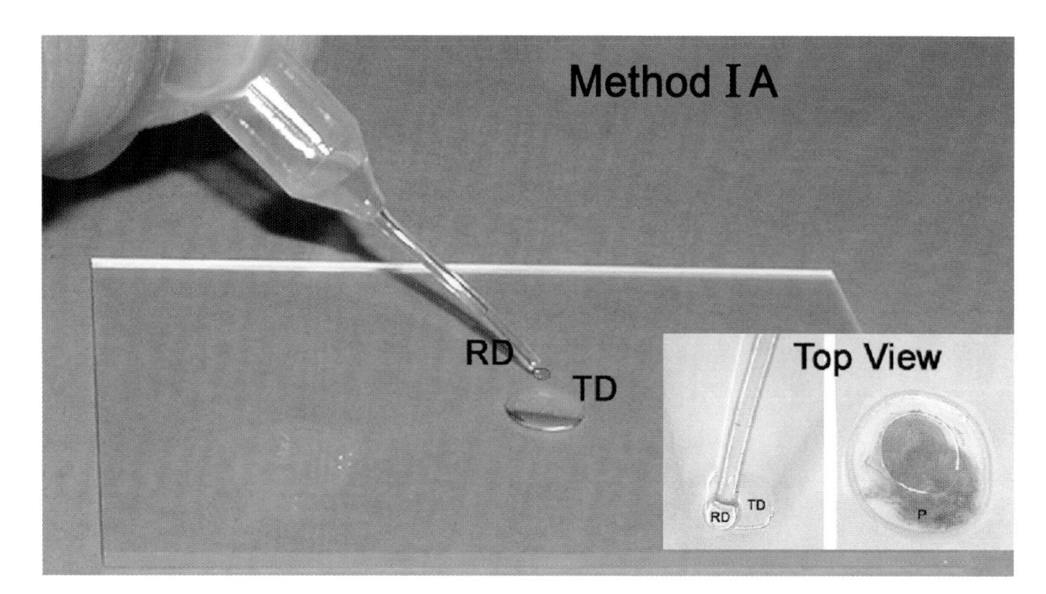

FIGURE 4.8 Method IA of Chamot and Mason: The RD is placed directly into the TD.

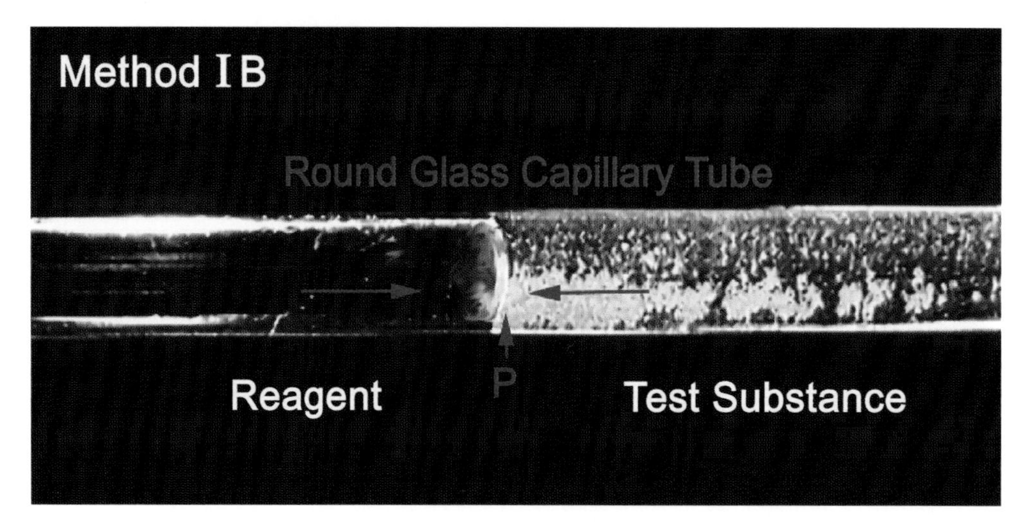

FIGURE 4.9 Method IB of Chamot and Mason: The reagent and test substance are placed into a 1-mm diameter capillary tube by capillary action and allowed to react. The precipitate (P) is viewed with a PLM.

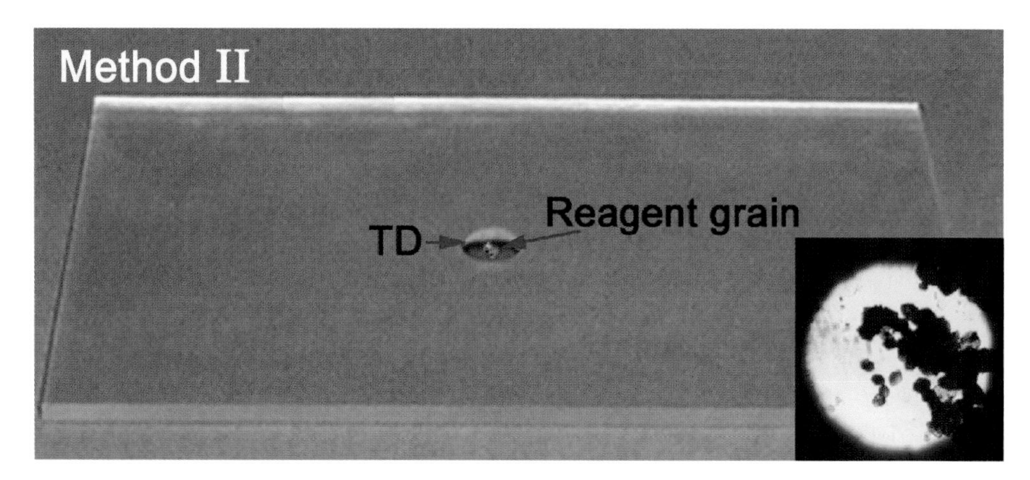

FIGURE 4.10 Method II of Chamot and Mason: A small fragment of the solid reagent is placed directly into the test solution. The resulting reaction is viewed with a PLM. The inset is iron III reacting with a grain of squaric acid.

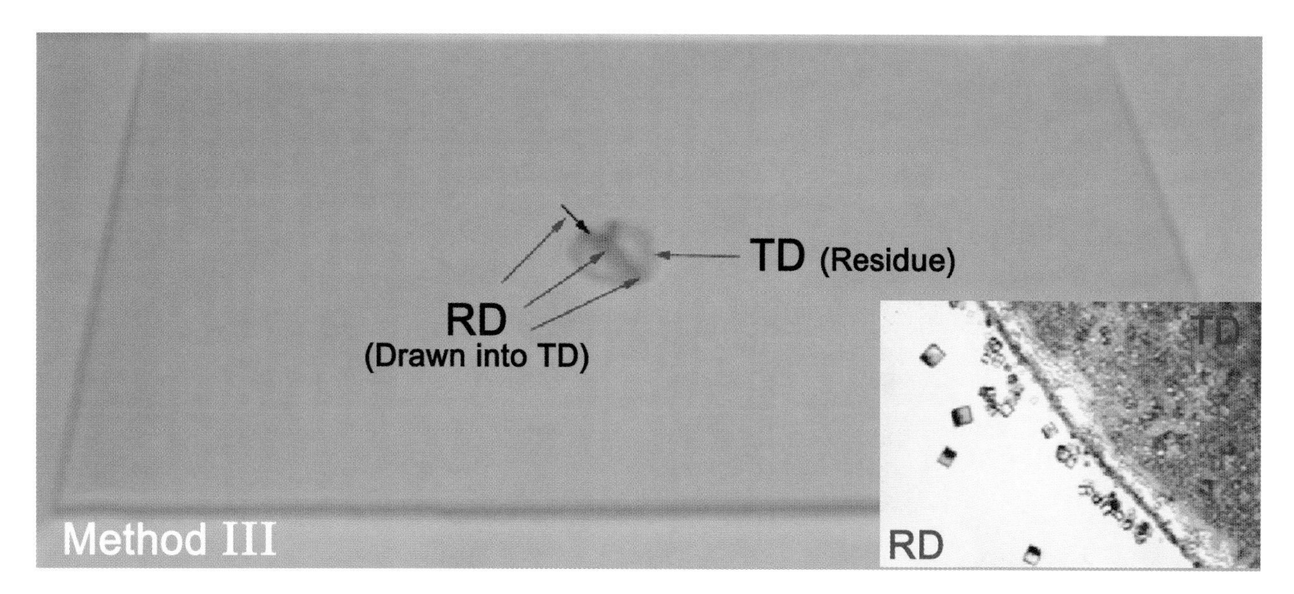

FIGURE 4.11 Method III of Chamot and Mason: The RD is drawn directly into a dried residue of TD. The inset is a saturated solution of squaric acid (RD) drawn across a dried residue of zinc white. The barrel-shaped zinc squarate crystals formed along the interface between the reagent solution and dried residue.

5 Identification and Comparison of Human Hair

Hair examinations and comparisons conducted by forensic scientists often provide important investigative and associative information. Human hair has been used in forensic investigations for nearly a century. Reports in the literature about the utilization of human and animal hairs encountered in forensic casework abound.[1–20] Human and animal hairs are encountered frequently in works of art, textiles, and tapestries. Hair has also been used as a reinforcement fiber in mortars and plasters. The microscopic identification and comparison of human hair are based on its physical morphology. This chapter discusses typical procedures used to examine, identify, and compare human hair.

Protocols describing the general morphology of human hair (and its interpersonal and intrapersonal variations) have been in common use by forensic examiners for almost a century. The sources of data used in the preparation of the protocol in Table 5.1 were the published literature and the authors' own work.[21–30] Photomicrographs depicting most of the microscopic traits of human hair cited in the protocol appear in Appendix A. Table 5.2 is a human hair data sheet intended to make the collection, tabulation, and evaluation of data somewhat easier. After collection and tabulation, the data can be used to determine the somatic region a questioned (Q) hair came from, the geographical origin of a person from whom a Q hair originated, and a common origin.

EXAMINATION

After preliminary examination, a hair specimen should be mounted in the appropriate medium, as outlined in Chapter 2. The internal and external identifying characteristics of human hair are observed readily when it is mounted in a medium with a refractive index (Melt Mount 1.539 or Cargille 1.540). Most mounting media with refractive indices ranging from 1.520 to 1.560 can be used successfully in hair examination.

The first task is to determine whether a Q hair originated from a human and not another mammal. Figure 5.1 depicts the primary anatomical regions of hair used in species identification: (1) the cuticle or outermost layer of hair composed of layers of overlapping scales; (2) the medulla or central canal of the hair (may be present or absent); and (3) the cortex or primary tissue. The cortex contains pigment granules, cortical fusi, and other morphological features. Figure 5.1 also illustrates a cast of the dominant scale pattern usually associated with hair of human origin. When necessary for identification purposes, a hair can be isolated, cast in Melt Mount 1.539 as detailed in Chapter 2, and then remounted in Melt Mount 1.539 for further study.

Human hairs to be examined typically come from the head or pubic regions of the body. However, hairs originating from other body areas such as the face and limbs may also be encountered and examiners should become familiar with the morphologies of all types of human hairs. Lists of the various morphological characteristics used to determine the somatic origin of human hair are in the cited literature. Figures 5.2 through 5.6 illustrate the morphologies of five distinct categories of human hair. Figures 5.2 and 5.3 show the primary physical characteristics used in the identification of human head and pubic hairs. It should be pointed out that these illustrations are only generalizations. Other configurations can and often do occur; for example, other anatomical regions of human hair, such as the tip and shaft, can vary as well.

SOMATIC ORIGIN

The regional origin of the person from whom a Q hair originated can often be determined from the morphological traits of the hair. The following information was compiled from the cited literature and the authors' research. Figures 5.7 through 5.9 illustrate differences in commonly encountered hairs.

REGIONAL ORIGIN

Regional origin is usually determined from head hair. Information about acquired characteristics present in questioned or known hair specimens can be very useful information to an analyst trying to associate a questioned hair specimen (QHS) to the source of a known hair standard (KHS). Acquired traits often strengthen the value of an association between a QHS and KHS. Evidence of a rare disease, peculiar treatment, mechanical damage, parasites, and other factors can frequently be used to further individualize a QHS to a given individual.

Because acquired characteristics appear throughout a lifetime, archeologists and anthropologists can often obtain important information about an individual's nutritional health, living environment, hygiene, and cultural practices from hairs. Photomicrographs of the acquired characteristics listed in Table 5.3 and most traits listed in the human hair protocol can be found in Appendix A. The information recorded in Table 5.2 may be used to help establish the somatic origin, regional origin, and source of a QHS. A glossary of terms is included in this chapter to aid readers in their studies of hair analysis. The appearances of hairs affected by a few diseases encountered in casework situations are shown in Figure 5.10.

Finally, before a positive association between a QHS and a KHS is made, a microscopic examination and comparison of the questioned and known hair specimens must be conducted on a high-quality transmitted light comparison microscope. The known standards should consist of at least 100 hairs when possible, root to tip, removed from all regions of the particular area of the body, e.g., hairs from the temporal, frontal, and parietal regions should be examined when a QHS is from the scalp. Both the questioned and known hair specimens should be treated identically and examined visually with a stereomicroscope and bright-field or polarized light microscope and then observed side-by-side on a comparison microscope. All data should be recorded in the examiner's notes for future reference. Examples of a human hair comparison are shown in Figure 5.11.

GLOSSARY OF HAIR TERMS

Amorphous Having no defined pattern or shape; usually refers to medullary configuration.

Anagen Synthesis phase of hair cycle. An anagen root comes from the active growth phase of a hair follicle during the growth cycle. The root from a pulled anagen hair is elongated, may have a root sheath, and is normally pigmented.

Buckling The appearance of caving in of the hair shaft; normally present in pubic hair.

Catagen Step-down phase of hair growth. The period between the active (anagen) growth phase and the resting (telogen) phase. A catagen root may be club-shaped and may have a meager amount of pigment and a dried root sheath.

Caucasoid Anthropological term designating one of the major groups of human beings, i.e., the people of Europe.

Cellular Displaying a definite form, pattern, or shape; usually refers to the medullary configuration.

Characteristic Microscopic or macroscopic feature of a hair.

Color Hue of a hair as judged under reflected and transmitted light.

Comparison Examination of questioned and known hair specimens in order to associate or disassociate them from a given individual.

Convolution Twisting of a hair shaft.

Cortex Primary anatomical region of a hair between the cuticle and medullary regions.

Cortical fusi Small spaces within a hair shaft that appear as tiny dark specks.

Cross-sectional shape Shape of a hair shaft when cut at right angles to its longitudinal axis.

Cuticle Outermost region of a hair composed of layers of overlapping scales.

Distal end End of a hair furthest from the root.

Eumelanin Brown pigment that provides the brown to black hues of human hair.

Fusiform Spindle-shaped gap found in a hair shaft.

Hair Fine, cylindrical fiber growing from the skin of a mammal.

Imbricate Scale pattern whose edges overlap in a wavy pattern.

Keratin Sulfur-containing fibrous protein forming the chemical basis for keratinized epidermal tissues such as hair.

Known A sample intended to be representative of a particular body area of a specific person or animal.

Looped cuticle Condition in which the distal edges of the cuticle scales are curved toward the hair shaft.

Macroscopic Characteristic large enough to be perceived by the unaided human eye or under low magnification.

Medial region Portion of a hair between the proximal and distal ends.

Medulla Anatomical region of a hair normally found in the center of the cortex.

Medullary configuration Form taken by medullary cells between the proximal and distal ends of a hair shaft.

Melanin Pigment providing hair color.

Mongoloid Anthropological term designating one of the major groups of human beings from Asia; includes the Inuit peoples and Amerindians.

Monilethrix A hair disorder that produces periodic nodes along the lengths of hairs with intervening, unmedullated, tapering constrictions.

Negroid Anthropological term designating one of the major groups of human beings, i.e., the peoples of Africa.

Ovoid bodies Oval- or round-shaped pigmented bodies usually found in the hair cortex.

Parasitic insects A parasite lives on another organism to obtain some benefit. Lice are parasitic insects found on humans. Types include head lice *(Pediculus humanus capitis)*, body and clothing lice *(Pediculus humanus corporis)*, and crab lice *(Phthirus pubis)*. The lice eggs or nits may attach to hair shafts.

Peripheral region The portion of the hair toward the outermost areas, distant from the medullary region, including the cuticle and outer areas of the cortex.

Pheomelanin Naturally occurring reddish brown to yellow melanin-based pigment; the pigment in human and other hair that provides the blonde to red color.

Pigment density Relative abundance of pigment granules in the hair cortex, as judged by their microscopic appearance.

Pigment distribution Pattern of pigment granules about the central axis of the hair shaft (uniform, peripheral, one-sided, random, or central).

Pili annulati Hair disorder resulting in ringed or banded hair; alternating bright and dark bands appear in the hair shafts.

Pili torti Hair disorder indicated by hairs that are flattened and twisted 180 degrees along their axes; the condition is found at irregular intervals along the shafts.

Proximal end End of the hair nearest the root end.

Questioned Sample of unknown origin.

Regional origin Originating from one of the three major groups of human beings (caucasoid, mongoloid, and negroid) as defined by anthropologists.

Root Structure at the proximal end of a hair extending from the hair follicle.

Scales Tiny plate-like structures made of keratin that protect the hair.

Telogen Last phase of the hair growth cycle during which the hair root becomes bulbous so it can be shed easily from the follicle.

Tip Most distal end of a hair shaft.

Translucent Able to transmit light but with sufficient diffusion to allow perception of an object.

Trichology Study of hair.

Trichonodosis Condition characterized by apparent or actual knotting of hair.

Trichoptilosis Disease condition characterized by longitudinal splitting or fraying of the hair shaft.

Trichorrhexis nodosa Disease or condition in which the hair breaks off at node-like structures formed primarily by over-bleaching and/or mechanical damage.

REFERENCES

1. von Hofman, E., *Lehrbuch der Gerichtlichen Medizin*, Vienna, 1898.
2. Marx, H., *Ein Beitrag zur Identitatsfrage bei der forensicschen Haaruntersuchung*, *Arch. Kriminol.*, 23, 57, 1906.
3. Gross, H., *Criminal Investigation*, adapted from *System Der Kriminalistik*, by J. C. Adams, London, Sweet & Maxwell, 1924, p. 144.
4. Locard, E., L' analyse des poussieres en criminalistique, *Rev. Int. Crim.*, 1, 176, 1929.
5. Glaister, J., *A Study of Hairs and Wools Belonging to the Mammalian Group of Animals, Including a Special Study of Human, Considered from the Medicolegal Aspect*, Publication 2, Cairo, MISR Press, 1931.
6. Söderman, H. and O'Connell, J.J., *Modern Criminal Investigation*, New York, Funk & Wagnalls, 1935, p. 188.
7. Smith, S. and Glaister, J., *Recent Advances in Forensic Medicine*, 2nd ed., Philadelphia, Blakiston's Son & Co., 1939, p. 118.
8. Gamble, L.H. and Kirk, P.L., Human hair studies. II. Scale counts, *J. Crim. Law Criminol.*, 31, 627, 1940.
9. O'Hara, C.H. and Osterburg, J.W., *An Introduction to Criminalistics*, New York, Macmillan, 1949.
10. Kirk, P.L., *Crime Investigation*, New York, Interscience, 1953, p. 152.
11. Gonzales, T.A., *Legal Medicine Pathology and Toxicology*, New York, Appleton, 1954.
12. Niyogi, S.K., A study of human hairs in forensic work, *J. Forens. Med.*, 9, 27, 1962.
13. Glaister, J., Hairs and fibers, *Criminologist*, 4, 23, 1969.
14. Hicks, J.W., Microscopy of Hair, Issue 2, Washington, D.C., U.S. Government Printing Office, 1977.
15. McCrone, W.C., Particle analysis in the crime laboratory, in *The Particle Atlas*, Vol. 5, McCrone, W.C., Delly, J.G., and Palenik, S.J., Eds., Ann Arbor, MI, Ann Arbor Science Publishers, 1979, p. 1379.
16. Bisbing, R.E., The forensic identification and association of human hair, in *Forensic Science Handbook*, Saferstein, R., Ed., Englewood Cliffs, NJ, Prentice-Hall, 1982, p. 184.
17. Deadman, H.A., Fiber evidence and the Wayne Williams trial, *FBI Law Enforce. Bull.*, March 1984, p. 13; May 1984, p. 10.
18. Gaudette, B.D., Fibre evidence, *RCMP Gazette*, 47, 18, 1985.
19. Petraco, N., The occurrence of trace evidence in one examiner's casework, *J. Forens. Sci.*, 30, 486, 1985.
20. Petraco, N., Trace evidence: the invisible witness, *J. Forens. Sci.*, 31, 321, 1986.
21. Glaister, J., *A Study of Hairs and Wool*, Publication 2, Cairo, MISR Press, 1931, p. 155.
22. Gaudette, B.D. and Keeping, E.S., An attempt at determining probabilities in human scalp hair comparison, *J. Forens. Sci.*, 19, 1974, pp. 599–606.
23. Gaudette, B.D., Probabilities and human pubic hair comparisons, *J. Forens. Sci.*, 21, 514, 1976.
24. Hicks, J.W., Microscopy of Hair, Issue 2, Washington, D.C., Federal Bureau of Investigation, 1977, p. 7.
25. Shaffer, S.A., A protocol for the examination of hair evidence, *Microscope*, 30, 151, 1982.
26. Bisbing, R.E., Forensic identification and association of human hair, in *Forensic Science Handbook*, Saferstein, R., Ed., Englewood Cliffs, NJ, Prentice-Hall, 1982, p. 184.

27. Strauss, M.A.T., Forensic characterization of human hair, *Microscope*, 31, 15, 1983.
28. Petraco, N., Protocol for human hair comparison in Chapter 37, in *Forensic Science,* Wecht, C., Ed., New York, Matthew Bender, 1984, p. 37A-8.
29. Ogle, R.R., Jr. and Fox, S., *Atlas of Human Hair,* Boca Raton, FL, CRC Press, 1999.
30. Robertson, J., Forensic and microscopic examination of human hair, in *Forensic Hair Examinatio*n, Robertson, J., Ed., London, Taylor & Francis, 1999, p. 79.
31. *Human Hair Guidelines*, SWGMAT, in press.
32. DeForest, P.R., Shankles, B., Sacher, R.L., and Petraco, N., Meltmount® 1.539 as a mounting medium for hair, *Microscope*, 35, 249, 1987.

TABLE 5.1
Human Hair Protocol

Macroscopic Features

Color (Reflected): White Gray Blonde Red Brown Black Other

Shaft form: Straight Arched Wavy Curly Twisted Tightly coiled Crimped

Shaft length, range in cm ____________________

Shaft texture: Fine Medium Coarse

Microscopic Characteristics

(Transmitted): Color: Colorless Blonde Red Brown Black Other ____________________

Natural pigmentation:

Pigment size: Fine Coarse

Pigment aggregation: Streaky Clumpy Patchy

Pigment aggregate size: Small Medium Large

Pigment density: Sparse Medium Heavy Opaque

Pigment distribution: Uniform Peripheral One-sided Random Other ____________________

Treatments: Dyed Rinsed Bleached Spray dye Coating dye Lightener Other ____________________

Structure:

Shaft characteristics: Diameter, range in µm ____________________

X-S shape: Round Oval Triangular Oblate Other ____________________

Shaft configurations: Buckling Convoluting Shouldering Invaginated Undulating Splitting Regular

Five Anatomical Regions and their Manifestations

Medulla:	**Cuticle:**	**Cortex:**	**Distal End:**	**Proximal End:**
❑ Present	❑ Present	❑ Cellular texture	❑ Tapered	❑ No Root
❑ Absent	❑ Absent	❑ Coarse	❑ Abraded	❑ Telogen
❑ Continuous	❑ Ovoid bodies	❑ Medium	❑ Square cut	❑ Catagen
❑ Discontinuous	❑ Size	❑ Fine	❑ Angular cut	❑ Anagen
❑ Fragmented	❑ Distribution	❑ Ovoid bodies	❑ Frayed	❑ Root tag
❑ Opaque	❑ Outer cuticle margin	❑ Size	❑ Split	❑ Root band
❑ Translucent	❑ Flattened	❑ Distribution	❑ Crushed	❑ Skeletal
❑ Amorphous	❑ Smooth	❑ Abundance	❑ Broken	
❑ Cellular	❑ Serrated	❑ Cortical fusi	❑ Burned	
❑ Other, i.e., doubled	❑ Cracked	❑ Size	❑ Other	
	❑ Looped	❑ Shape		
	❑ Other	❑ Distribution		
	❑ Inner cuticle margin			
	❑ Distinct			
	❑ Indistinct			
	❑ Other			

TABLE 5.2
Data Sheet for Human Hair Protocol[a]

Macroscopic (Gross) Characteristics:

1. Length ____________________
2. R.L. color ____________________
3. Shaft shape ____________________
4. Texture ____________________

Microscopic Characteristics:

A. Cuticle

1. Margin ____________________
2. Distribution ____________________
3. Shape ____________________
4. Color ____________________
5. Thickness ____________________

B. Medullary Configuration

1. Amorphous/opaque ____________________
2. Amorphous/translucent ____________________
3. Cellular/opaque ____________________
4. Cellular/translucent ____________________
5. Distribution ____________________
6. Diameter thickness, μm ____________________
7. Absent ____________________

C. Cortex

1. T.L. color ____________________
2. X-S shape ____________________
3. Pigment shape ____________________
4. Pigment distribution ____________________
5. Pigment density ____________________
6. Shaft diameter range ____________________
7. Shaft variation ____________________
8. Thickness change ____________________
9. Cortical fusi ____________________
10. Cortical fusi distribution ____________________
11. Cortical damage ____________________
12. Oddities ____________________
13. Foreign debris ____________________

D. Proximal End

1. Tip shape ____________________

E. Distal End

1. Root structure ____________________
2. Root shape ____________________
3. Root end ____________________
4. Growth phase ____________________

[a] See Table 5.1.

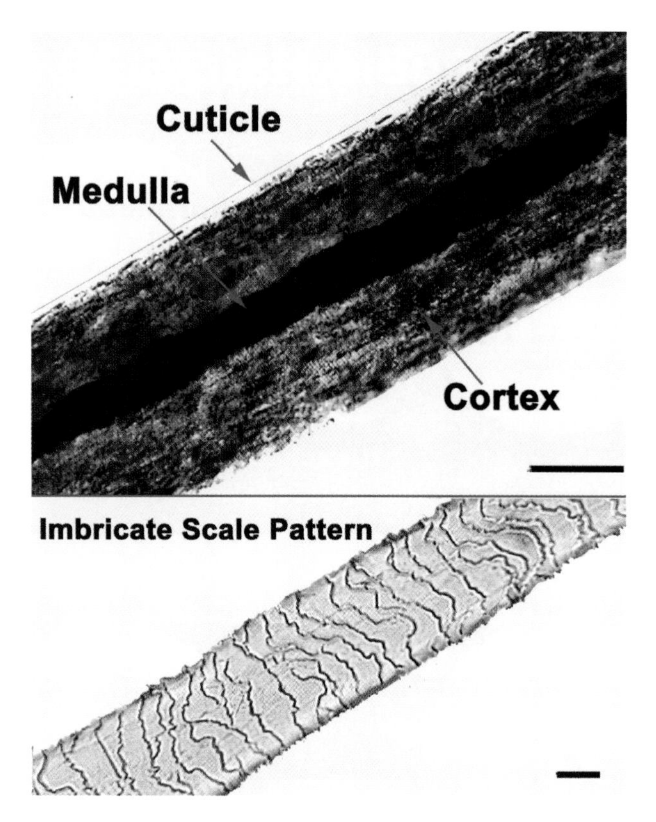

FIGURE 5.1 Top: the three primary anatomical regions of hair used in species identification: the cuticle (outermost layer composed of layers of overlapping scales); the medulla (central canal); and cortex (primary tissue). Bottom: typical imbricated scale pattern. The black scale is equal to 40 μm, unless otherwise noted.

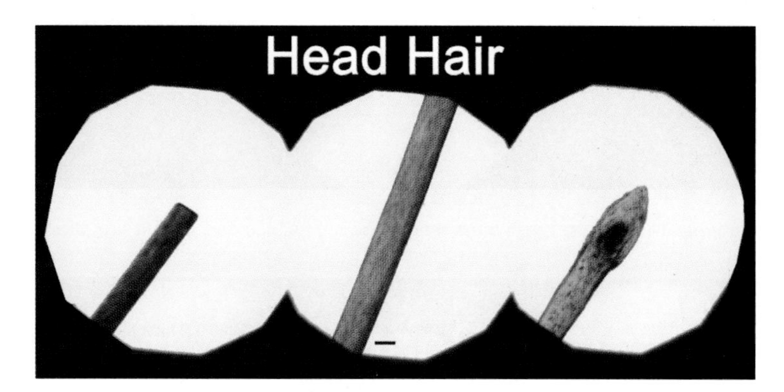

FIGURE 5.2 Appearance of a typical Eastern European shed scalp hair.

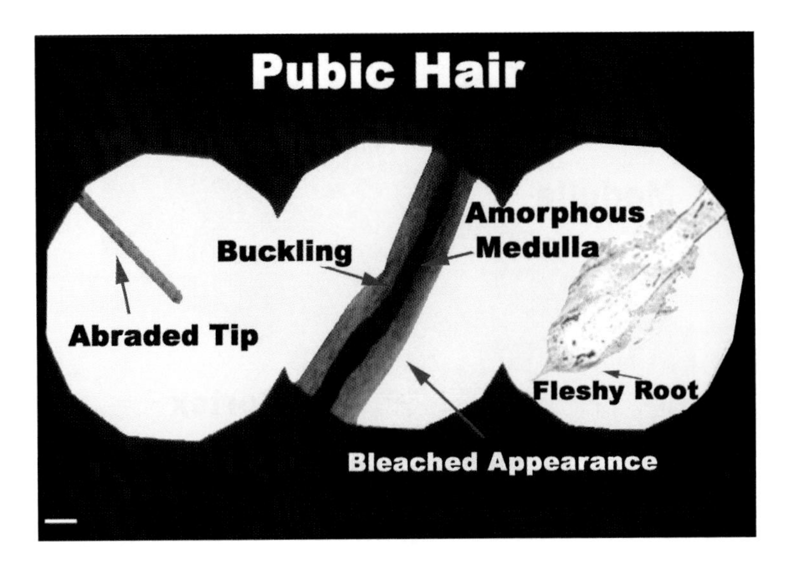

FIGURE 5.3 Appearance of a typical pubic hair.

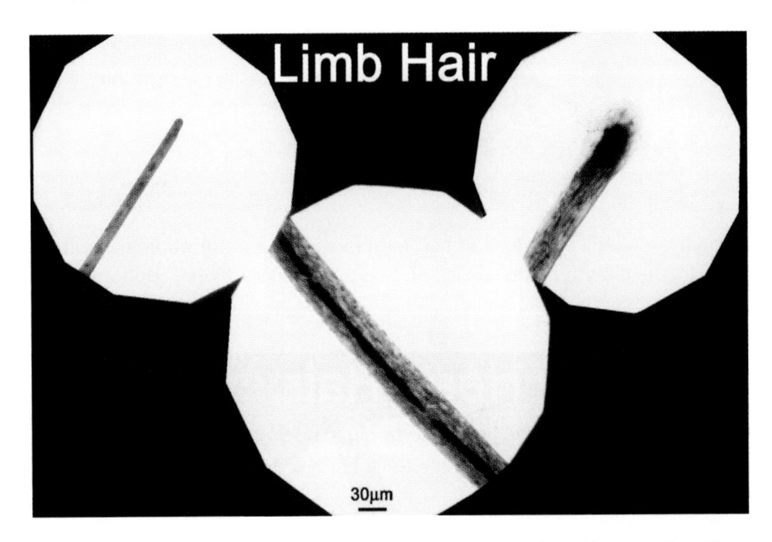

FIGURE 5.4 Appearance of a typical limb hair. Limb hairs tend to be arc shaped and have a fine diameter.

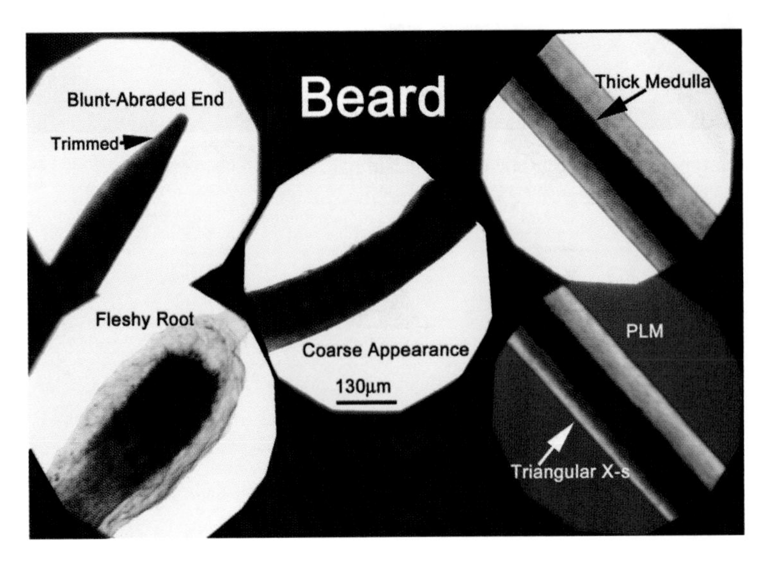

FIGURE 5.5 Appearance of a typical facial hair.

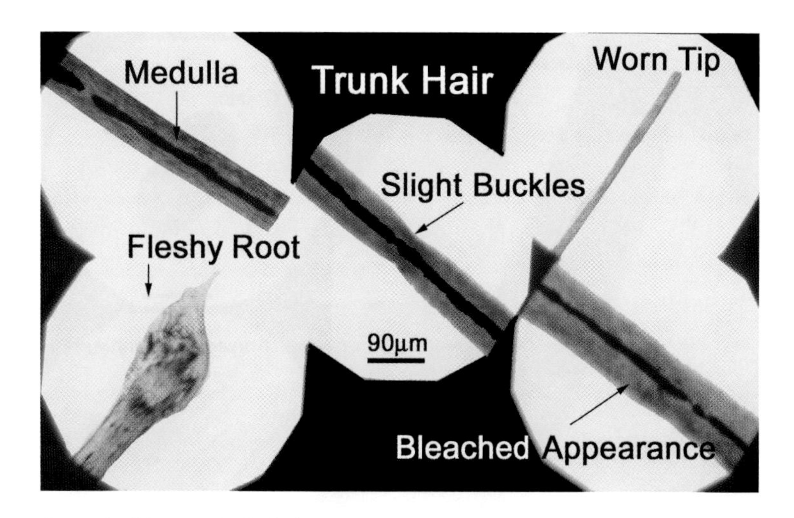

FIGURE 5.6 Appearance of a typical human body hair.

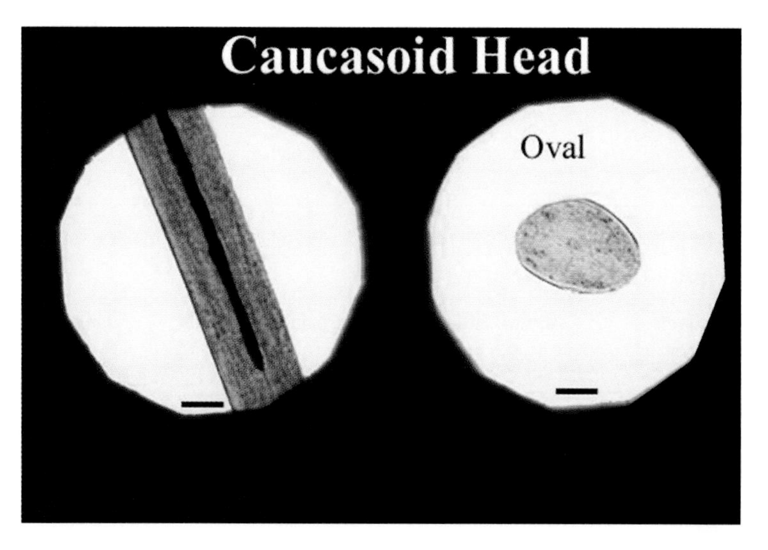

FIGURE 5.7 Eastern European head hair.

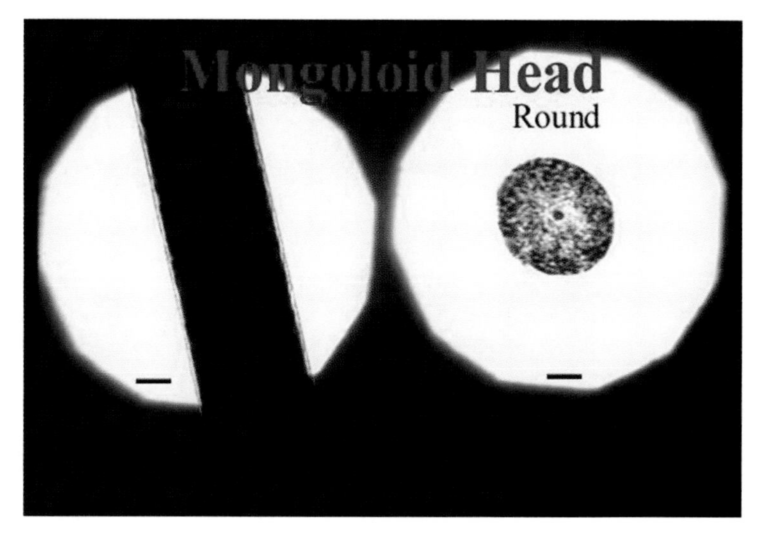

FIGURE 5.8 East Asian head hair.

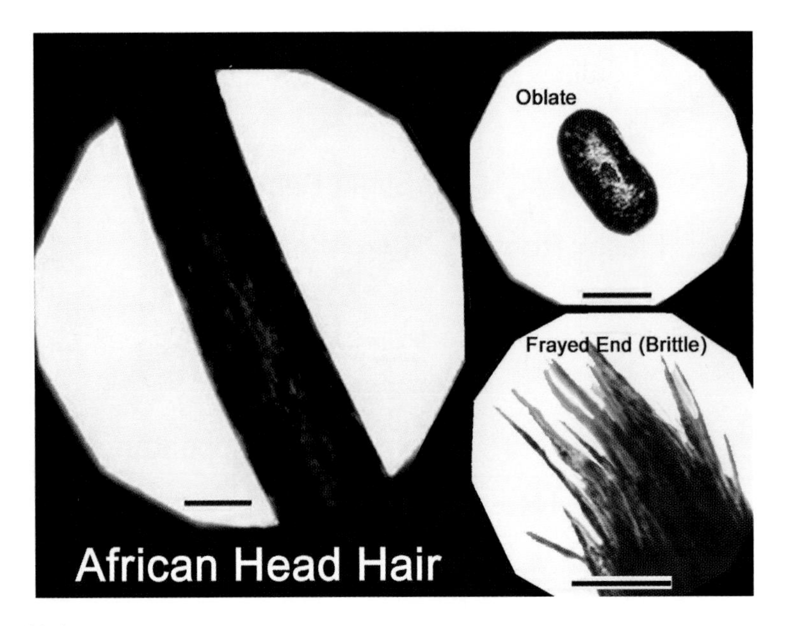

FIGURE 5.9 African head hair.

TABLE 5.3
Acquired Characteristics

Artifacts	Abnormalities	Artificial Treatments	Damage
Lice	Hair casts	Hair spray	Environmental
Nits	Pili annulati	Hair gel	Mechanical
Mold/fungus	Trichoschisis	Permanent	Burned
Bite marks	Monilethrix	Hair cosmetic	Crushed
Debris	Trichorrhexis nodosa	Other	Glass cut
Blood	Trichorrhexis invaginati		Broken
Fibers	Pili torti		Frayed
Other	Trichonodosis		Twisted
	Cartilage hair hypoplasia		Tangled
			Other

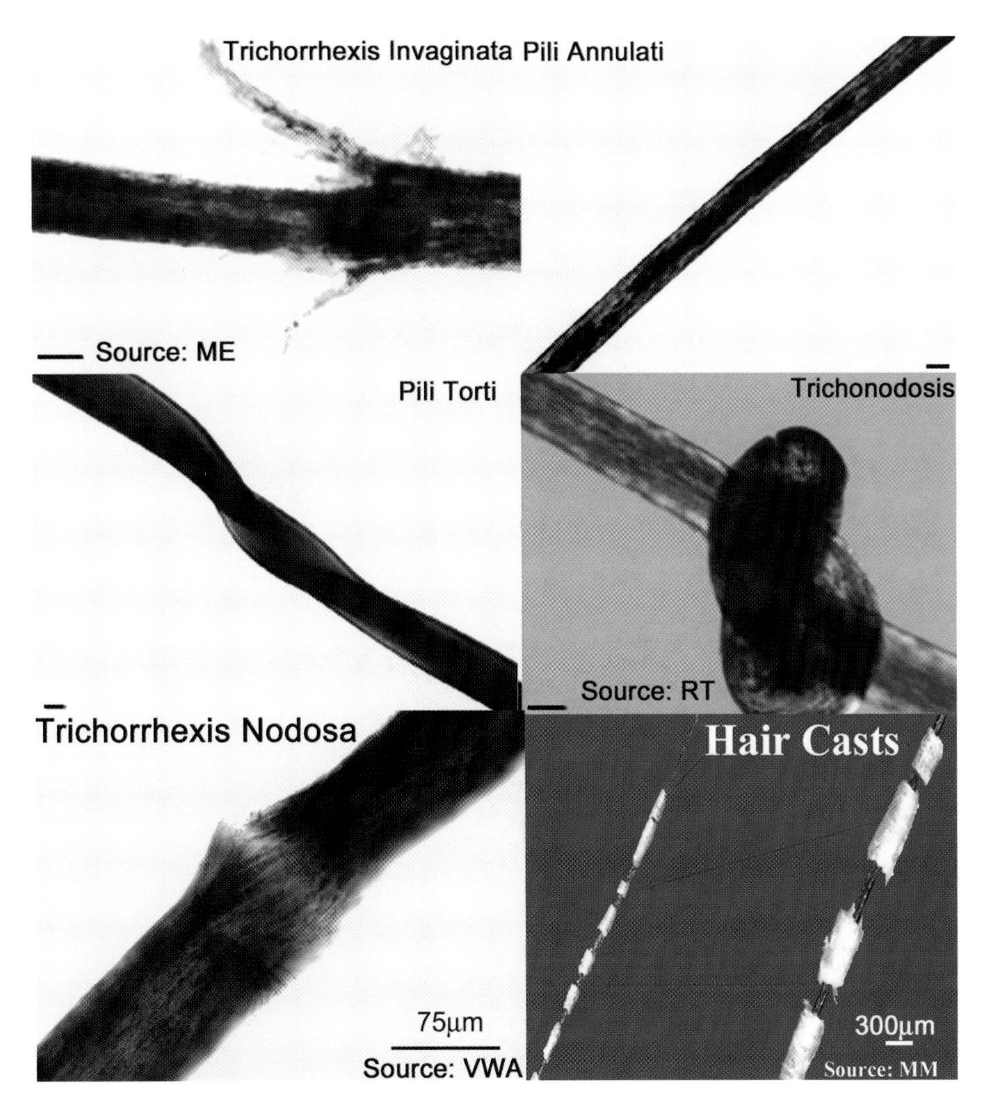

FIGURE 5.10 Six hair conditions seen in casework. All photomicrographs are of questioned or known head hair specimens. The black scale is equal to 40 µm, unless otherwise noted.

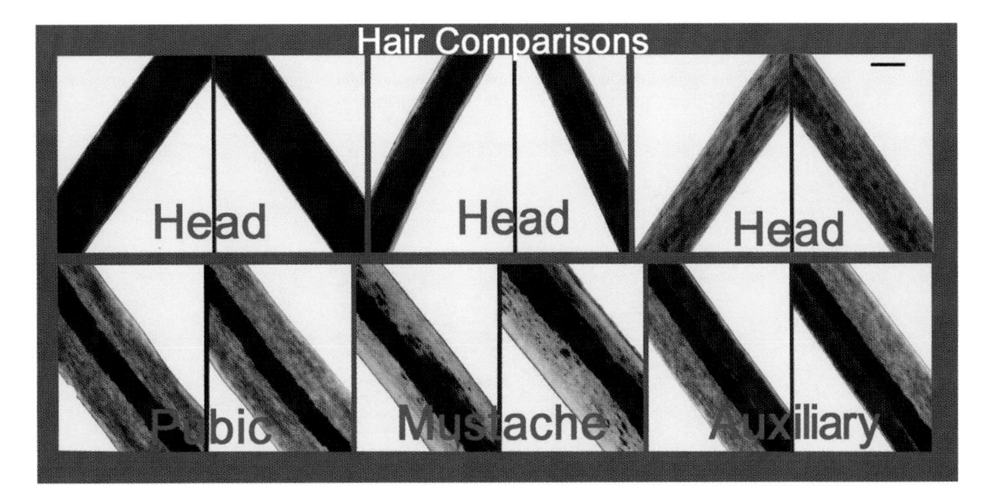

FIGURE 5.11 Photomicrographs of six hair comparisons from casework. The known hairs are on the left; the questioned hairs are on the right.

6 Animal Hair Identification

Animal hair is often encountered in forensic casework. Hairs shed by domestic pets such as cats, dogs, ferrets, and hamsters are often found on clothing or in dust specimens. Hair from pet grooming finds its way into dust. Animal hair originating from clothing and other textile items made from animal hair or fur can become airborne and become incorporated into dust at a given location. Animal hairs used to manufacture wearing apparel, draperies, textiles, and carpets are easily transferred between people, places, and things and for that reason often play firmly established roles in forensic investigations.[1–10] Information provided by hair evidence is often used to associate people, reconstruct events, and solve crimes.

Art historians, conservators, archeologists, and anthropologists also encounter various categories of mammalian hair in their work. Animal hairs used in building materials, textiles, clothing, costumes, tapestries, and other decorative arts are found in articles from ancient and modern cultures around the globe.[11–17] The clues provided by hair evidence can be used by forensic scientists, conservators, archeologists, anthropologists, and other investigators to solve many types of scientific inquiries. Three types of animal hairs are encountered: (1) fine undercoat hairs such as wool and fillers; (2) guard hairs normally characteristic of a family or species of mammal; and (3) large hairs that resemble guard hairs (tactile hairs, for example).

This chapter concentrates on the identification of guard hairs. However, species of questioned animal hairs taken from commercial textiles can be identified on the basis of the microscopic appearance of the fur or underhairs and a microscopist should be familiar with many different types of animal hairs. The animal hairs encountered in casework come from three basic sources: domestic animals; commercial fur animals (clothing and textiles); and wild animals. Identification within these basic categories is based primarily on macroscopic and microscopic morphological features that will be discussed in the next section. A comprehensive discussion of guard hair morphology can be found in the literature.[18–25]

MACROSCOPIC AND MICROSCOPIC EXAMINATIONS

Macroscopic Examination

A questioned guard hair must be isolated from its matrix sample and examined visually and with a stereomicroscope. The following characteristics should be noted:

1. Reflected light colors
2. Color band pattern
3. Length of shaft
4. Shaft shape including presence or absence of a shield.

Figure 6.1 shows common dorsal guard hair shaft shapes with shields.

Microscopic Examination

The three primary anatomical regions common to all mammalian hair and thus important in their classification and identification are the cuticle (outer protective layer), cortex (main body), and medulla (central canal). See Figure 6.2. Microscopic examination of animal hairs should include observation of the following characteristics:

1. Cuticle scale patterns
2. Medulla configuration and medullary index
3. Pigmented areas of cortex — colors and banding
4. Shaft shape
5. Cross-sectional shape
6. Root morphology

Cuticle

A cuticle is composed of many layers of overlapping scales covering the entire hair shaft. A.B. Wildman and other researchers described basic scale patterns in detail. For additional study, readers are again referred to the literature.[18–26] The scale patterns exhibited along a guard hair shaft are useful for identifying certain characteristics. Scale patterns should be determined before a hair is mounted permanently.

Many techniques for replication of hair scale patterns have been published. The authors used the methods cited in Chapter 2 to produce the specimens shown in Figures 6.3 through 6.6. The size, shape and pattern of arrangement along the hair shaft are useful criteria for animal hair identification because many species have distinctive scale patterns. Figures 6.3 through 6.6 are photomicrographs of the primary scale patterns used in the identification of mature guard hairs. After the principal cuticle pattern is determined, the guard hair should be preserved in a permanent mounting medium, as described in Chapter 2.

Medulla

The medullary configurations exhibited by guard hairs can be useful in determining the species from which a questioned hair originated. Some species have quite distinct medullae. Rabbit hair is a good example. Its medulla is in the form of a multiserial ladder that resembles a corncob. Figure 6.7 consists of photomicrographs of basic medullary configurations. The specimens were mounted in Cargille Melt Mount (refractive index of 1.539 for the sodium D line) or Permount (refractive index after setting of about 1.523 for the sodium D line).

The medullary index (MI) is a measure of the ratio of the overall diameter of a hair shaft to the diameter of the medulla. Most mammals except humans have MI values between 0.50 and 0.90. Some species have quite large MI measurements, for example, rabbits. The MI for a rabbit guard hair is at least 0.90. This feature, along with the corncob appearance of the medulla, indicates that a questioned guard hair most likely originated from a rabbit or hare. Figure 6.8 demonstrates the computation of MI. The measurement is useful for determining the species of mammal from which a questioned hair originated.

Cortex

The cortex is the main body of a mammalian hair composed of spindle-shaped keratinized fibril cells arranged along the long axis and held together by a sulfur-rich protein matrix. Dispersed within the cortical region are pigment granules, ovoid bodies, and air spaces known as cortical fusi. A comprehensive discussion of hair physiology can be found in a work edited by Robertson.[27]

The cross-sectional shape of a guard hair can be helpful in determining the species of mammal from which a questioned guard hair originated. Rabbit hair is again a good example. See Figure 6.9. Note the ribbon-shaped cross-section and several rows of medullary cells that indicate rabbit or hare guard hair.

Color and Color Banding

Examination of the colored and colorless areas includes observing lack of pigment, density of pigment, reflected and transmitted light colors, whether or not color is banded along the shaft, and the pattern of banding. In addition, any artifacts such as ovoid bodies, cortical fusi, damage, and treatment should be noted.

Shaft and Cross-Sectional Shapes

Observations should include overall shape. Tapering or strictures found on the shaft should be noted. Cross-sectional shape characteristics should be observed and noted. See dorsal guard hairs shown in Figure 6.1.

Root Morphology

Root morphology can also be an indicator of species. For example, dog hair roots are usually spade-shaped, while cat hair roots have frayed tips.

Species Determination and Comparisons

All the data describing macroscopic and microscopic features of the questioned hair specimens should be noted on an animal hair data sheet (see Table 6.1). The observations noted on the data sheet should be compared with the data in Figure 6.10 and Appendix B to determine preliminary identification of species. For positive identification after preliminary identification, the hair should be compared with reference standards, the data in Appendix B, published references, and illustrations in animal hair atlases.

Questioned animal hairs can be compared to hairs from a known source in the same way human hair comparisons are conducted. As with human hairs, a good standard sample from the animal is required. It should include, if possible, at least 25 hairs representing all colors and sample hairs from all areas of the body.

It is not possible to completely cover all the information needed to identify the species of animal hairs in this chapter. Readers are referred to the reference list for materials for further study.

REFERENCES

1. Gross, H., *Criminal Investigation*, adapted from *System Der Kriminalistik*, Adams, J.C., London, Sweet & Maxwell, 1924, p. 131.
2. Locard, E., The analysis of dust traces, *Am. J. Police Sci.*, 1, 276, 1930.
3. Söderman, H. and Fontell, E., *Handbok I: Kriminalteknik*, Stockholm, 1930, p. 534.
4. Glaister, J., *A Study of Hairs and Wools Belonging to the Mammalian Group of Animals, Including a Special Study of Human Hair Considered from Medicolegal Aspects*, Cairo, MISR Press, 1931.
5. Smith, S. and Glaister, J., *Recent Advances in Forensic Medicine*, 2nd ed., Philadelphia, Blakiston's Son, 1939, p. 118.
6. Kirk, P.L., *Crime Investigation*, New York, Interscience, 1953, p. 152.
7. Hicks, J.W., Microscopy of Hair, Issue 2, Washington, D.C., U.S. Government Printing Office, 1977.
8. McCrone, W.C., Particle analysis in the crime laboratory, in *The Particle Atlas*, Vol. 5, McCrone, W.C., Delly, J.G., and Pelanik, S.J., Eds., Ann Arbor, MI, Ann Arbor Science Publishers, 1979, p. 138.
9. Sato, H., Yoshino, M., and Seta, S., Macroscopical and microscopical studies of mammalian hairs with special reference to morphological differences, *Rep. Natl. Res. Inst. Police Sci.*, 33, 116, 1980.
10. Petraco, N., The occurrence of trace evidence in one examiner's casework, *J. Forens. Sci.*, 30, 485, 1985.
11. Cennini, C., *The Craftsman's Handbook*, Thompson, D.V., Jr., Trans., New York, Dover, 1954.
12. Gettens, R.J. and Stout, G.L., *Painting Materials: A Short Encyclopedia*, New York, Dover, 1962.

13. Minor, M. and Minor, N., *The American Indian Craft Book*, Lincoln, University of Nebraska, 1978.
14. Miller, R.S., *Art of the Andes from Chavin to Inca*, New York, Thames & Hudson, 1995.
15. Miller, M.E., *The Art of Mesoamerica from Olmec to Aztec*, New York, Thames & Hudson, 1996.
16. Blier, S.P., *The Royal Arts of Africa: Majesty of Form*, New York, Abrams, 1998.
17. Lucas, A. and Harris, J.R., *Ancient Egyptian Materials and Industries*, New York, Dover, 1999.
18. Glaister, J., *A Study of Hairs and Wools Belonging to the Mammalian Group of Animals, Including a Special Study of Human Hair, Considered from Medicolegal Aspects*, Cairo, MISR Press, 1931.
19. Hausman, L.H., Structural characteristics of the hair of mammals, *Am. Naturalist*, 54, 496, 1920.
20. Brown, F.M., The microscopy of mammalian hair for anthropologists, *Proc. Am. Philos. Soc.*, 85, 250, 1942.
21. Wildman, A.B., *Microscopy of Animal Textile Fibres*, Leeds, WIRA, 1954.
22. Adoryan, A.S. and Kolenosky, G.B., *A Manual for the Identification of Hairs of Selected Ontario Mammals*, Wildlife Research Report 90, Ontario, Canadian Department of Lands and Forests, 1969.
23. Moore, T.D., Spence, L.E., Dugnolle, C.E., and Hepworth, W.G., *Identification of the Dorsal Guard Hairs of Some Mammals of Wyoming*, Cheyenne, 1974.
24. Brunner, H. and Conan, B.J., *The Identification of Mammalian Hair*, Melbourne, Lukata Press, 1974.
25. Appleyard, H.M., *Guide to the Identification of Animal Fibres*, 2nd ed., Leeds, WIRA, 1978.
26. Moore, T.D., Spence, L.E., Dugnolle, C.E., and Hepworth, W.G., *Identification of the Dorsal Guard Hairs of Some Mammals of Wyoming*, Cheyenne, 1974, p. 317.
27. Robertson, J., Ed., *Forensic Examination of Hair*, London, Taylor & Francis, 1999, p. 156.

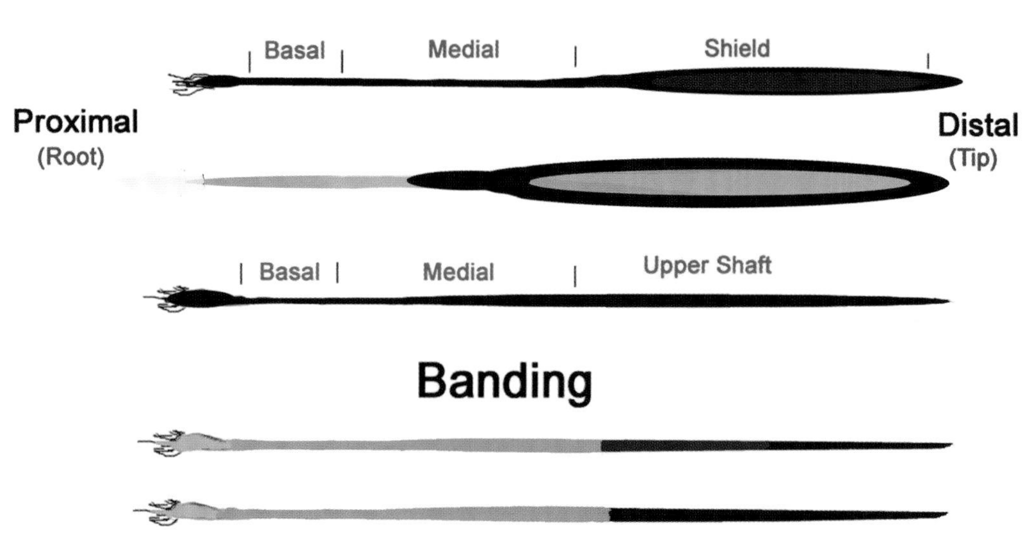

FIGURE 6.1 Typical shapes, banding, and nomenclature of guard hairs. A comprehensive discussion of guard hair shapes and color banding can be found in the work of Moore et al.[26]

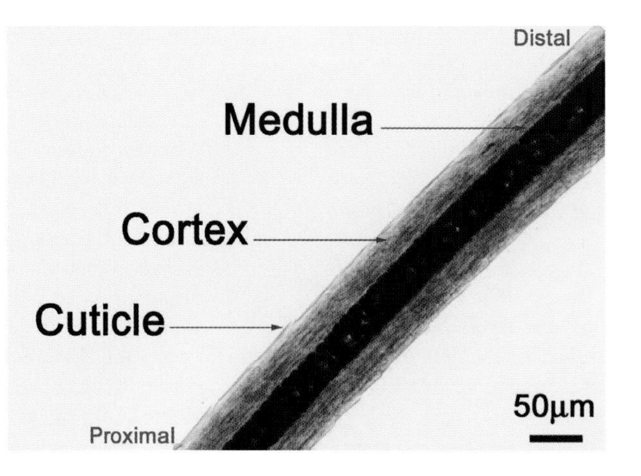

FIGURE 6.2 Three anatomical regions of mammalian hair: cuticle, cortex, and medulla. The proximal or root end and the distal or tip end are also depicted.

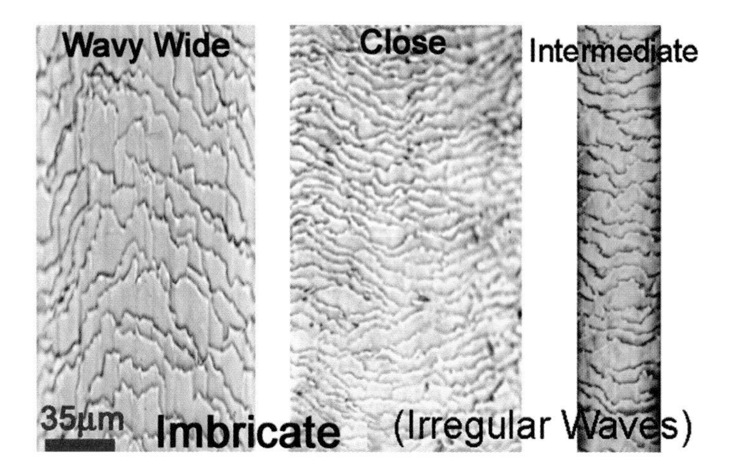

FIGURE 6.3 Imbricate or wavy scale patterns.

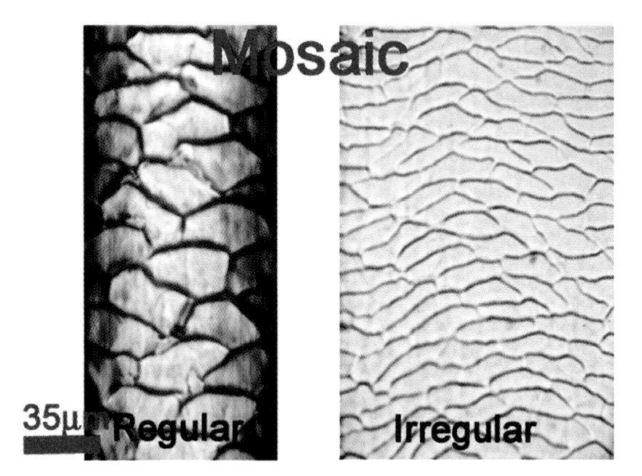

FIGURE 6.4 Mosaic or tile-shaped scale patterns.

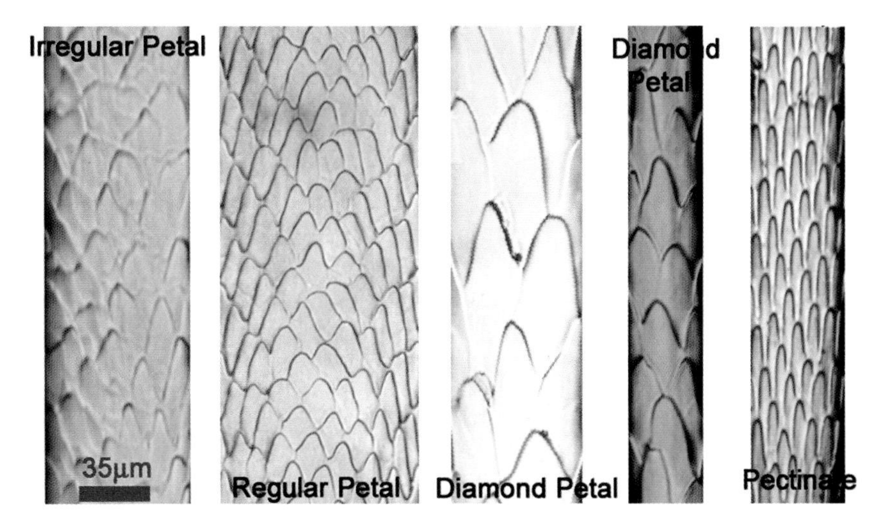

FIGURE 6.5 Petaloid-shaped scale patterns: irregular, regular, diamond, and pectinate.

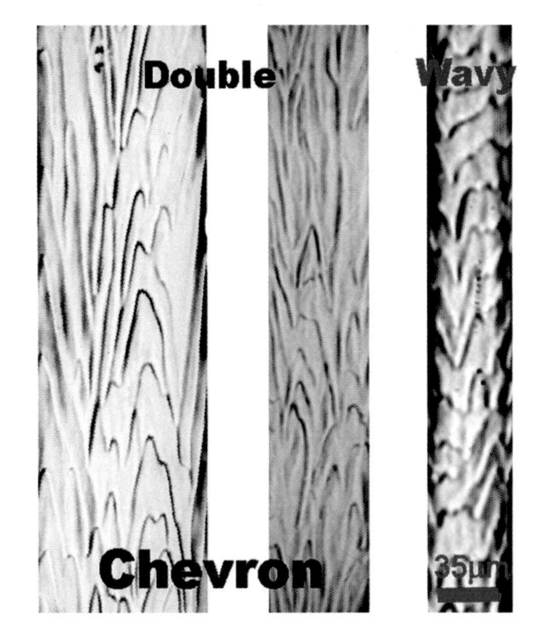

FIGURE 6.6 Chevron-shaped scale patterns.

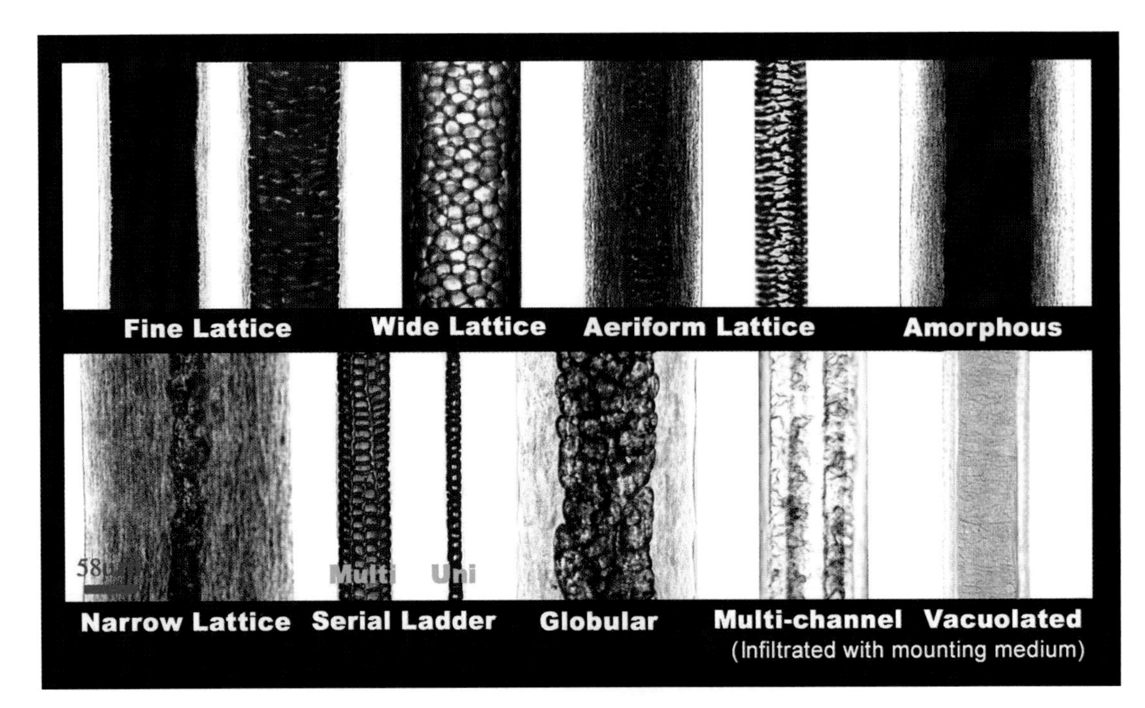

FIGURE 6.7 Basic medullary configurations found in mature mammalian guard hairs.

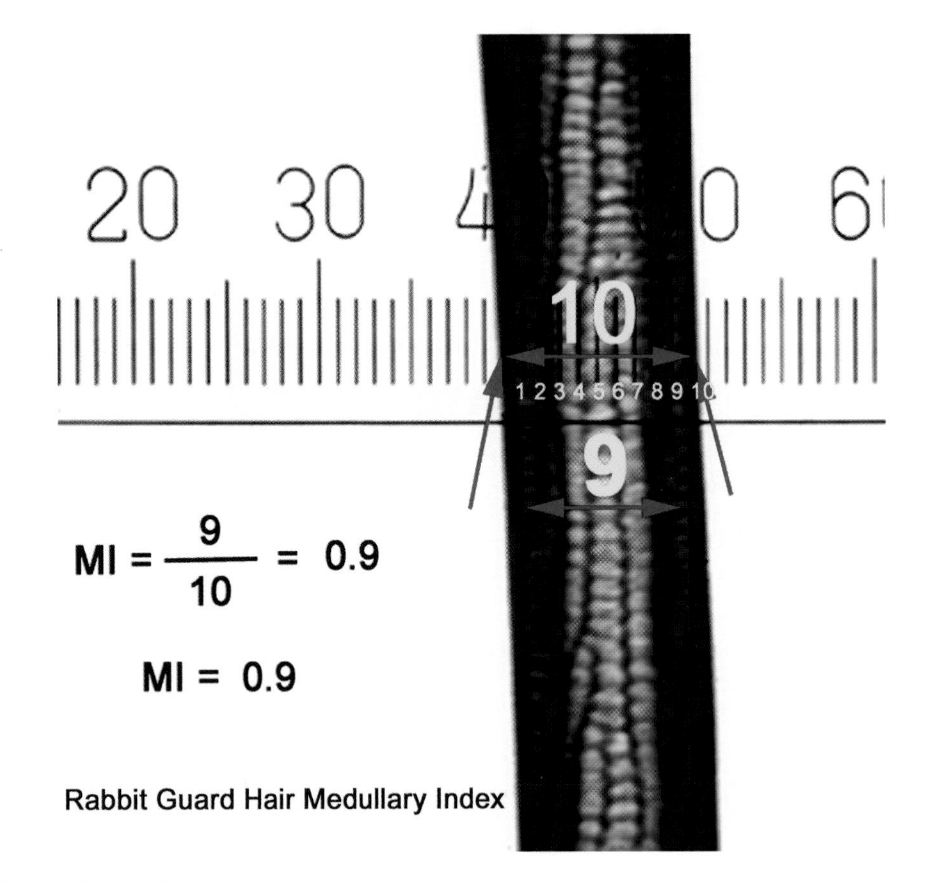

FIGURE 6.8 Computation of MI.

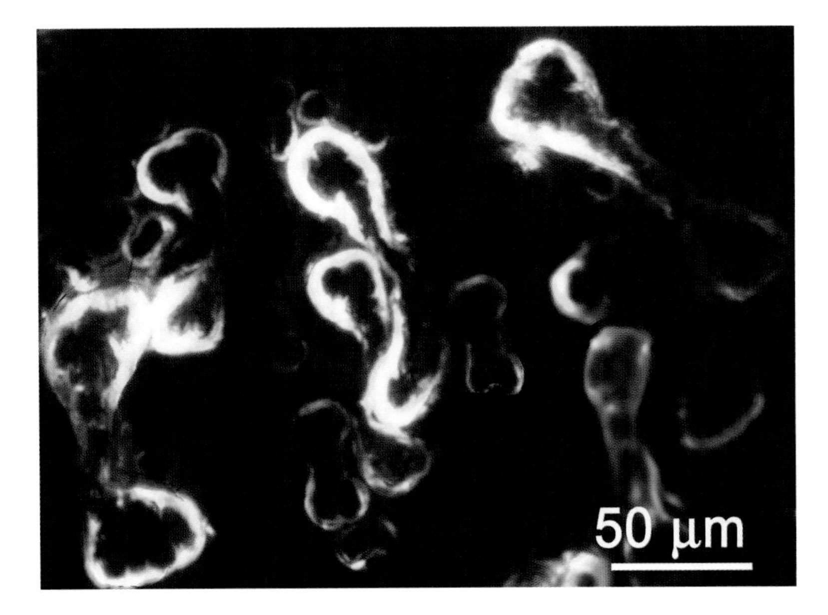

FIGURE 6.9 Cross-sectional appearance of rabbit guard hairs.

TABLE 6.1
Hair Data Sheet

Insert appropriate response in blank or circle appropriate description.

Classification of hair: Guard Wool Underhair Down Other

Cortex:

Shape of hair: Straight Curly Wavy

Length of Shaft in mm ______________________________

Maximum______________ Minimum______________ Average ______________

Color: Reflected____________________ Transmitted ____________________

Single color ______________Multicolored ______________ Banded______________

Describe banding from tip end to root ______________________________

____

Pigment density and distribution (i.e., heavy toward cuticle) ______________________

____

Shaft diameter in µm______________________________

Root shape ______________________________

Tip shape ______________________________

Medulla:

Medulla: Absent Present

Medullary configuration ______________________________

Changes along shaft (describe from tip to root end):______________________

____

Medullary index (medulla diameter/shaft diameter) = ______________________

Cuticle:

Scale patterns along shaft (describe from tip to root end): ______________________

____

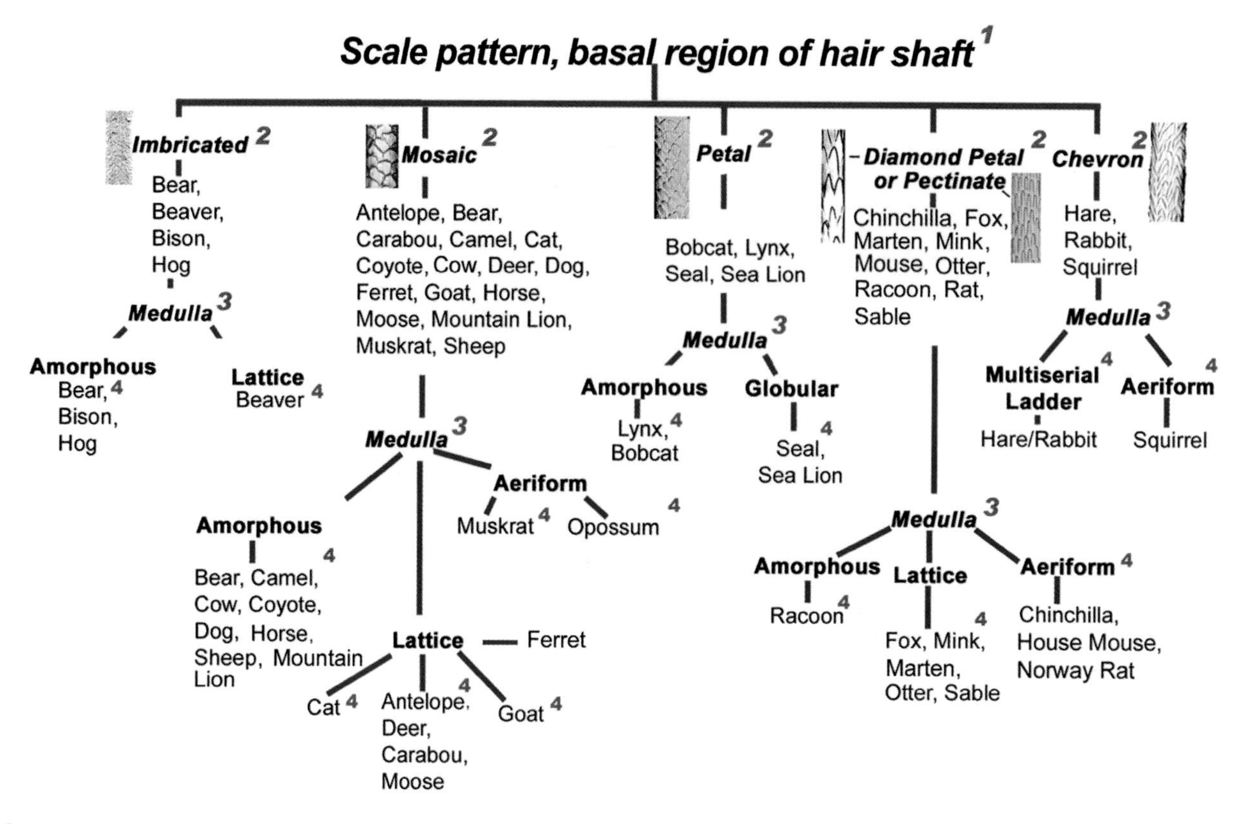

1 Principal scale pattern in the basal region of mature guard hair.

2 Imbricated scale patterns: **close, intermediate, wide; mosaic scale patterns: irregular and regular**
Chevron scale patterns: single and double; and petal scale patterns: regular and irregular.

3 Primary medulla type exhibited in upper shaft or shield area of mature guard hairs.

4 See Appendix B for identifying characteristics.

FIGURE 6.10 Flow chart for preliminary identification of mammalian hair species based on the principal scale pattern found at the basal (near the root) region and the primary medullary configuration of a mature guard hair.

7 Synthetic Fiber Identification

Today's enormous production and worldwide availability of textile fibers mean the average person invariably finds himself surrounded by items manufactured from synthetic fibers. As a result, people often touch many types of textile fibers during their daily routines. The need to have frequent contact with clothing items, rugs, draperies, furniture, and vehicle interiors ensures that we all come into contact with vast numbers of synthetic and natural fibers every day. The important role textile fibers play in forensic science investigations has been clearly established in the literature.[1–20] Forensic scientists often use textiles and fibers to associate people, places, and things involved in incidents and help reconstruct events.

People throughout the ages have found themselves in constant contact with textiles and the fibers of which they are composed. Natural fibers in the form of animal hair, wool, vegetable fibers, wood (paper) fibers, and silk are often encountered in works of art, textiles, and tapestries. Vegetable fibers and hair from people and animals have been used as a reinforcement fibers in mortars and plasters. Consequently, practitioners of disciplines concerned with studies of past and present cultures, textile materials used in everyday lives, and art and artifacts must be able to identify textile fibers.

Forensic scientists, art historians and conservators, textile conservators, architectural conservators, archeologists, and anthropologists all require straightforward methods that allow them to quickly and accurately identify textile fibers. Polarized light microscopy provides such a method.

Textile fibers fall into many categories. Natural fibers include human and animal hair, insect fibers, vegetable fibers, and mineral fibers. A large number of commercially manufactured fibers contain different genera of natural and synthetic polymers. Human and animal hairs were discussed in Chapters 5 and 6. Vegetable and insect fibers will be discussed in Chapter 8. The identification of manufactured natural and synthetic polymers and mineral fibers will serve as the focus of this chapter. Table 7.1 lists the natural and synthetic polymer fibers to be included in this discussion.

PRELIMINARY EXAMINATION PROCEDURE

After sorting with a stereomicroscope, an aliquot of the questioned fiber is mounted in either Permount or a Cargille oil with a refractive index (RI) of 1.525 for the sodium D line at 25°C. Temporary mounts are usually prepared with Cargille oil. Permanent mounts are generally prepared with Permount. However, unless a slide is sealed, Permount can degrade by oxidizing, thereby destroying the specimen. To prevent the Permount from oxidizing after drying, the preparation should be sealed with liquid electrical tape as shown in Figure 7.1.

The next step is to examine the specimen with a polarized light microscope using plane-polarized light (PPL) at 25 to 500× magnification. Fong reported a procedure for identifying synthetic fibers in a single mounting medium with an RI of 1.525. Petraco followed with a scheme involving Permount and/or Cargille oil, and later devised a method using Melt Mount 1.539.[14,16,26,27] (As a point of interest, the authors studied the RIs of various batches of dried Permount using traditional RI immersion methods, GRIM II apparatus, and phase-contrast microscopy as originally presented by Ojena and De Forest).[21] The RI range of dried Permount is between 1.524 through 1.526 (average is 1.525).

The first step is determining whether a fiber is natural or synthetic. If it is natural, determine whether its origin is animal or vegetable; if man-made, determine whether the material is a natural or synthetic polymer or mineral. Refer to Chapters 5, 6, and 8 for information about fibers classified as natural (human or animal hair or vegetable material). This chapter covers microscopy of synthetic fibers.

MAN-MADE NATURAL AND SYNTHETIC FIBERS

After examining a preparation, a microanalyst will usually observe several fibers. He or she must visually single out a fiber and make certain observations. Information concerning fiber morphology is collected first, then the relative refractive index (RRI) of the N$\parallel$ and N$_\perp$ directions of the fiber as they compare to the RI of the mounting medium are obtained by using the Becke line (BL) method under PPL. The elongated axis of a fiber analyzed by the BL method is aligned parallel to the vibrational (preferred) direction of the polarizer. The movement of the BL when the microscope focus is raised is noted. The BL moves toward the medium of higher RI under these conditions.

The elongated axis of the fiber is aligned perpendicular to the preferred direction of the polarizer and the movement of the BL in this orientation is noted. See Figure 7.2 for illustrations of fiber orientation and Becke line movement.

The subject fiber is then observed between crossed polars (CPs). If the fiber is isotropic, it will remain dark (extinct) when rotated between CPs. If anisotropic, it will appear to turn on (display interference colors) four times and turn off (go extinct) four times as it is rotated 360° (see Figure 7.3).

The degree of retardation exhibited by an anisotropic fiber depends on its generic class. Acetate and acrylic fibers demonstrate low-order interference colors (ICs). Rayon and olefin fibers exhibit first- and second-order ICs, while aramid, nylon, and polyester fibers show higher-order ICs, as shown in Figure 7.4.

The amount of retardation an anisotropic fiber exhibits as it is rotated between CPs is determined by using an interference chart, and if necessary, appropriate compensators. First, the microscopist rotates the fiber until it exhibits its brightest IC, normally 45° off extinction, then looks for the corresponding retardation color on an interference chart, as shown in Figure 7.5.

Fiber birefringence (Bi) can be estimated with the following method:

1. Measure specimen diameter in micrometers with a calibrated ocular micrometer, as outlined in Chapter 3.
2. Locate the fiber thickness value (t = 40 μm) on an interference chart.
3. Locate the fiber retardation value due to the IC displayed by that thickness.
4. Trace a left-to-right line from t until it intersects the bottom-to-top line traced from r.
5. From the intersection of t and r, trace a diagonal line adjacent to or along the closest diagonal line running from the origin until it intersects the chart's outer perimeter.
6. Estimate fiber birefringence at the point at which it intersects the Bi scale (see Figure 7.6).

Another important optical property of an anisotropic material is its sign of elongation (SE). The SE for a synthetic fiber is the relationship between a fiber's length or long axis as its relates to its RIs. If the highest RI of a fiber is along its length (Nll), its SE is said to be positive. If the highest RI is along a fiber's width ($N_{\perp}$), its SE is said to be negative. Figure 7.7 depicts the procedure for determining SE of a synthetic fiber with a fixed compensator. A Sénarmont compensator is used to obtain a precise retardation reading for anisotropic fibers that exhibit low order interference colors. A variable compensator, i.e., a quartz wedge, is used normally to determine the SE and order of retardation for a fiber that exhibits high-order interference colors. Variable compensators are useful because they enable a microscopist to add or subtract several orders of compensation to or from a specimen during an examination. An excellent explanation of the use of variable compensators and Sénarmont compensators can be found in Bloss.[36]

Other comparative information about fiber appearance (dulling agent, optical properties, pleochroism; Figure 7.8) and other factors should be collected and recorded in an examiner's notes on a fiber data sheet (Table 7.2). The following data should be compiled for the effective identification and comparison of an unknown man-made fiber:

1. Physical properties
   Reflected color
   Length (mm)
   Length (staple or continuous)
   Appearance (crimped or straight)
   Transparency
   Transmitted color
   Coloration (pigment or organic dye)
   Dyeing process
   Thickness (μm)
   Cross-sectional shape
   Longitudinal morphology (smooth, striated, cross-hatched, etc.)
   Surface texture
   Fish eyes (present or absent); shape if present ________________
   Dulling agent (present or absent)
   Extent of dulling
   Twists/crimps (present or absent)
   Other treatment (e.g., melting)
   Other manufacturing artifacts (e.g., gas bubbles)
   Foreign artifacts (e.g., blood, pollen, etc.)

2. Optical properties
   Isotropic/anisotropic
   Degree of relief
   Interference colors (ICs)
   Thickness (μm) of fiber portion causing observed IC
   Retardation (nm)
   Estimated birefringence
   Sign of elongation
   Relative refractive indices for Nll and $N_{\perp}$ directions using PPL
   Pleochroism

To determine the generic classification of an unknown fiber, compare the data collected and tabulated in Table 7.2 to information in Table 7.3 and Figure 7.9. Each class of fiber specimen is identified in the same manner. If a comparison of fibers is desired, the questioned and known specimens can be compared side by side on a comparison microscope. Appendix C is an atlas of synthetic fibers. Most classes of synthetic fibers are pictured in both longitudinal and cross-sectional views. In addition, data tables, flow charts, and photomicrographs covering other aspects of synthetic fiber identification such as modification ratios of rug fibers, solubility testing, and use of interference filters to determine order of retardation can be found in Appendix C.

REFERENCES

1. Locard, E., The analysis of dust traces. II, *Am. J. Police Sci.*, 1, 405, 1930.
2. Locard, E., The analysis of dust traces. III, *Am. J. Police Sci.*, 1, 496, 1930.
3. O'Neill, M.E., Police microanalysis. II. Textile fibers, *J. Am. Inst. Crim. Law Criminol.*, 25, 835, 1935.
4. Burd, D.Q. and Kirk, P.L., Clothing fibers as evidence, *J. Am. Inst. Crim. Law Criminol.*, 32, 353, 1942.
5. Frei-Sulzer, M., Coloured fibres in criminal investigations with special reference to natural fibres, in *Methods of Forensic Science,* Vol. 4, Curry, A.S., Ed., New York, Interscience, 1965, p.141.
6. Longhetti, A. and Roche, G.W., Microscopic identification of man-made fibers from the criminalistics point of view, *J. Forens. Sci.*, 3, 303, 1958.
7. Mitchell, E.J. and Holland, D., An unusual case of identification of transferred fibres, *J. Forens. Sci.*, 19, 23, 1979.
8. Grieve, M.C., The role of fibers in forensic science examinations, *J. Forens. Sci.*, 28, 877, 1983.
9. Fong, W., Fiber evidence: laboratory methods and observations from casework, *J. Forens. Sci.*, 29, 55, 1984.
10. Deadman, H.A., Fiber evidence and the Wayne Williams trial, *FBI Law Enforcement Bull.*, March 1984, p. 13; May 1984, p. 10.
11. Gaudette, B.D., Fibre evidence, *RCMP Gazette*, 47, 18, 1985.
12. Petraco, N., The occurrence of trace evidence in one examiner's casework, *J. Forens Sci.*, 30, 485, 1985.
13. Petraco, N., Trace evidence: the invisible witness, *J. Forens. Sci.*, 31, 321, 1986.
14. Fong, W., Rapid microscopic identification of synthetic fibers in a single liquid mount, *J. Forens. Sci.*, 27, 257, 1982.
15. Petraco, N., DeForest, P.R., and Harris, H., New approach to the microscopical examination and comparison of synthetic fibers encountered in forensic science cases, *J. Forens. Sci.*, 25, 571, 1980.
16. Gaudette, B.D., The forensic aspects of textile fiber examination, in *Forensic Science Handbook*, Vol. 2, Saferstein, R., Ed., Englewood Cliffs, NJ, Prentice-Hall, 1988, chap. 5.
17. De Forest, P.R., Gaensslen, R.E., and Lee, H.C., *Forensic Science: An Introduction to Criminalistics*, New York, McGraw-Hill, 1983.
18. Gorski, A. and McCrone, W.C., Birefringence of fibers, *Microscope*, 46, 3, 1998.
19. Grieve, M.C., Fibers and forensic science: new ideas, developments, and techniques, *Forens. Sci. Rev.*, 6, 59, 1994.
20. Robertson, J., Forensic examination of fibres: protocols and approaches: an overview, in *Forensic Examination of Fibres,* Robertson, J., Ed., Chichester, Ellis Horwood, 1992.
21. Ojena, S.M. and De Forest, P.R., Precise refractive index determinations by the immersion method using phase contrast microscopy and the Mettler hotstage, *J. Forens. Sci.*, 12, 315, 1972.
22. Hudson, P.B., Clapp, A.C., and Kness, D., *Joseph's Introductory Textile Science*, 6th ed., New York, HBJ College Publishers, 1993.
23. McCrone, W.C. and Delly, J.D., *The Particle Atlas*, 2nd ed., Ann Arbor, MI, Ann Arbor Science Publishers, 1973.
24. Moncrief, R.W., *Man-Made Fibers*, 6th ed., London, Newnes-Butterworths, 1975.
25. Petraco, N., A Guide to the Identification of Synthetic Fibers in Textile Materials, 1985 (unpublished work).
26. Petraco, N., A guide to the rapid screening, identification, and comparison of synthetic fibers in dust samples, *J. Forens. Sci.*, 32, 768, 1987.
27. Petraco, N. and De Forest, P.R., A guide to the analysis of forensic dust specimens, in *Forensic Science Handbook*, Vol. 3, Saferstein, R., Ed., Englewood Cliffs, NJ, Prentice-Hall, 1993, chap. 2.
28. Robertson, J., Ed., *Forensic Examination of Fibers*, New York, Ellis Horwood, 1992.
29. Robertson, J. and Grieve, M., Eds., *Forensic Examination of Fibers*, London, Taylor & Francis, 1999.
30. Palenik, S.J., Chapter 7, Microscopical examination of fibers, in *Forensic Examination of Fibers*, Robertson, J. and Grieve, M., Eds., Philadelphia, Taylor & Francis, 1999, chap. 7.
31. Reference Collection of Synthetic Fibers, Technical Data, Washington, D.C., U.S. Department of Commerce, January 1983 and January 1984.
32. *Identification of Textile Materials*, Manchester, U.K., Textile Institute, 1970.
33. *Man-Made Fibers Fact Book*, New York, Man-Made Fiber Producers Association, Inc., 1978.
34. *Technical Manual of the American Association of Textile Chemists and Colorists,* Vol. 53, 1977.
35. *Forensic Fiber Examination Guidelines*, Washington, D.C., Fiber Subgroup, TWGMAT, 1998.
36. Bloss, F.D., *Optical Crystallography,* Washington, D.C., Mineral Society of America, 1999, p. 109.

TABLE 7.1
Classes of Synthetic and Mineral Fibers

Generic Class	Fiber-Forming Polymer (% by weight)	Polymer Source
Acetates: diacetate and triacetate	Cellulose acetate 100%; 92% of OH groups are acetylated	Natural
Acrylic	Acrylonitrile monomers; at least 85%	Synthetic
Modified acrylics (modacrylics):	Acrylonitrile monomers; at least 35 to 85% plus various monomers, i.e., vinyl chloride	Synthetic
Dynel®		
SEF®		
Verel®		
Fiberglass	Silicone dioxide	Natural
Mineral wool	Hydrated magnesium silicates	
Chrysotile asbestos[a]	Mineral	
Metallic fibers	Metal foils or metal/plastic plies	
Rayon Viscose	Regenerated cellulose	Natural
High-tenacity viscose rayon	Regenerated cellulose	Natural
Lyocel	Regenerated cellulose	Natural
Nylon 6, 6.6	Polyamide monomers; <85% of amide linkages are bonded to two aromatic rings	Synthetic
Aramid	Polyamide monomers; at least 85% of amide linkages are bonded to two aromatic rings	Synthetic
Kevlar®		
Nomex®		
Olefins	At least 85% propylene monomers or ethylene monomers	Synthetic
Polyester	At least 85% PET, PCDT, PBT	Synthetic
Saran	At least 80% polyvinylidene	Synthetic
Spandex (Lycra®)	At least 85% urethane	Synthetic

Source: Forensic Fiber Guidelines, Washington, D.C., Fiber Subgroup, Technical Working Group for Materials Analysis, January 1998.

[a] Primary form of asbestos used in the past by the building industry.

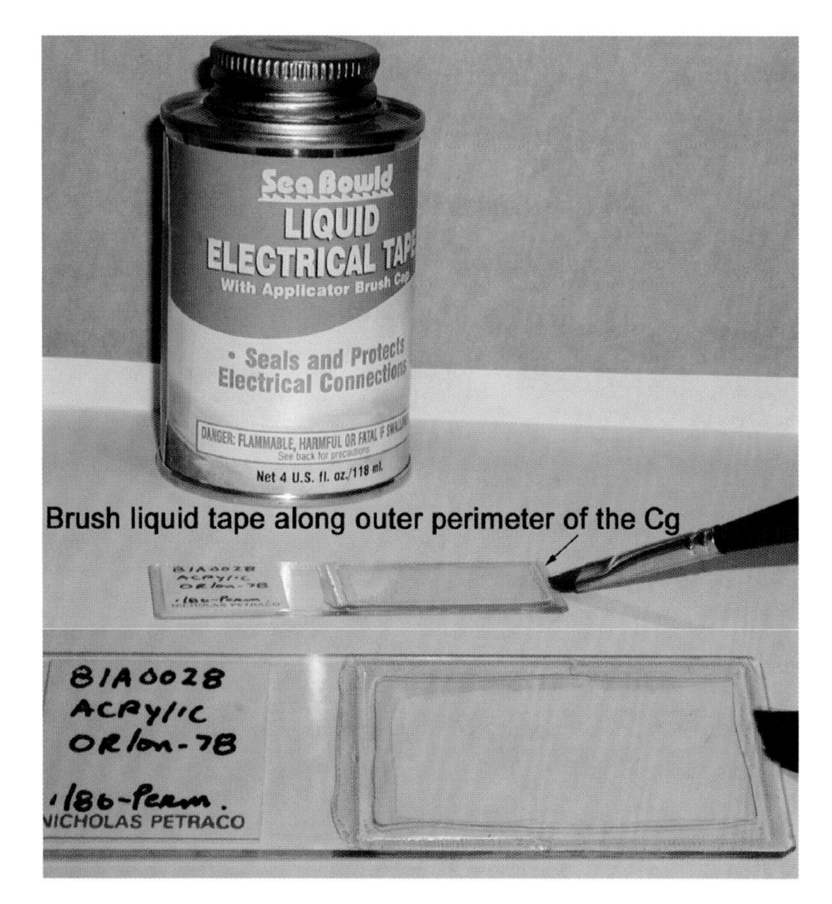

FIGURE 7.1 Edges of the cover glass are sealed with liquid electrical tape to prevent the oxidation of Permount.

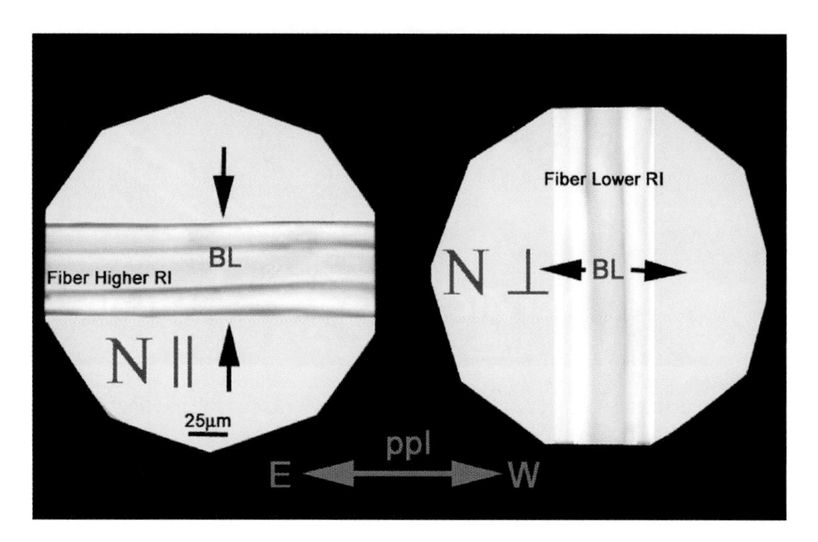

FIGURE 7.2 Edges of the cover glass are sealed with liquid electrical tape to prevent the oxidation of Permount.

FIGURE 7.3 Appearance of an anisotropic fiber as it is rotated 360 degrees between CPs.

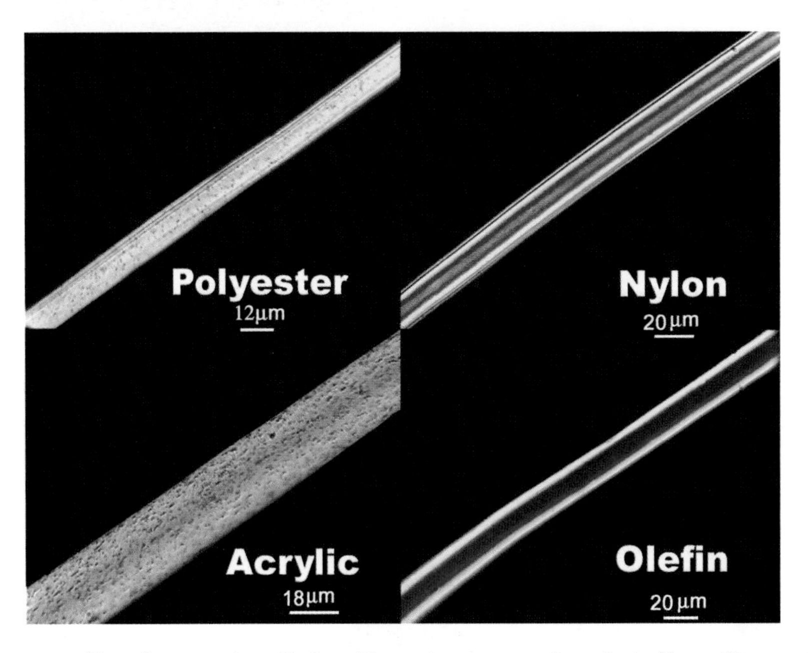

FIGURE 7.4 The appearance of interference colors displayed by various genera of synthetic fibers. Changes in thickness and cross-sectional shape greatly influence a synthetic fiber's appearance between crossed polars.

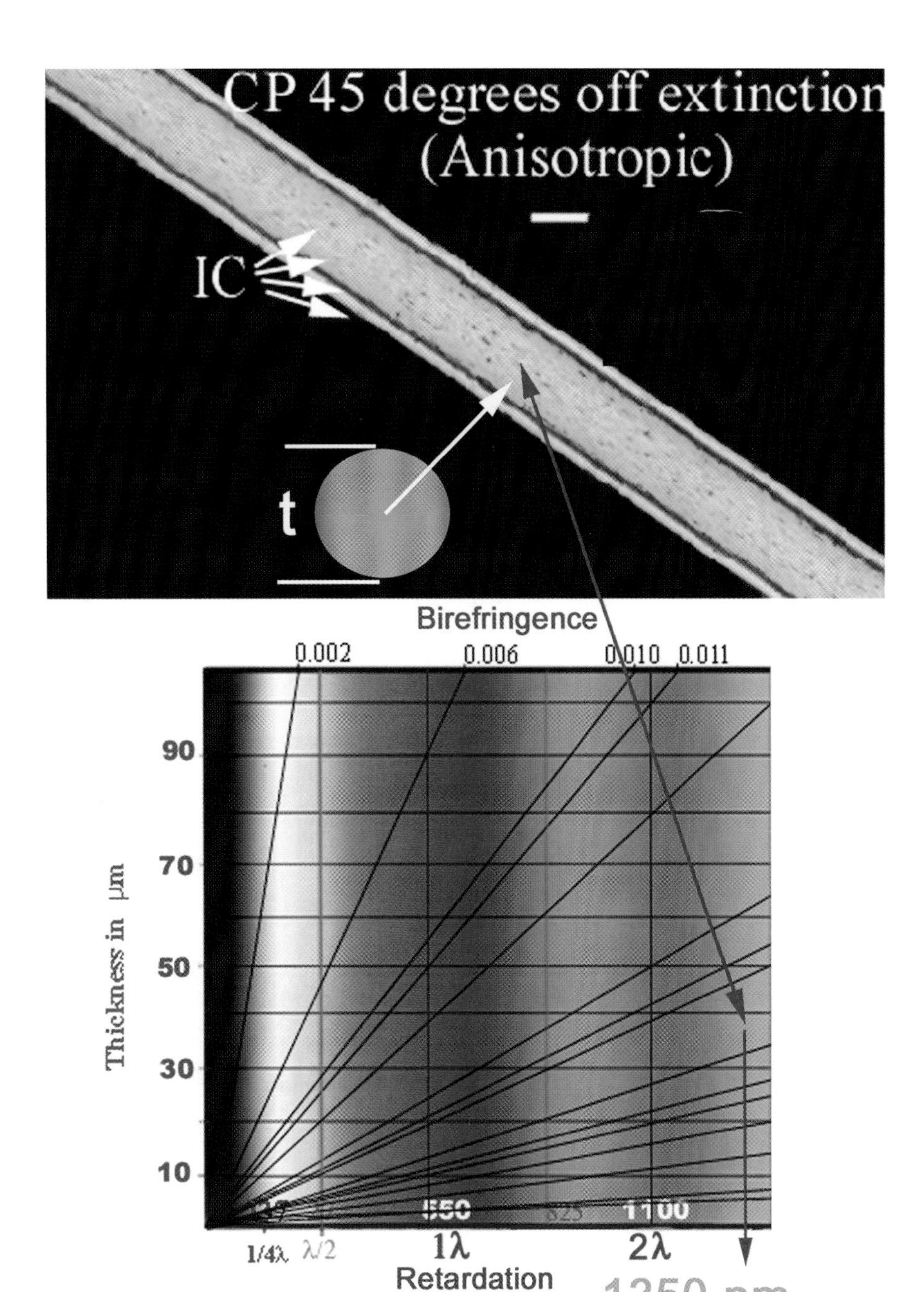

FIGURE 7.5 Computing the degree of retardation a fiber specimen exhibits when viewed between crossed polars. The fiber is placed at the position where it exhibits its brightest IC, normally 45 degrees of extinction; then the IC displayed by the fiber at its thickest point (t = 40 μm) is observed and the corresponding color is found on an interference chart. The amount of retardation in nanometers is read from the interference chart. In this case, the retardation is 1350 nm.

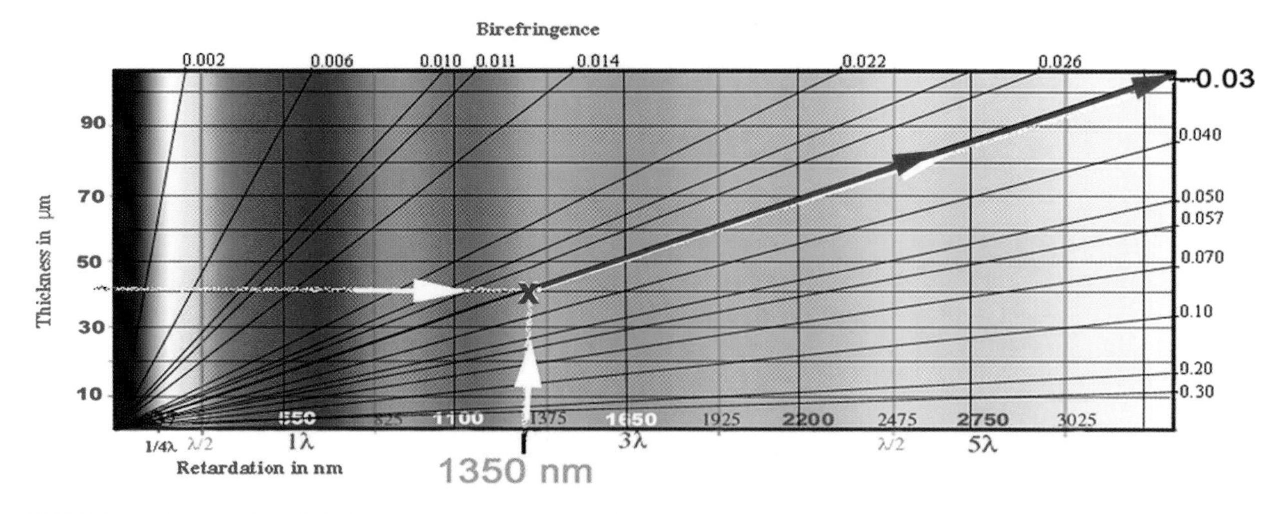

FIGURE 7.6 Computation of birefringence: First measure specimen diameter (t) in micrometers with a calibrated ocular micrometer. Next, on an interference chart, locate the t-value of the fiber (40 µm), and r-value (1350 nm). Trace a line from t (left to right) until it intersects the line being traced from r (bottom to top). Finally, from the intersection of both lines, trace a diagonal line adjacent to or along the intersecting black diagonal line coming from the origin of the interference chart and then estimate the fiber birefringence at the intersection of the diagonal line and the birefringence scale along the outer perimeter of the interference chart (0.030).

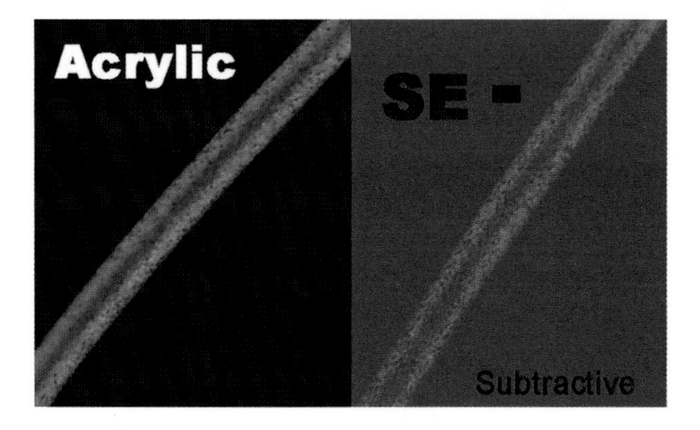

With the fiber specimen placed between CP, parallel to the z-ray of the compensator, and at 45 degrees off extinction, the full wave compensator is inserted along the PLM's OA. If subtractive retardation occurs, the fiber has a negative SE; if, on the other hand, additive retardation occurs, the fiber has a positive SE (see Chapter 1, Figures 1.15 to 1.22).

FIGURE 7.7 Determination of sign of elongation.

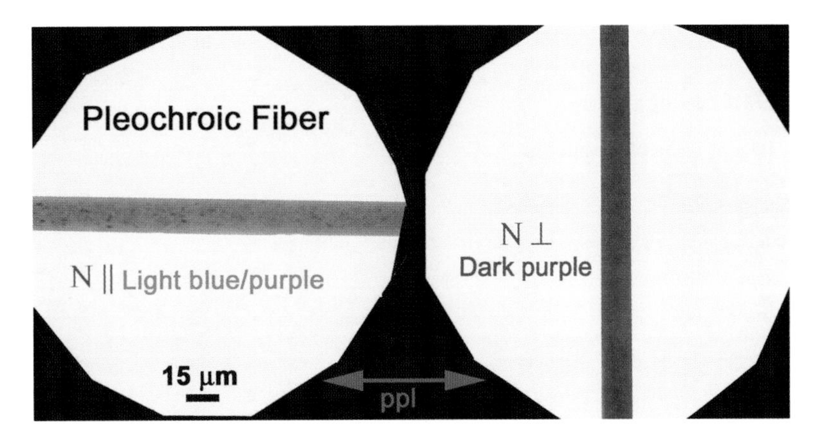

FIGURE 7.8 Pleochroism is the ability of a transparent material to exhibit up to three different colors when viewed with plane-polarized light along different axes. Simply stated, it is a change in color as a material is rotated in plane-polarized light. The specimen depicted is a synthetic fiber. Colors are recorded for both the N|| and $N_{\perp}$ directions.

TABLE 7.2
Generic Classes of Common Synthetic Fibers[22–35]

Generic Class	RRI N|| and N⊥	Birefringence	Sign Elongation	Typical X-S/ Comments
Diacetate	N|| and N⊥ < 1.525	0.002 – 0-.005	+	Serrated
Triacetate	N|| and N⊥ < 1.525	Near 0.0	+/–	Serrated
Acrylic	N|| and N⊥ <1.525 N⊥ near 1.525	0.001 – 0.006	–	Dogbone, bean, peanut, mushroom, oval, round
Modacrylic	N|| and N⊥ >1.525	0.001 – 0.005	+/–	Irregular, ribbon unusual shape, rough texture, fish eyes
Chrysotile asbestos	N|| and N⊥ >1.525	0.004 – 0.016	+	Very fine fibers, usually bundled, sticky
Carbon	opaque	—	—	Ovoid
Aramid	N|| and N⊥ >>>1.525	0.120 – 0.710	+	Round
Nylon 6, 6.6	N|| >>1.525 N⊥ near 1.525	0.049 – 0.063	+	Round, trilobal, tetralobal
Glass wool and mineral wool	Isotropic Range: 1.520 – 1.570	—	—	Round, continuous lengths; various colored resins attached; irregular shapes; conchoidal fractures; broken ends
Metallic	Opaque	—	—	Flat, ribbon; Cu°, Ag°, Al°, Au°; usually plastic coating
Polypropylene	N|| > 1.525 & N⊥ < 1.525 or N|| and N⊥ > 1.525	0.028 – 0.034	+	Round, trilobal
Polyethylene	N|| >> 1.525 N⊥ < close 1.525	0.050 – 0.052	+	Ovoid
Polyester	N|| >> 1.525 N⊥ > 1.525	0.098 – 0.175	+	Round, donut, propeller, polygonal, trilobal
Viscose rayon	N|| > 1.525 N⊥ < 1.525	0.020 – 0.028	+	Serrated, multilobal
High tenacity rayon	N|| > 1.525 N⊥ < 1.525	0.035 – 0.039	+	Serrated, multilobal
Polynosic rayon	N|| > 1.525 N⊥ < 1.525	0.041 – 0.043 0.044	+ +	Serrated, multilobal
Vinyl	N|| > 1.525 N⊥ < 1.525	0.025 – 0.030	+	Dogbone
Vinyon	N|| and N⊥ ≤1.525 or near it	0.000 – 0.005	+	Round, dogbone

TABLE 7.3
Fiber Data Sheet

(Circle item or write observations)

Morphology

Longitudinal morphology: smooth striated irregular other ____________

Cross-sectional shape____________

Lobe diameter (μm)____________

Reflected color____________

Transparency ____________

Transmitted color____________

Length (mm) ____________

Surface texture____________

Staple length or continuous____________

Crimped or straight ____________

Coloration: pigment organic dye

Dye process ____________

Treatment: twisted crimped melted

Dulling agent extent and type ____________

Other manufactured artifacts (gas bubbles, fish eyes, etc.) ____________

Foreign artifacts (blood, pollen, etc.) ____________

Optical Data

RRI (refractive index relative to medium (Permount 1.525 or other mounting medium with known RI):

N parallel (N_{Π}) above below near equal to ____________

N perpendicular ($N_{\perp}$) above below near equal to ____________

Crossed polars: isotropic anisotropic

Interference colors____________

Estimated retardation (nm) ____________

Estimated birefringence (retardation/thickness) ____________

Sign of elongation ____________

Dyed color ____________

Dulling agent: none bright semi-dull dull type____________

Pleochroism: no yes

Extinction____________

UV fluorescence____________

Other information ____________

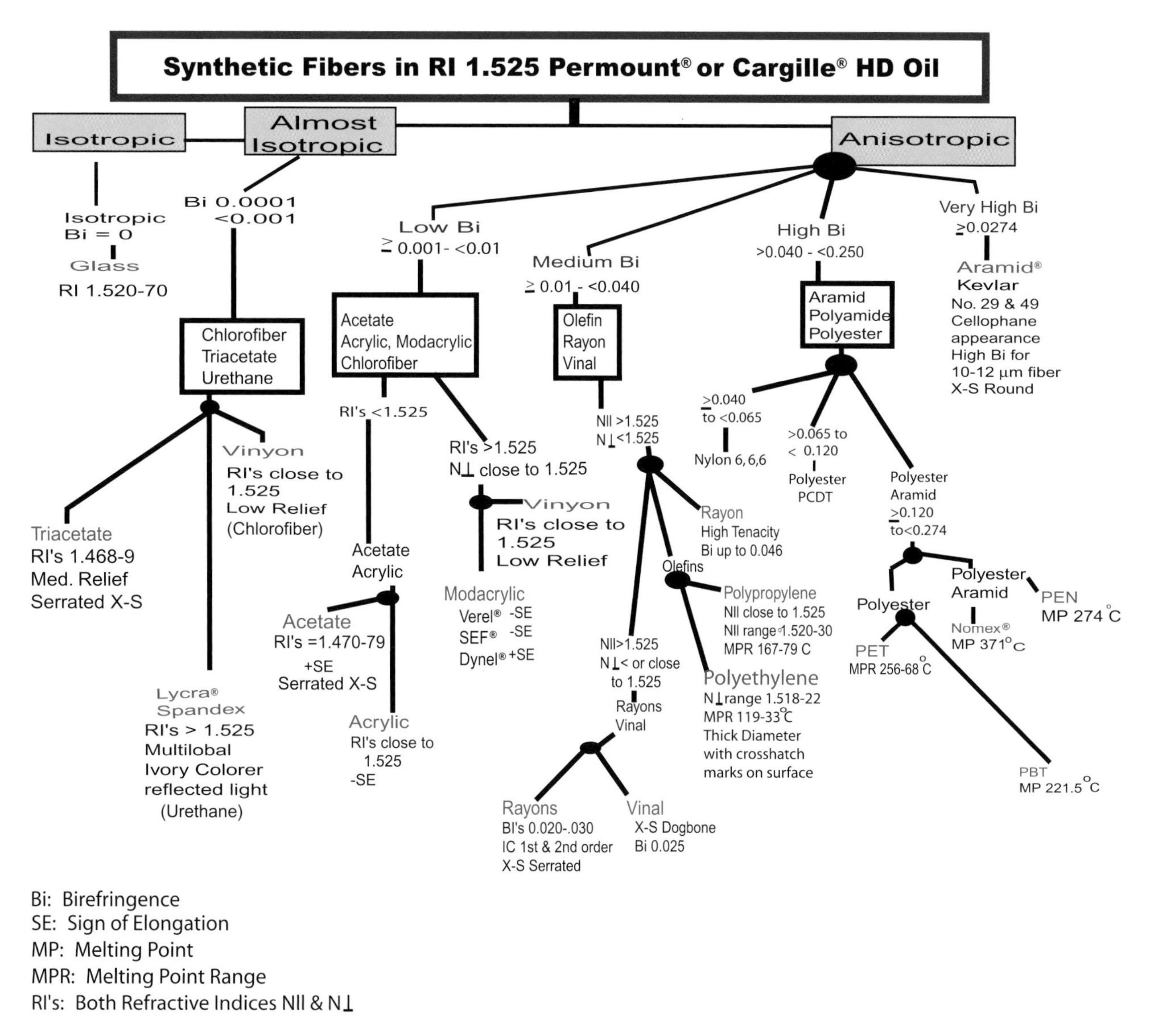

FIGURE 7.9 Synthetic fiber flow chart for mounting medium with RI 1.525.

8 Natural Fiber Identification

Much archaeological evidence indicates that people have utilized natural fibers since the end of the Stone Age. Key archaeological sites of Neolithic towns and settlements produced textiles, ropes, and other items made of natural fibers. Additionally, animal and human hair, wool, vegetable fibers, wood fibers, and silk are often encountered in works of art, paper used for writing, drawing and painting, ceremonial items, clothing, tools and weapons, furniture and decorative items, tapestries used for insulation and decoration, fibers as reinforcement for mortar and plaster, and for many other diverse functions. Consequently, practitioners of the disciplines concerned with studying past and present cultures must be able to identify textile fibers.

Forensic scientists, art historians, art conservators, textile conservators, architectural conservators, archaeologists, chemists, and anthropologists all need straightforward methods that allow them to quickly and accurately identify and categorize natural fibers. Light microscopy, in particular PLM, provides just such a method.

Natural fibers fall into four categories: mammalian hairs, vegetable fibers, insect fibers (silk), and mineral fibers. Human and animal hairs and mineral fibers were covered in earlier chapters. Natural fibers are normally identified and categorized primarily on the basis of their microscopic anatomies. Features such as cell structure, cell shape, cell size, cross-markings, crystal shapes, lumen size, length and width, and other factors are studied to determine classification. Identifying various types of natural fibers on the bases of these features along with other physical characteristics and optical properties will be the primary focus of this chapter. Table 8.1 lists the natural fibers to be covered in this chapter.

PRELIMINARY EXAMINATION PROCEDURE

Mount an aliquot of the questioned fiber specimen in Cargille Melt Mount 1.539 or 1.540 HD (high dispersion) oil as described in Chapter 3. Temporary mounts are normally prepared with Cargille oil and permanent mounts with Melt Mount. Examine the specimen microscopically under a polarized light microscope using plane-polarized light (PPL) at magnifications from 25 to 500×. Determine whether the fiber is natural or synthetic. If it is a human hair, refer to Chapter 5. If it is an animal hair, refer to Chapter 6. See Chapter 7 for details on man-made natural and synthetic polymers and minerals.

Identification of Natural Fibers

Plant, paper, insect, and wood fibers are typically identified on the basis of physical morphology. The literature abounds with morphological data and information about staining techniques, flow charts, and the like, all for use in the identification of natural fibers.[1–14] The following scheme was developed by the authors from information taken from the literature. The authors have used this scheme successfully in their casework.

Information about morphology is collected first. Next, the relative refractive index (RRI) of the N|| and $N_\perp$ directions of the fiber specimen as they compare to the RI of the mounting medium are determined by the Becké line (BL) method using plane-polarized light. The subject fiber is examined between crossed polars (CPs). If a fiber is isotropic, it will remain dark (extinct) when rotated between CPs; if anisotropic, it will appear bright and dark as it is rotated 360° between CPs. At the positions of maximum brightness, the fiber displays its highest order interference colors (ICs). The specimen's degree of retardation is computed in the maximum brightness position by using an interference color chart and compensation as discussed in Chapters 1 and 7. Extinction of the fiber is noted; see Figure 8.1. All of the data describing the questioned natural fiber should be noted on a data sheet as shown in Table 8.1.

Next, the modified Herzog test is utilized to determine a plant fiber's molecular orientation.[14] The plant fiber specimen is observed between CP at its east ↔ west extinction position. A full-wave (550 nm) fixed compensator is inserted at a 45° angle to the preferred directions of the crossed polars. If an additive effect occurs (fiber turns slightly blue) the fiber is said to have an S-twist; if a subtractive effect occurs (fiber turns slightly yellow) the fiber is said to have a Z-twist. Figure 8.2 depicts this test for both S- and Z-twist situations. Record the results on the data sheet (Table 8.1).

Finally, the dispersion colors exhibited by the subject fiber are determined with a central stop and the following Cargille HD oils. A short review of the central stop dispersion staining technique can be found in Chapter 3 of this text. The natural fiber is mounted in HD oil with an IR of 1.585, and placed in its east ↔ west orientation. The fiber is observed with east ↔ west oriented plane-polarized light and a 10× central stop objective. The dispersion colors exhibited by the fiber are reworded on Table 8.1. The procedure can be repeated using Cargille HD oils having RIs of 1.590 and 1.565 for sodium D-line (see Figure 8.3).[14]

Other features such as the presence of crystals, vessel elements, spiral thickenings and the like can also be important identifying characteristics. Occasionally, these features can be detected, *in situ*, while examining longitudinal mounts of fiber specimens as shown in Figures 8.4 through 8.6.

Unfortunately, in order to see most natural fiber specimens' crystals, the specimen must be ashed in a muffle furnace. Ashing is a destructive method, and should only be carried out when there is an excess of the questioned fiber specimen. A procedure for preparing ash mounts of natural fibers is shown in Figure 8.7. Figure 8.8 depicts two ash specimens prepared in the described manner. One specimen is manila ash, and the other specimen is sisal ash. These two crystal forms are often used to differentiate between sisal and manila plant fibers.

Over the millennia, mankind has utilized wood and wood by-products for a myriad of purposes. Consequently, specimens composed of wood such as pieces of furniture, sculptures, artifacts, tools, clubs, canes, sawdust, wood fragments, and paper are examined on a routine basis by criminalists, chemists and conservators. Before the mid-nineteenth century, all paper was made primarily from Mitsumata inner tree bark (paper mulberry), old rags, or other vegetable fibers. Today, more than 90% of all paper is made primarily from wood pulp. The type of wood pulp(s) used to make a piece of questioned paper can be determined by the identification of the basic wood cells present in the specimen.[16-19]

Many schemes have been published for the identification of wood and paper-making fibers.[2-11,13] Consequently, only a brief discussion of wood's anatomy will be given in this text. Instead, the emphasis will be placed on the preparation of paper specimens; for microscopic examination the serious reader is referred to the literature.

The xylem or wood fraction of hardwoods and softwoods is the source of the raw material used to manufacture paper. The fundamental building block of wood is the cell. A piece of wood is composed of an immense number of basic wood cells, composed of cellulosic microfibrils, and surrounded by a void-filling matrix material known as hemicellulose. This is all held together by a natural adhesive polymer known as lignin. There are three primary classes of cells in wood. They are parenchyma, vessel elements, and fibers. Some identifying features of wood cells are depicted in Figures 8.9 and 8.10.[16,17]

Specimens of wood pulp (obtained from the maceration of wood)[18] or specimens of paper are prepared for microscopic examination by cutting them into small pieces and placing them into a beaker containing hot, distilled water. A glass rod is used to mix the wood fibers with the hot water. The beaker is covered and the specimen is gently heated to a boil.

The beaker is removed from the hotplate and allowed to cool. The mixture is placed into a small blending jar, which is half filled with very hot (98°C) distilled water. The specimen is blended for 1 minute until a slurry forms. Next, the slurry of water and wood fibers is decanted into a simple strainer. Wash the wood fibers with cold running water before removing them from the strainer (see Figure 8.11). The wood fibers are dyed and mounted as illustrated in Figure 8.12.

To determine the classification of an unknown natural fiber, the collected data in Table 8.1 should be compared against the information in Table A8.2 and Figures 8.1, 8.4, 8.6, and 8.8 through 8.10, Figure 8.13 (fiber flow chart), the natural fiber appendix at the end of this chapter, known standards, and the published literature. Each class of fiber specimen is identified in the same manner. If a comparison of the fibers is desired, the questioned and known specimens can be compared side by side on a comparison microscope.

REFERENCES

1. Willard, M.L., *Chemical Microscopy Laboratory Outline*, Ann Arbor, MI, Edwards Brothers, 1952.
2. Heyn, A.N.J., *Fiber Microscopy*, New York, Interscience, 1953.
3. Harris, M., Ed., *Handbook of Textile Fibers*, Washington, D.C., Harris Research Laboratories, 1954.
4. Carpenter, C.H. and Leney, L., *Papermaking Fibers*, Syracuse, NY, Syracuse University, 1963.
5. Britt, K.W., Ed., *Handbook of Pulp and Paper Technology*, New York, Reinhold Publishing, 1964.
6. *Identification of Textile Materials*, Manchester, Textile Institute, 1970.
7. McCrone, W.C. and Delly, J.D., *The Particle Atlas*, 2nd ed., Ann Arbor, MI, Ann Arbor Science Publishers, 1973.
8. Core, H.A., Côté, W.A., and Day, A.C., *Wood Structure and Identification*, 2nd ed., Syracuse, NY, Syracuse University Press, 1980.
9. Côté, W.A., Ed., *Papermaking Fibers*, Syracuse, NY, Syracuse University Press, 1980.
10. Schaffer, E., Fiber identification in ethnological textile artifacts, *Stud. Conserv.*, 26, 119, 1981.
11. Purham, R.A. and Gray, R.L., *The Practical Identification of Wood Pulp Fibers*, Atlanta, Tappi Press, 1982.
12. Catlings, D. and Grayson, J., *Identification of Vegetable Fibres*, London, Chapman & Hall, 1982.
13. Hoadley, R.B., *Identifying Wood*, Newtown, CT, Taunton Press, 1990.
14. Van Hoven, H., Dispersion Staining: An Aid to Plant Fiber Identification, presented at the annual meeting of NEAFS, Springfield, MA, October 1993.
15. Bloss, F.D., *Optical Crystallography*, Washington, D.C., Mineral Society of America, 1999, p. 51.
16. Côté, W.A., Ed., *Papermaking Fibers*, Syracuse, NY, Syracuse University Press, 1980, p. xix.
17. Core, H.A., Côté, W.A., and Day, A.C., *Wood Structure and Identification*, 2nd ed., Syracuse, NY, Syracuse University Press, 1980, p. 25.
18. Purham, R.A. and Gray, R.L., *The Practical Identification of Wood Pulp Fibers*, Atlanta, Tappi Press, 1982, p. 195.

TABLE 8.1
Natural Fiber Data Sheet

Morphology

Longitudinal morphology (smooth, striated, cross-hatched, etc.) ____________________

Cross-section shape ____________________

Diameter of single fiber (μm) ____________________

Surface texture ____________________

Crystals ____________________

Spiral elements ____________________

Vessel elements ____________________

Parenchyma ____________________

Fibers ____________________

Optical Data

Crossed polars: isotropic anisotropic

Refractive index relative to mounting medium (Melt Mount 1.539, Cargille 1.540 high-dispersion oil, or other mounting medium with known RI):

N parallel ($N_{||}$) above below near equal to ____________________

N perpendicular ($N_{\perp}$) above below near equal to ____________________

Interference colors ____________________

Estimated retardation (nm) ____________________

Estimated birefringence (retardation/thickness) ____________________

Results of modified Herzog test ____________________

Sign of elongation ____________________

Extinction displayed ____________________

Dispersion staining colors (central stop) in appropriate Cargille oil:[14]

Cargille 1.585 high-dispersion oil ____________________

Cargille 1.590 high-dispersion oil ____________________

Cargille 1.565 high-dispersion oil ____________________

Other information ____________________

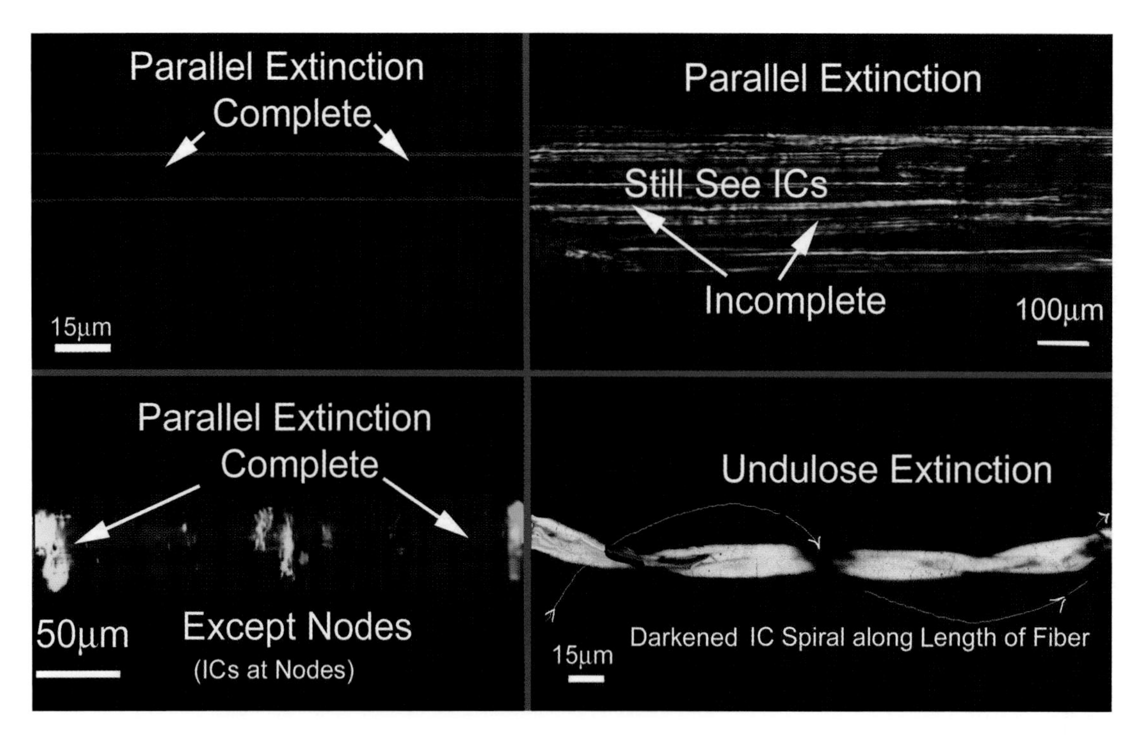

FIGURE 8.1 Major varieties of extinction exhibited by natural fibers.

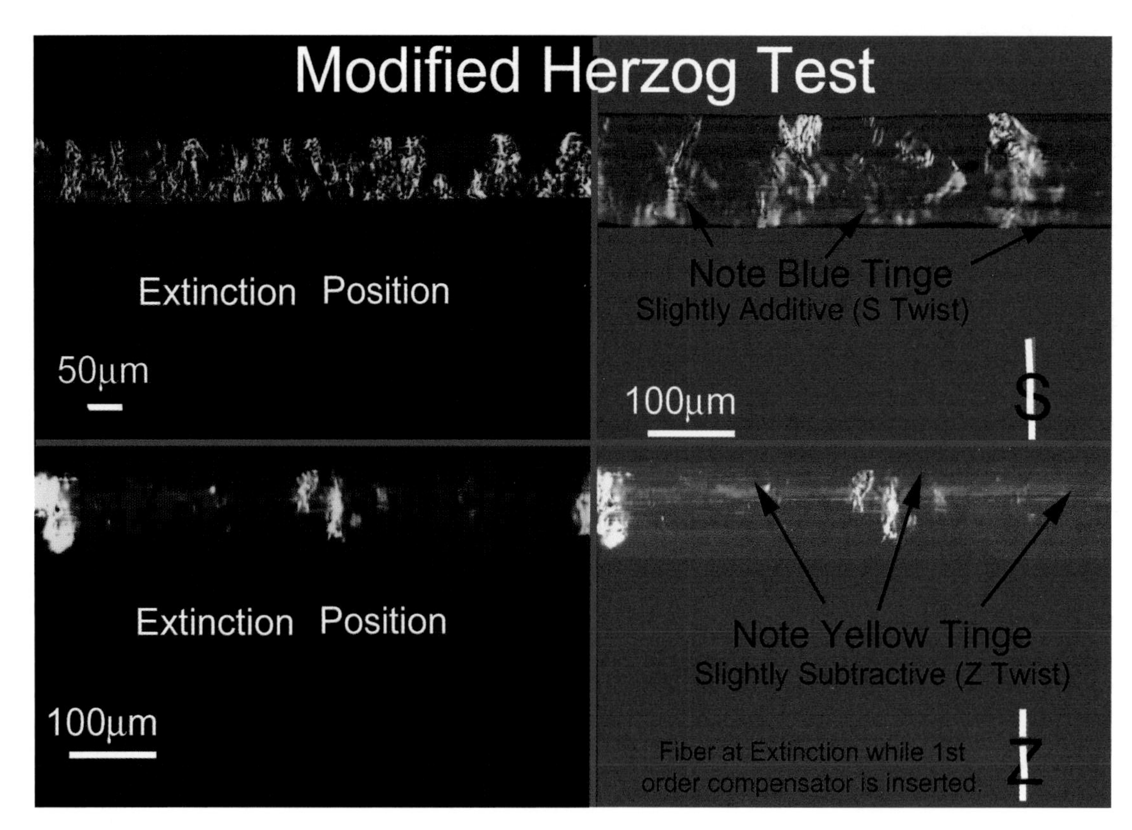

FIGURE 8.2 S- and Z-twist-oriented fibers.[14]

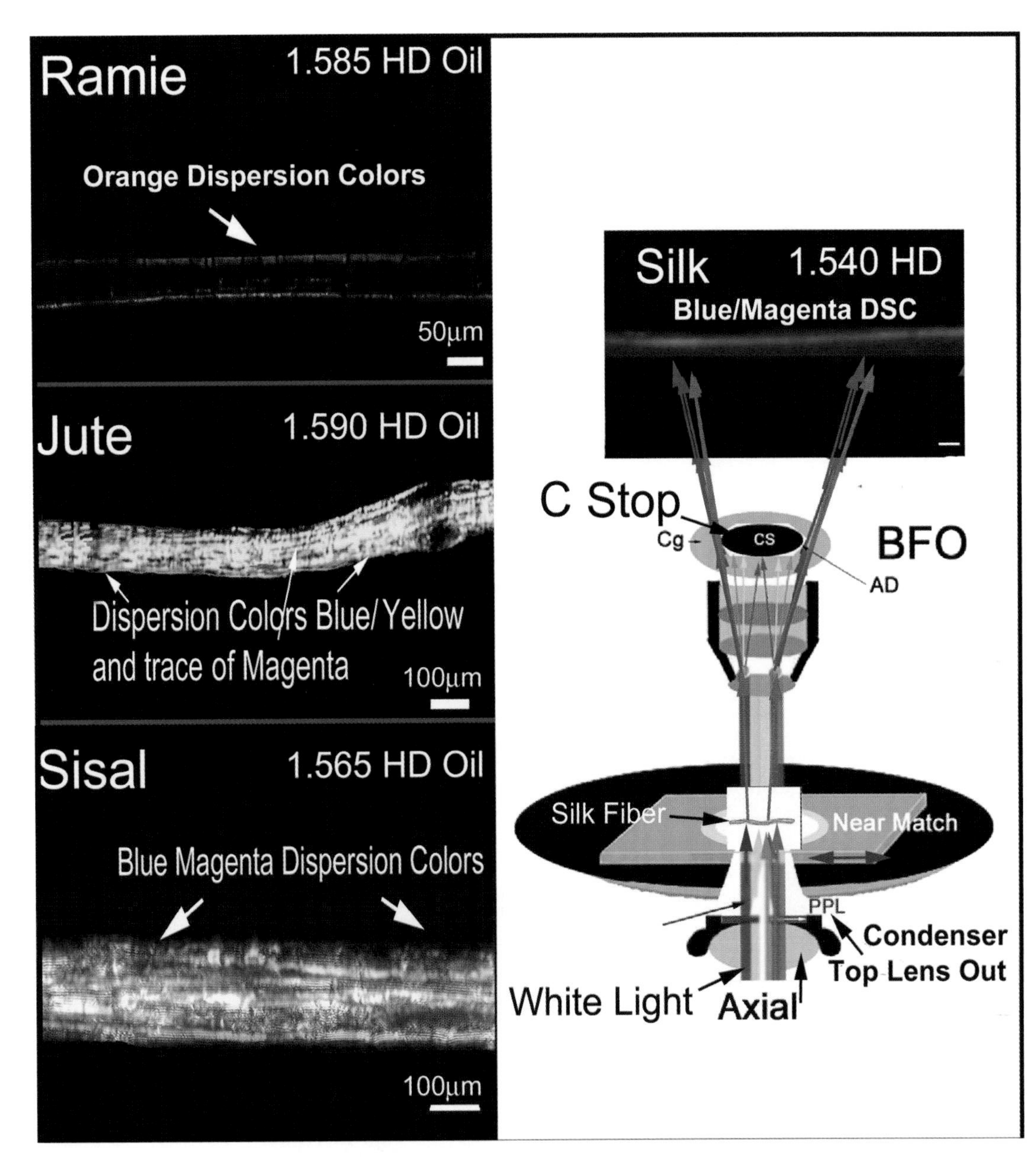

FIGURE 8.3 Dispersion staining procedure for natural fibers utilizing Cargille HD oils and a 10× central stop objective.[14,15]

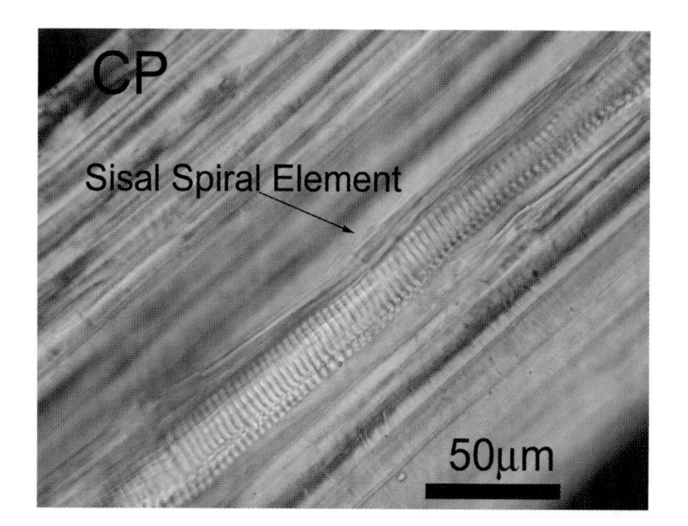

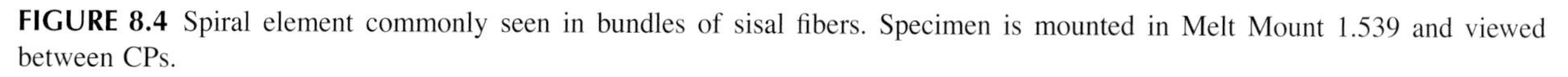
FIGURE 8.4 Spiral element commonly seen in bundles of sisal fibers. Specimen is mounted in Melt Mount 1.539 and viewed between CPs.

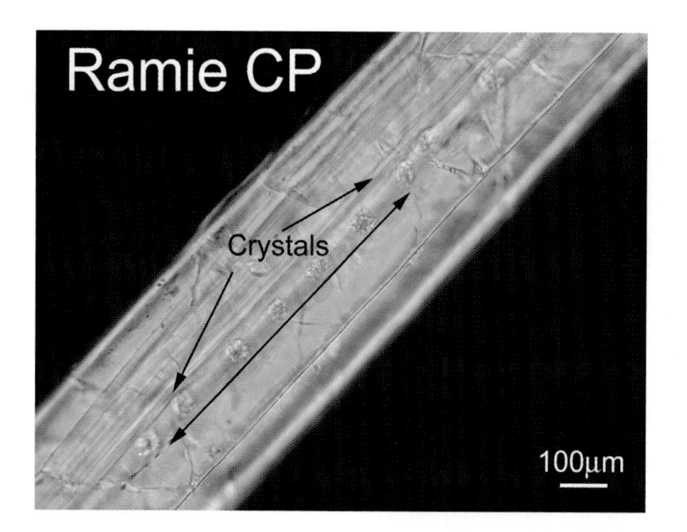

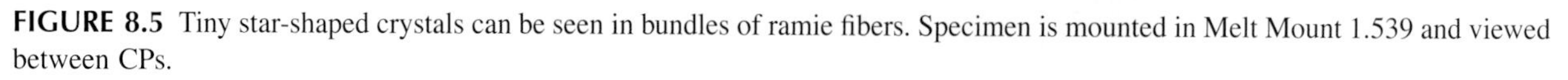
FIGURE 8.5 Tiny star-shaped crystals can be seen in bundles of ramie fibers. Specimen is mounted in Melt Mount 1.539 and viewed between CPs.

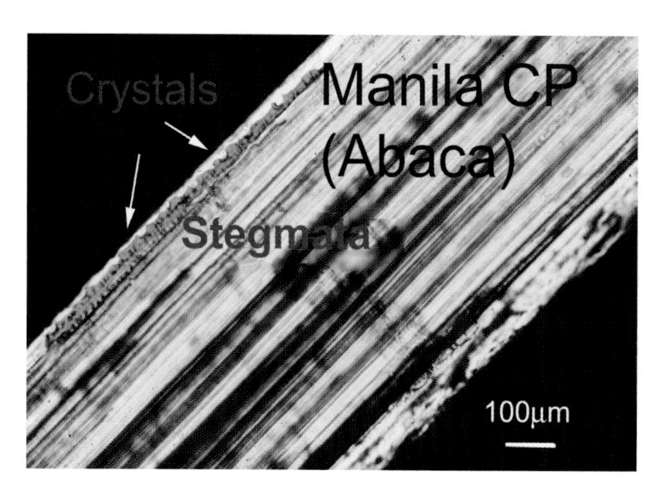

FIGURE 8.6 A string of rectangular-shaped silica stigmata seen in abaca or manila fiber bundles. Specimen is mounted in Melt Mount 1.539 and viewed between CPs.

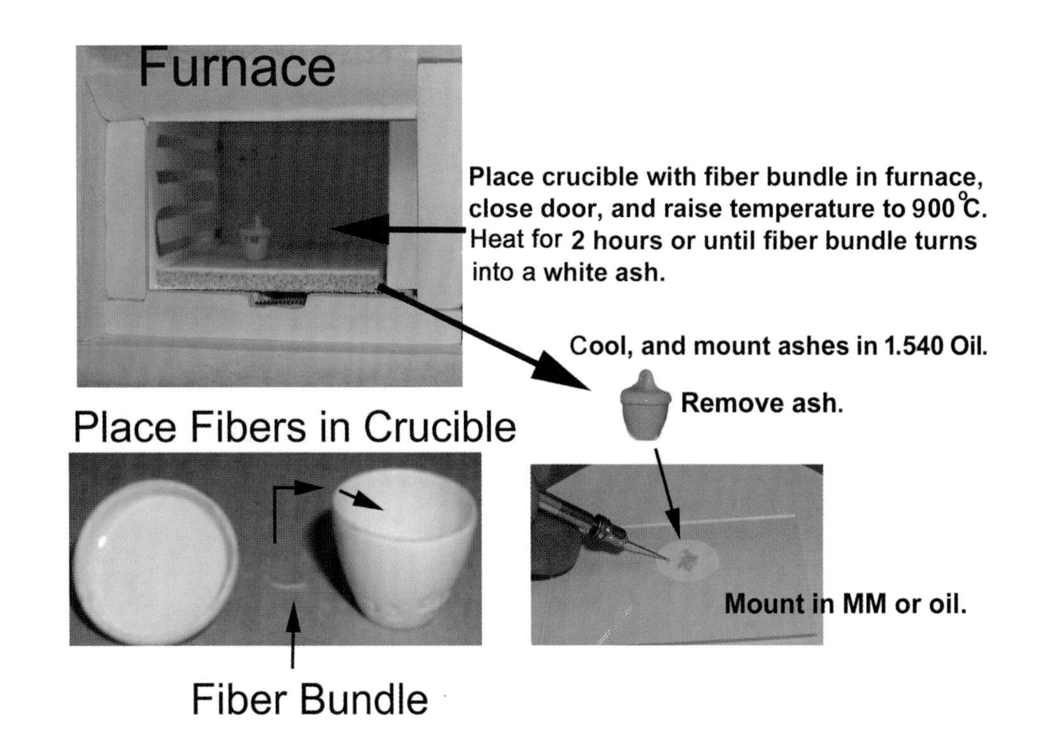

FIGURE 8.7 Procedure for ashing a natural fiber specimen. A small specimen is placed into a micro-sized porcelain crucible. The crucible is covered and placed into a muffle furnace and heated at 900° about 1 hour until a white ash is formed. After cooling, the crucible is removed from the furnace, and the ash is mounted in Cargille Melt Mount or a high-dispersion oil.

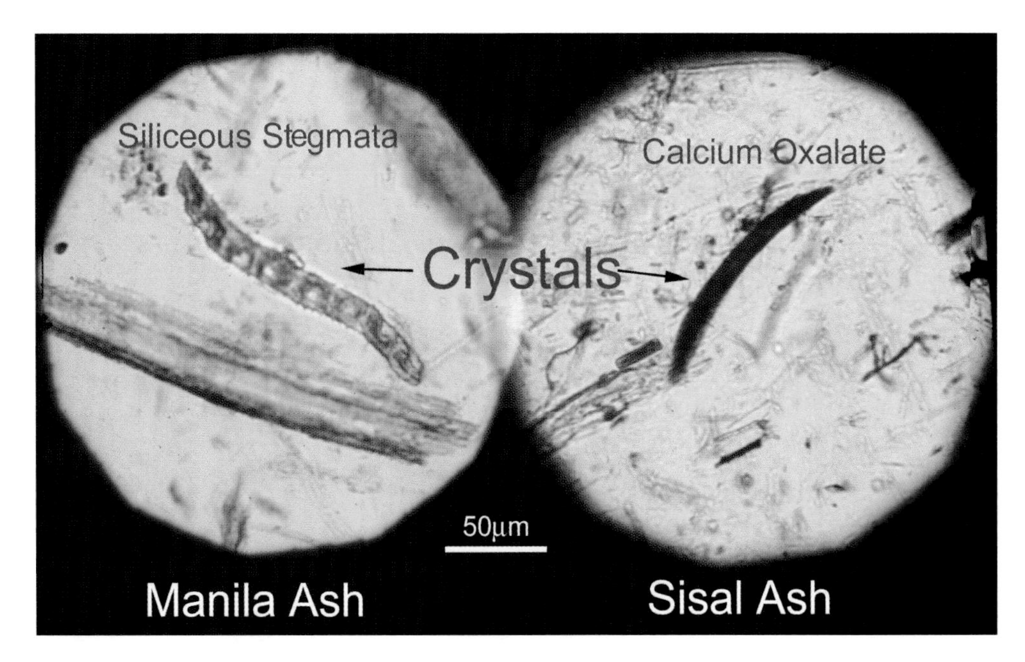

FIGURE 8.8 Left: crystals common to manila. Right: crystals common to sisal.

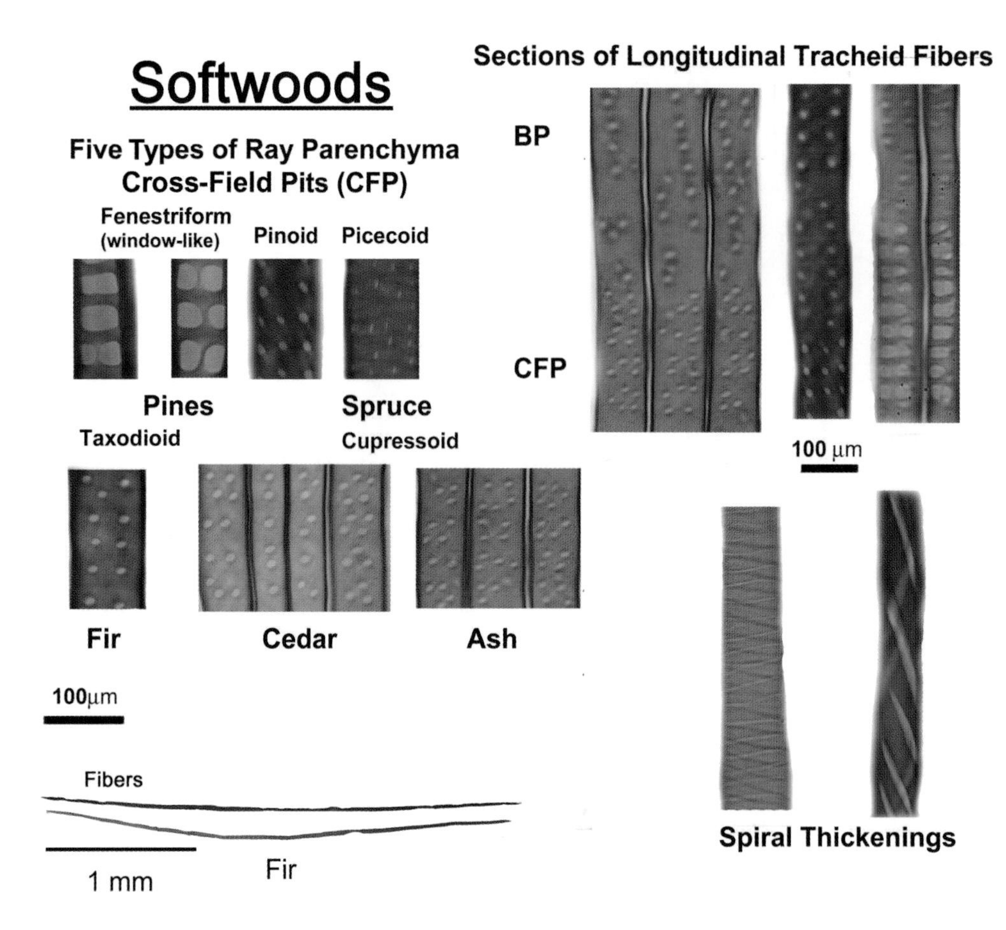

FIGURE 8.9 Assorted species of softwoods and some of their identifying features.

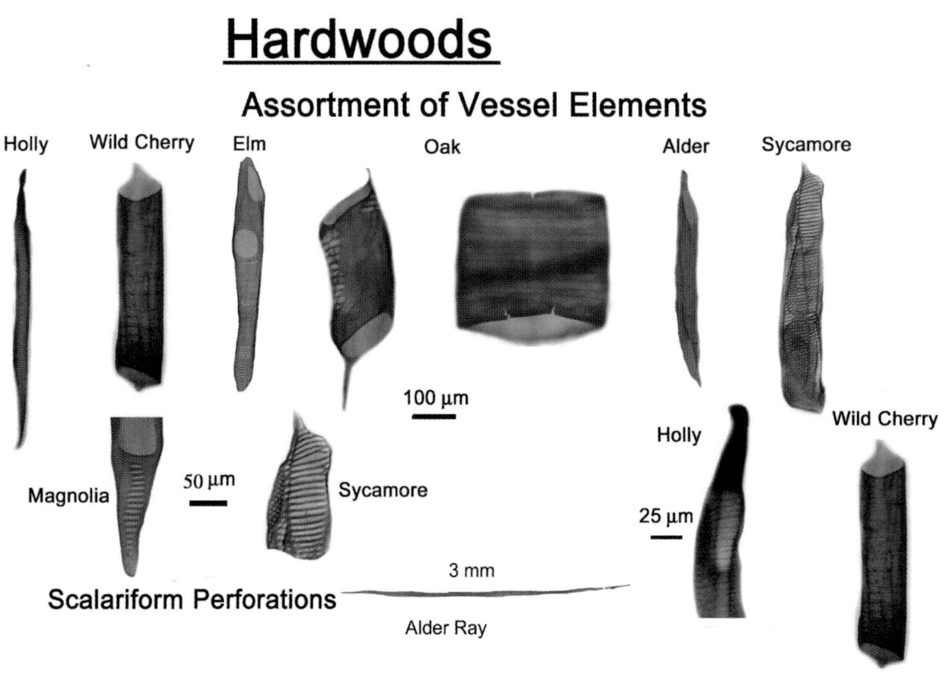

FIGURE 8.10 Assorted species of hardwoods and some of their identifying features.

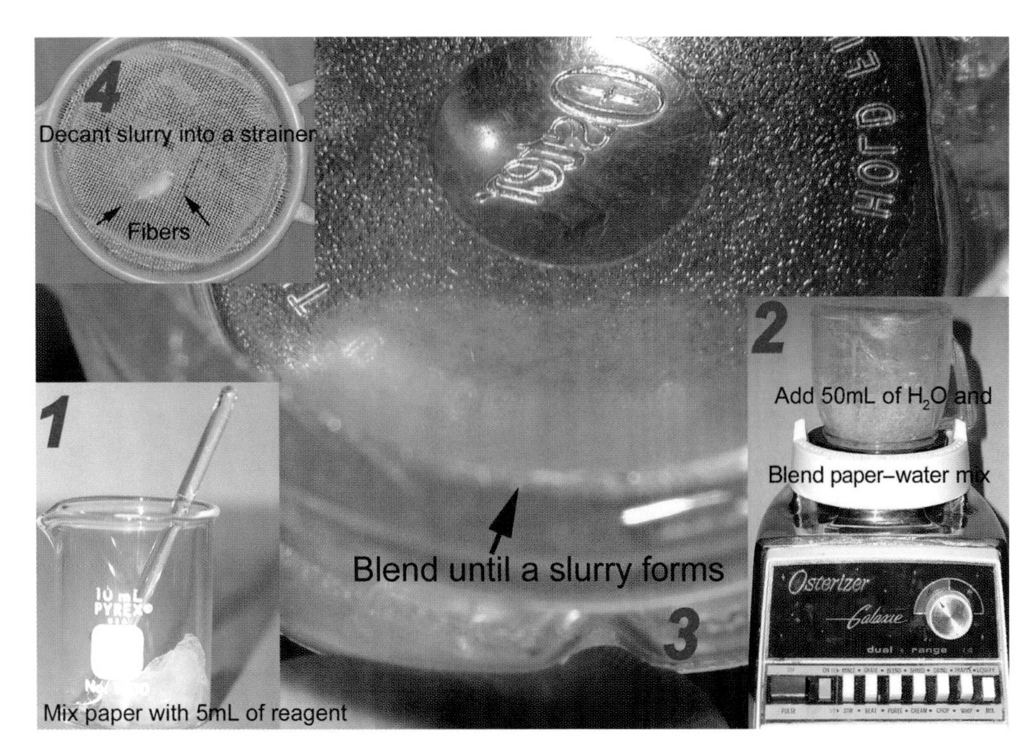

FIGURE 8.11 Maceration of paper: (1) place paper specimen in a small beaker with 5 mL of very hot distilled water; agitate with a glass rod. Cover the beaker and heat gently to a boil. Remove the beaker from the hotplate and allow it to cool; (2) place the mixture into a small blending jar half filled with very hot (98°C) distilled water; (3) blend 1 minute until a slurry forms; (4) decant the slurry of water and paper fibers into a simple strainer; wash the fibers with cold running water before removing them from the strainer; place in a storage container filled with ethanol until needed.

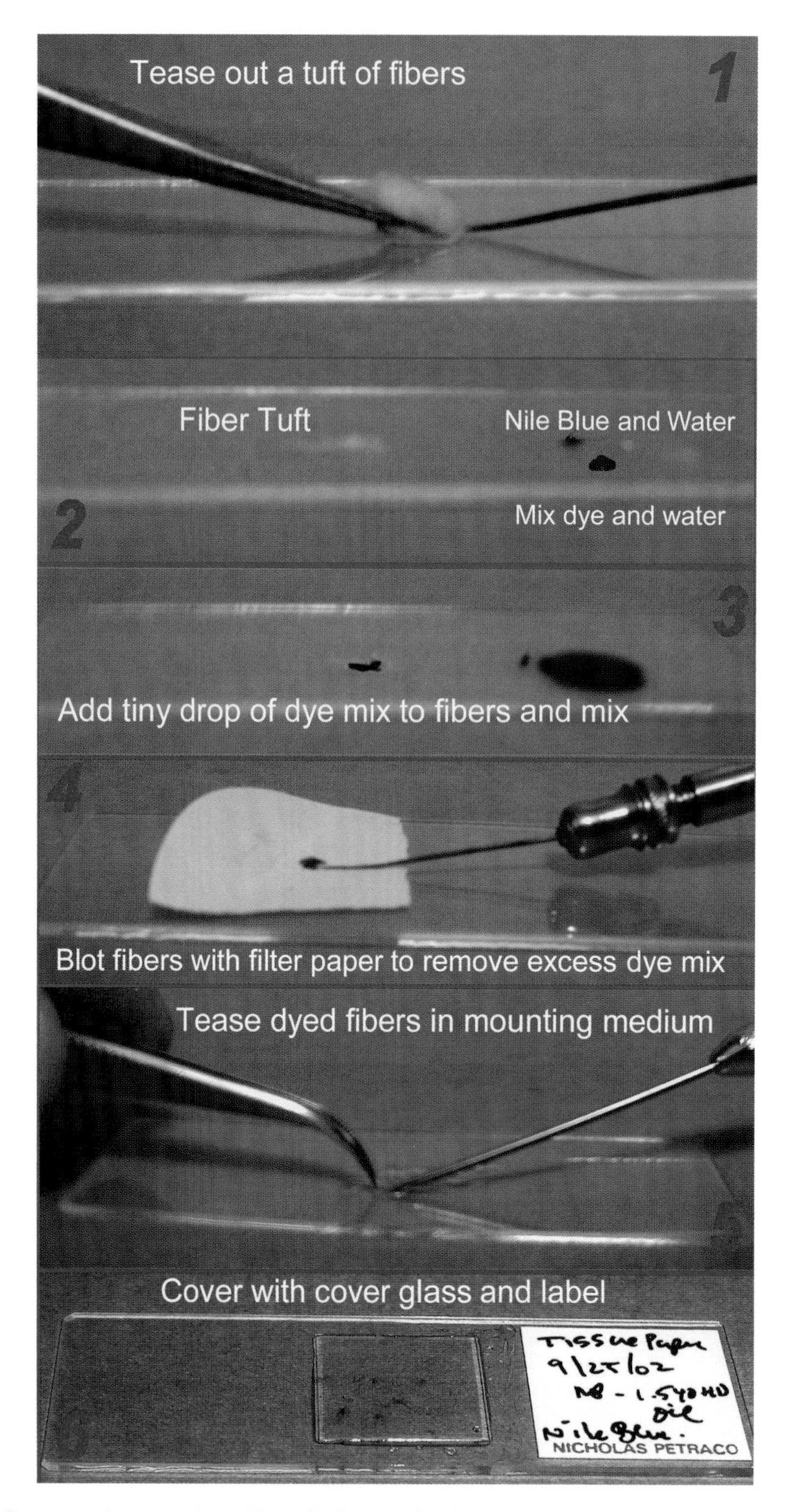

FIGURE 8.12 Paper fiber mounting procedure: (1) a tuft of paper fiber is removed from the storage container and teased out; (2) a tiny speck of dry dye (Nile blue) is placed on the slide next to a drop of water; (3) the dye is mixed and added to the fibers; (4) the dyed fibers are blotted with filter paper; (5) the dyed fibers are dried on a hot plate and then teased in mounting medium; (6) a cover glass is placed over the preparation and the slide is labeled.

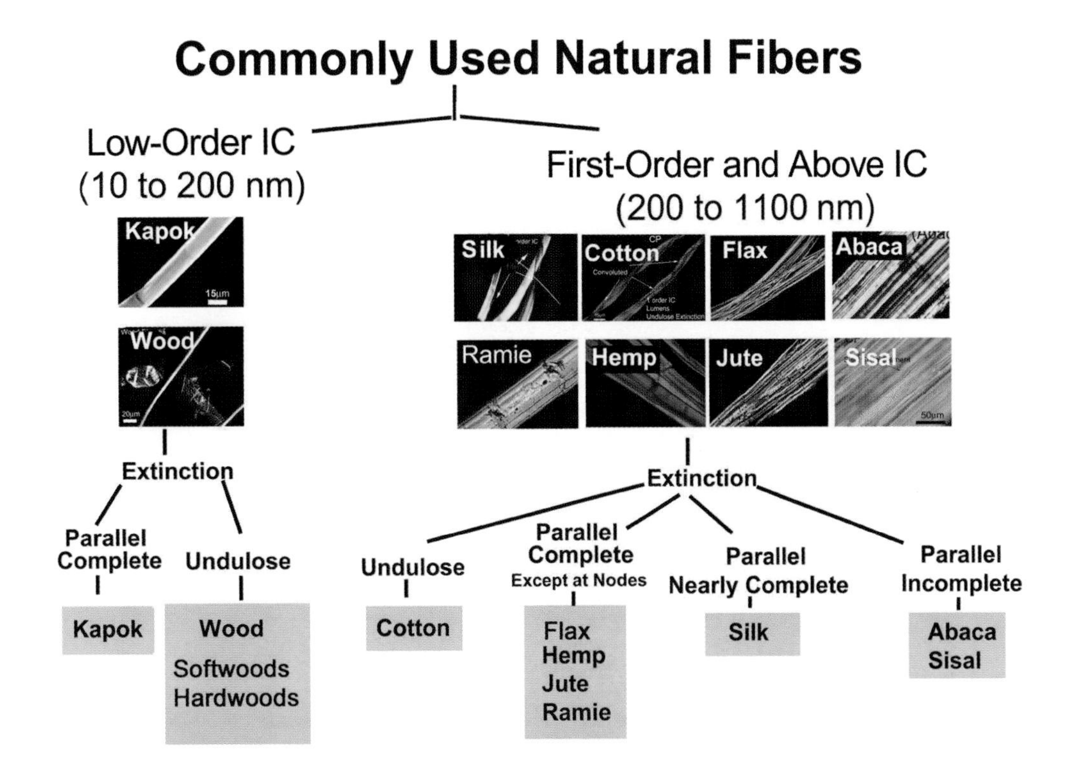

FIGURE 8.13 Flow chart of common natural fibers mounted in Melt Mount 1.539 as they appear between CPs at 45° off extinction and at extinction.

Natural Fiber Appendix

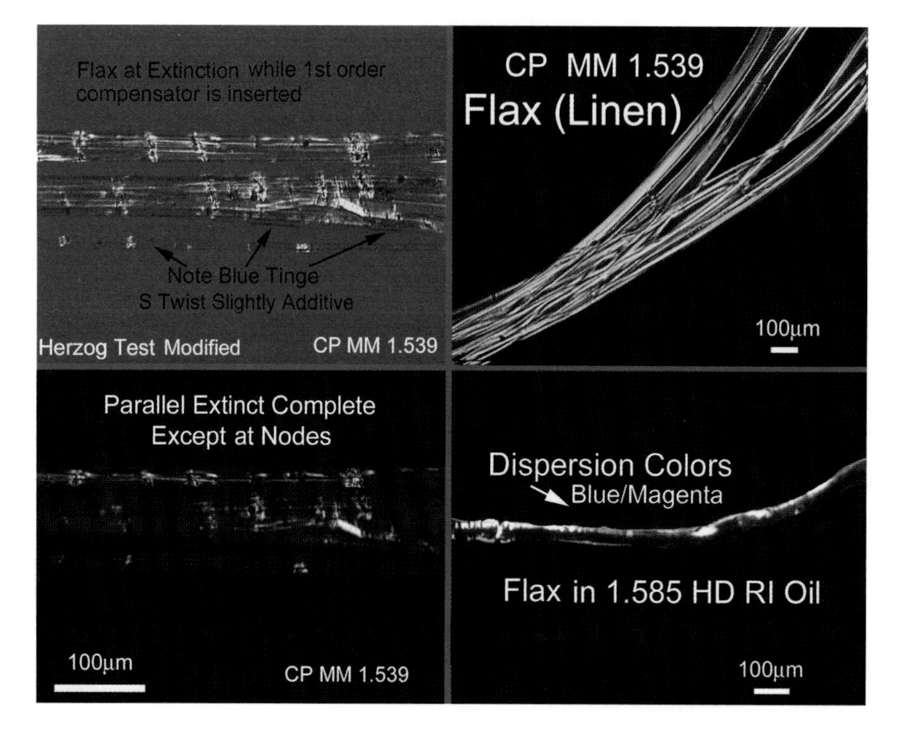

FIGURE A8.1 Flax fibers as they appear in Melt Mount 1.539 and viewed between CPs (top right); at extinction (bottom left); during Herzog test (top left); and central stop dispersion staining colors in designated Cargille oil (bottom right).

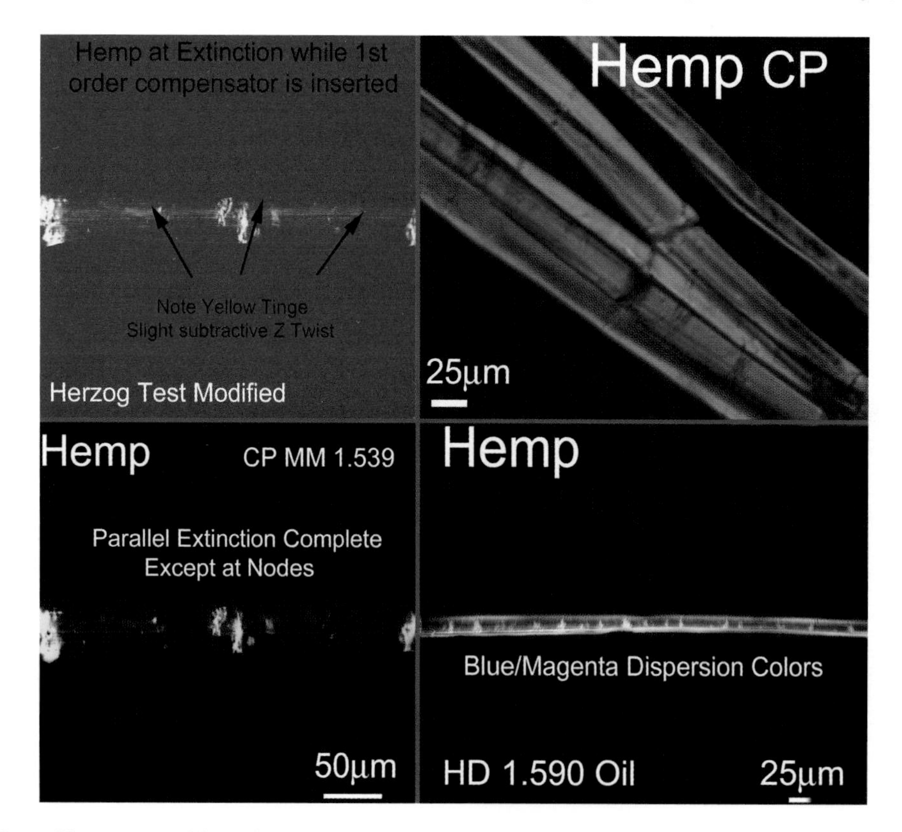

FIGURE A.8.2 Hemp fibers mounted in Melt Mount 1.539 and viewed between CPs (top right); at extinction (bottom left); during Herzog test (top left); and central stop dispersion staining colors in designated Cargille oil (bottom right).

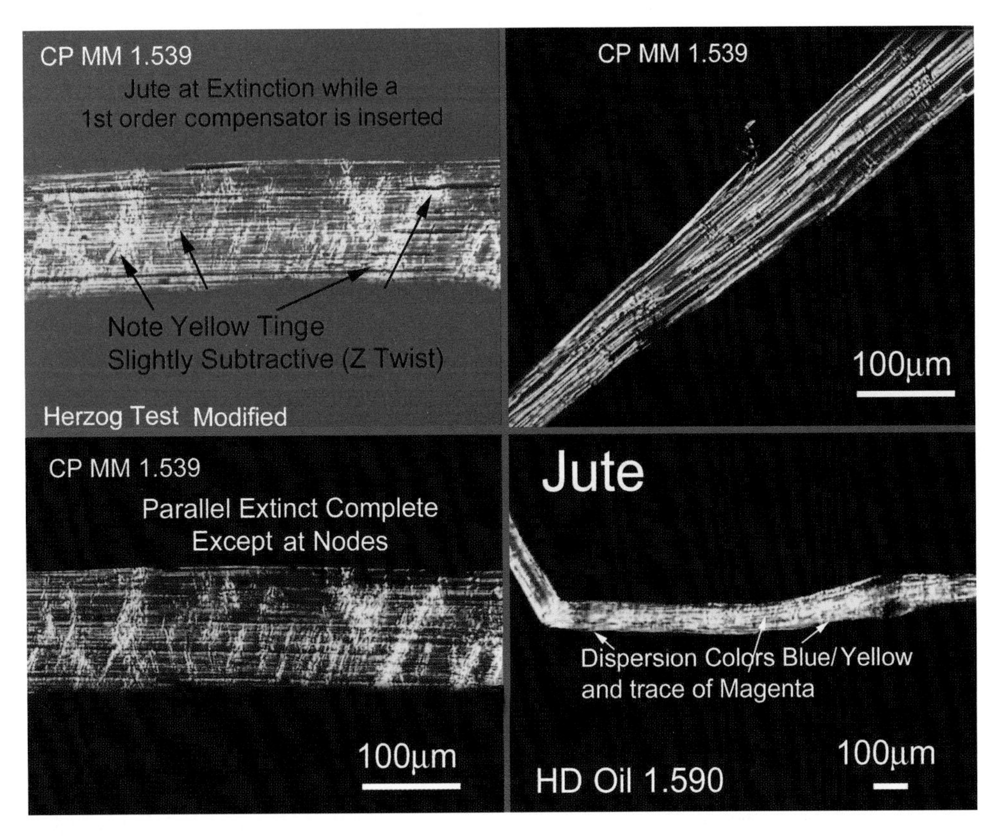

FIGURE A8.3 Jute fibers mounted in Melt Mount 1.539 and viewed between CPs (top right); at extinction (bottom left); during Herzog test (top left); and central stop dispersion staining colors in designated Cargille oil (bottom right).

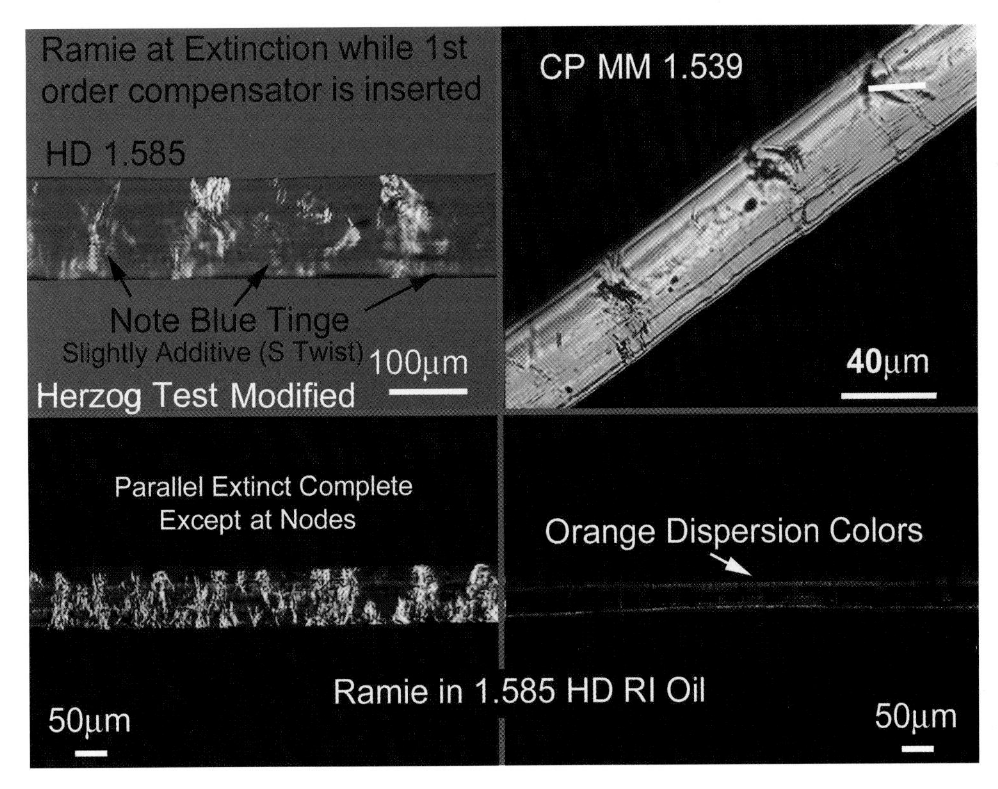

FIGURE A8.4 Ramie fibers mounted in Melt Mount 1.539 and viewed between CPs (top right); at extinction (bottom left); during Herzog test (top left); and central stop dispersion staining colors in designated Cargille oil (bottom right).

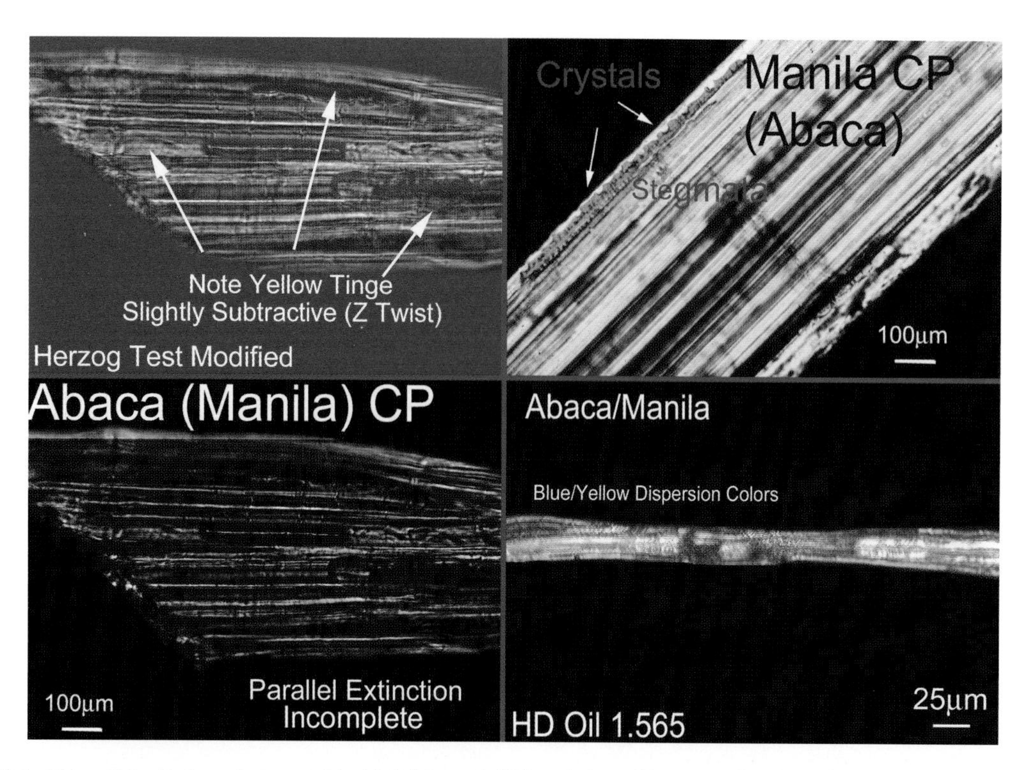

FIGURE A8.5 Abaca (Manila hemp) mounted in Melt Mount 1.539 and viewed between CPs (top right); at extinction (bottom left); during Herzog test (top left); and central stop dispersion staining colors in designated Cargille oil (bottom right).

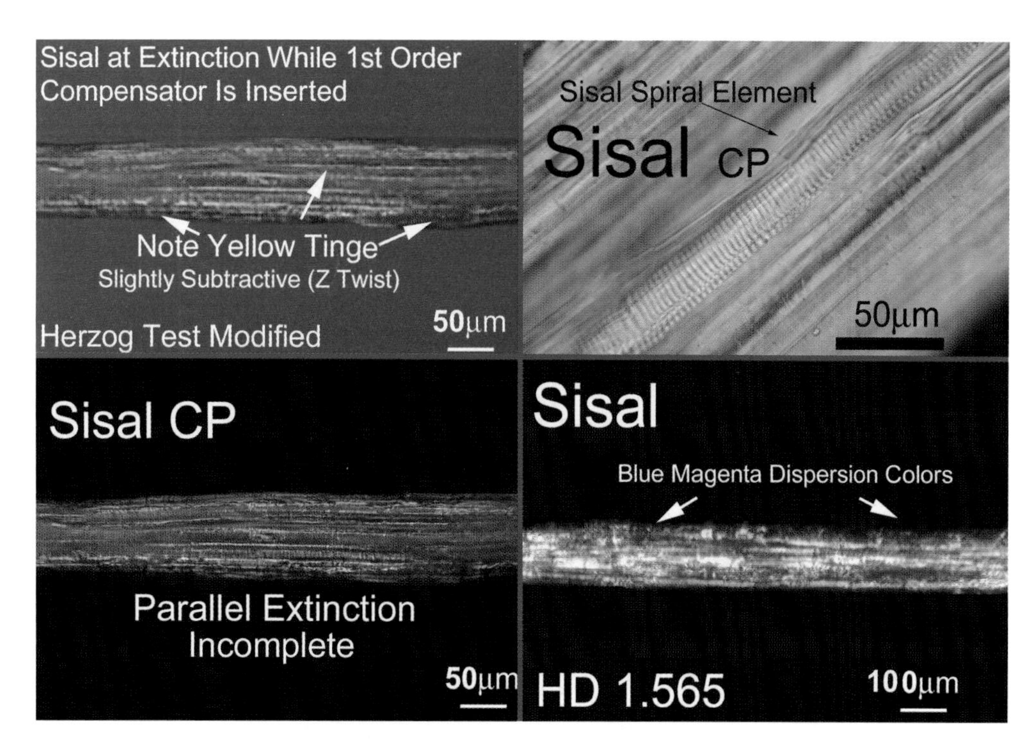

FIGURE A8.6A Sisal fibers mounted in Melt Mount 1.539 and viewed between CPs (top right); at extinction (bottom left); during Herzog test (top left); and central stop dispersion staining colors in designated Cargille oil (bottom right).

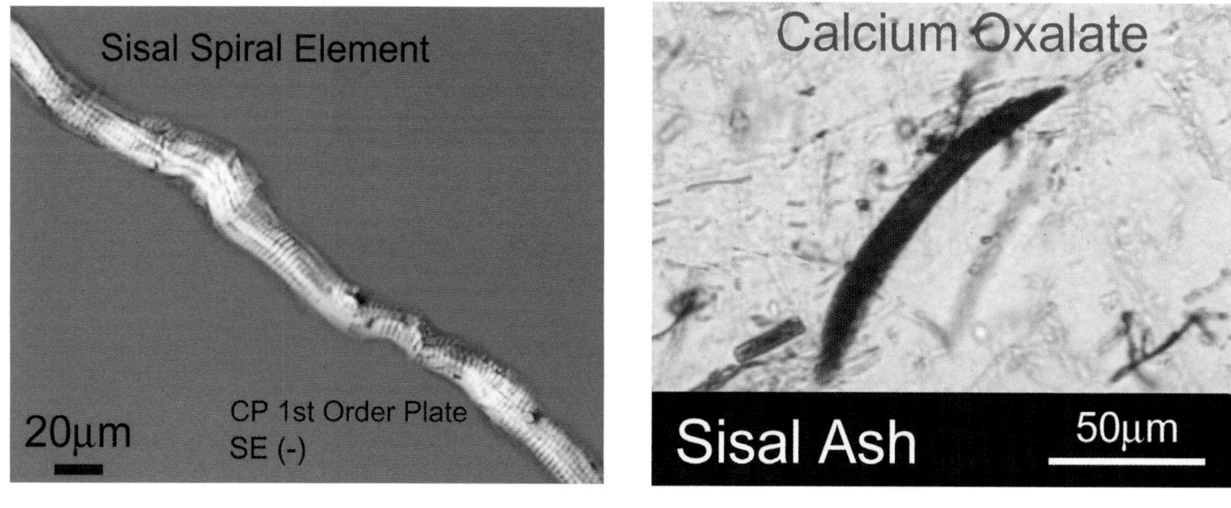

FIGURE A8.6B Sisal spiral element (looks like a long spring).

FIGURE A8.6C Sisal crystal.

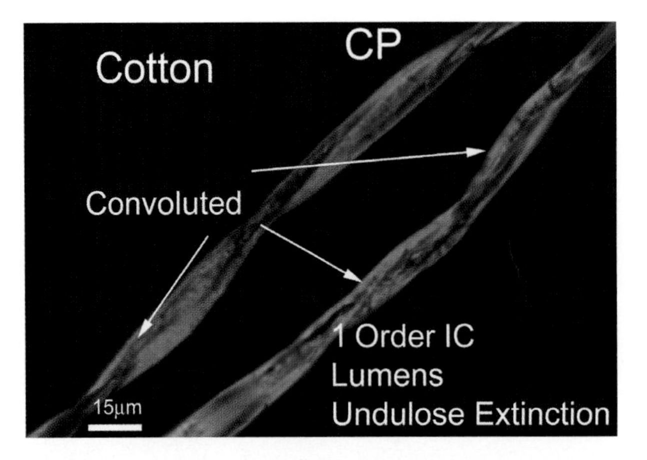

FIGURE A8.7A Cotton fibers mounted in Melt Mount 1.539 and viewed between CPs.

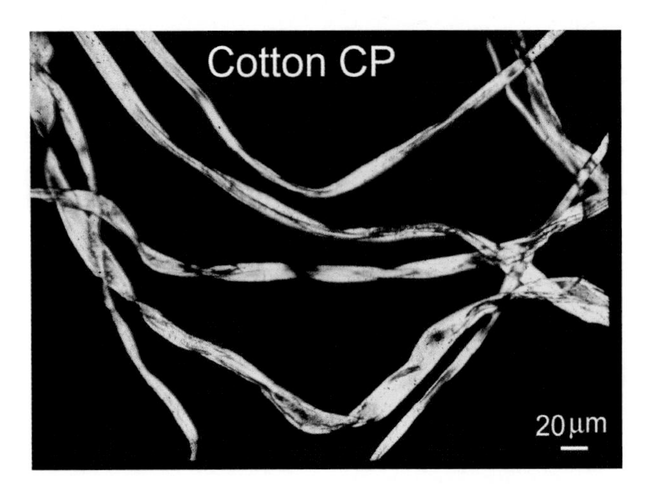

FIGURE A8.7B Cotton fibers mounted in Melt Mount 1.539 and viewed between CPs.

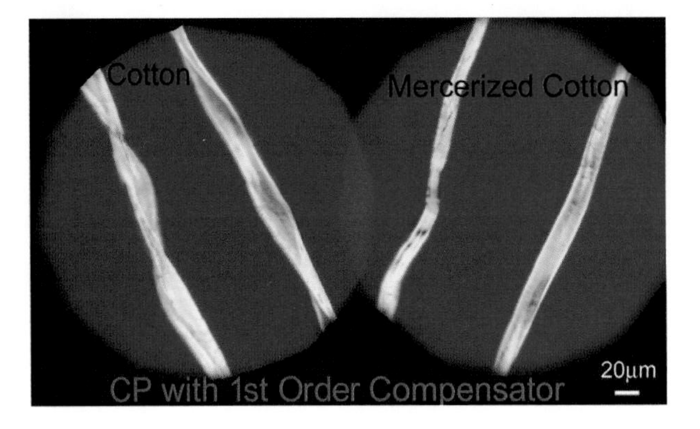

FIGURE A8.7C Chemically treated (mercerized) cotton has fewer convolutions and a finer diameter than untreated cotton.

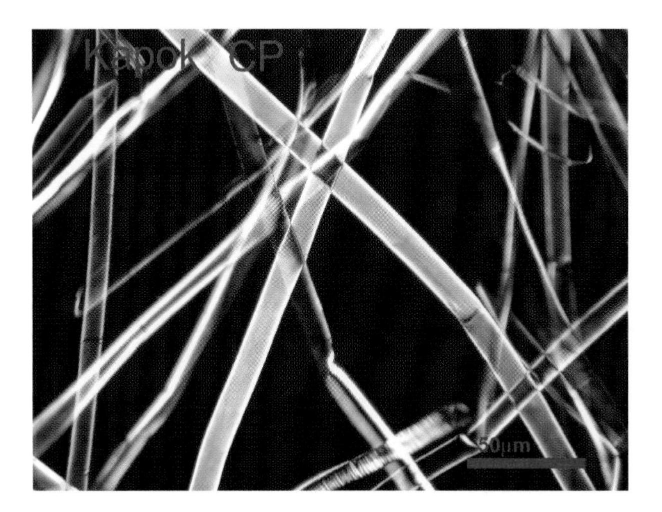

FIGURE A8.8A Kapok fibers mounted in Melt Mount 1.539 and viewed between CPs.

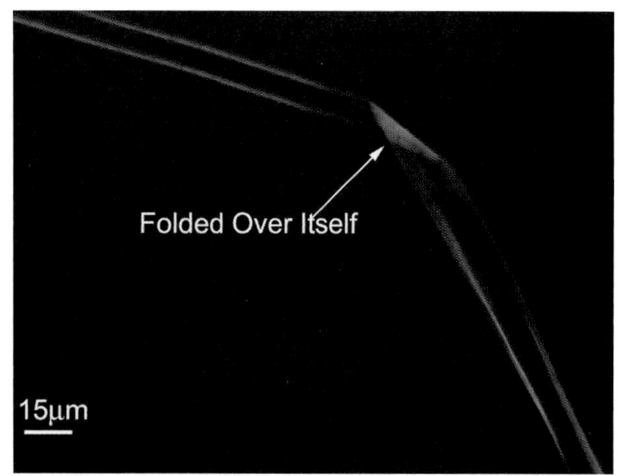

FIGURE A8.8C Kapok folds easily over itself.

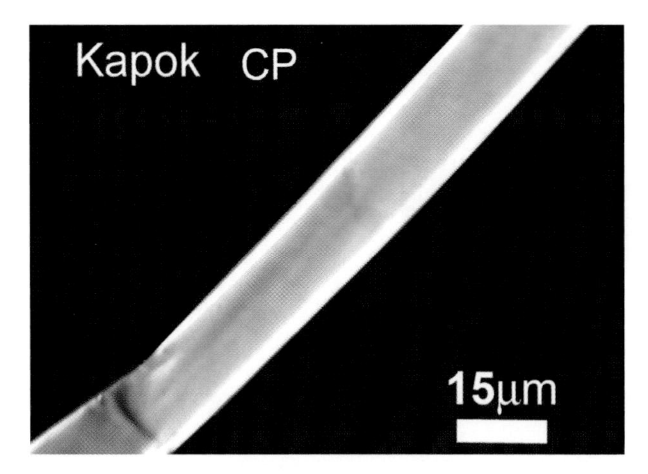

FIGURE A8.8B Kapok fiber (enlarged).

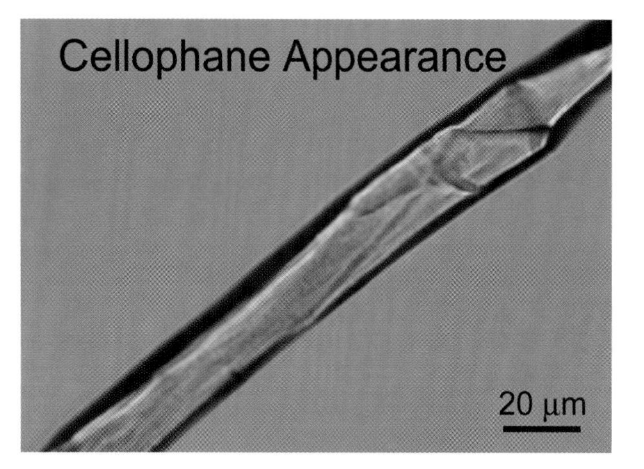

FIGURE A8.8D Kapok sometimes looks like cellophane (CP 10° off extinction).

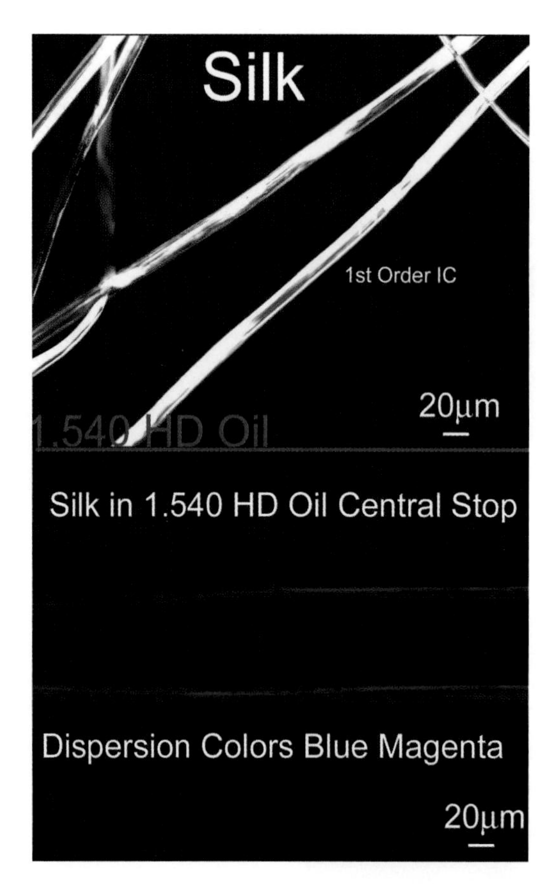

FIGURE A8.9A Tussah (wild silk) mounted in Melt Mount 1.539 and viewed between CPs (top); viewed with central stop dispersion staining objective and mounted in Cargille 1.540 high-dispersion oil (bottom).

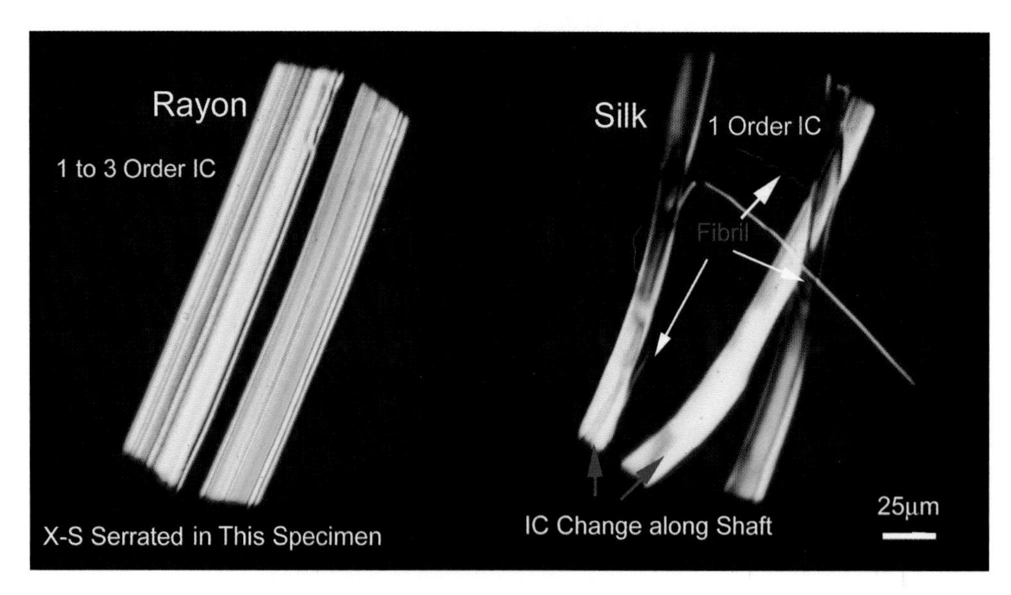

FIGURE A8.9B Cultivated silk fibers mounted in Melt Mount 1.539 and viewed between CPs (right) and compared with viscose rayon (left). Note fibril fibers present in silk (usually 1 to 2 μm in diameter).

TABLE A8.1
Natural Fibers Discussed in Text

Source of Natural Fiber	Common Name
Leaf	Abaca (Manila); sisal
Seed	Cotton
Bast (stem)	Flax (linen); hemp; jute; ramie
Fruit	Kapok
Secondary xylem (wood): coniferous (softwood); non-coniferous (hardwood)	Large numbers of families and species; paper fibers
Silkworm	Wild or tussah silk; cultivated and spun silk

TABLE A8.2
Natural Fiber Reference Data[1–14]

Natural Fiber	Extinction	Herzog Test	Dispersion Staining Colors	Structured Cells and Elements	Crystals Shape
Abaca (Manila)	Parallel; incomplete	Z-twist	1.565 HD oil; blue/yellow	—	Rectangular (backbone)
Sisal	Parallel; incomplete	Z-twist	1.565 HD oil; blue/magenta	Spiral element	Banana
Cotton	Undulose	—	—	Undulating convolution Lumens	—
Flax (linen)	Parallel; complete except at nodes	S-twist	1.585 HD oil; blue/magenta	—	None
Hemp	Parallel; complete except at nodes	Z-twist	1.590 HD oil; blue/magenta	—	Star
Jute	Parallel; complete except at nodes	Z-twist	1.590 HD oil; blue/yellow; trace magenta; magenta	—	Rhombic
Ramie	Parallel; complete except at nodes	S-twist	1.585 HD oil; orange	—	Star
Kapok	Parallel; incomplete	—	—	Folds common; cellophane appearance	—
Secondary xylem (wood): coniferous (softwood) non-coniferous (hardwood)	Undulose	—	—	Softwoods: cross-field pitting; ray tracheids (RTs); RT dentates; spiral thickenings Hardwoods: vessel elements; spiral thickenings; scalariform plates; tracheids; ray pitting	—
Cultivated silk (fine); wild or tussah silk (coarse)	Parallel; complete	—	1.540 HD oil; blue/magenta	Triangular X-S; tiny, fine fibrils coming from main fiber; low first-order IC uneven along fiber length	—

9 Textile Examination

Items of clothing and decorative arts, remnants of cloth, strands of thread and yarn, impressions of textiles, whole and broken buttons, pieces of rope and cordage, and pieces of cloth tape are only a few of the items encountered in museum studies, at crime scenes, and at archeological sites.[1–5,8,15–29] Forensic scientists, archeologists, museum scientists, and conservators are often asked to examine, identify, and compare these types of specimens to reconstruct events, study cultures and lifestyles of ancient peoples, determine how textiles were damaged; maintain, repair, and clean artifacts; and differentiate authentic and forged items.

Figure 9.1 depicts a typical forensic textile case. The arresting officer requested that a piece of questioned terrycloth collected at the scene of an attempted arson be compared to a known piece of terrycloth found in a suspect's auto to determine whether they had a common origin. This chapter discusses introductory methods for identifying, examining, and comparing a wide range of textiles. Textile science is a very complex subject and concerned readers are referred to the literature for further study.

Most textile materials are constructed from various types of synthetic or natural fibers that are twisted, untwisted, or spun together to form threads or yarns that are woven or knitted into fabric. Thus, a study of textile examination should begin with a working knowledge of the physical, optical, and chemical properties of synthetic and natural fibers and how such fibers are used in the construction of threads and yarns. Information about the identification of synthetic and natural fibers can be found in Chapters 5 through 8. The basic structures of fibers and how they are woven or knitted into textile materials will be the subjects of this chapter. Interested readers are referred to the literature for further information.[6–14]

STRUCTURES OF FIBERS

The structure of a fiber determines, to a large part, the appearance and behavior of the fabric it forms. Attributes such as length, shape, surface structure, treatment, diameter thickness, and denier are some of the primary factors that effect fabric manifestation and behavior.

Fibers are manufactured into two primary forms: staple lengths and continuous filaments. Staple fibers can be anywhere from several millimeters to many centimeters long. Filament fibers can be several kilometers in length (see Figures 9.2 and 9.3). Filament fibers are manufactured by extruding polymer melts or solutions containing dissolved polymers through the holes of a spinneret directly into a stream of air or a solvent bath, thereby producing a continuous fiber filament that is spun directly onto a spool. This process, known as spinning, is illustrated in Figure 9.4.

The cross-sectional shape of a fiber determines many characteristics of the fiber and the yarns and fabrics made from them. A multisurfaced fiber such as the star-shaped example depicted in Figure 9.2 has many more surfaces to reflect light than a round fiber. Thus, a star-shaped cross section produces a more lustrous and radiant fabric than a round fiber. In addition, a star-shaped cross-section will scatter more light and thus mask soiling or staining of the fabric made from it more readily than a fiber with a round or irregular cross-sectional shape.

Transverse shape influences surface texture. A round or flat shape causes a fiber to feel smooth; a serrated or multilobed shape causes a fiber to feel rough. It follows that any fabric made from fibers will feel smooth or rough, depending on its constituent fiber composition. Photomicrographs of fiber cross-sectional shapes appear in Appendix C.

YARN STRUCTURES

Yarn is a general term for continuous lengths of textile fibers (continuous filament or staple lengths) twisted or spun together into thread suitable for knitting, weaving, or otherwise intertwining to form a fabric. Yarn is produced in several forms:

1. **Spun yarn** — A number of cut or staple-length fibers held together by twisting and spinning
2. **Zero-twist yarn** — A number of filaments placed together without twisting; often held by crimping
3. A number of filaments held together with a degree of twist
4. **Monofilament** — A single filament with or without twist
5. Narrow strips of material, such as paper, plastic, or foil, with or without twist, intended for use in construction of a textile

Yarns are categorized into two primary groups: filament and spun. Filament yarns are made of continuous lengths of fiber strands, untwisted (usually crimped) or twisted, so that they stick together to form long continuous threads or yarns. Spun yarns are made of short staple lengths of fibers twisted or spun together so that they adhere to each other to form a thread or yarn. Filament and spun yarns can be differentiated simply by untwisting the yarn. Filament yarns unravel into long strands of fibers, while spun yarns unravel into short staple lengths of fiber that can be easily pulled apart (Figure 9.5).

In addition to determining whether fibers are composed of filaments or staple lengths, other physical properties play roles in the comparison of yarns. Since yarns are made by twisting parallel fibers together, the twist direction of a yarn is usually determined during comparison. To determine twist direction, simply hold a yarn and rotate it in a clockwise direction. If the yarn unravels, it has a Z-twist. If it becomes tighter, it has an S-twist (see Figures 9.5 through 9.7). Table 9.1 is a textile data sheet used for recording observations.

Other physical properties to compare are smoothness or roughness (texture); whether a yarn is monofilament or multifilament; the number of twists per inch (TPI); the number of single strands; threads, or yarns composing a ply; the number of fibers making up each ply, strand, or thread; the twist direction of each ply, strand, or thread; the angle of twist; whether the yarn is composed of a blend of fibers; and whether it exhibits pilling or other unusual features. Some of these properties are shown in Figures 9.3 and 9.5 through 9.8. The examination of yarn specimens is normally carried out with a stereomicroscope and fine tweezers. The specimen should be unraveled slowly while being viewed under low power and its construction should be noted, as depicted in Figure 9.9. All data should be recorded (see Table 9.1).

FABRIC STRUCTURES

Woven fabrics are made by intertwining two arrays of textile yarns placed at a 90° angle. The lengthwise or warp yarns should be parallel to the length of the fabric, while the filler or weft yarns are interlaced at right angles to the warp yarns. The warp yarns run the length of the fabric and the weft yarns run the width of the fabric from selvage to selvage. Each strand of filler yarn is called a pick.

A loom is a device used to weave fabrics. The essential parts are the warp beam, harness, heddles, reed, shuttle, bobbin, weft or filling yarn, and fabric roll. During weaving, the warp yarns must be mounted parallel onto the warp or loom beam. Each warp yarn is passed through the harness and strung through the appropriate heddle and reed. The harness picks up and lowers the warp yarns. The shuttle that carries a bobbin holding filler fibers inserts the filler yarns into the warp yarns that are kept separated by the reed. As each thread of filler yarn is inserted into the warp yarns, the reed pushes the loose filler yarn or pick to the edge of the completed fabric. This cycle is repeated thousands of times to produce woven fabric. The completed fabric is slowly wound onto the roll of cloth (see Figure 9.10).

The basic weaves are plain, twill, and satin. The plain-weave pattern is the most common and therefore is utilized most of the time. The plain weave consists of alternate placement of one warp yarn over a filler yarn and then under the adjacent filler yarn for the entire length of the fabric (Figure 9.11). Figure 9.12 illustrates a plain weave design.

Common variations of the plain weave are the rib and basket weaves. Ribs are formed on the fabric surface because one yarn is thicker than the opposing yarn (rib weave; Figure 9.13). A group of ends or warp yarns is alternatively passed over and under a group of picks or filler yarns (basket weave; Figure 9.14).

The twill weave is another common textile design. The appearance of diagonal lines in a fabric typifies twill-weave textiles. The many variations of the twill weave include regular twill, steep twill, reclining twill, right-hand twill, left-hand twill, broken twill, and balanced twill. All variations concern the angle of the diagonal and the direction of the warp yarn (twill angle). Denim is one of the most common twill-weave fabrics (Figure 9.15). Denim is classified as regular twill because its twill angle is 45°. Figure 9.16 shows several twill designs.

The satin weave is another basic weave. The face of a satin-weave fabric is composed almost entirely of warp or filler floats produced in the repeat of the weave. Satin weaves appear smooth and shiny. Figure 9.17 shows an example of a satin weave.

Knit fabrics are produced by interlooping (as a series of loops) one or more yarns. The two major types of knitted fabrics are warp and weft knits. Vertical columns in a knit fabric are known as wales. In the warp knit, the yarn runs parallel to the length of the fabric. In the weft knit, it runs perpendicular to the length of the fabric. Figure 9.18 shows a typical weft knit. Knits tend to stretch easily and usually recover from wrinkling quickly. Another type of fabric that stretches easily is crepe. Crepes are lightweight fabrics that have crinkly texture surfaces (Figure 9.19).

FABRIC ANALYSIS

A fabric analysis is normally performed with a linen tester for counting threads, a protractor for measuring angles, a stereomicroscope, fine needles, and tweezers (Figure 9.20). The steps are as follows:

1. Determine overall appearance of fabric (color, luster, texture).

2. Determine the fiber composition:
   a. Is the yarn composed of synthetic or natural fibers?

b. If synthetic, what is the generic classification?
c. If natural, is the material hair, wool, or vegetable fiber?
d. Is the yarn made of a single fiber type or is it a blend?
e. How many plies? Single or multiple-ply yarn?
f. Is the yarn composed of tow, filament, or staple lengths?
g. How are fibers held together? Are they crimped, twisted, other?

3. Determine warp direction.
   a. Selvage runs in warp direction.
   b. Reed marks run parallel to warp.
   c. Warp yarns usually have higher twists.
   d. Warp yarns are normally finer in diameter.
   e. Stripes usually run in warp direction.
   f. Fabric is harder to tear in the warp direction.

4. Determine weft direction.
   a. Novelty yarns are usually in filling direction.
   b. Bulkier yarns are in the filling direction.
   c. Ridges usually run in weft direction.
   d. Weft direction usually has more stretch.

5. Determine face side of fabric.

6. Determine basic type of weave (plain, basket, twill, satin).
   a. Note fabric finish (print design, dyed, mechanical texturing, coating).
   b. Hold the fabric face-up with the warp yarns running toward you.
   c. Pull out one end of the warp yarn with a tweezer and determine how it is woven from the first pick down.
   d. Continue examining the fabric until you are certain that you have completed one repeat (Figure 9.20).
   e. As a check, make sure you examine and determine several repeats of the weave.

7. Prepare a graphic representation of the fabric. All collected data and observations should be noted on a textile data sheet.

8. Use the data compiled on the data sheet to examine, identify, and compare questioned and known textile specimens.

Figure 9.1 shows a jigsaw match of two pieces of cloth. In addition, data tabulated by an analyst showed that the piece of cloth from the suspect's auto and the piece from the crime scene were consistent in construction, and in physical, optical, and chemical properties.

TEXTILE GLOSSARY[14]

Back side Rear or posterior side of a fabric.
Beating up Final operation performed by the loom during weaving in which the last filler yarn inserted by the shuttle is beat or pushed into position against the finished fabric by the reed.
Cord Object produced when two or more fine continuous filaments or staple lengths of fibrous materials are spun, twisted, or otherwise plied together into a continuous length of material.
Crimp Wave-like bend in tow and filament yarn; the number of waves or crimps per unit length can help characterize yarns and fabrics.
Crimping Process of imparting wave-like bends to tow or filament yarns.
End Individual warp yarn.
Face side Front or anterior side of a fabric.
Filament Long continuous length of fiber.
Float Portion of a warp or filling yarn that extends over two or more neighboring picks or ends.
Hand The way a fiber, yarn or fabric feels when handled.
Novelty yarn Yarn that exhibits a special effect.
Pick Single filling or weft thread or yarn.
Pilling Formation of tufts of small rounded balls of fibers on a fabric surface.
Ply Number of single strands, threads, or yarns twisted together to form a plied yarn.
Reed Comb-like device on a loom that separates warp yarns and also beats or pushes newly woven filler yarn against the finished fabric by a process known as beating up.
Reed mark Defect composed of light and heavy streaks in a woven fabric; may be caused by damaged yarns or damaged or worn reeds; always runs parallel to warp direction.
Repeat Weave design covered by a single unit of a pattern; a design repeated over and over again to form the fabric.
Selvage Narrow edge of a woven fabric that runs parallel to the warp thread or yarn.
Spinneret Disc having many tiny holes, used in the production of filament fibers.
Spinning Process used in the manufacture of fibers; a polymer melt or solution of dissolved polymer is extruded through the holes of a spinneret disc directly into a stream of air or a solvent bath, thereby producing a long continuous filament that is spun directly onto a spool.
Spool Cylinder upon which filament yarn is wound.
Spun yarn Yarn composed of staple lengths of fibers normally held together by twisting.
Staple Fiber whose length is measured in millimeters or centimeters.
Strand Single fiber, filament, or monofilament.

Thread Slim, strong strand of yarn or cord made for knitting, weaving and sewing; most threads are made by plying and twisting yarns.

Tow Untwisted strands of continuous filaments not held together by twisting (zero-twist yarns).

Twisted Designation of a clockwise (S) or counter-clockwise (Z) orientation of fibers, filaments, plies, or cords when they are intertwined to produce cord, rope, thread, or yarn.

Untwisted Yarn not held together by twisting (zero-twist yarn).

Wales Vertical columns of stitches present in knits.

Warp Lengthwise thread or yarn that runs parallel to a textile length and perpendicular to its filler thread or yarn.

Weave Pattern of intertwined warp (lengthwise) and weft (filler) yarns used in fabric construction.

Weft Yarn running from selvage to selvage at right angles to the warp or lengthwise yarn; filler yarn.

Yarn Common term for a continuous length of filament, yarn, or fine cord used in production of textile fabrics.

REFERENCES

1. Söderman, H. and O'Connell, J.J., *Modern Criminal Investigation*, New York, Funk & Wagnalls, 1935.
2. O'Hara, C.H. and Osterburg, J.W., *Introduction to Criminalistics*, New York, Macmillan, 1949.
3. Kirk, P.L., *Crime Investigation*, New York, Interscience, 1953.
4. Svensson, A. and Wendel, O., *Techniques of Crime Scene Investigation*, 1st ed., New York, Elsevier, 1974.
5. Deadman, H.A., Fiber evidence and the Wayne Williams trial, *FBI Law Enforce. Bull.*, March 1984, p. 13; May 1984, p. 10.
6. Press, J.J., Ed., *Man-Made Textile Encyclopedia*, New York, Interscience, 1959.
7. *Identification of Textile Materials*, Manchester, U.K., Textile Institute, 1970.
8. *Man-Made Fibers Fact Book*, Washington, D.C., Man-Made Fiber Producers Association, Inc., 1978.
9. *Technical Manual*, Vol. 53, Research Triangle Park, NC, American Association of Textile Chemists and Colorists, 1977.
10. Schaffer, E., Fiber identification in ethnological textile artifacts, *Stud. Conserv.*, 26, 119, 1981.
11. *Dictionary of Fiber and Textile Technology*, 6th ed., Chatham, NJ, Hoechst Celanese Corporation, 1990.
12. Goutmann, M., *Fabric Structure: Basic Weave Design*, 2nd ed., Philadelphia, Philadelphia College of Textiles and Science, 1986.
13. Zielinski, S.A., *Master Weaver Library*, 22 vols., Quebec, Canada, Nilus Leclerc, 1979.
14. Minor, M. and Minor, N., *The American Indian Craft Book*, Lincoln, University of Nebraska Press, 1978.
15. De Forest, P.R., Gaensslen, R.E., and Lee, H.C., *Forensic Science: An Introduction to Criminalistics*, New York, McGraw-Hill, 1983, p. 217.
16. Robertson, J., Ed., *Forensic Examination of Fibers*, New York, Ellis Horwood, 1992.
17. Miller, R.S., *Art of the Andes from Chavin to Inca*, New York, Thames & Hudson, 1995.
18. Miller, M.E., *The Art of Mesoamerica*, New York, Thames & Hudson, 1996.
19. Blier, S.P., *The Royal Arts of Africa*, New York, Harry N. Abrams, 1998.
20. Elsner, J., *Imperial Rome and Christian Triumph*, New York, Oxford University Press, 1998.
21. Clunas, C., *Art in China*, New York, Oxford University Press, 1998.
22. Shaw, H., *Dress and Decoration of the Middle Ages*, Cobb, CA, First Glance Books, 1998.
23. Aston, M. and Taylor, T., *The Atlas of Archaeology*, New York, DK Publishing, 1998.
24. Lucas, A. and Harris, J.R., *Ancient Egyptian Materials and Industry*, Mineola, NY, Dover Press, 1999.
25. Robertson, J. and Grieve, M., Eds., *Forensic Examination of Fibers*, London, Taylor & Francis, 1999.
26. Taupin, J.M., Adolf, F.P., and Robertson, J., Examination of damage to textiles, in *Forensic Examination of Fibres*, 2nd ed., Robertson, J. and Grieve, M., Eds., London, Taylor & Francis, 1999, p. 65.
27. Taupin, J.M., Arrow damage to textiles: analysis of clothing and bedding in two crossbow deaths, *J. Forens. Sci.*, 43, 205, 1998.
28. Adolf, F.P., The structure of textiles: introduction to the basics, in *Forensic Examination of Fibres*, 2nd ed., Robertson, J. and Grieve, M., Eds., London, Taylor & Francis, p. 33.
29. Wiggins, K.G., Ropes and cordage, in *Forensic Examination of Fibres*, 2nd ed., Robertson, J. and Grieve, M., Eds., London, Taylor & Francis, 1999, p. 55.

FIGURE 9.1 Comparison of a piece of terrycloth collected at the scene of an attempted arson and a known piece of terrycloth found in a suspect's auto to determine whether they have a common origin.

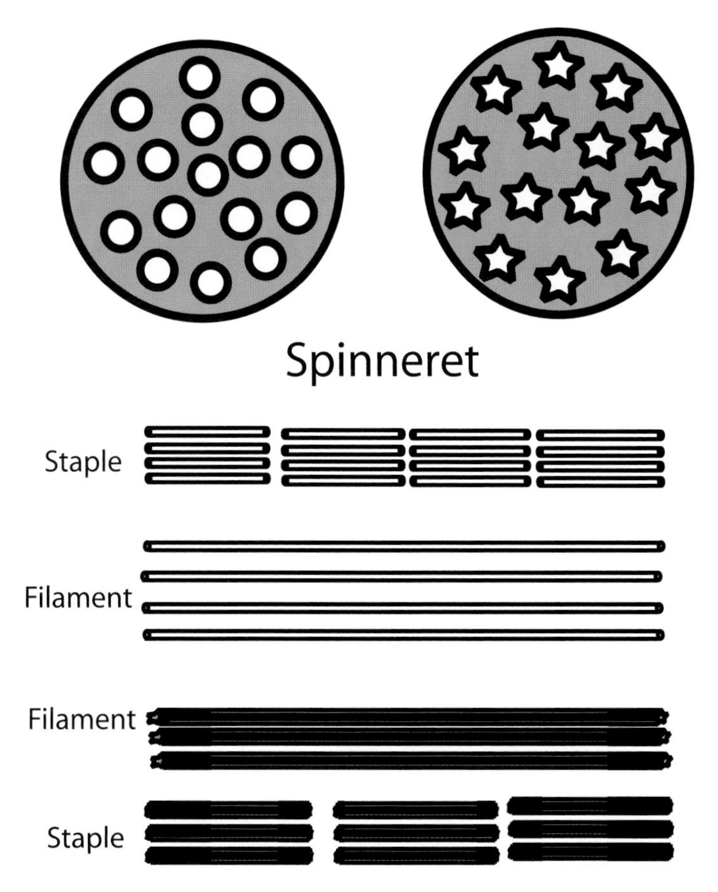

FIGURE 9.2 Two spinneret discs with different-shaped holes. The filament yarns produced by these spinnerets are cut into various lengths to produce staple fibers.

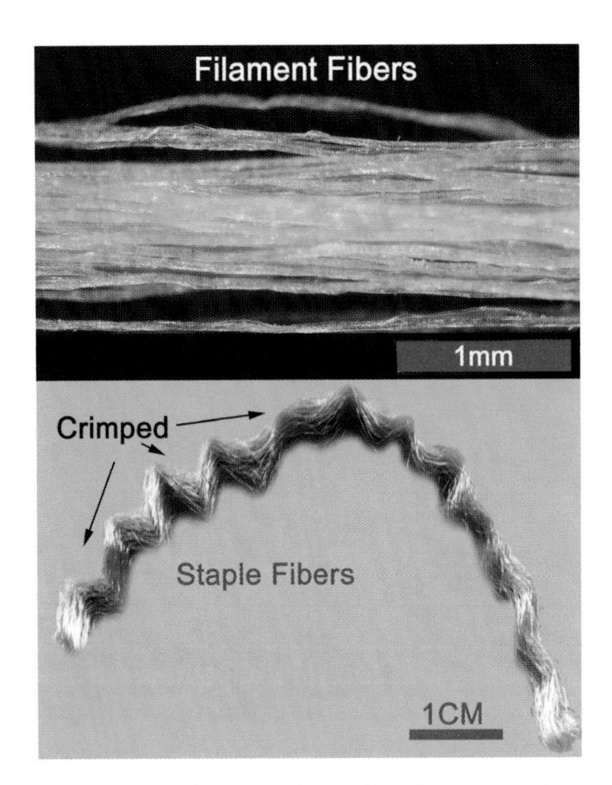

FIGURE 9.3 Two primary fiber types: continuous filament and staple length. Crimping helps hold untwisted fibers together and increases their volume and insulating properties.

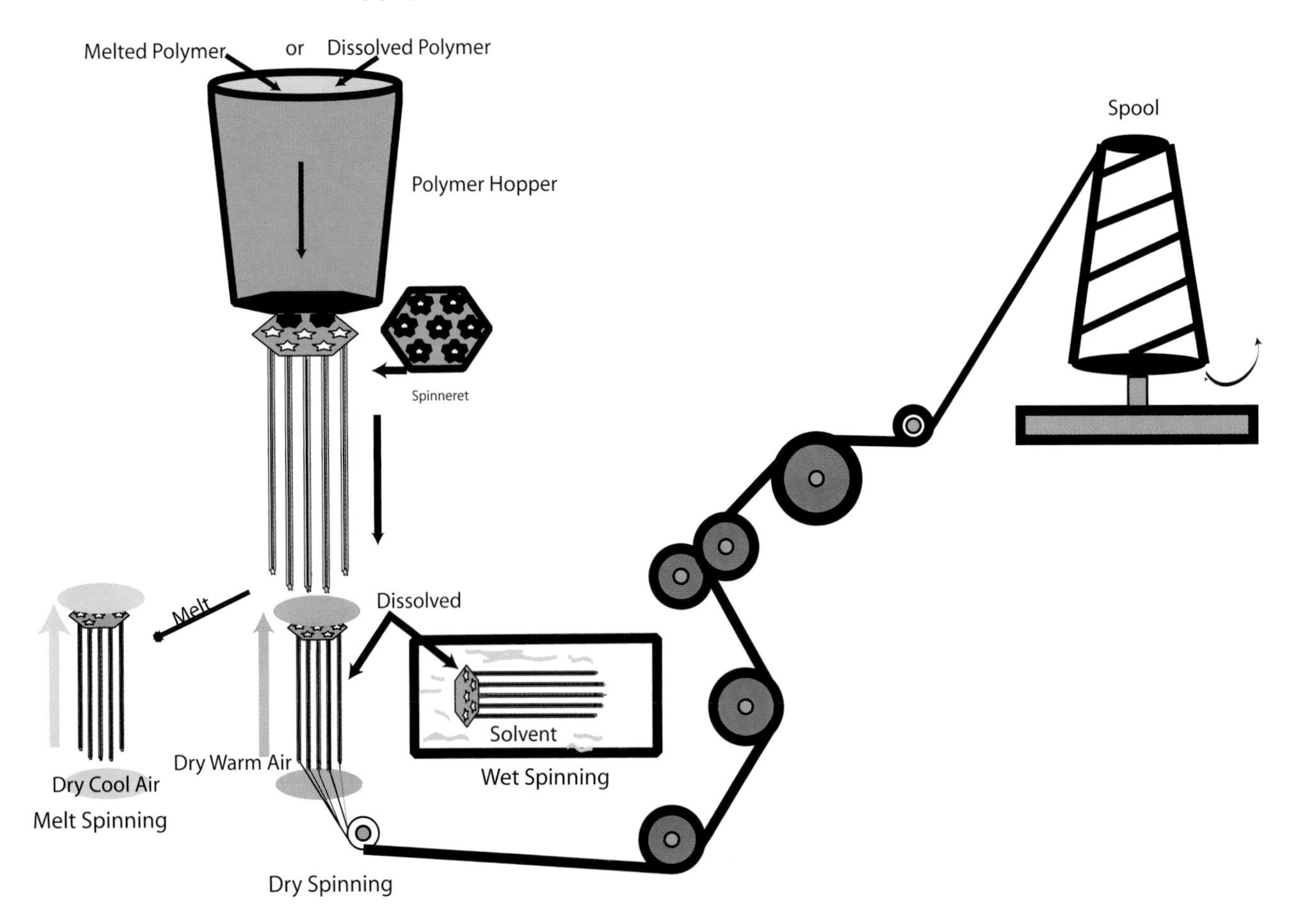

FIGURE 9.4 Spinning process used to produce synthetic filament yarns.

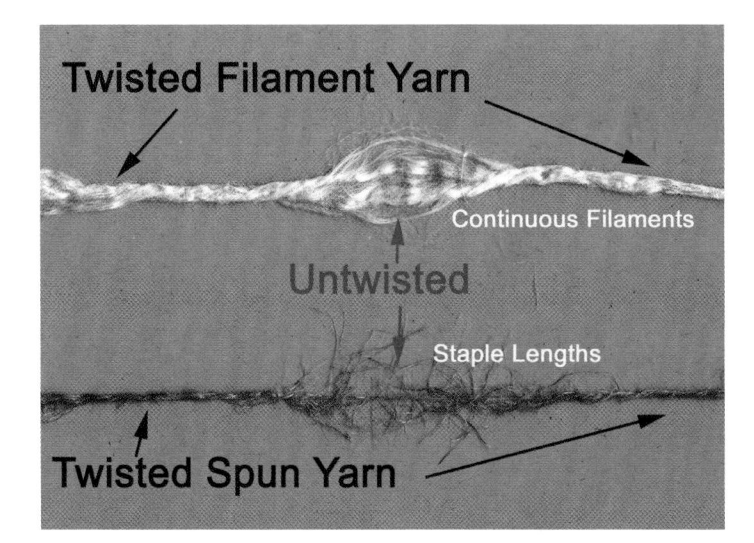

FIGURE 9.5 Filament and spun yarns. To differentiate between filament and spun yarns, untwist the yarn between the fingers. A filament yarn will unravel into long strands of fibers; a spun yarn will unravel into short staple lengths of fiber that can be easily pulled apart.

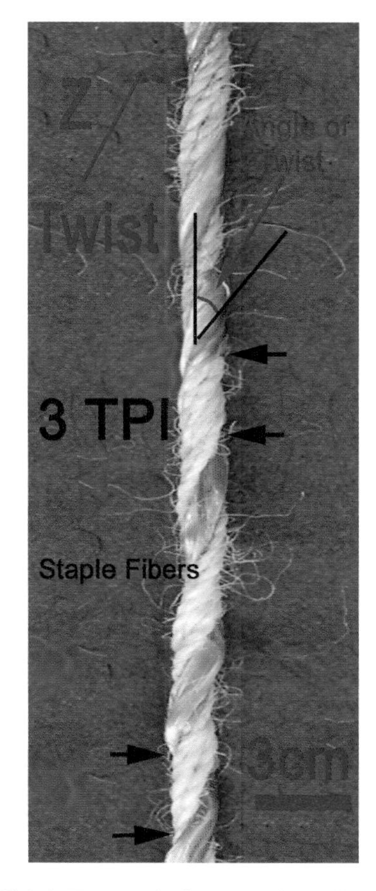

FIGURE 9.6 Yarn made from multiple plies of short staple-length fibers held together with a Z-twist. The number of twists per inch (TPI) and the angle of twist are illustrated.

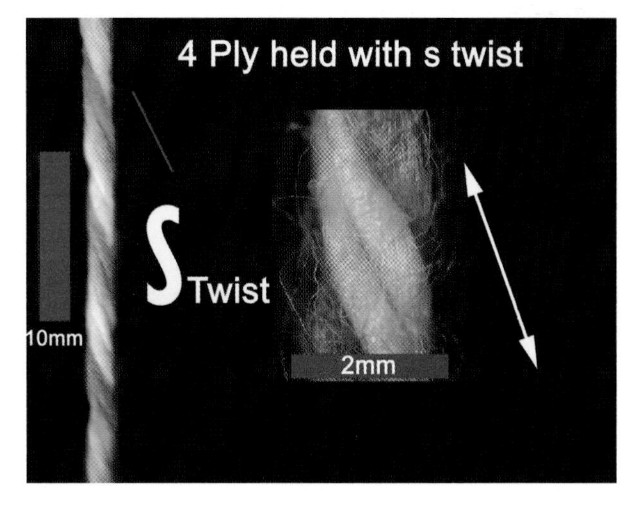

FIGURE 9.7 Four-ply yarn made from short staple-length fibers held together with an S-twist.

TABLE 9.1
Textile Data Sheet
Examine both warp and weft yarns and record all observations

Specimen No. **Observations**

Fiber Class	Genera	
Overall Appearance	Luster	
	Texture	
	Color	
	Surface	
	Crimps/inch	
	Other	
Fiber Type	Staple filament	
Yarn Type	Single ply	
	Cord	
	Combed	
	Worsted	
	Stretch	
	Spun	
	Other	
Twists/inch		
Fiber/thread count		
Pile	Uncut/cut	
	W	
	V	
	Other	
Knit	Warp or weft	
	Plain stitch	
	Purl stitch	
	Miss stitch	
	Tuck stitch	
Weave pattern	Face side	
	Back side	
Printed/dyed design		

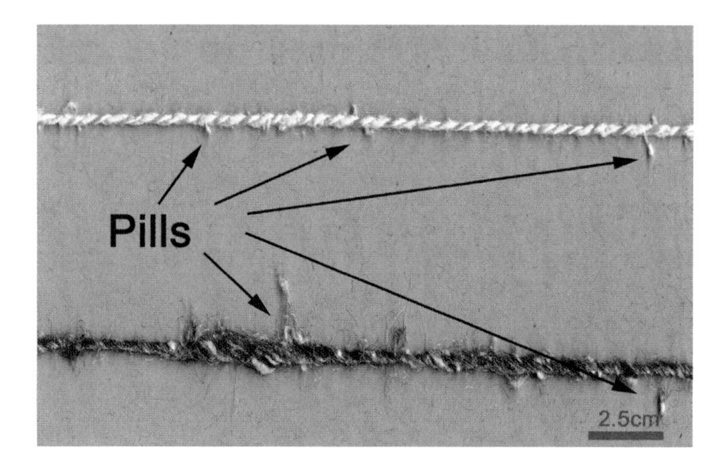

FIGURE 9.8 Two specialty yarns exhibiting pilling.

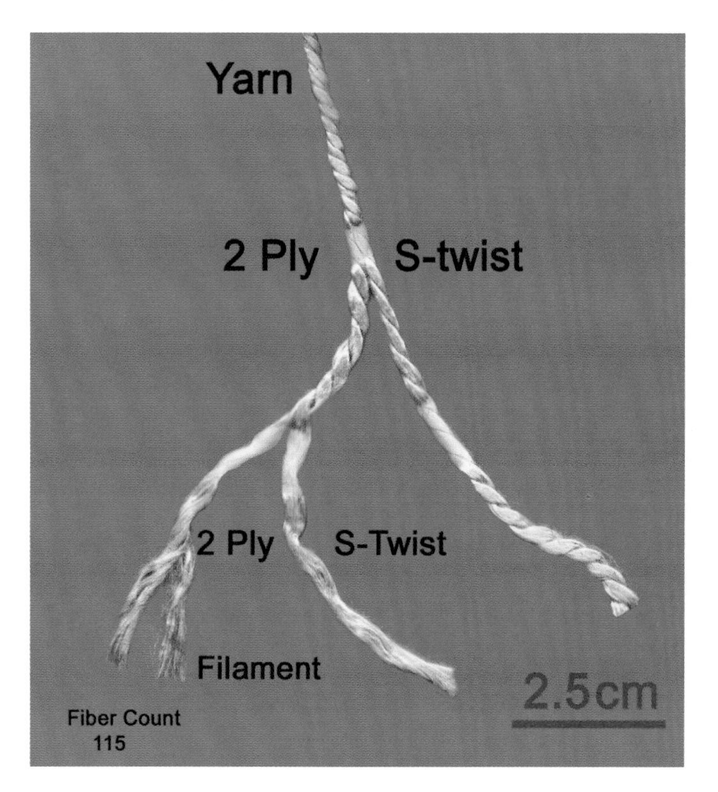

FIGURE 9.9 Breakdown examination of a four-ply continuous multifilament yarn held together with S-twists using a stereomicroscope.

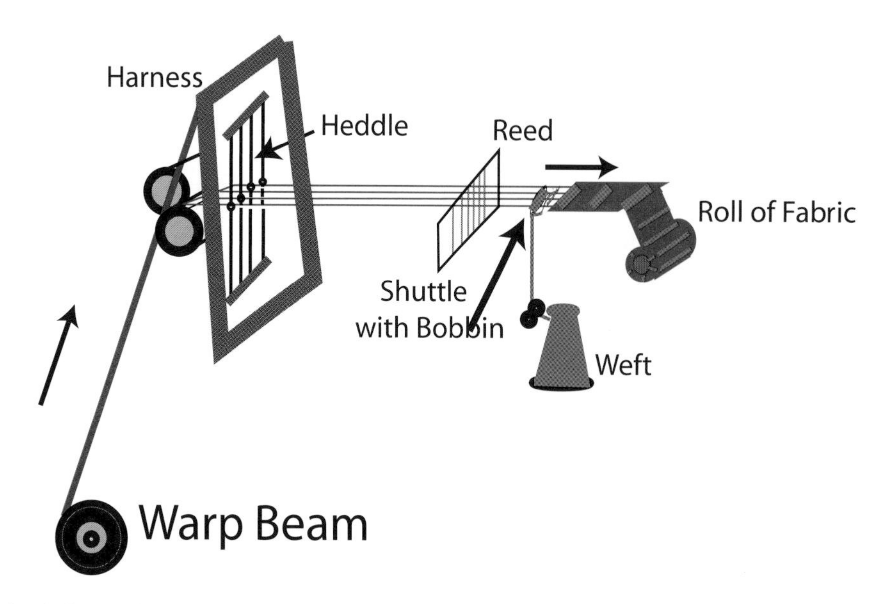

FIGURE 9.10 A simple loom with its essential parts.

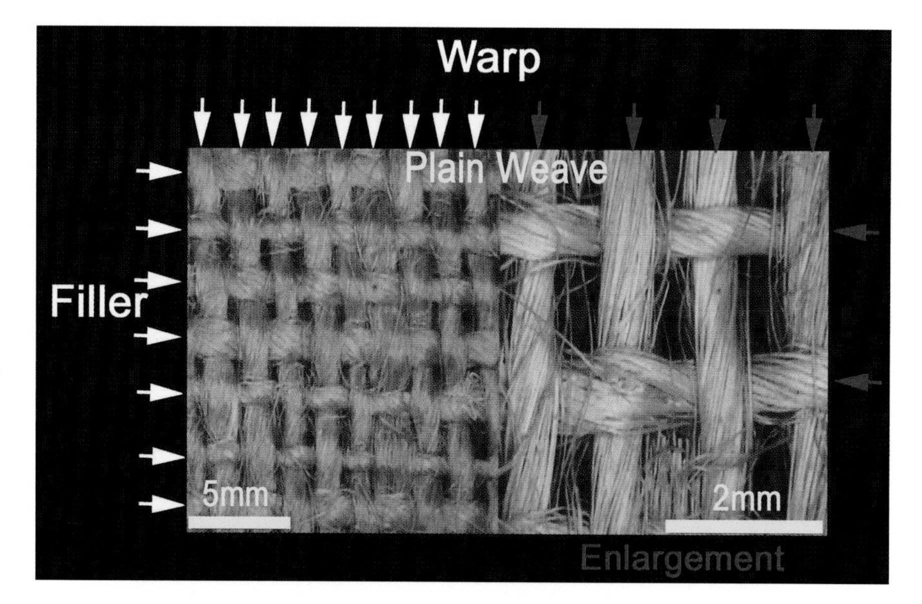

FIGURE 9.11 Plain weave design. Each warp yarn is placed over one filler yarn and then under the next or adjacent filler yarn for the entire length of the fabric.

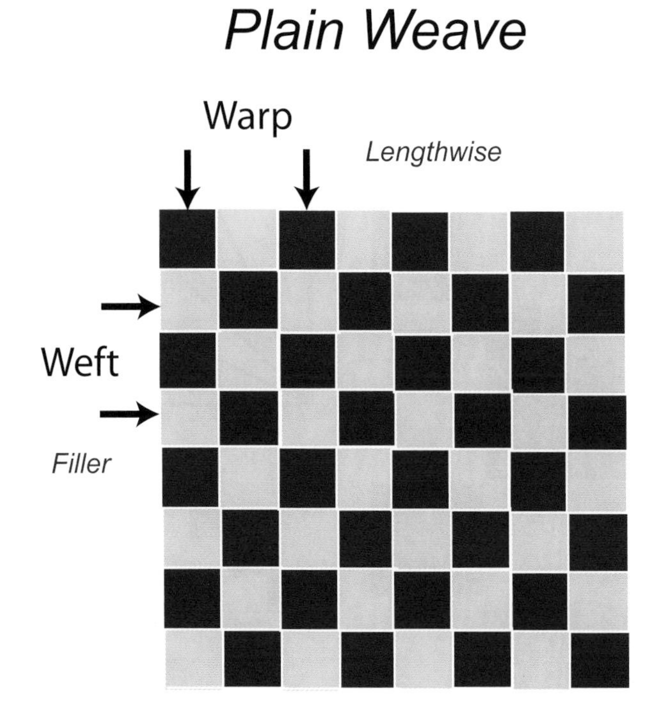

FIGURE 9.12 Graphic representation of a plain weave. The blue squares symbolize warp yarns; yellow squares signify filler yarns.

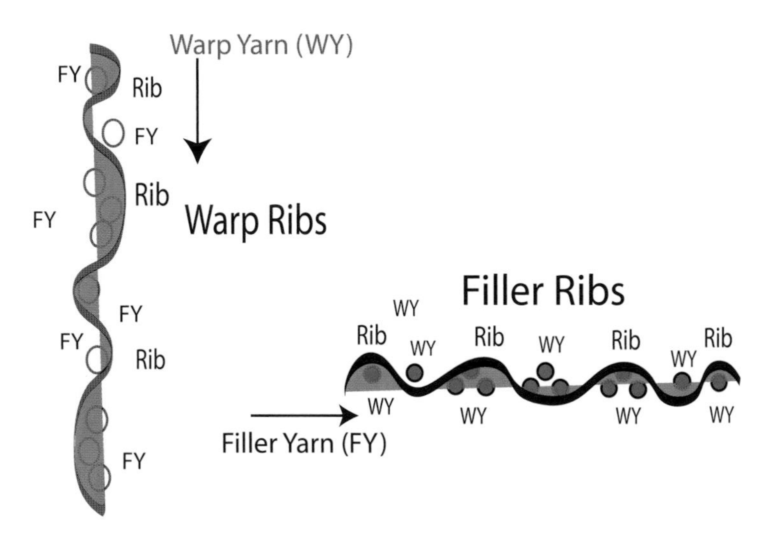

FIGURE 9.13 Cross section of warp and filler rib weave patterns.

FIGURE 9.14 Basket weave. Each group of warp yarns is passed over and under a group of filler yarns. The repetition continues for the entire length of the fabric.

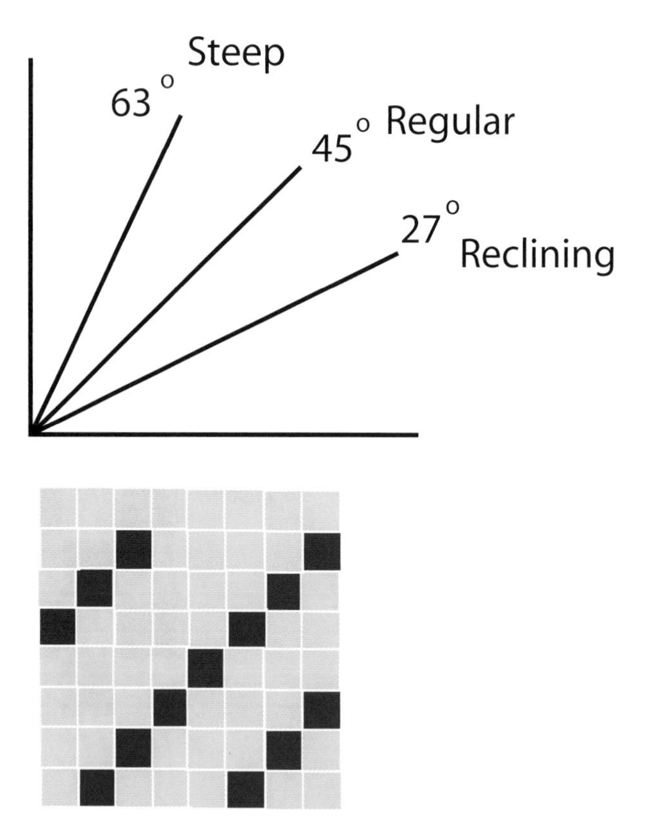

FIGURE 9.16 Several twill patterns.

FIGURE 9.15 Specimen of denim twill with a 45° twill angle.

FIGURE 9.17 Typical satin weave fabric.

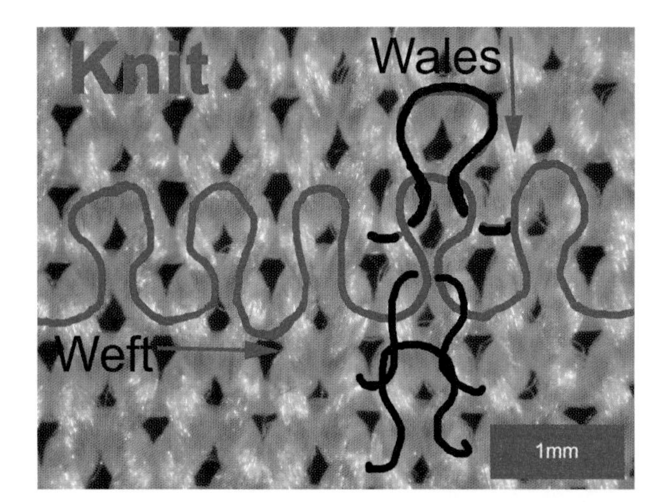

FIGURE 9.18 Typical weft knit. One thread runs across the fabric width. Vertical columns of stitches are known as wales.

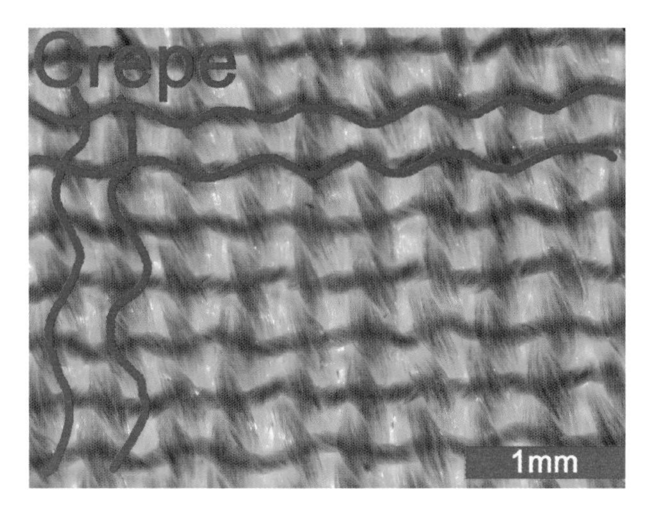

FIGURE 9.19 Typical crepe fabric.

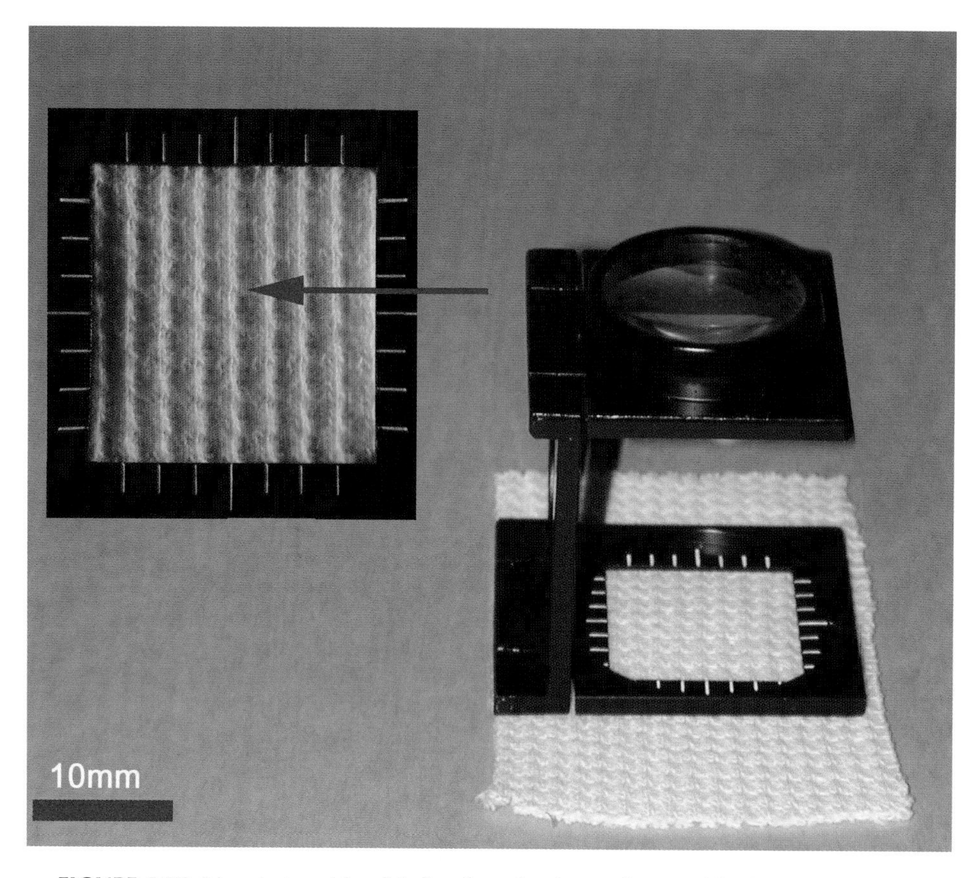

FIGURE 9.20 Liner tester with a 1 in.^{2} scale and at least a 5× magnification lens.

10 Paint Examination

Paint and paint-like materials have been used throughout recorded history. Various mixtures of coloring agents, fluid materials, and binding substances have been utilized for recording, expression, decoration, and surface protection for millennia. Scientific investigators frequently encounter paint specimens. Forensic scientists, chemists, and conservators examine and identify miniscule traces of pigment and paint (see Figure 10.1). Most paint analyses today are conducted with highly sophisticated and costly instrumentation. However, inexpensive, timely, simple, and irrefutable analytical tests and procedures that can be used on minute traces of paint are still required. This chapter will concentrate on classical, less complex, but proven microchemical methods for analyzing paints and pigments.

Paint samples submitted to forensic and conservation analytical laboratories can play important roles in investigations and prosecutions. These analyses and comparisons for common origin are distinguished from tests performed by industrial laboratories by the sizes of samples submitted for characterization. These paint samples are not pristine; they are subjected to uncontrolled environmental and collection effects.

Paint sample analysis is required in a number of situations, for example: (1) investigating vehicular accidents in which contact between two objects must be established; (2) investigating hit-and-run incidents in which make, model, and color of an unknown vehicle must be determined; (3) architectural paint exhibits related to restoration projects or investigations of crimes against property; (4) duplicating and testing paints and lacquers used in the decorative arts; (5) authentication studies and restoration of artistic works; (6) testing fingernail polishes and other coating materials used in the cosmetics industry; and (7) archeological research involving paints used by ancient peoples for protection, decoration, and personal adornment.

PAINT USE AND COMPOSITION

Paints are applied for protective value, aesthetic purposes, or both. For purpose of this discussion, *paint* includes a range of materials, from thin, translucent stains to heavy, opaque films. A brief treatment of these topics of interest to the investigator and analyst is appropriate. A complete discussion of the formulations, manufacturing steps, methods of application, properties, uses, and analyses of paint films is beyond the scope of this work and readers are directed elsewhere for such information.[1–2]

Paint is composed of three principal components. The vehicle is the binder that holds all the components together and is usually a polymeric material consisting of natural or synthetic resins. The binder can form a surface film in a number of ways. When the film forms by the simple evaporation of the solvent system of the liquid, a paint is normally classified as a lacquer. One characteristic of lacquers is their ability to redissolve when subjected to many organic solvents.

When a film is formed by chemical cross-linkages of a number of components, a paint is usually referred to as an enamel. The cross-linkages can be initiated by elevated temperature, oxidation by exposure to oxygen in air, chemical reactions of components of special initiators, or a combination of these factors. Latex paints form films through the coalescence of dispersed latex particles after water loss. These working definitions are not exclusive; combinations of film-forming mechanisms are common. Table 10.1 lists common binders and vehicles.

Pigments supply paint with color, hue, and saturation. Blues and greens are predominantly organic; whites, yellows, and reds are inorganic. This is not a strict rule, and crossovers and mixtures are common in modern formulations. Pigments are expensive and manufacturers seek to minimize their use to control costs. Table 10.2 lists common inorganic and organic pigments.

Extenders are generally less expensive inorganic materials that, when added to paint, increase its solid content and, as a result, its opacity and hiding ability. Certain extenders provide other advantages. Titanium dioxide, known for its hiding capacity, can fill the roles of pigment and extender. Table 10.3 lists common pigments and extenders.

PAINT EXAMINATION

After numerous applications, the film of paint covering a surface becomes thicker. It consists of numerous layers of paint and layers of soot, pollution, and dirt from the environment that collect on the surface between successive applications. In addition, evidence of chemical oxidation, discoloration, bleeding, erosion due to weathering and sunlight, and fibers or hairs from brushes used to apply the paint can be present in paint chips. All this information can be useful in forensic investigations, archeological studies, corrosion studies, fine art authentications, and conservation studies.[7,8,11,12]

Paint film investigations can be based on a number of physical and chemical characteristics.[1–8,10–13,15,17,19–38] The size and shape of an exhibit, its surface condition, color, layer sequence, and thickness are physical attributes that can be readily assessed by macroscopic and microscopic examinations.

The chemical compositions of paint components can be evaluated individually or in combination by a number of microchemical, microscopical, and instrumental methods. The sizes of paint specimens, conditions of the samples, and the information sought will guide the analyst in choosing methods and techniques to employ for the physical and chemical analysis of paint evidence.

In forensic science casework, the first part of paint examination should be an attempt to establish a jigsaw fit of edges or a match of surface striae of questioned and known samples. The quantity and quality of the characteristics that match should be sufficient to establish uniqueness. These examinations are generally conducted macroscopically, using an illuminated desk magnifier, stereomicroscope at a low range of magnification, and reflected light illumination at various incident angles.

If a physical match is not attained, the layer structure order, color, thickness, and other details should be documented. Some manipulation of the sample may be necessary to achieve the collection of needed information in sufficient detail. Angle cuts and thin sectioning with a clean (new) scalpel blade can easily reveal the layer structure. It may be necessary to embed the sample in a resin and employ microtome techniques to obtain high-quality thin sections. Embedded samples make it possible to grind and polish samples so that fine physical details such as pigment size and distribution can be evaluated by higher-resolution microscopy.

In most cases, an initial step in the analysis of paint is documentation of physical characteristics. Before any chemical testing is conducted on a pigment or paint specimen, it should be examined under a stereomicroscope. If the specimen appears homogeneous, a tiny aliquot should be preserved in Melt Mount 1.539 or 1.540 high-dispersion oil for polarized light microscopy (PLM) examination. If the specimen appears to be a mixture, its components should be isolated under a stereomicroscope and each constituent should be readied for PLM examination. Finally, if the specimen consists of paint chips, cross-sections should be prepared as described in Chapter 2. If the paint chips are minute, cross-sections can be prepared via the procedure shown in Figure 10.2. Figure 10.3 depicts a cross-section of a tiny, single-layered paint flake.

In any examination involving paint and coating materials, specimen color is always of primary importance in identifying the pigments used in manufacture. Another vital factor is layer sequencing. The authors have found that placing a tiny droplet of water to act as a tiny magnifying lens on the surface of a paint flake or across the layers of a cross-section of a paint chip usually makes it easier to determine colors and examine or measure its layer structure. This procedure is also useful when comparing the specimen color to Munsell® color chips (Figures 10.4 and 10.5). Fluorescence microscopy is often useful in determining layer structures, especially on multi-layered white architectural paints with clear lacquers (Figures 10.6 and 10.7).

If sample size is sufficient, destructive tests based on chemical reactions can serve as sources of additional data. The dissolution, swelling, or generation of colors with various solvents or reagents is informative about the possible identities of resins, pigments, and extenders. These tests can be performed in a porcelain spot plate, small disposable test tube, or glass microscope slide. Observing the tests with a microscope allows them to be performed successfully on micro-sized samples. Because the pigments and extenders found in paint are ground to such a small size, they can be identified via PLM and microchemical methods.[39–56] Among the major advantages of these testing procedures, they are (1) relatively inexpensive to perform; (2) very sensitive and require only nanogram-sized specimens; (3) proven and documented in the literature; and (4) specific for certain ions, elements, compounds, and molecules.

Components of paint specimens are isolated and prepared for PLM examination as described in Chapter 2. The specimen is examined and data concerning physical, crystalline, and optical properties are collected and documented as noted in Chapter 3. A paint specimen data sheet is prepared for each sample (Table 10.4).

After basic information about physical properties is compiled, PLM assessment and chemical characterization of each specimen can commence. The component of interest can be isolated and examined by employing the optical, microchemical, and microcrystalline methods discussed in Chapters 1 through 4, the basic procedures covered in Appendix D, and the cited literature. The data recorded on a data sheet can be used to characterize and identify the specimen. Guerra Paint and Pigments in New York City (510 East 13th Street, 212-529-0628) is a great source for hard-to-find paints, pigments, extenders, additives, and resins.

Most modern paint specimens are composed of synthetic resins, binders, and organic dyes usually identified with infrared microspectroscopy (IMS). A brief discussion of IMS is in order. However, a complete discussion of IMS is beyond the scope of this text and interested readers are referred to the literature.[57]

IMS is routinely employed for paint analysis. Transmission measurements can be obtained on thinly sliced or rolled samples and by compressing the paint in a diamond cell as shown in Figure 10.8. Attenuated total reflectance objectives allow for the collection of spectra from the surface of a paint sample without the need to prepare a thin specimen. Reflection data vary slightly from transmission data, but successful comparisons, interpretations, and database searches can be carried out. The infrared microscope has a limited spectral range because of the detectors employed and the optics available, but is applicable to the analysis of the major organic components in paint. If data on the inorganic constituents are desired, additional tests with extended range spectrometer optics are required. An IMS spectrum and database search of damar, a classical resin, is illustrated in Figure 10.9.

The application of microscopy and microtechniques to the examination of tiny paint specimens has many advantages. It should be apparent that paint examination, although complex, can be achieved with these fast, proven, low-cost methods. It is hoped that this work will encourage the use of microchemical methods and PLM in everyday casework.

REFERENCES

1. Forensic Paint Analysis and Comparison Guidelines, Washington, D.C., Scientific Working Group for Materials Analysis, U.S. Department of Justice, 1998.
2. *Standards E1610 through 1695*, West Conshohocken, PA, American Society for Testing and Materials.
3. Toch, M., *Paint, Paintings, and Restoration,* New York, Van Nostrand, 1931.
4. Cennini, C.A., *Il Libro dell Arte (The Craftsman's Handbook*, 15th Century), Thompson, D.V., Jr., Transl. (1933), New York, Dover Publications, 1954.
5. Cellini, B., *The Treatises of Benvenuto Cellini on Goldsmithing and Sculpture*, Thompson, D.V., Jr., Transl., Mineola, NY, Dover Publications, 1960.
6. Massey, R., *Formulas for Painters,* New York, Watson-Guptill, 1967.
7. Gettens, R.J. and Stout, G.L., *Painting Materials: A Short Encyclopedia*, New York, Dover Publications, 1966.
8. Crown, D.A., *The Forensic Examination of Paints and Pigments*, Springfield, Il, Charles C Thomas, 1968.
9. Adrosko, R.J., *Natural Dyes and Home Dyeing*, New York, Dover Publications, 1971.
10. Mayer, R., *The Artist's Handbook of Materials and Techniques*, 5th ed., New York, Viking, 1991.
11. McCrone, W.C., Delly, J.G., and Palenik, S.J., Eds., *The Particle Atlas*, Vol. 2, 1973; Vol. 5, 1979, Ann Arbor, MI, Ann Arbor Science Publishers.
12. Thornton, J.I., Forensic paint examination, in *Forensic Science Handbook*, Saferstein, R., Ed., Englewood Cliffs, NJ, Prentice-Hall, 1982, chap. 10.
13. Ryland, S.G., Infrared microspectroscopy of forensic paint evidence, in *Practical Guide to Infrared Microspectroscopy*, Humecki, H.J., Ed., New York, Marcel Dekker, 1995, chap. 6.
14. Derrick, M.R., Infrared microspectroscopy in the analysis of cultural artifacts, in *Practical Guide to Infrared Microspectroscopy*, Humecki, H.J., Ed., New York, Marcel Dekker, 1995, chap. 8.
15. *Paint and Coatings: Testing Manual of the Gardner–Sward Handbook*, 14th ed., Koleske, J.V., Ed., West Conshohocken, PA, American Society for Testing and Materials, 1995.
16. Miller, M.E., *The Art of Mesoamerica from Olmec to Aztec,* 2nd ed., New York, Thames & Hudson, 1996.
17. Buxbaum, G., Ed., *Industrial Inorganic Pigments,* 2nd ed., New York, Wiley-VCH, 1998.
18. Garcia, M., *Couleurs Végétales, Teintures, Pigments et Encres*, Paris, Édisud, 2002.
19. Forensic Paint Analysis and Comparison Guidelines by Paint Subgroup, Washington, D.C., Scientific Working Group for Materials Analysis, U.S. Department of Justice, May 2000.
20. De Wild, M.A., *The Scientific Examination of Pictures*, translated from the Dutch, London, G. Bell, 1929.
21. Gettens, R.J., Cross-sectioning of paint films, *Tech. Stud. Fine Arts*, 5, 18, 1938.
22. Fogg Museum, *Technical Studies in the Field of the Fine Arts,* Cambridge, Harvard University, 1932–1942.
23. Stout G.L., *The Care of Paintings*, New York, Columbia University Press, 1948.
24. Bradley, M.C., *The Treatment of Pictures*, Cambridge, MA, Art Technology, 1950.
25. *Studies in Conservation*, London, National Gallery, published quarterly since 1952.
26. Kirk, P.L., *Crime Investigation*, New York, Interscience, 1953, p. 257.
27. Keck, C.K., *How to Take Care of Your Pictures*, New York, Museum of Modern Art and Brooklyn Museum, 1954.
28. Gettens, R.J. and Upsilon, B.M., *Abstracts of Technical Studies in Art and Archaeology 1943–1952*, Washington, D.C., Freer Gallery of Art, 1955.
29. *Art and Archaeology Technical Abstracts*, New York University, 1955 through 1965.
30. Nickolls, L.C., The identification of stains of nonbiological origin, in *Methods of Forensic Science*, Vol. 1, Lundquist, F., Ed., New York, Interscience, 1962, p. 335.
31. Butler, M.H., Application of the polarized microscope in the conservation of paintings and other works of art, *Microscope*, 21, 101, 1973.
32. Bromelle and Smith, Eds. *Conservation and Restoration of Pictorial Art*, London, Butterworths, 1976.
33. Ryland, S.G. and Kopec, R.J., The evidential value of automobile paint chips, *J. Forens. Sci.*, 24, 64, 1981.
34. Laing, D.K., Dudley, R.J., Home, J.M., and Isaacs, M.D.J., The discrimination of small fragments of household gloss paint by microspectrophotometry, *Forens. Sci. Int.*, 20, 199, 1982.
35. Suzuki, E.M., Infrared spectra of U.S. automobile original topcoats (1974–1989): differentiation and identification based on acrylonitrile and ferrocyanide C ≡ N stretching absorptions, *JFSCA*, 41, 376, 1996.
36. McCrone, W.C. and Markowski, E., A protocol for authentication of paintings, *Microscope*, 47, 135, 1999.
37. Welsh, F.S., Analyzing historical wallpapers: fibers and pigments, *Microscope*, 49, 35, 2001.
38. Smith, K.J. and Barabe, J.G., Raman analysis of watercolor pigments, *Microscope*, 49, 159, 2001.

39. Chamot, E.M. and Mason, C.W., *Handbook of Chemical Microscopy*, Vol. 1, New York, John Wiley & Sons, 1930.
40. Chamot, E.M. and Mason, C.W., *Handbook of Chemical Microscopy*, Vol. 2, New York, John Wiley & Sons, 1931.
41. *The Merck Index*, 5th ed., Rahway, NJ, Merck & Co., Inc., 1940.
42. Vesce, V.C., *Classification and Microscopic Identification of Organic Pigments*, Mattiello, J.J., Ed., New York, John Wiley & Sons, 1942.
43. Fiegl, F., *Qualitative Analysis by Spot Tests: Inorganic and Organic Applications*, 3rd ed., New York, Elsevier, 1946.
44. Schaefer, H.F., *Microscopy for Chemists*, New York, Van Nostrand, 1953.
45. McCrone, W.C., *Fusion Methods in Chemical Microscopy,* New York, Interscience, 1957.
46. Bloss, F.D., *An Introduction to the Methods of Optical Crystallography*, New York, Holt, Rinehart & Winston, 1961.
47. Schneider, F.L., *Qualitative Organic Microanalysis,* New York, Academic Press, 1964.
48. Fulton, C.C., *Modern Microcrystal Tests for Drugs*, New York, Interscience, 1969.
49. Fiegl, F., *Spot Tests in Inorganic Analysis*, 6th ed., Amsterdam, Elsevier, 1972.
50. Stevens, R.E., Squaric acid: a novel reagent in chemical microscopy, *Microscope*, 22, 63, 1974.
51. Wills, W.F., Jr. and Whitman, V.L., Extended use of squaric acid as a reagent in chemical microscopy, *Microscope*, 25, 1, 1977.
52. Delly, J.G., Microchemical tests for selected cations, *Microscope*, 37, 139–166, 1989.
53. Wills, W.F., Jr., Squaric acid revisited, *Microscope*, 38, 169, 1990.
54. Jungreis, E., *Spot Test Analysis: Clinical, Environmental, Forensic, and Geochemical Applications*, 2nd ed., New York, John Wiley & Sons, 1997.
55. Schaeffer, H.F., *Microscopy for Chemists*, New York, Van Nostrand, 1953.
56. Feigl, F., *Qualitative Analysis by Spot Tests*, 3rd ed., New York, Elsevier, 1946.
57. Humecki, H.J., Ed., *Practical Guide to Infrared Microspectroscopy*, New York, Marcel Dekker, 1995.

FIGURE 10.1 Six paint specimens used in casework: (1) paint smear on a screwdriver used to enter the scene of a double homicide; (2) paint smear found in a World Trade Center Ground Zero dust specimen; (3) specks of paint from a ricocheted lead bullet found at the scene of a homicide; (4) cross-section of an architectural paint chip used in a crime scene reconstruction; (5) known automobile paint flakes used during a homicide investigation; and (6) cross-section of a paint chip removed during restoration of the Statue of Liberty.

TABLE 10.1
Common Binder Materials[3–18]

Natural Binder (Circa)	Synthetic Binder (Circa)
Natural resins: damar, amber, elemi, copal, rosin, mastic, sandarac, shellac (antiquity)	Phenolic resins (early 20th century)
Oils and waxes (antiquity)	Nitrocellulose (1920s)
Temper (1600 B.C.)	Vinyl acetate (1925)
Linseed oil (130 A.D.)	Alkyd resins (1927)
Bituminous coatings (antiquity)	Melamine–formaldehyde resins (1930s)
Animal glues (antiquity)	Urethane resins (1937)
Casein (antiquity)	Vinyl resins (1930s)
Gums (antiquity)	Acrylics (1930s)
Whitewash (1700s)	Methacrylate (1930s)
Tung oil (1900s)	Urethane (late 1930s)
Zein (1940s)	Styrene–butadiene (1940s)
Safflower oil (recent)	Silicone oils (1940s)
	Polyester (1940s)
	Polyamide resins (1950s)
	Alkyd–melamine (1950s)
	Epoxy resins (1950s)

TABLE 10.2
Common Inorganic and Natural Organic Pigments[3–23]

Color	Pigment/Mineral/Compound	Circa	Tests/Ions	Method/Reagent/Comments
White	Anhydrite — $CaSO_4$		Ca	MI/5% H_2SO_4
	Gypsum — $CaSO_4.2H_2O$		Ca	MI/oxalic Acid (A)
	Bone white[a] — $Ca_3(PO_4)_2$	Middle ages	Ca, PO_4^{-3}	MI/(PO_4^{-3}), $AgNO_3$
	Lead white — $Pb(OH)_2{\cdot}Pb(CO_3)$	8th century[c]	CO_2 (↑), Pb	MI-MIII/KI, dilute HCl (gas ↑)
	Titanium white (TiO_2)[a]		Ti	MII chromotropic A
	Whiting, limestone — $CaCO_3$[a]		Ca	MI/5% H_2SO_4 and dilute HCl (CO_2 gas ↑)
	Zinc white — ZnO		Zn	MI or MII/$K_2Hg(SCN)_4$ squaric acid (SA)
Black	Bone black[c]	Antiquity	PO_4^{-3}	MI/(PO_4^{-3}), $AgSO_4$
	Graphite[b]		C	Metallic luster Rλ, greasy texture
	Lamp black[b] (carbon soot)	Antiquity		Tiny black balls
	Mars black, magnetite — (Fe_3O_4)	Middle 19th century	Fe	Magnetic, opaque, iron filings
	Wood charcoal[c]	Antiquity		Wood cellular structures
Brown	Raw umber, goethite — Fe_2O_3 + 5–20% MnO_2, burnt umber — Calx RU[d]	Middle ages; before 15th century	Fe, Mn	MI or MII/SA, $K_4Fe(CN)_6$ or KSCN
	Raw sienna, goethite — Fe_2O_3 + <1% MnO_2, burnt sienna — Calx RS[d]	Middle ages before 15th century	Fe, Mn	MI or MII/SA, $K_4Fe(CN)_6$ or KSCN
	Van Dyke brown	Antiquity	Al	90% organic matter, aluminon (AL)
Blue/green	Cobalt Blue — $CoAL_2O_4$	1820	Co, Al	MI or MII/(Co)& (Zn) $K_2Hg(SCN)_4$, (Al) Cs_2SO_4, AL
	Cobalt Green — CoO·ZnO	1835	Co, Zn	
	Terre vert complex silicate/Fe, Al, Mg, K (mica group of minerals)	Antiquity	K, Fe, Mg, Al	MI or MII/SA, for K, Mg, UA for Mg, glauconite, celadonite
	Egyptian blue/Pompeian blue, calcium copper silicate	Antiquity	Ca, Cu	MI/(Ca) 5% H_2SO_4, (Cu) SA and $K_2Hg(SCN)_4$
	Cerulean blue, $CoO{\cdot}nSnO_2$	1870	Co, Sn	MI/(Co) $K_2Hg(SCN)_4$, metal/Zn
	Maya blue (blue-green)	600 A.D.	Al	MI clay spherlites close to Melt Mount

TABLE 10.2 (CONTINUED)
Common Inorganic and Natural Organic Pigments[3–23]

Color	Pigment/Mineral/Compound	Circa	Tests/Ions	Method/Reagent/Comments
	Azurite blue/malachite green $2CuCO_3.Cu(OH)_2$ 2 or 3 for malachite	Antiquity	Cu	(Cu) SA and $K_2Hg(SCN)_4$, any test for Cu
	Verdigris — $CuC_2H_3O_2{\cdot}2Cu(OH)_2$	Antiquity	Cu	Test for Cu, $C_2H_3O_2^-$, pleochroic
	Verditer $2CuCO_3{\cdot}Cu(OH)_2$ (artificial)	Middle ages	Cu	Any test for Cu
	Manganese blue — $BaMn_4$ + $BaSO_4$	1935	Ba, Mn	MI, $K_2Hg(SCN)_4$ for (Mn) SA and 5% H_2SO_4 for Ba
	Smalt (cobalt glass)	Antiquity	Co	Fuse, dissolve, MI $K_2Hg(SCN)_4$
	Chrome oxide — Cr_2O_3	1797	Cr	Fuse, MI, test Cr with $AgNO_3$
	Viridian $Cr_2O_3.2H_20$	1838	Cr	Fuse, MI, Test Cr with $AgNO_3$
	Lapis lazuli/ultramarine blue (natural)	Antiquity	Na, Ca, Al	Fuse, dissolve, MI test for Na
	Ultramarine blue (synthetic)	1826	Na, Al	Uranyl Acetate (UA), ZnUA, SA For Ca or Al, pyrite in natural
	Indigo (vegetable dye)	Antiquity	Organic	Sublimates, fine particle <1 µm
	Prussian blue	1704	Fe	MII SA, replacement with Zn dust
	Phthalocyanine blue or green	1935	Organic	Sublimation/hanging drop H_2SO_4
	Sap green (organic)	Middle ages	Organic	Dye from ripened buckthorn berries
Yellow	Saffron crocus (golden yellow)	Roman epoch	Organic	Plant cellular structures
	Orpiment — As_2S_3 (natural mineral)	Antiquity	As	MI, As with silver nitrate
	Realgar — As_2S_2 (natural mineral)	Antiquity	As	MI, As with silver nitrate
	Indian yellow Mg Lake euxanthic acid	India	Mg	MI, test for Mg with UA
	Cadmium yellow — CdS	1829	Cd	MI, Cd with $K_2Hg(SCN)_4$ and CsCl
	Litharge/massicot — PbO	Antiquity	Pb	MII, Pb with KI and $K_2Hg(SCN)_4$
	Gamboge	15th century	Organic	Tree gum resin soluble in alcohol
	Yellow ochre $Fe_2O_3{\cdot}xH_20$	Antiquity	Fe	MII SA or with KSCN
	Naples yellow ($Pb_3[SbO_4]_2$)	Antiquity	Pb, Sb	MI, Pb with KI, Sb with CsCl
	Zinc Yellow — $ZnCrO_4$	1850	Zn, Cr	Test for Zn and Cr
Purple	Mauve	1856	Organic	Reddish violet crystals
	Cochineal carminic acid	1523	Organic	Soluble in Melt Mount
	Mars violet	Mid-19th century	Fe	Artificial ochres, test for Fe
	Cobalt violet	1860	Co, PO_4	Test for Co and PO_4
Red	Alizarin	1868	Organic	Turns purple in dilute NaOH
	Rose madder	Antiquity	Organic	Turns purple in dilute NaOH; fluoresces yellow-red with purpurin
	Dragon's blood	Greek epoch	Organic	Soluble in alcohol
	Red lead — Pb_3O_4	Antiquity	Pb	Test for Pb
	Cadmium red	1910	Cd, S, Se	Test for Cd
	Cadmium red + $BaSO_4$, Lithopone	1926	Cd, S, Se	Test for Cd and Ba
	Hematite — Fe_2O_3 (mineral)	Middle ages	Fe	Test for Fe
	Red Ochre — Fe_2O_3	Antiquity	Fe	Test for Fe
	Cinnabar — HgS (mineral)	Antiquity	Hg, S	MI, test for Hg with $AgNO_3$
	Vermilion — HgS (synthetic)	8th and 9th centuries	Hg, S	MI, test for Hg with $AgNO_3$
Metallic	Gold — Au, Silver — Ag, Copper — Cu, Tin — SN, Iron — Fe	Antiquity	Au, Ag, Cu, Sn, Fe	Replacement of metal ions with zinc (Zn) dust

[a] Ground.

[b] Elemental.

[c] Charcoal.

[d] Calcined (heated or roasted) to calx.

Note: See references 39 to 56, Chapter 4, and Appendix D for testing procedures.

TABLE 10.3
Common Extenders and Additives

Extender	Compound	Circa
Barytes or blanc fixe	$BaSO_4$	Early19th century
Bentonite	$Al_2O_3{\cdot}4SiO_2{\cdot}2H_2O$	—
China clay/kaolinite	$Al_2SiO_5(OH)_4$	Antiquity
Diatomaceous or infusorial earth	SiO_2	—
Chalk, limestone	$CaCO_3$	8th century
Quartz	SiO_2	—
Anhydrite	$CaSO_4$	—
Micas	Complex silicates	—
Talc	$MgO{\cdot}SiO_2{\cdot}2H_2O$	—
Wollastonite	$CaSiO_3$	Late 20th century
Glass beads	Silicone dioxide	Late 20th century
Ceramic spheres	Clays	Late 20th century
Plastic specks	Melamine	Late 20th century
Polypropylene powder	Olefin	Late 20th century
Tire rubber powder	Neoprene	Late 20th century
Waxes	Beeswax, carnauba	Antiquity
Gum arabic	Resin	Antiquity
Bitumen	Organic coal tars	Antiquity

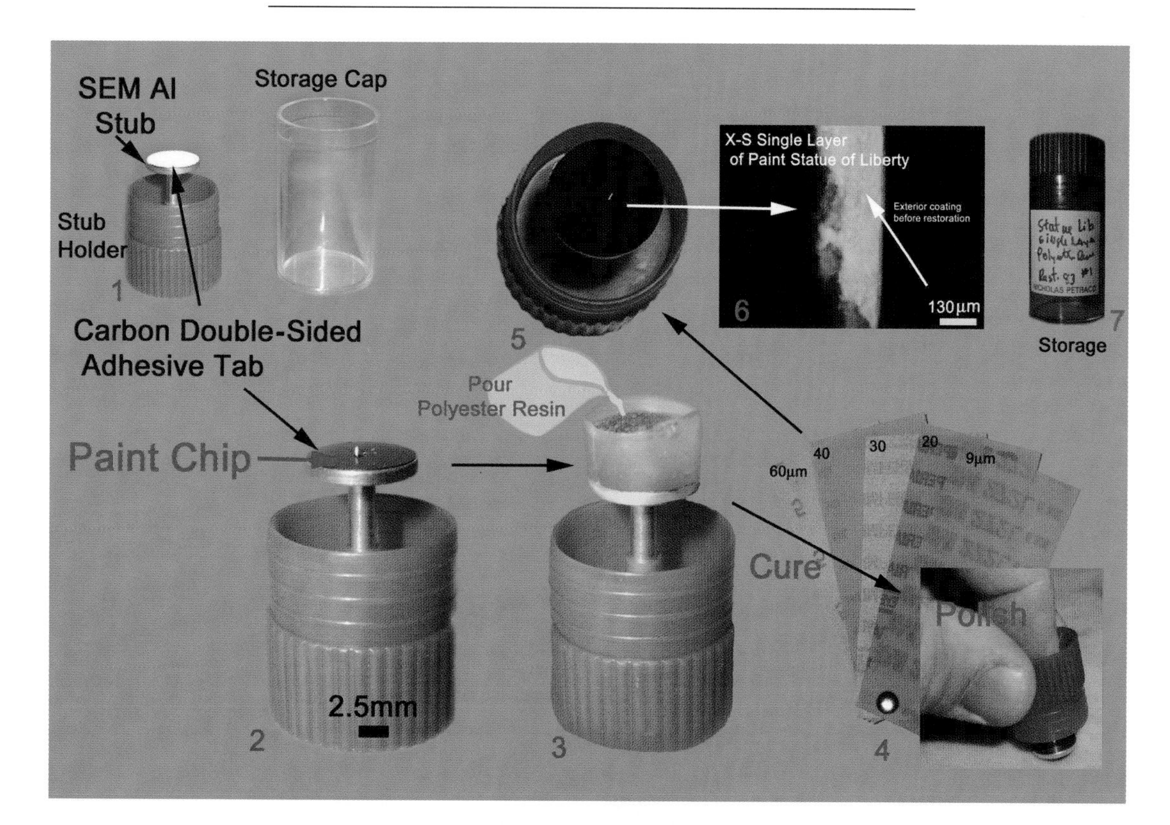

FIGURE 10.2 Cross-sectioning of minute paint flakes: (1) A 1-cm-diameter aluminum SEM stub with a carbon adhesive tab attached is placed into a stub holder; (2) the protective white cover of the carbon tab is removed and the paint flake is placed onto the stub (use stereomicroscope); (3) a piece of cellophane tape is applied around the aluminum stub to form a well; 1 to 2 ml of the desired casting resin is poured into the well and allowed to cure; and (4) the preparation is polished with a series of 3M® wet/dry abrasive papers (60, 40, 30, 20, and 9 μm) and a gem polishing cloth; (5) the specimen is viewed with a stereomicroscope; (6) examined; and (7) secured for storage or subsequent use.

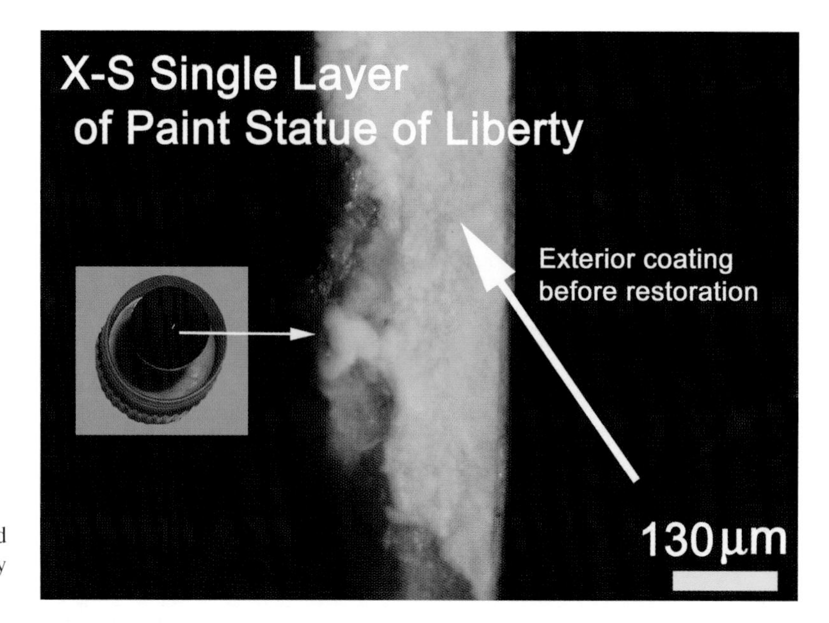

FIGURE 10.3 Single-layer paint chip removed from the exterior coating of the Statue of Liberty during restoration.

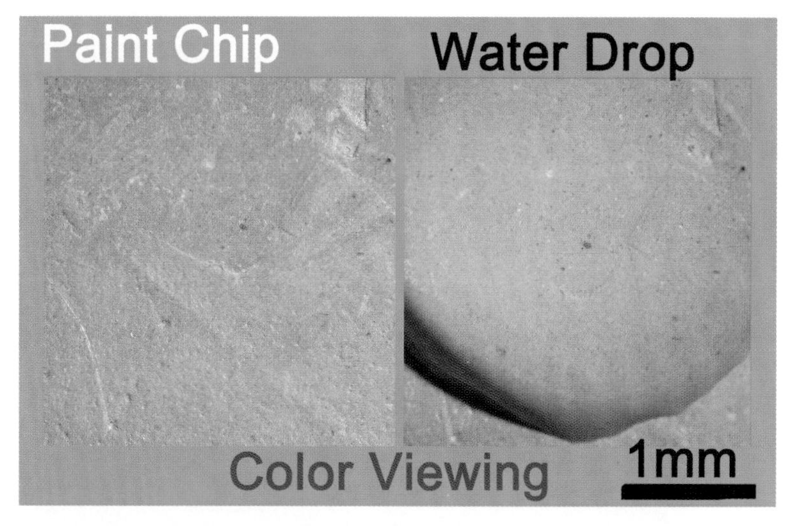

FIGURE 10.4 Left: paint flake without water droplet. Right: paint flake with water droplet. This procedure is also useful when comparing a specimen to Munsell color chips.

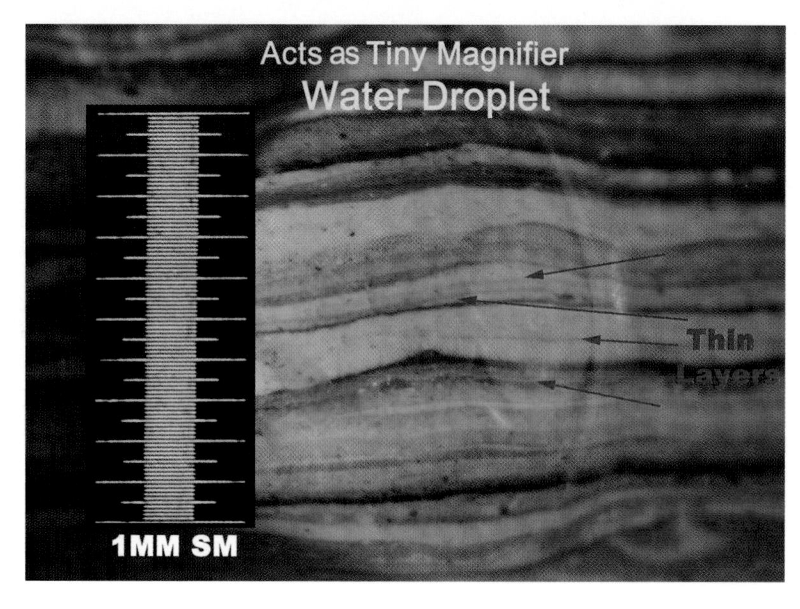

FIGURE 10.5 Cross-sectional preparation of paint chip made with polyester resin and discussed in Chapter 2. A tiny water droplet has been added to aid in layer sequencing and measuring layer thickness. Inset: a 1-mm stage micrometer with 100 equal divisions; each division equals 10 μm.

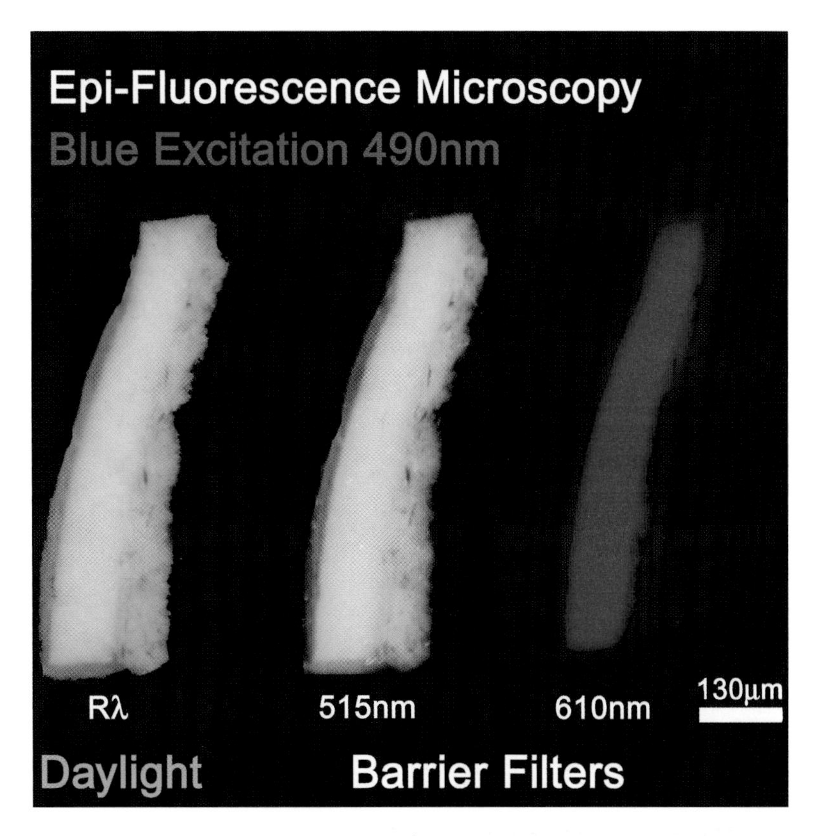

FIGURE 10.6 Left: white layer of paint under reflected light (Rλ). Center: same layer with epifluorescence illumination, 490-nm excitation source, and 515-nm barrier filter. Right: appearance with 610-nm barrier filter.

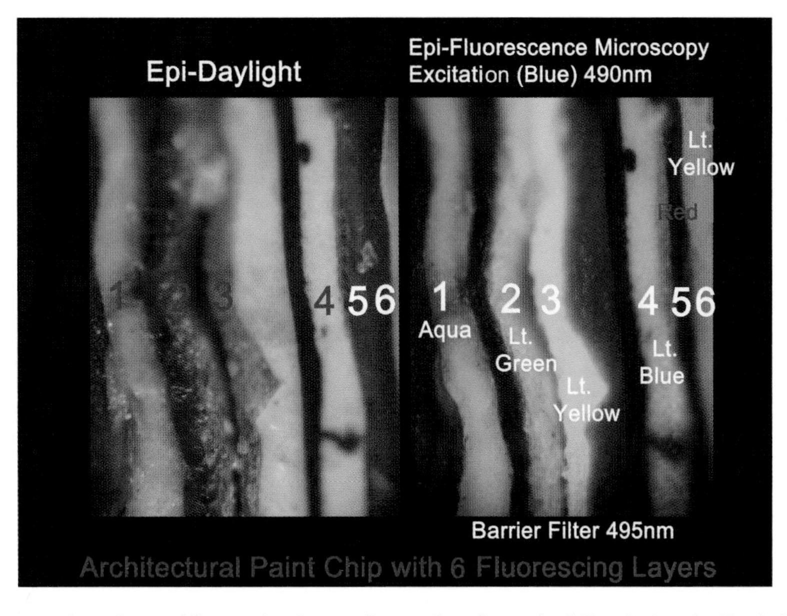

FIGURE 10.7 Left: cross-section of an architectural paint specimen taken from a building during the South Street Seaport restoration in lower Manhattan, viewed under Rλ. Right: same cross-section viewed with epifluorescence illumination, a 490-nm λ excitation source, and a 495-nm barrier filter. Note the six fluorescing layers. This information is very useful for determining origins of questioned and known paint chips, and identifying paint components.

TABLE 10.4
Paint and Pigment Data Sheet

File or Case No. ____________________

Specimen Source ____________________ Sample No. ____________________

Color: transmitted light ____________________ reflected light ____________________

Cross-section layer sequence ____________________

Layer thickness (μm) ____________________

Appearance in Melt Mount 1.539 or high-dispersion oil ____________________

Appearance at other RI levels (i.e., 1.66) ____________________

Degree of relief displayed by specimen in ____________________

Becke line appearance and movement ____________________

Pigment, extender or binder appearance ____________________

in transmitted light ____________________ reflected light ____________________

other ____________________

Transparency: transparent ________ opalescent ________ opaque ________

Crystalline shape ____________________

Sketch:

Crystalline type: system __________ amorphous ________ habit __________

twinning ____________________ lamellar ____________________

Surface texture or appearance ____________________

Particle size (μm) ____________________ Thickness (μm) ____________________

Particle distribution ____________________

Pleochroic color(s) ____________________

Crossed polars: isotropic ____________________ anisotropic ____________________

Extinction: parallel _____ symmetrical _____ oblique _____ undulose _____

Estimated retardation (nm) __________ Estimated birefringence __________

Solubility ____________________

FIGURE 10.8 Utilization of IMS in the analysis of minute paint and pigment specimens. A minute aliquot of specimen is removed, placed onto a diamond cell, and flattened. The diamond cell is then removed from its holder and placed on the microscope stage. The sample is viewed with the microscope and aligned with the sampling window and a spectrum is collected.

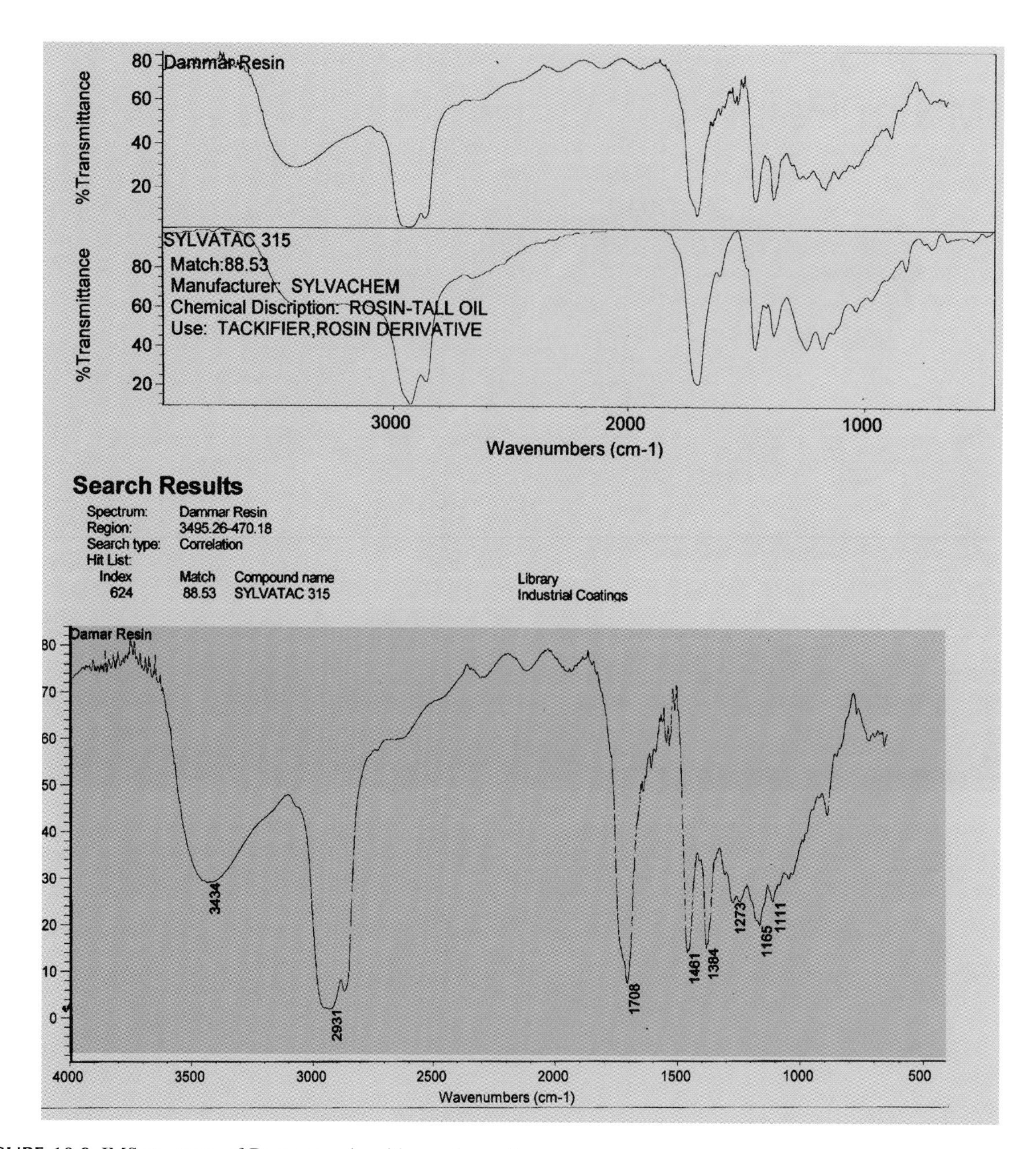

FIGURE 10.9 IMS spectrum of Dammar resin with search results.

11 Soil and Mineral Examination

The formation of soil is a long and complex process. Soil generally is composed of all the unconsolidated materials above the bedrock of a given geological region.[1] Thus, its intrinsically complex nature provides means for its characterization, classification, and comparison. While the nature of soil is complex, most soil samples contain only four or five minerals combined with vegetation and miscellaneous debris. Murray, et al. point out that although thousands of minerals exist in nature, only about twenty are common in soil specimens and only three to five of the twenty are found in a given soil sample.[2]

If an analytical method allowed an analyst to identify and quantify the 20 most prevalent mineral species found in soil on sight in a few seconds, the comparison of forensic soil specimens would be much easier and a good deal more significant. A method that meets these criteria was published by W.J. Graves, a Canadian forensic geologist.[3]

Although the identification and comparison of the ratios of all the materials present in soil has been of longstanding interest to the forensic science community, many forensic laboratories refuse to do soil casework.[4–8] In the authors' opinions, the primary reason for the lack of widespread use of this excellent procedure is the erroneous belief that soil, because of its ubiquitous consistency within a given geologic region, is not a valuable form of associative physical evidence. A recent article on the usefulness of close-proximity soil evidence concluded that soil samples can yield viable information for identification and possible individualization in forensic soil casework.[9] The authors have used this method with slight modifications in soil cases for over two decades with great success.

METHOD

The method developed and published by Graves was adopted with certain modifications and published by N. Petraco.[8] The primary changes are the substitution of Cargille's Melt Mount (MM) ND 1.539 at 25°C for Cargille refractive index oil ND 1.540 at 25°C, examination of a narrower sieve-fraction range, use of a smaller cover glass, replacement of a cross-hair micrometer with a Becket disc, and modification of the tabulation sheet to suit the user's needs.

Initial Observation

Each soil specimen is opened separately, weighed accurately, and examined visually in daylight and with a stereomicroscope. The overall wet or dry color observed after drying in an oven at 100°C of each sample is noted. All the data are recorded on a data tabulation sheet (Figure 11.1). Artifacts such as fibers, hairs, paint chips, glass fragments, and so on are removed after documentation; each artifact is sorted, placed into a container and stored. The artifacts are examined carefully, classified, and preserved for individualization of questioned and known soil specimens.

As suggested by Graves, a representative sample (50 to 100 mg) of each soil sample is obtained and sifted with a set of stainless steel sieves.[8] If available, a microsample splitter should be used to obtain a nonbiased aliquot. The authors recommend a sieve mesh (M) range as follows: 40, 80, 120, 140, 170, and 200 M. Separate sets of sieves must be available for processing different questioned and known samples. The soil fraction trapped between the 170 and 200 mesh sieves is weighed and divided into three equal aliquots: one for color matching, one for PLM examination, and one for density gradient comparison (optional). Depending on the soil type or grain size, different sieve fractions may be selected for examinations.

Density gradients, if prepared, are made with Clerici's reagent (CR) and distilled water (DS). Clerici's reagent is an aqueous, saturated solution of a thallium–malonate–formate (TMF) salt in distilled water.[22] Each gradient is composed of 11 layers starting with undiluted saturated CR, followed by a 9:1 mixture of CR/DS, subsequent CR/DS mixtures of 8:2, 7:3, 6:4, 5:5, 4:6, 3:7, 2:8, and 1:9 ratios, and finally pure DS. The density range of a gradient prepared in this manner has been determined (using standard mineral chips) to be 4.05 g/ml to 1.00 g/ml at 20°C. Another way to determine the density of each layer composing each gradient is to use the following formula* (ambient temperature must be taken into account):

$$\text{Density of Mixture (DM)} = \frac{D_{CR}V_{CR} + D_{DS}V_{DS}}{V_{CR} + V_{DS}}$$

where D_{CR} = density of Clerici's solution, V_{CR} = volume of Clerici's solution, D_{DS} = density of distilled water, and V_{DS} = volume of distilled water.

Figure 11.2 depicts a CR/DS density gradient comparison of known and questioned soil specimens recovered in a criminal investigation. The remaining sieve fractions are weighed, packaged, and stored for possible examination. The weights are recorded on a data sheet (Figure 11.1).

* Adopted from Kirk, P., *Density and Refractive Index*, 1951.

The following tests are conducted for each questioned and known specimen.

Color Matching

Isolate the aliquot for color matching in its own soil sample cell. Determine its color by using a Munsell® soil color chart and color system as described in References 11 and 17 and shown in Figure 11.3. Record all findings on the data sheet.

Mounting in Melt Mount 1.539

Place a standard 1 × 3-in. glass microscope slide on a hotplate and warm. Two to three drops of warmed (65 to 70°C) Melt Mount (MM) 1.539 ND at 25°C are placed on one end of the heated slide. Remove the slide from the hotplate. Spread a thin, even layer of MM over the end of the microscope slide with another glass slide and allow the microscope slide to cool to room temperature.

A 1- to 2-mg portion of each homogeneous aliquot of the fraction trapped between the 170- to 200-M sieves is carefully weighed out on an analytical balance. The entire aliquot is spread evenly over the cooled MM. A 25-mm square No.1 1/2 cover glass is then placed on top of the MM soil preparation. The microscope slide is then placed on the warm hotplate and a 5-gm weight is placed on top of the cover slip.

As the MM liquefies, the mineral grains will be sandwiched between the microscope slide and the cover glass, thus embedding the mineral grains and other particulates in the MM. Upon cooling, a permanent mount ranging in thickness from around 74 to 88 μm of each specimen will be available for PLM examination. If it is necessary to know the exact thickness of the specimen along the microscope's optic axis, a more precise technique using the micrometer located on the fine-adjustment focusing knob found on most high-quality polarized light microscopes is described in Chapter 3.

The thickness value is important because it will be used with the retardation approximation to estimate the birefringence (EBi) of the unknown grain, which will in turn be used to help identify the unknown mineral grain. If possible, at least three slides of each sample should be prepared and made available for PLM examination. A different clean sable-haired brush must be used for each specimen. Figure 11.4 is a flow chart showing common minerals and related materials seen in casework soil specimens.

Mineral Identification

When using PLM for the identification and comparison of mineral grains in soil specimens, the ability to ascertain optical properties such as birefringence (Bi), relief (degree of shadowing), refractive index (RI), dispersion of light, Becké line (a white halo around the edge of the particle if polychromatic light is used), movement, and other parameters is vital to a successful analysis.[12] Explanations of these phenomena can be found in Chapters 1 and 3.

Each mineral grain in a soil specimen is identified on the basis of physical appearance and optical properties. The estimated birefringence is determined by comparing the retardation colors exhibited when observed between crossed polars and thickness measurement with data on a Michel–Lévy interference color chart (Figure 11.5). Various types of fixed and variable compensators and monochromatic interference filters are available to help determine the precise retardation color exhibited by each mineral present in a soil specimen. Accurate retardation data can be used with mineral thickness data to estimate birefringence.

The importance of correct thickness and retardation assessments cannot be overstated. The identity of each mineral grain is determined by referring to the literature, a set of mineral standards mounted in MM 1.539, published atlases, and interference charts. Other optical and morphological features such as color, pleochroism, extinction angle, sign of elongation, interference figures, shape, crystal system, crystal habit, twinning, conchoidal fractures, cleavage, interfacial angles, and chemical composition can be used to identify each grain in a soil specimen.[3,13–15] Several common minerals that exhibit these features are depicted in the mineral atlas at the end of this chapter.

Other types of materials in addition to mineral grains are often found in soil samples, for example, peat, plant parts, insects and insect parts, glass fragments, starch grains, diatoms, feathers, hairs, and fibers, to name only a few. These materials can be very useful in soil analysis and comparison. If positive identification of a mineral grain is required, the grain should be isolated and identified via a dispersion staining scheme developed for mineralogical soil analysis or the spindle-stage method.[19, 20]

Serious readers should become familiar with the optical and morphological natures of these substances (Table 11.1). With a little work and practice, a microscopist should be able to identify each particle encountered after a few seconds of examination with a PLM.

Counting Procedure

The authors routinely use a polarized light microscope fitted with a Whipple micrometer disc in its right 10× ocular in combination with a plain 4× strain-free achromatic objective to perform quantitative particle counts. The total magnification is varied from 40 to 500× as needed.

The specimen slide is placed on a rotating circular stage set on 0°. The slide can be held on the microscope stage with clips, a low-profile mechanical stage, or by hand. The soil specimen mounted in MM is viewed simultaneously with both transmitted light illumination, crossed polars with the analyzer set 10° off extinct, and top illumination provided by a standard stereomicroscope incident illuminator. Several counts of distinct portions of the specimen slide are made. The microscope slide is moved from right to left. Each examined section is marked to avoid duplication. All particles within the Whipple disc boundary are identified by eye and counted before proceeding to the next field of observation. At least 300 to 500 particles are counted on each slide and counts are recorded on the data sheet. Three slides of each soil sample are examined for a total particle count range of 900 to 1500 particles per soil sample. A separate counting sheet is prepared for each slide counted. The microscope settings and illuminations are varied frequently as needed for identification and counting purposes. The percentage of each particle class present in a given soil specimen is computed from the collected data. The results are compared for both questioned and known samples.

DISCUSSION

After pertinent data for each questioned and known soil specimen are obtained and recorded, the collected data are used to make the final comparisons and interpretations. The likelihood of common origin is assessed for each questioned and known specimen. Information such as wet and dry color, particle size distribution, mineral profile, artifact profile, and so on is used in each evaluation and an accurate soil analysis data sheet should be prepared for each specimen. The data sheets can be useful when preparing or presenting court exhibits and testimony.

REFERENCES

1. Jenny, H., *Factors of Soil Formation: A System of Quantitative Pedology*, New York, McGraw-Hill, 1941.
2. Murray, R.C. and Tedrow, J.C.F., *Forensic Geology*, New Brunswick, NJ, Rutgers University Press, 1975, p. 35.
3. Graves, W.J., Mineralogical soil classification technique for the forensic scientist, *JFSCA*, 24, 323, 1979.
4. Murray, R.C. and Tedrow, J.C.F., *Forensic Geology*, New Brunswick, NJ, Rutgers University Press, 1975.
5. Dudley, R.J., Particle size analysis of soils and its use in forensic science, *J. Forens. Sci. Soc.*, 16, 219, 1976.
6. McCrone, W.C., Particle analysis in the crime laboratory, in *The Particle Atlas*, Ann Arbor , MI, Ann Arbor Science Publishers, 1979, p. 1379.
7. McCrone, W.C., Soil comparison and identification of constituents, *Microscope*, 30, 17, 1982.
8. Petraco, N., Microscopic analysis of mineral grains in forensic soil analysis, *ALBYBL*, Part 1, April 1994; Part 2, September 1994.
9. Junger, E.P., Assessing the unique characteristics of close-proximity soil samples: just how useful is soil evidence? *JFSCA*, 41, 27, 1996.
10. Murray, R.C. and Tedrow, J.C.F., *Forensic Geology*, 2nd ed., Englewood Cliffs, NJ, Prentice-Hall, 1992.
11. Antoci, P.R. and Petraco, N., Technique for comparing soil colors in the forensic laboratory, *JFSCA*, 38, 437, 1993.
12. McCrone, W.C., McCrone, L.B., and Delly, J.G., *Polarized Light Microscopy*, Ann Arbor, MI, Ann Arbor Science Publishers, 1978.
13. Allen, R.M., *Practical Refractometry by Means of the Microscope*, Cedar Grove, NJ, Cargille Laboratories, 1954.
14. Bloss, F.D., *An Introduction to the Methods of Optical Crystallography*, New York, Holt, Rinehart & Winston, 1961.
15. McCrone, W.C. and Delly, J.G., *The Particle Atlas*, Ann Arbor, MI, Ann Arbor Science Publishers, 1973, p. 2.
16. Fleischer, M., Wilcox, R.E., and Matzko, J.J., Microscopic Determination of Nonopaque Minerals, U.S. Geological Survey Bulletin 1627, Washington, D.C., U.S. Government Printing Office, 1984.
17. *Munsell Soil Color Charts and Book of Color*, Baltimore, Munsell Color Company, 1954.
18. Chamot, E.M. and Mason, C.W., *Handbook of Chemical Microscopy*, 2nd ed., Vol. 2, New York, John Wiley & Sons, 1957.
19. Fraysier, H.D. and Van Hoven, H., A simple mineralogical soil analysis method using dispersion staining, *Microscope*, 40, 107, 1992.
20. Bloss, F.D., *The Spindle Stage: Principles and Practice*, London, Cambridge University Press, 1981.
21. Bates, R.L. and Jackson, J.A., Eds., *Dictionary of Geological Terms*, 3rd ed., New York, Doubleday, 1984.
22. Mange, M.A. and Maurer, H.F.W., *Heavy Minerals in Colour*, London, Chapman & Hall, 1992.
23. Petraco, N. and Kubic, T., A density gradient technique for use in forensic soil analysis, *JFSCA*, 45, 872, 2000.

Soil Data Tabulation Sheet[1,3–8,10,12,17,19,21,22]

File no. ______________________ Date ______________________

Type of investigation______________________________________

Soil sample no. __________ Aliquot no. __________ Slide no. __________

Sample condition ____

Overall soil color: wet ______________ dry ______________

Substances removed ____

Sieve fraction weight: 40 M ____ 80 M _____ 120 M _____ 140 M _____ 170 M _____ 200 M_____

Munsell color system: hue (H) ______ value (V)________ chroma (C)_____ Final: H V/C[11,17]________

Wet ______________________ Dry______________________

FIGURE 11.1 Soil analysis tabulation and data sheet customized for the geographical region where the author works and lives.

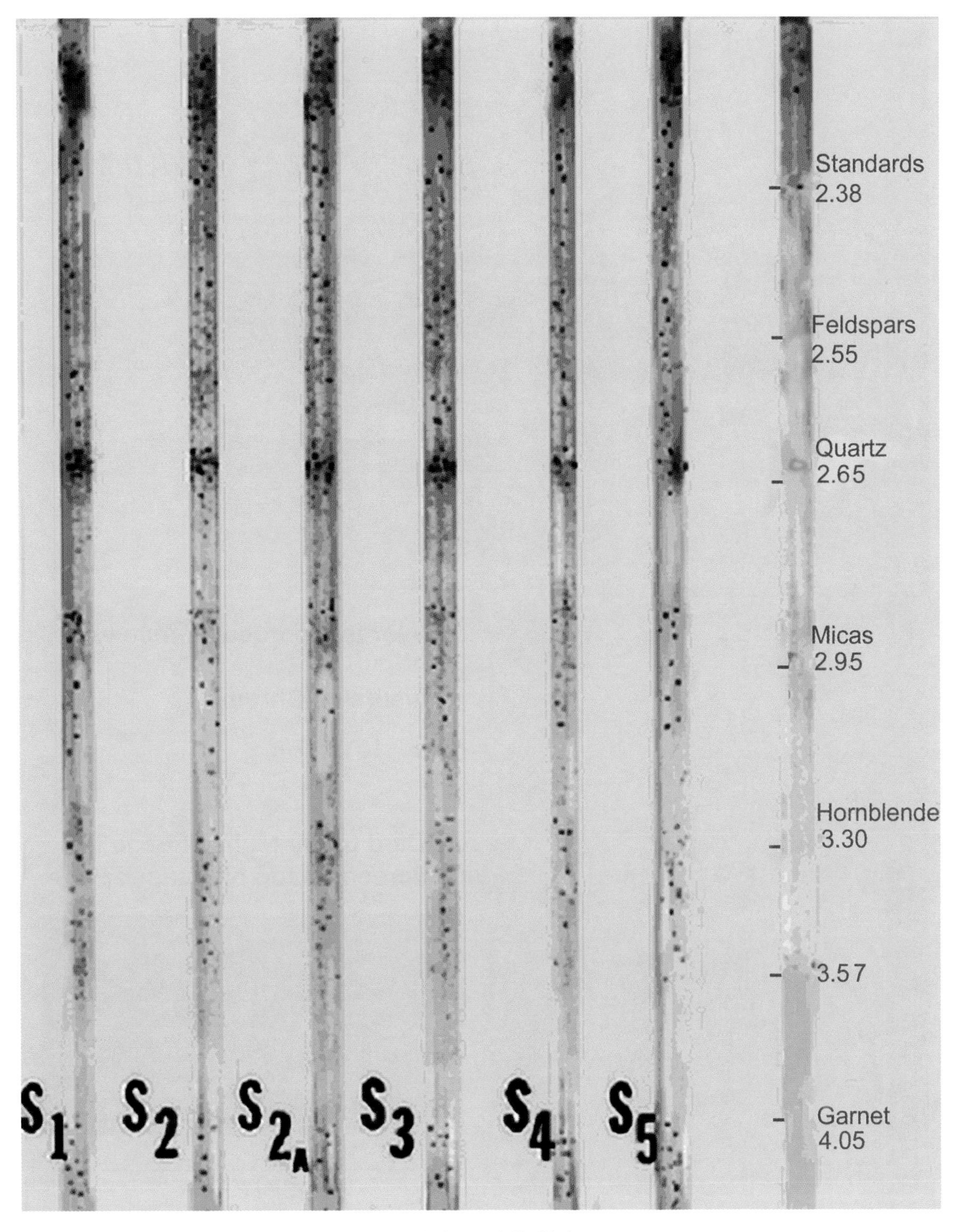

FIGURE 11.2 Density gradient prepared with Clerci's solution and distilled water.

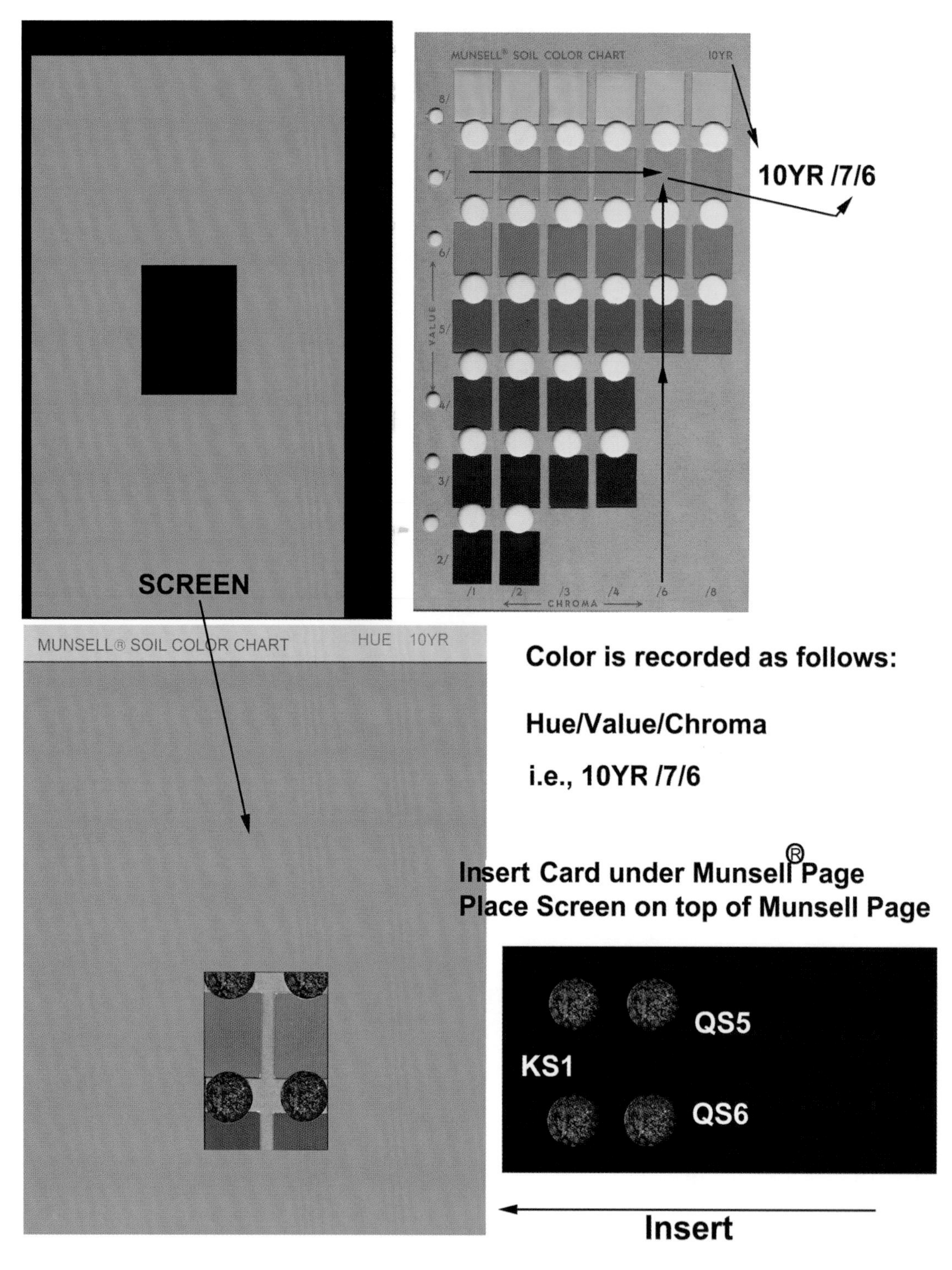

FIGURE 11.3 Part of a Munsell soil color chart system.

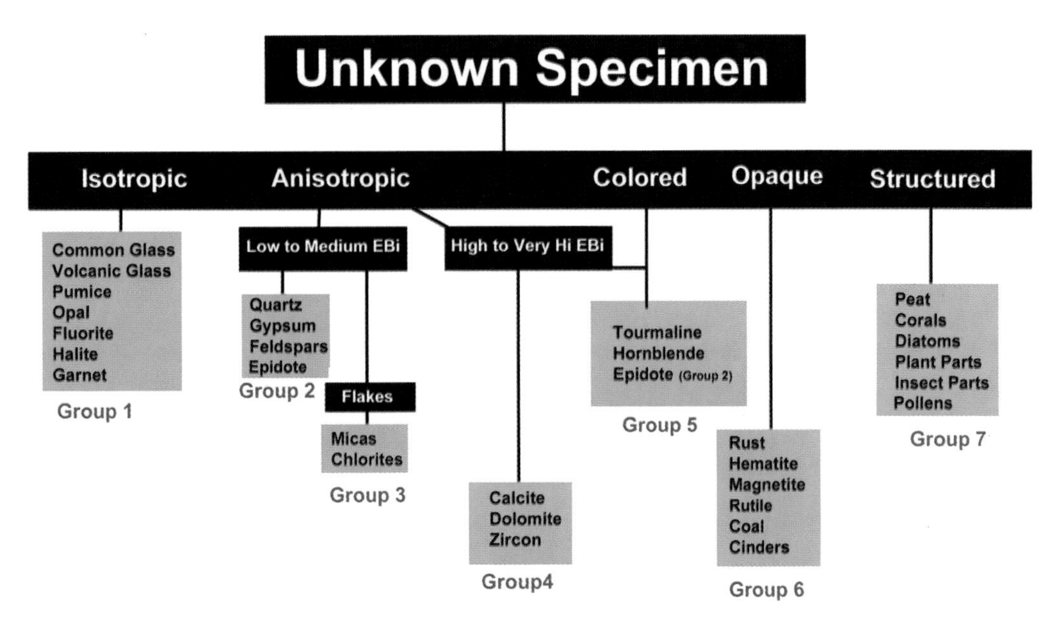

FIGURE 11.4 Flow chart for PLM identification of common minerals and related substances in soil specimens. See Table 11.1 and the mineral atlas in this chapter for descriptions and depictions of the minerals and related materials as they appear when mounted in MM 1.539.

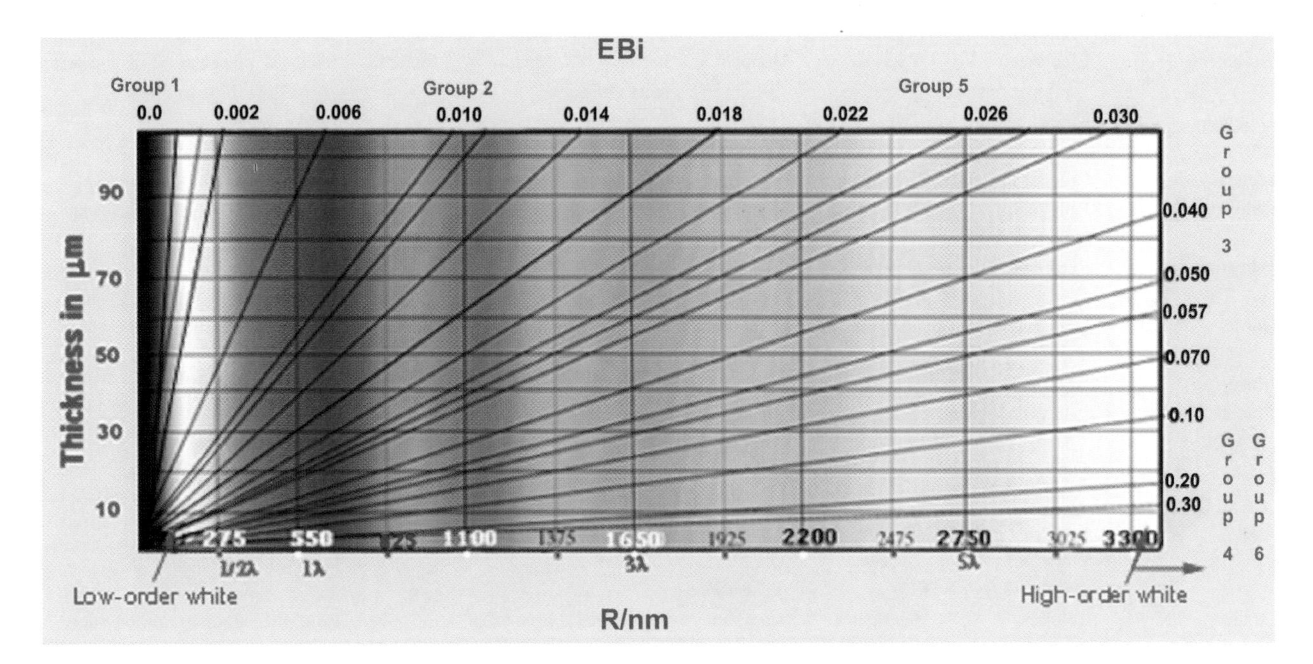

FIGURE 11.5 Michel–Lévy interference chart for rapid identification of a mineral by estimating its birefringence (EBi). See Figure 11.4 for group identification, and Table 11.1 and Figures 11.6 through 11.14 for identifying mineral characteristics.

TABLE 11.1
Minerals and Related Materials Encountered in Soil Specimens[1,3,4–8,10,12,17,19,21,22]

Specimen	Color	Transparency	Crystal Form	Cleavage/ Fracture	Relief	Est. Bire-fringence	Remarks
Glass	Green, amber, brown, colorless	Transparent	Amorphous	Conchoidal	Very low to high	—	Sharp edges; RI range 1.510 to 0.580; poss. strain Bi
Quartz — SiO_2	Colorless	Transparent	Hexagonal	Conchoidal	Low	+0.009	RIs >MM; 1 to 2° IC; inclusion common; uniaxial +
Calcite — $CaCO_3$	Colorless	Transparent	Triagonal	Perfect, rhombic	Very high	+0.172	Rhombic crystals; Hi IC like dolomite
Gypsum/plaster of Paris $CaSO_4 \cdot 2H_2O$	Colorless to white	Transparent to opaque	Rhombic	Perfect	Low to medium	+0.13	RIs <MM; tiny particles; surface rough
Halite — NaCl	Colorless to white	Transparent to opaque	Cubic	Perfect, cubic	Very low	—	Salt, RI = 1.544; sol. in water; RI close to MM
Feldspar	Colorless to pink	Transparent	Monoclinic or triclinic	Perfect to good	Low	+0.005 to 0.011	Twin lamellae; albite RIs ≤MM; labradorite >MM; microcline <MM; orthoclase <MM
Mica	Colorless, gray, green, yellow-brown	Transparent	Monoclinic	Perfect, plate	Medium	–0.015 to 0.046	Pleochroic; multilayered; biaxial
Garnet	Colorless to pink	Transparent	Cubic	Conchoidal	High	—	Isotropic; RI >>MM; strain Bi
Zircon — $ZrSiO_4$	Colorless to amber, pink	Transparent to opaque	Tetragonal	—	Medium to very high	0.04 to 0.065	RI >>MM; may be opaque; rounded; color varies
Tourmaline	Colorless to yellow-brown	Transparent to opaque	Triagonal	Rectangular	Medium to high	+0.02 to 0.036	Very pleochroic; colors vary (yellow, red, blue, green, pink, blue-green)
Hornblende (Ferro)	Blue-green to yellow-brown	Transparent	Monoclinic	Perfect, rectangular	Medium	+0.023	RI >MM; pleochroism strong; green to yellow; blue to green
Dolomite	Colorless, white to brown	Transparent to opaque	Triagonal	Perfect	Very high	0.184	Looks like calcite
Fluorite — CaF_2	Colorless to violet	Transparent	Isometric	Conchoidal	High	—	Color varies; RI <<<MM
Chlorites	Pale yellow to green	Transparent	Monoclinic	Flakes	Low	0.01	Pleochroic; dark gray in reflected λ; difficult to ID
Hematite	Red-orange	Opaque	Triagonal	Aggregates	Very high	0.23 to 0.28	Color varies; sol. in HCl; nonmagnetic
Rutile	Orange to red	Near opaque	Tetragonal	Near perfect	Very high	0.28	Color varies; yellow to brown
Magnetite	Gray-black	Opaque	Cubic	Uneven	—	—	Magnetic; metallic luster
Epidote	Pale yellow-green	Transparent	Monoclinic	Perfect	Low to medium	0.014	Rough surface; inclusions; RI >>MM
Pyrite — FeS_2	Yellow	Opaque	Cubic	Conchoidal to uneven	—	—	Fools' gold; sol. in HNO_3
Coal	Black	Opaque	—	Conchoidal	—	—	Lustrous; brittle
Cinders	White, brown, black	Opaque	Glassy appearance	Globular	—	—	Brittle; hard; clusters
Pumice/perlite	Colorless	Transparent	Amorphous	Conchoidal	Low to medium	—	RI <MM, strain Bi.; air bubbles; perlite; heated pumice
Obsidian	Varies	Transparent	Amorphous	Conchoidal	Low to medium	—	RI <MM; strain Bi.; air bubbles; volcanic glass
Diatoms, corals, shells	Colorless to black	Opaque	—	—	—	—	Fine structures; varied forms; corals vary in color: white, pink, red, black
Peat	Green, brown, yellow	Transparent to opaque	Fragments	—	Medium	Low	Cellular structure; plant fibers; wood fragments

Note: Specimens described as they appear in Melt Mount 1.539.

TABLE 11.2

Mineral/ Specimen	Color	EBi	Relief	Remarks	Specimen Number						
					1	2	3	4	5	6	7
Quartz											
Feldspar											
Calcite											
Dolomite											
Garnet											
Fluorite											
Halite											
Muscovite											
Vermiculite											
Biotite											
Phlogopite											
Gypsum											
Tourmaline											
Epidote											
Chlorite											
Rutile											
Limonite											
Zircon											
Hematite											
Magnetite											
Obsidian											
Pumice											
Opal											
Hornblende											
Cinders											
Coal											
Glass chips											
Diatoms											
Shells, corals											
Peat											
Pollen											
Seeds											
Plant parts											
Others											
Totals											

MINERAL ATLAS

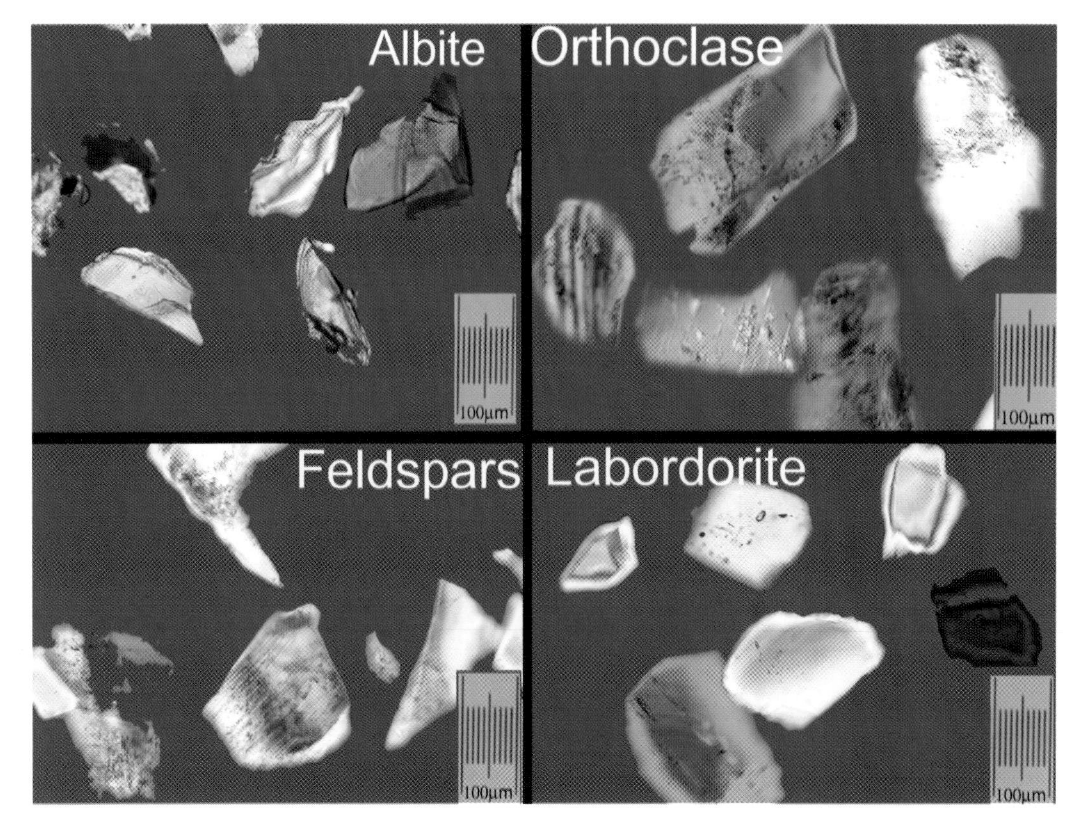

FIGURE 11.6 Common feldspars as they appear in MM 1.539, between crossed polars, at 10° off extinction.

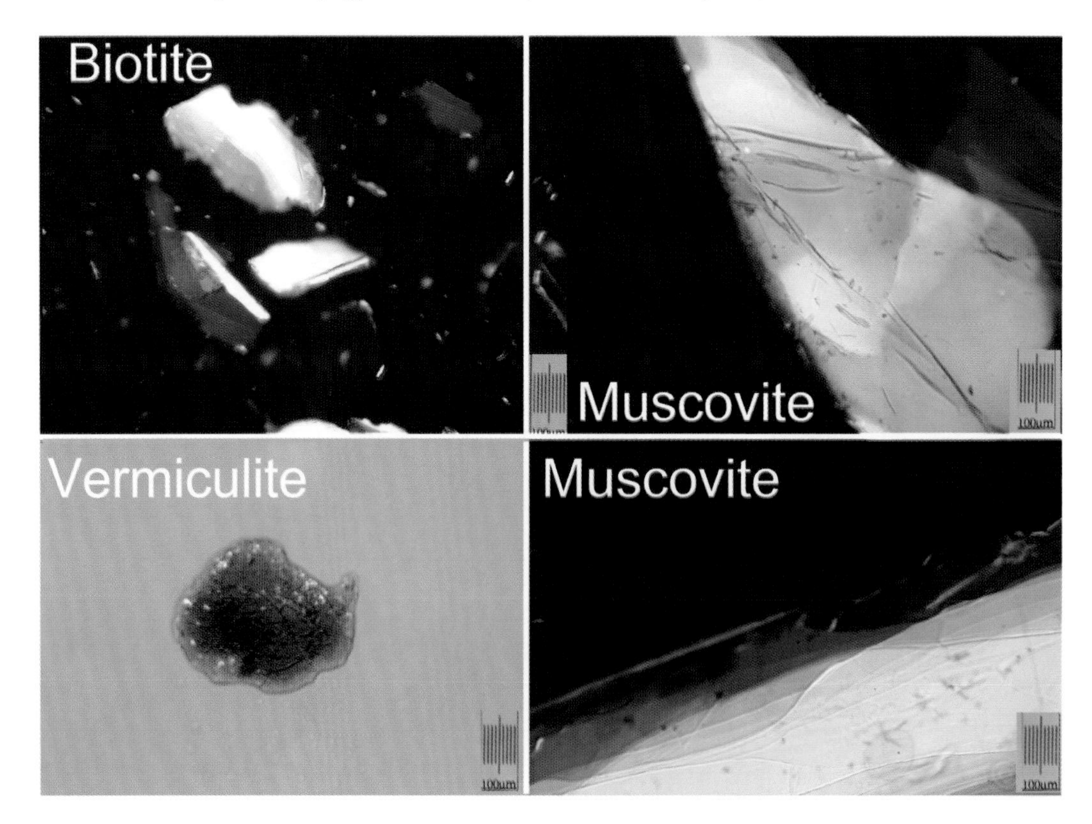

FIGURE 11.7 Common micas as they appear in MM 1.539 between crossed polars; vermiculite was photographed in plane-polarized light.

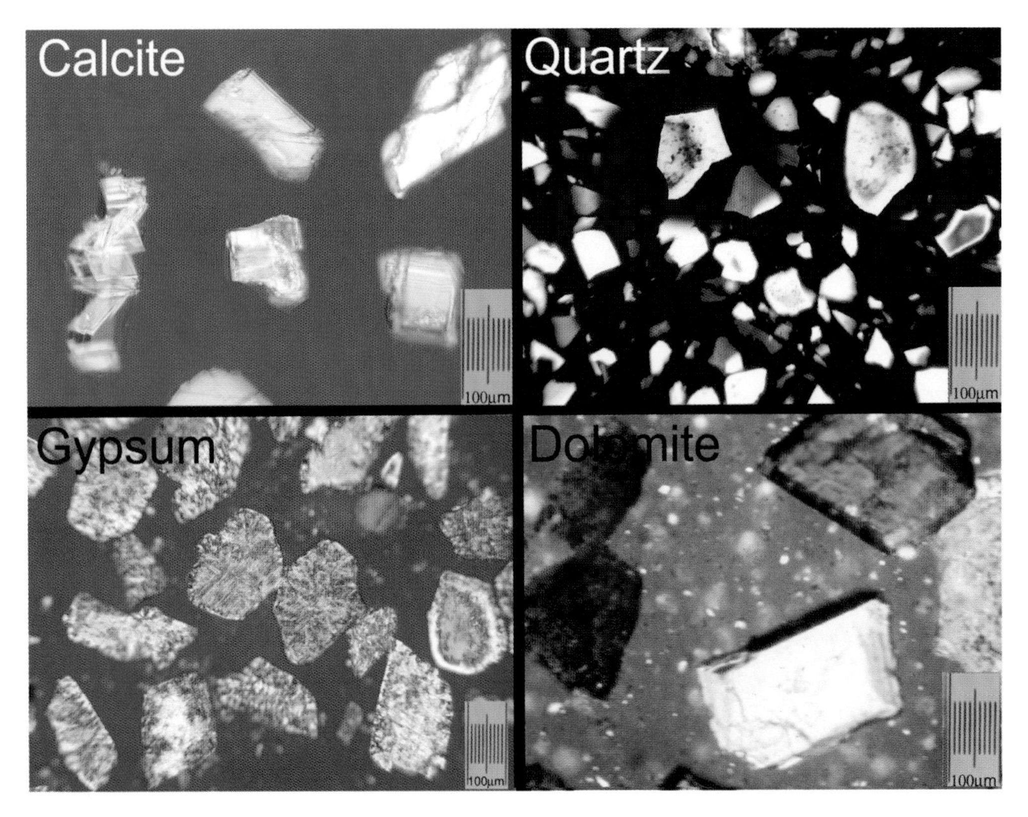

FIGURE 11.8 Common minerals as they appear in MM 1.539; quartz between crossed polars; calcite and gypsum at 10° off extinction; and dolomite between crossed polars with a first-order (550-nm) compensator.

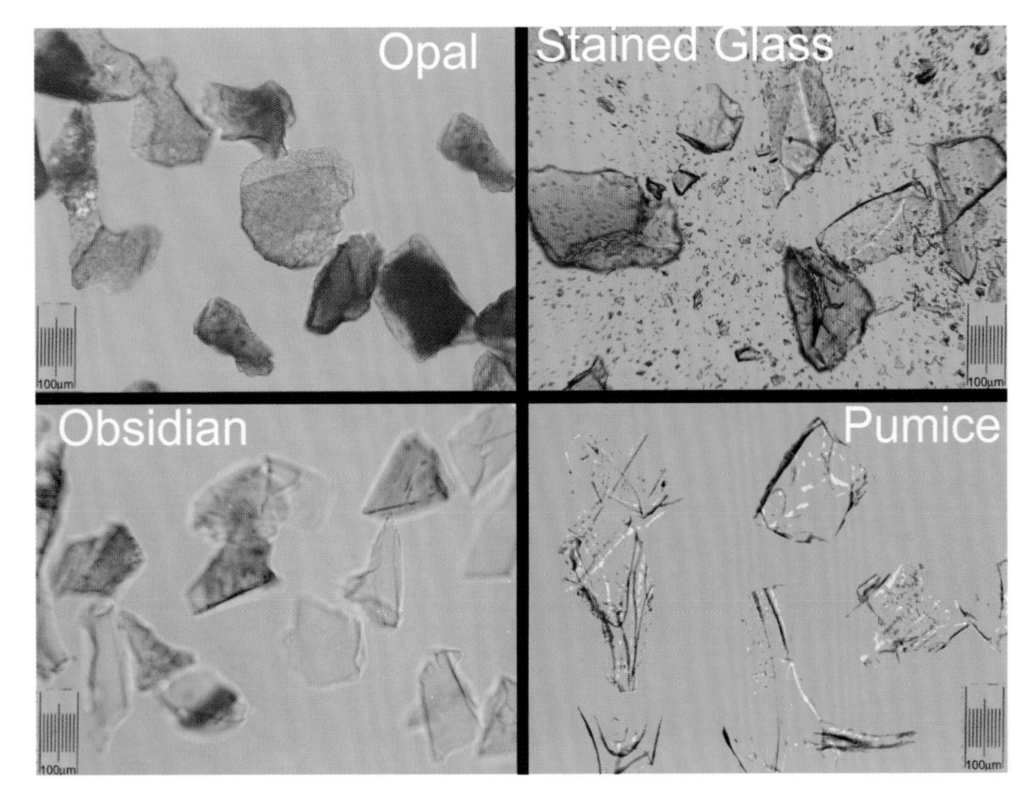

FIGURE 11.9 Common amorphous minerals and related substances as they appear in MM 1.539 in plane-polarized light. All are isotropic.

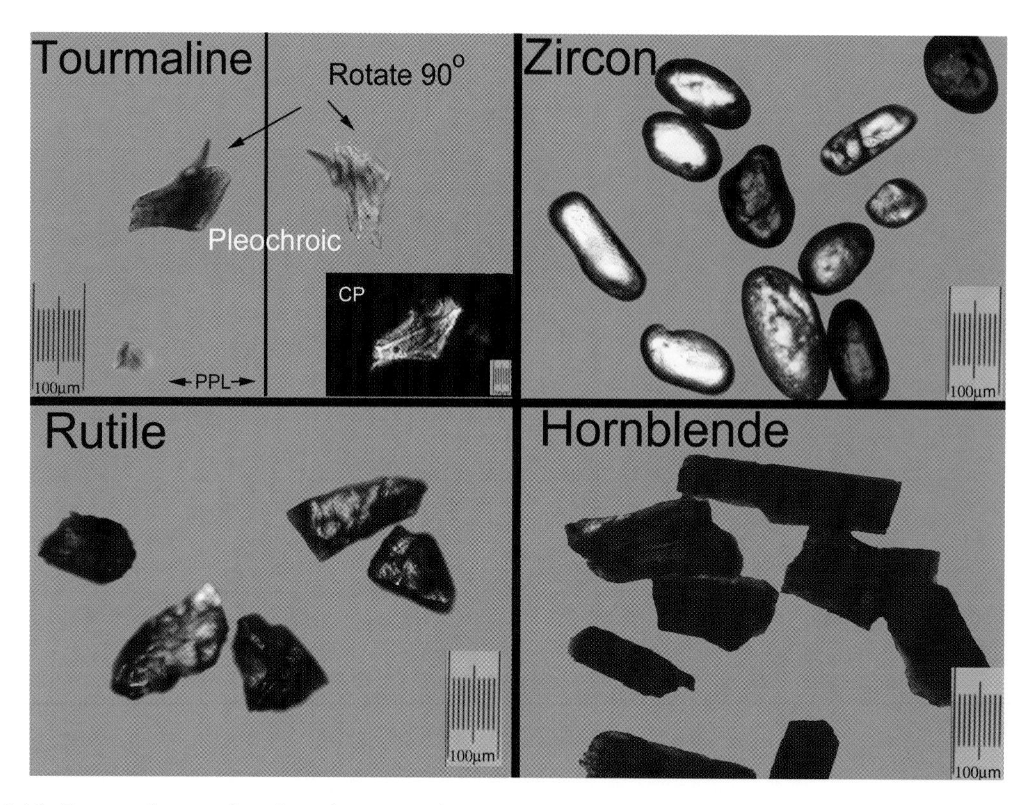

FIGURE 11.10 Common heavy minerals as they appear in MM 1.539, between crossed polars, at 10° off extinction.

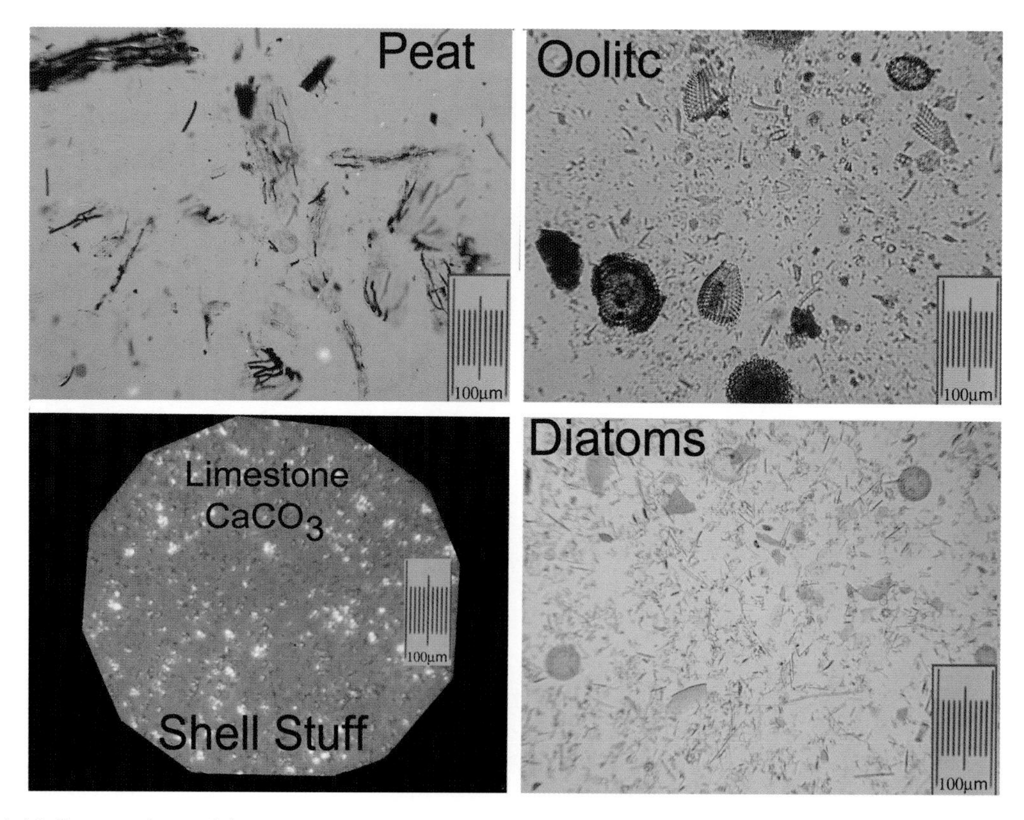

FIGURE 11.11 Structured materials common to soil they appear in MM 1.539 in plane-polarized light; limestone viewed between crossed polars at 10° off extinction.

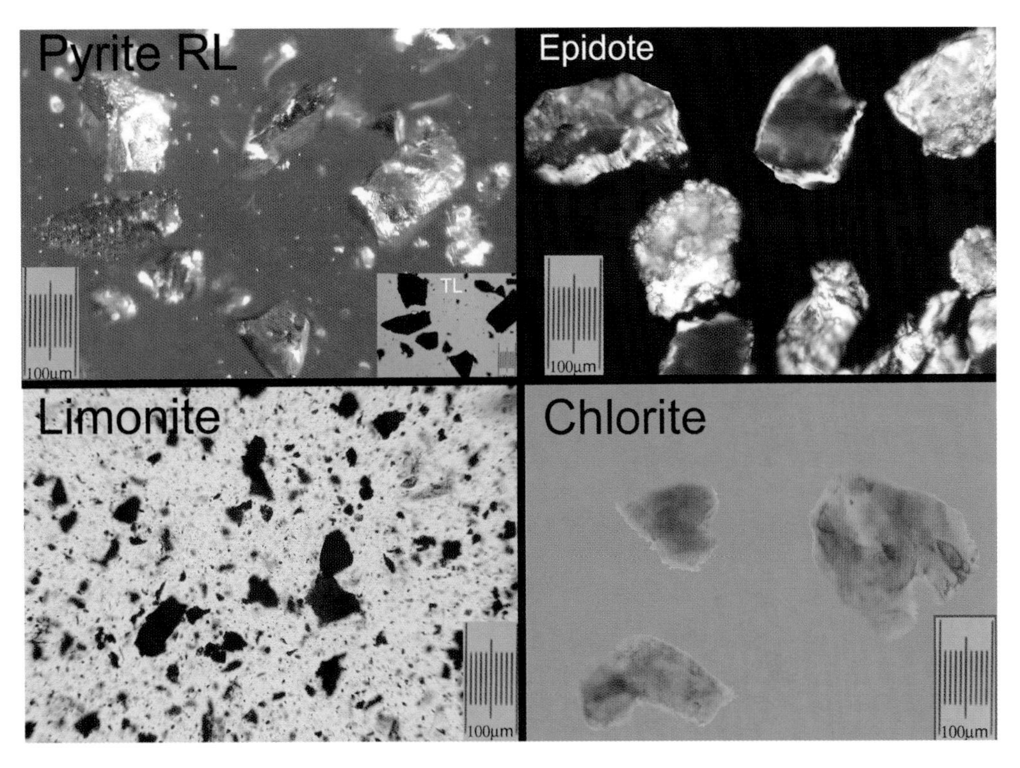

FIGURE 11.12 Miscellaneous materials common to soil as they appear in MM 1.539; pyrite with reflected light; epidote between crossed polars; limonite and chlorite in plane-polarized light.

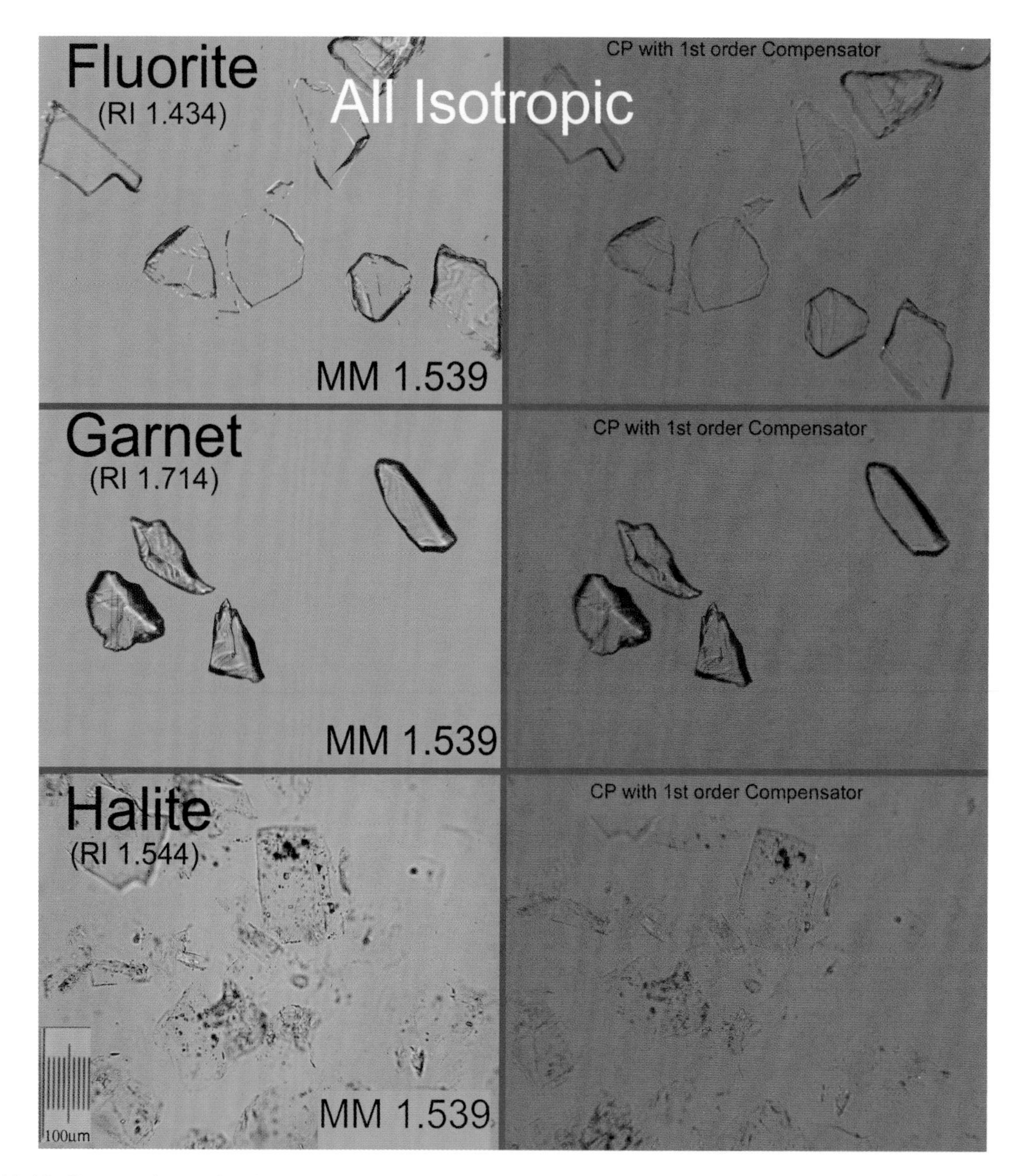

FIGURE 11.13 Common isotropic minerals. Left: fluorite, garnet, and halite (NaCl) as they appear in MM 1.539 with plane-polarized light. Right: Same minerals viewed between cross polars with a first-order compensator. Note no interference colors.

FIGURE 11.14 Common crystal systems represented by common materials mounted in MM 1.539 and viewed between crossed polars at 10° off extinction.

12 Gemstone Identification

Forensic scientists, museum scientists, and conservators are sometimes asked to identify semiprecious and precious gemstones and differentiate natural and synthetic stones. Although these tasks are best handled by professional gemologists, such as those certified by the Gemological Institute of America (GIA), it is important for scientists of other disciplines to have a general working knowledge of the procedures, equipment, and rationale employed in gemstone characterization and identification. The goal of this chapter is to provide this fundamental information.

True gemstones are fashioned from naturally occurring inorganic minerals. Natural minerals are composed of atoms or ion complexes in a formless matrix structure or in precise crystalline forms. Certain minerals have their atoms or ions arranged by nature and natural forces into precise crystalline forms or lattices known collectively as crystals. Minerals composed of randomly arranged atoms or ions have no distinct lattice structures and are therefore designated amorphous (without form). Natural glasses such as obsidian and naturally occurring minerals such as opals fall under the amorphous classification system. Naturally occurring minerals such as diamonds, zircons, morganite, topaz, kunzite, and kyanite come under one of the six primary categories of crystalline classifications (Figure 12.1). Crystallography is a complex subject beyond the scope of this book. Resolute readers are referred to the literature.[1–18]

Gems composed of organic materials have been used throughout the ages for personal adornment and in the decorative arts. Pearls and ivory originate from living organisms. Amber, coral, shells, bone, and jet originate from once-living organisms. Gems fashioned from these organic materials are known as organic gemstones, and they are prized as highly as gemstones made from inorganic minerals. Their pleasing forms, colors, varying transparencies, textures, lusters, and easy workability are only a few reasons these materials are considered equals to precious inorganic gemstones.

Mineralogists have developed their own nomenclature system used by them for centuries. They are classified by group, species, and variety. Gemologists use the same system to name and classify gemstones. A group of gemstones is defined as two or more gems similar in physical structure and properties. A distinct member of a given group is known as a species. Different members of a species that vary in color or other characteristics (i.e., inclusions) are known as varieties. Gemstone taxonomy is presented in Figure 12.2. Mineral groups and organic gemstone materials possess constant physical and optical properties. Mineralogists and gemologists utilize these properties to classify, characterize, and identify them.

PROCEDURES FOR CHARACTERIZING AND IDENTIFYING GEMSTONES

Preliminary Examination

An initial visual examination of the specimen with unaided eyes and a 10× loupe is the first step. The following physical and optical properties are noted:

Color
Clarity
Dispersion of light
Characteristic appearance
Luster
Habit
Interfacial angle
Cleavage

Next, the weight of a gemstone in carats is determined and its physical dimensions are measured with a micrometer as shown in Figure 12.3. The collected data are recorded on a gemstone data sheet (Figure 12.4). (*Note:* All data concerning the questioned red gemstone studied in this chapter are recorded on Figure 12.16.)

Microscopic Examination

The next step is conducting a macroscopic examination utilizing a stereomicroscope equipped with dark-field (DF) illumination. The gemstone should be cleaned with a lint-free cloth, placed in a stone holder, mounted on the microscope stage, and examined as demonstrated in Figures 12.5 and 12.6.

Information concerning physical characteristics revealed by the microscope and DF illumination, for example, inclusions, fracture data, color banding, color zoning, curved striae, and so on, is collected, described, sketched, or otherwise recorded on the gemstone data sheet (Figure 12.4).

Optical Examination

Determining optical properties including birefringence, refractive index (RI), dispersion, and pleochroic nature is vital to gemstone classification and identification. Chapters 1 and 3 review these terms. Gemologists use the equipment depicted in Figure 12.7 daily to determine these fundamental optical properties.

A polariscope is an instrument that determines whether a gemstone has one or more primary refractive indices (RIs). It is composed of a housing with a light at its base and two polarizing filters. The lower filter is called the polarizer; the upper is called the analyzer. The gemstone to be tested is placed between the crossed polarizing filters. While the examiner observes the specimen through the analyzer, the gemstone is rotated until it appears extinct (or dark). See A in the inset on the right in Figure 12.8. The gemstone is then rotated 90° (B in Figure 12.8). If the stone color changes from dark to light, it is birefringent or doubly refractive. If it remains extinct or dark, it is singly refractive or isotropic.

RIs are determined using a Duplex II refractometer. A minute drop of high RI liquid (1.81) is placed cautiously on the polished prism of the refractometer. The stone is gently positioned, table facet side down, onto the tiny drop of RI liquid and carefully moved to the prism center over the gray bar. RI readings are taken by looking through the eyepiece and observing the blue-green stripe at the end of the shadowed area. The shadow is produced by the refraction and reflection of the light entering the gemstone surface through the interface formed by the liquid and glass prism (Figures 12.9 and 12.10). Precise RI readings can be obtained by using monochromatic light as the illumination source. Collected data are recorded on a gemstone data sheet (Figure 12.4).

Another important and distinguishing feature of many natural and synthetic gemstones is their behavior when viewed under long- and short-wave ultraviolet (UV) radiation. The specimen gemstone is placed into a UV cabinet and the short-wave UV excitation source is activated. The short-wave lamp is turned off and the long-wave UV source is activated. Any fluorescence (a gemstone emits visible light as long as the excitation UV lamp is turned on) or phosphorescence (a stone continues to emit light after the UV source is turned off) is noted for each UV source. Figure 12.11 illustrates the fluorescence of a synthetic ruby. (**Caution:** UV light can damage your eyes. Never look directly at a UV light source. Always employ a protective shield or other protective eyewear when working with UV radiation!)

Anisotropic minerals often absorb different wavelengths of visible light along different crystalline axes, and thus transmit different colors of visible light in different orientations. A mineral exhibiting this optical property is said to be pleochroic. This phenomenon is best observed under plane-polarized light. The dichroscope depicted in Figure 12.12 is used to determine whether a gemstone is pleochroic. The gemstone is held with a stone holder and placed against the double polarizer calcite window of the device. The gemstone and apparatus are held up to a daylight source. The observer views the stone through the lens. If each window displays a different color, a gemstone is pleochroic. Observations should be noted on the Figure 12.4 data sheet.

Finally, information concerning crystal growth, optical phenomena such as asterism and chatoyancy, and the types of inclusions found inside a gemstone all aid in identification. Figures 12.13 through 12.15 portray some of these features and inclusions. Gemstones display literally hundreds of these distinct characteristics and serious readers are referred to the literature, particularly References 1, 3, 4, 6, 8, and 11.

GEMSTONE IDENTIFICATION

To determine a gemstone's classification, the collected data noted on the data sheet (Figure 12.4) should be compared to the data compiled on Table 12.1 and information in the literature. In addition, comparison of the questioned gemstone with known reference standards is recommended, especially if the gemstone is identified as originating from a particular mine location or geographical location, i.e., a Colombian emerald or African amethyst.

Figure 12.16 contains the data collected for the red gemstone shown in various figures throughout this chapter. The collected data are utilized as follows:

1. The deep red color suggests a number of possible groups: corundum, coral, diamond, cubic zirconium, hematite, yag, glass, jadite, tourmaline, beryl, spinel, garnet, or zircon.
2. Transparency characteristics eliminate three groups cited above. The remaining groups are corundum, diamond, cubic zirconium, yag, glass, tourmaline, beryl, spinel, garnet, and zircon.
3. Based on birefringence, six more groups are eliminated; corundum, tourmaline, beryl, and zircon remain.
4. RI values of 1.762 to 1.770 indicate corundum.
5. The questioned stone is pleochroic; that, too, indicates corundum.
6. Strong fluorescence under long and short UV radiation normally indicates synthetic corundum.
7. The inclusions (gas bubbles and curved striae) prove the material is synthetic corundum.
8. The conclusion is that the gemstone is a synthetic ruby.

REFERENCES

1. Schumann, W., *Gemstones of the World*, New York, Sterling Publishing, 1984.
2. Bates, R.L. and Jackson, J.A., Eds., *Dictionary of Geological Terms*, 3rd ed., New York, Doubleday, 1984.

3. Liddicoat, R.T., Jr., *Handbook of Gem Identification*, 12th ed., Santa Monica, CA, GIA, 1993.
4. Cipriani, C. and Borelli, A., *Simon & Schuster's Guide to Gems and Precious Stones*, Lyman, K., Ed., New York, Simon & Schuster, 1986.
5. Duda, R. and Lubo, R., *Minerals of the World*, 2nd ed., New York, Arch Cape Press, 1989.
6. Prinz, M., Harlow, G., and Peters, J., *Simon & Schuster's Guide to Rocks and Minerals*, New York, Simon & Schuster, 1978.
7. Holden, A. and Singer, P., *Crystals and Crystal Growing*, New York, Doubleday, 1960.
8. Pagel–Theisen, V., *Diamond Grading ABC: Handbook for Diamond Grading*, 11th ed., Brussels, Rubin & Son, 1993.
9. Fleischer, M., Wilcox, R.E., and Matzko, J.J., Microscopic Determination of the Nonopaque Minerals, U.S. Geological Survey Bulletin 1627, Washington, D.C., U.S. Government Printing Office, 1984.
10. Mange, M.A. and Maurer, H.F.W., *Heavy Minerals in Colour*, London, Chapman & Hall, 1992.
11. Gübelin, E.J. and Koivula, J.I., *Photoatlas of Inclusions in Gemstones*, 3rd ed., Zurich, ABC Editions, 1997.
12. McCrone, W.C., McCrone, L.B., and Delly, J.G., *Polarized Light Microscopy*, Ann Arbor, MI, Ann Arbor Science Publishers, 1978.
13. Allen, R.M., *Practical Refractometry by Means of the Microscope*, Cedar Grove, NJ, Cargille Laboratories, 1954.
14. Bloss, F.D., *An Introduction to the Methods of Optical Crystallography*, New York, Holt, Rinehart & Winston, 1961.
15. Bloss, F.D., *Optical Crystallography*, Monograph Series Publication 5, Washington, D.C., Mineral Society of America, 1999.
16. Wood, E.A., *Crystals and Light: An Introduction to Optical Crystallography*, New York, Van Nostrand, 1964.
17. Winchell, A.N., *Elements of Optical Mineralogy, An Introduction to Microscopic Petrography: Part I: Principles and Methods*, 5th ed., New York, John Wiley & Sons, 1949.
18. Stoiber, R.E. and Morse, S.A., *Crystal Identification with the Polarizing Microscope*, New York, Chapman & Hall, 1994.

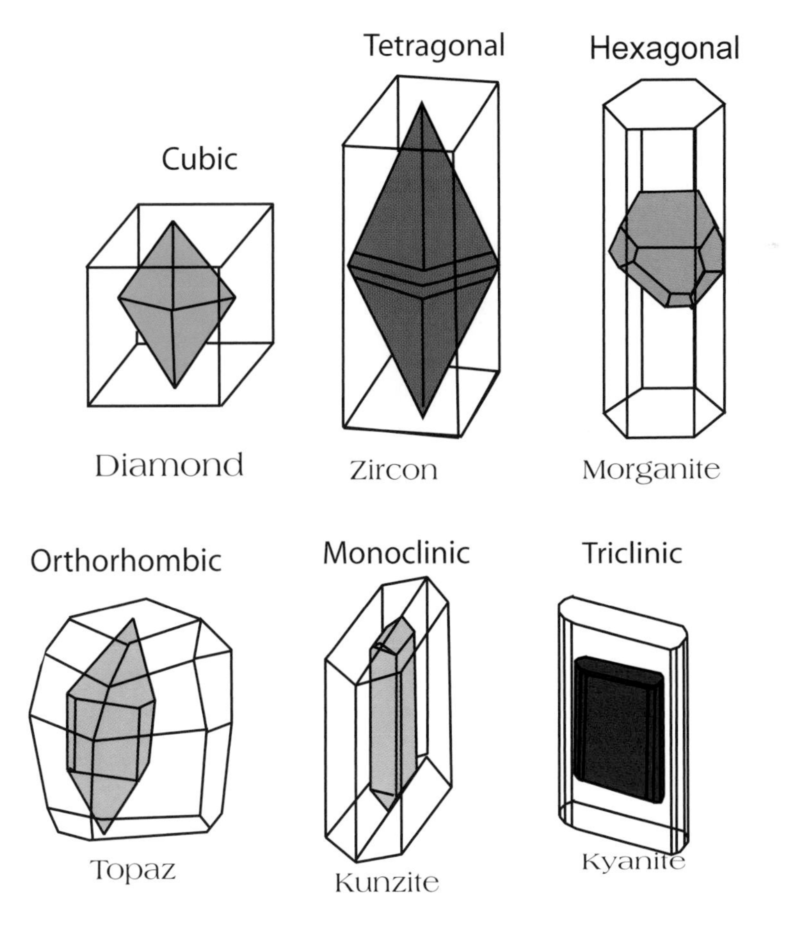

FIGURE 12.1 The six primary crystal systems used by crystallographers to describe crystalline substances. Inserted into each crystal form is a gemstone mineral representative of the crystal type depicted.

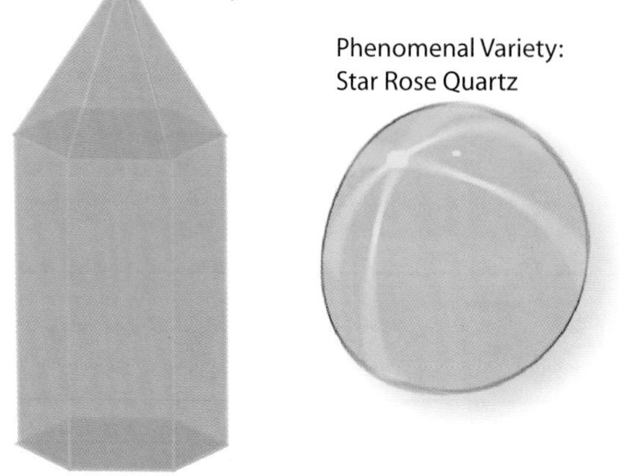

FIGURE 12.2 Classification of a common semiprecious gemstone known as rose quartz into its group, species, and two varieties.

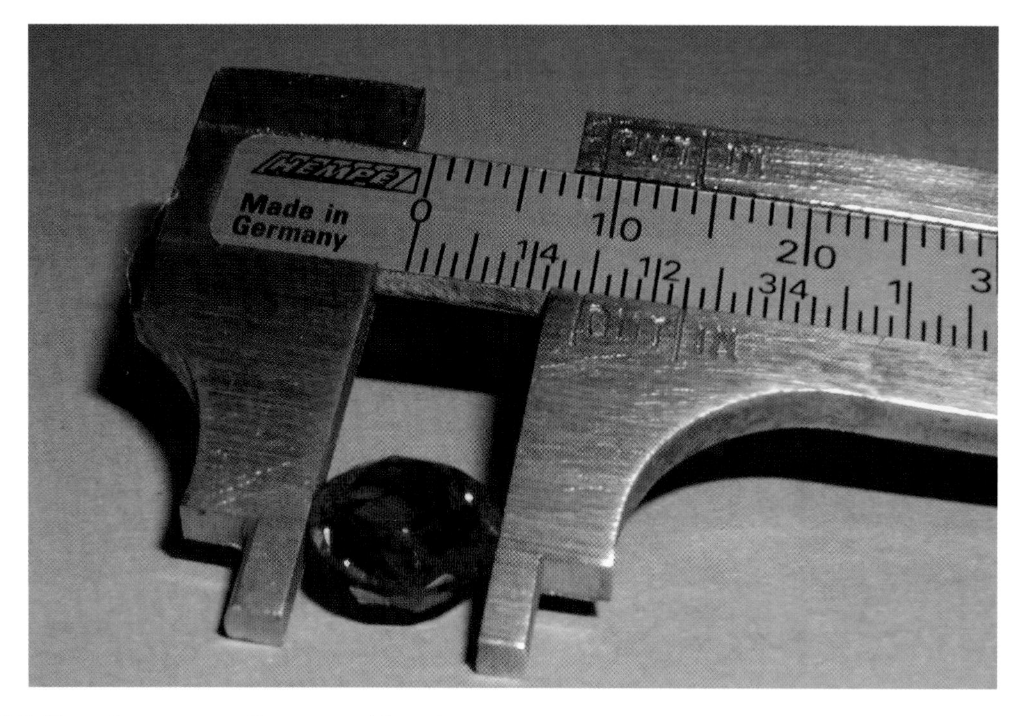

FIGURE 12.3 Measuring gemstone dimensions with a micrometer.

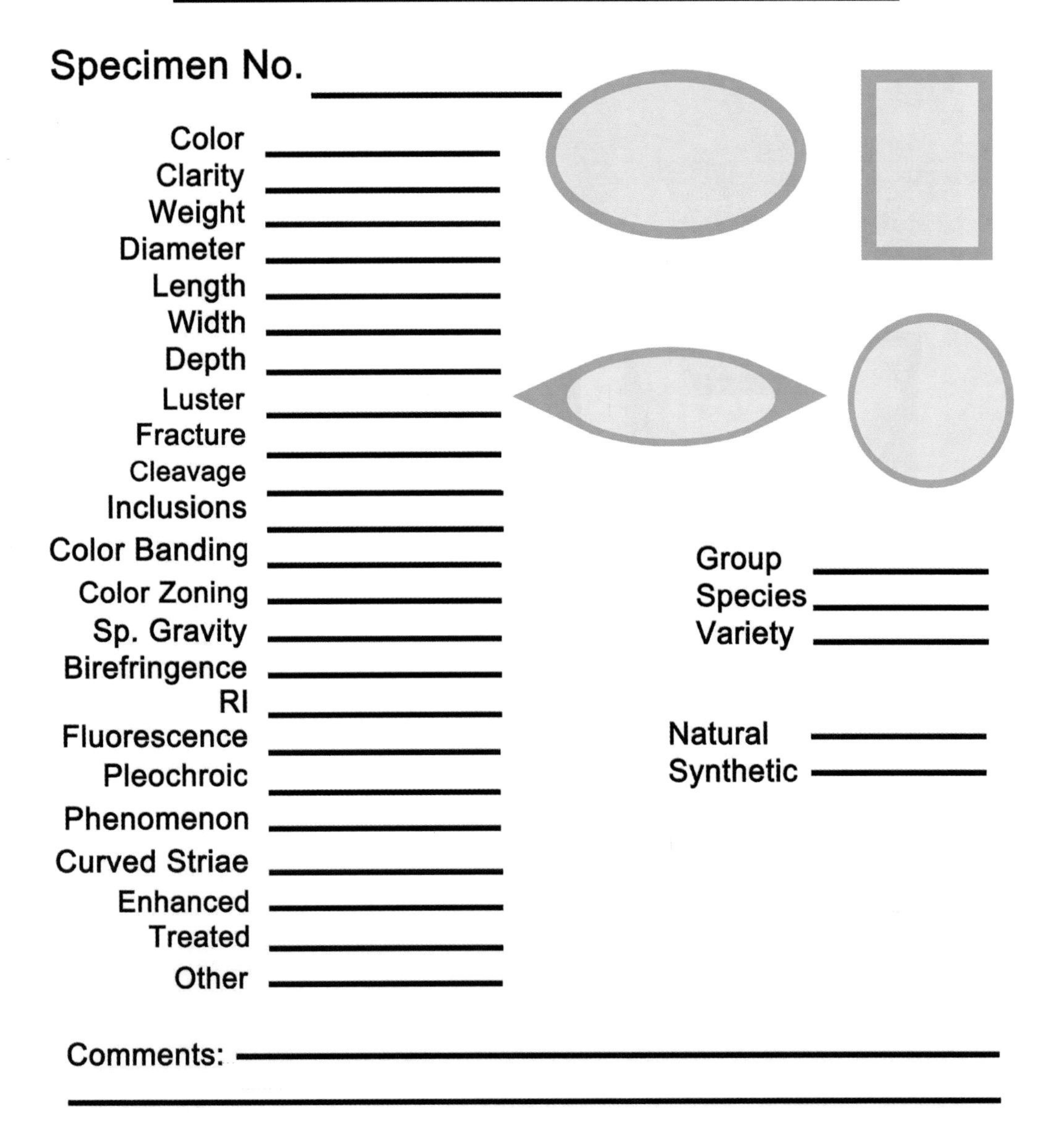

Gemstone Data/Examination Sheet

Specimen No. ____________

Color ____________
Clarity ____________
Weight ____________
Diameter ____________
Length ____________
Width ____________
Depth ____________
Luster ____________
Fracture ____________
Cleavage ____________
Inclusions ____________
Color Banding ____________
Color Zoning ____________
Sp. Gravity ____________
Birefringence ____________
RI ____________
Fluorescence ____________
Pleochroic ____________
Phenomenon ____________
Curved Striae ____________
Enhanced ____________
Treated ____________
Other ____________

Group ____________
Species ____________
Variety ____________

Natural ____________
Synthetic ____________

Comments: ____________________________________

FIGURE 12.4 Gemstone data sheet. Check off, circle, write, or sketch all responses.

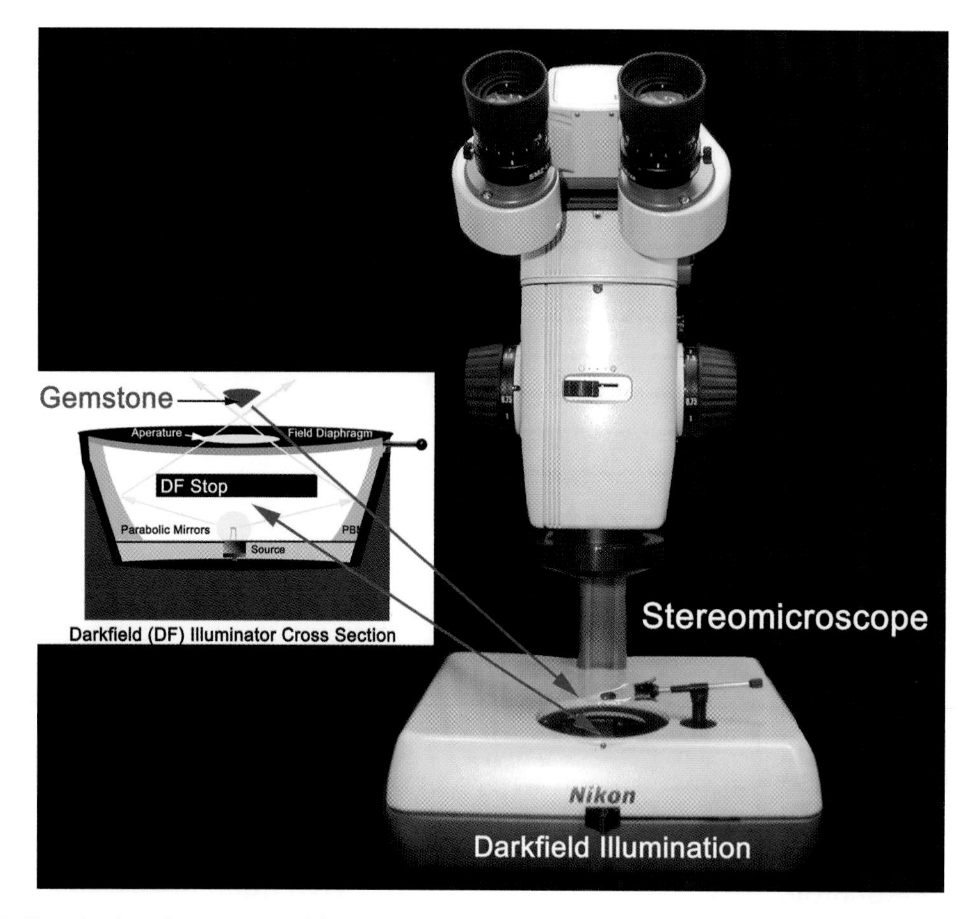

FIGURE 12.5 Examination of a gemstone with a stereomicroscope equipped with dark-field illumination.

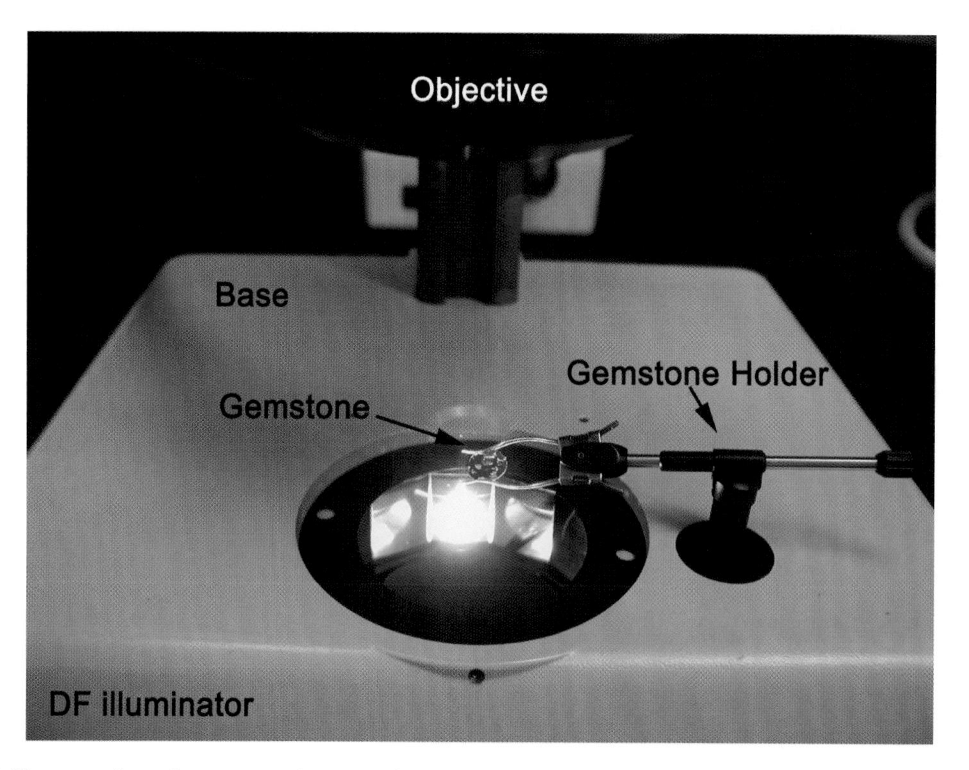

FIGURE 12.6 Close-up view of gemstone placement for viewing with a gemological microscope or a stereomicroscope with dark-field illumination.

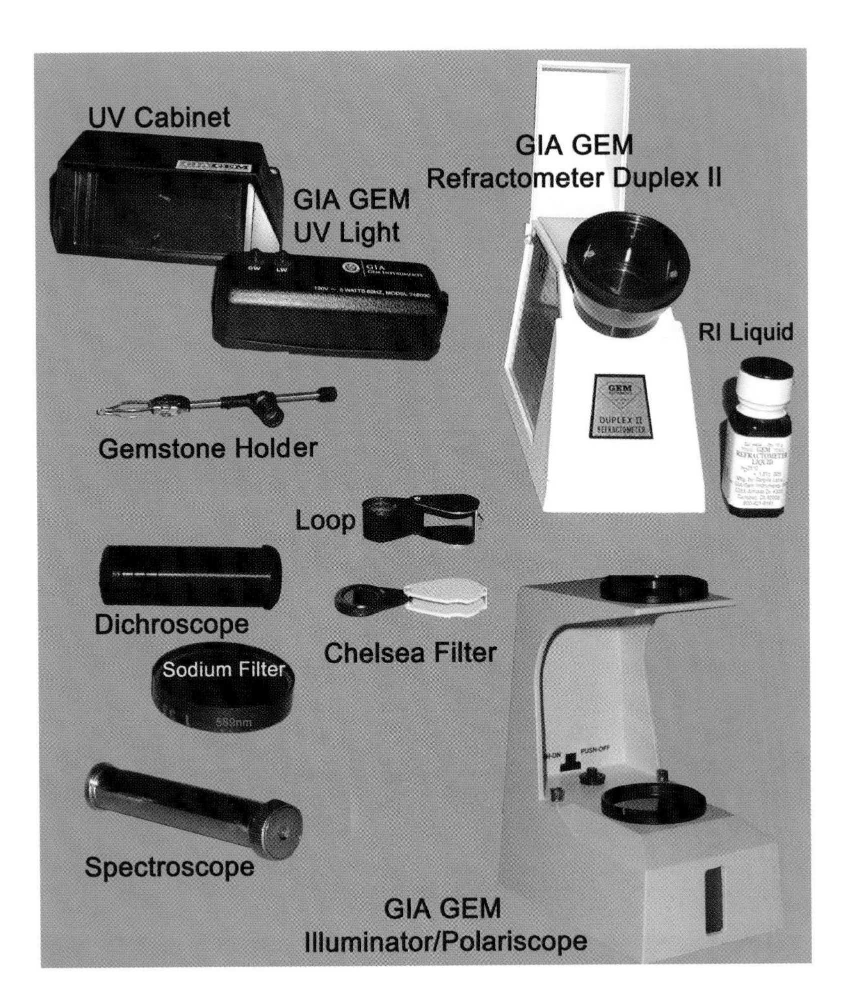

FIGURE 12.7 Basic equipment needed to determine physical and optical properties of gemstones.

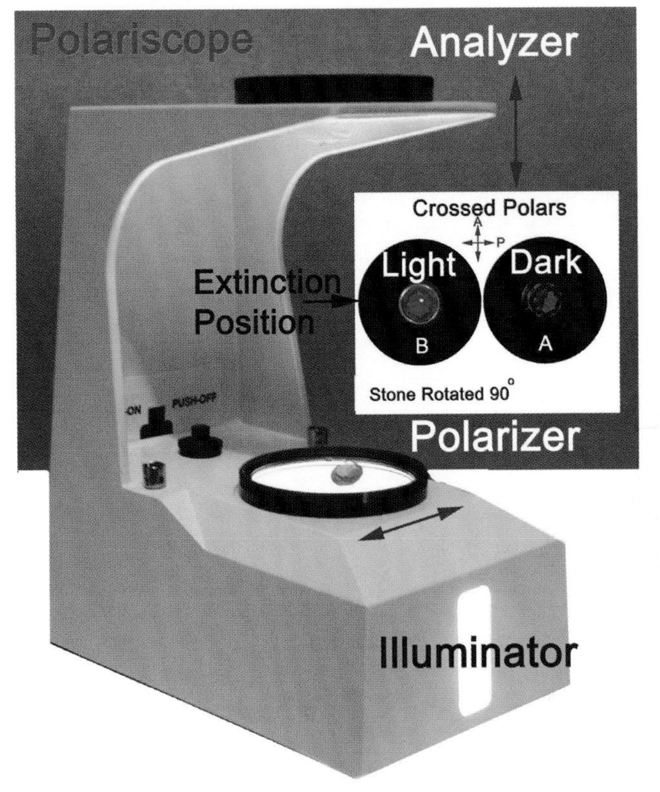

FIGURE 12.8 Procedure for determining whether a gemstone is singly or doubly refractive utilizes a polariscope. The gemstone is placed between the crossed polarizers set at the extinction position. At position A (right), the stone is rotated until it appears dark or black. The stone is rotated 90° (position B, left). If the color changes from dark to light, the stone is birefringent or doubly refractive. If the stone remains dark or extinct, it is singly refractive (isotropic).

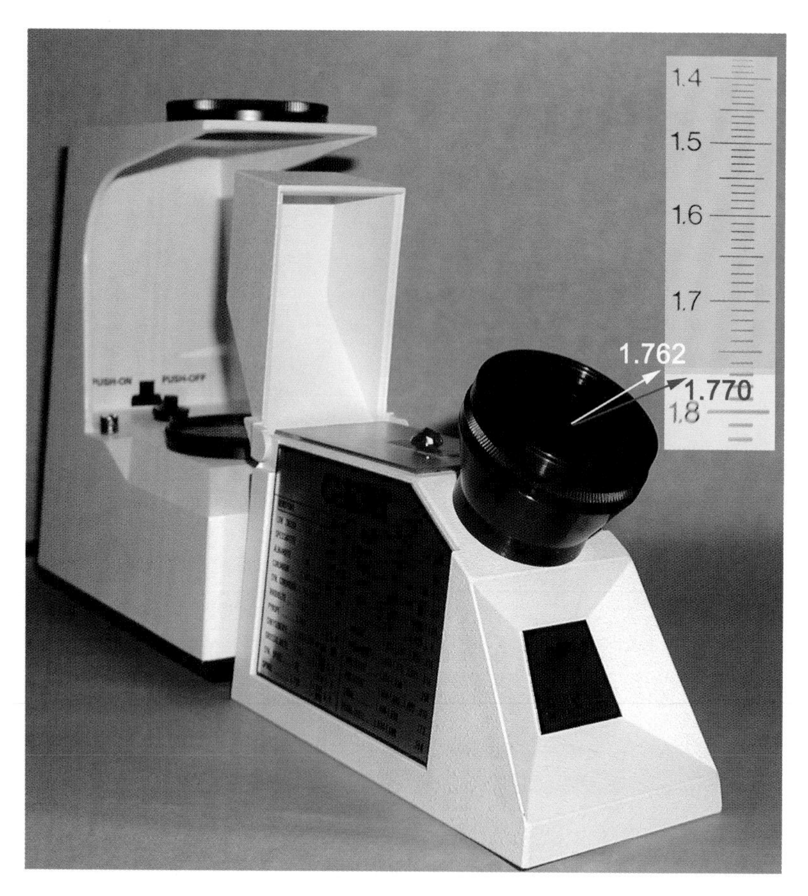

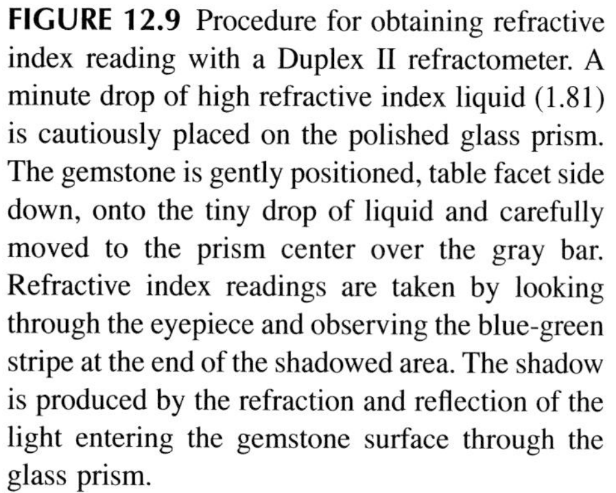

FIGURE 12.9 Procedure for obtaining refractive index reading with a Duplex II refractometer. A minute drop of high refractive index liquid (1.81) is cautiously placed on the polished glass prism. The gemstone is gently positioned, table facet side down, onto the tiny drop of liquid and carefully moved to the prism center over the gray bar. Refractive index readings are taken by looking through the eyepiece and observing the blue-green stripe at the end of the shadowed area. The shadow is produced by the refraction and reflection of the light entering the gemstone surface through the glass prism.

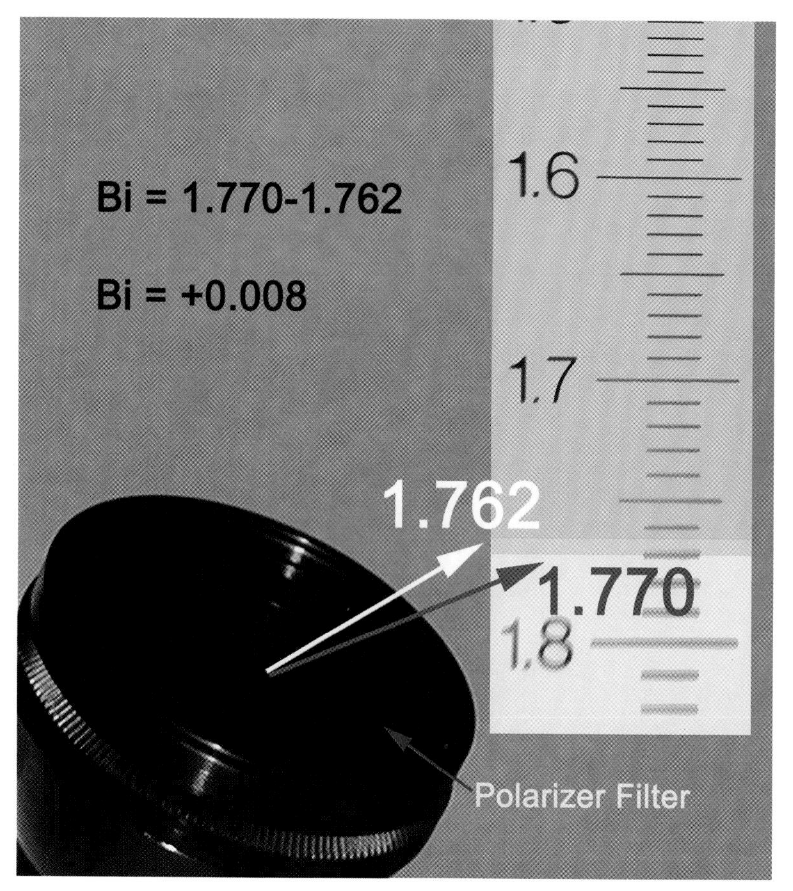

FIGURE 12.10 Close-up view of a refractive index scale and refractometer readings for the questioned red gemstone.

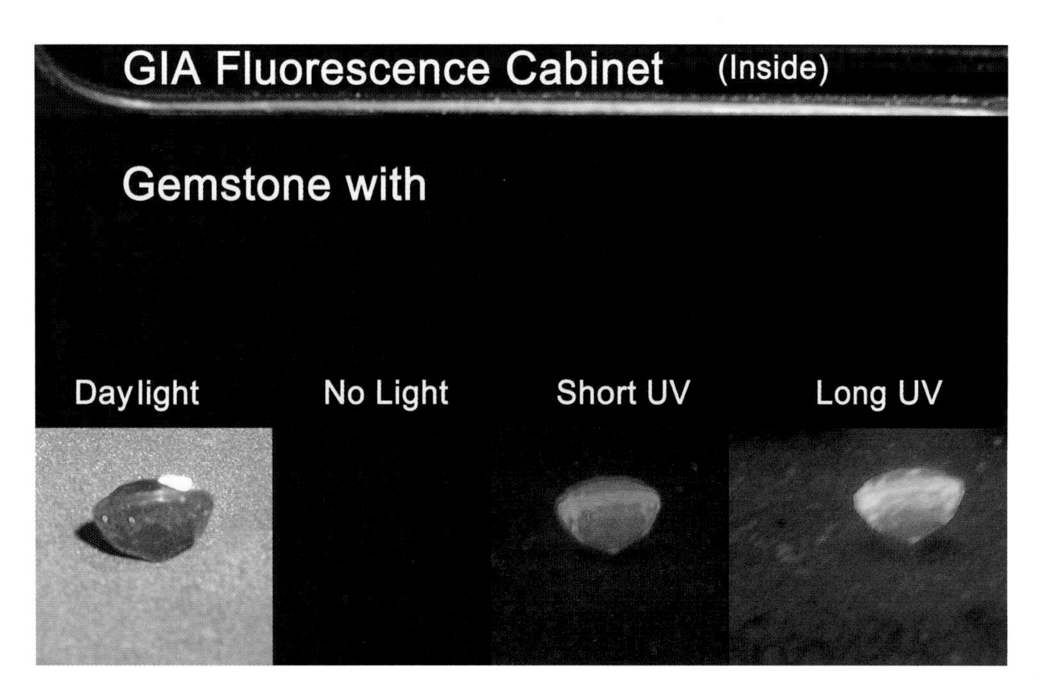

FIGURE 12.11 The appearance of the red specimen gemstone when viewed in daylight and with short and long UV light.

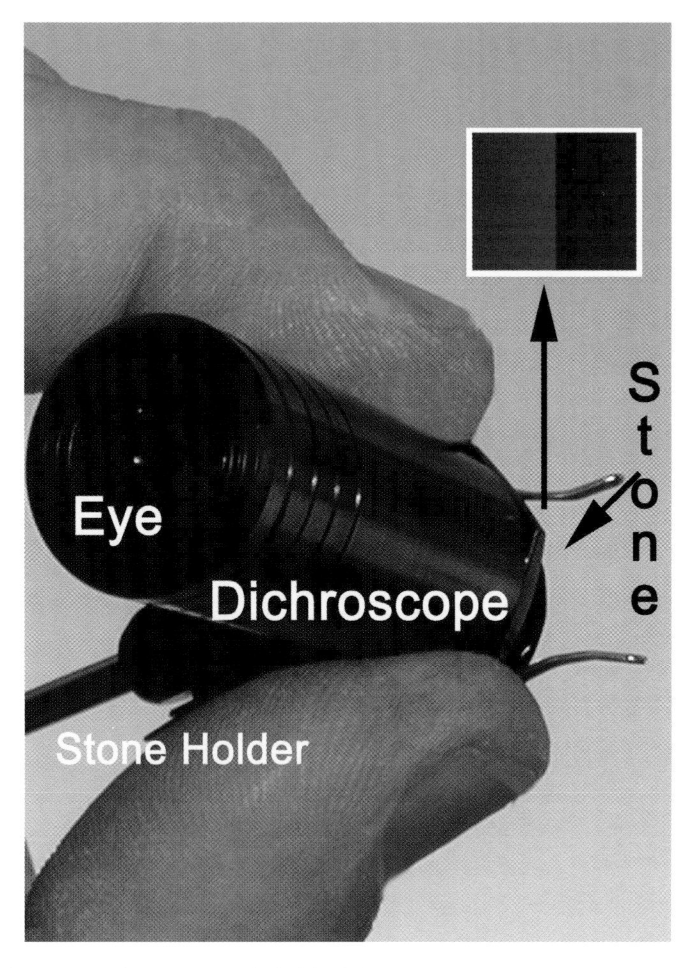

FIGURE 12.12 Use of a dichroscope. The gemstone is held with a stone holder and placed against the double polarizer calcite window. The gemstone and apparatus are held up to a daylight source. The observer views the stone through the lens. If each window displays a different color, the gemstone is pleochroic.

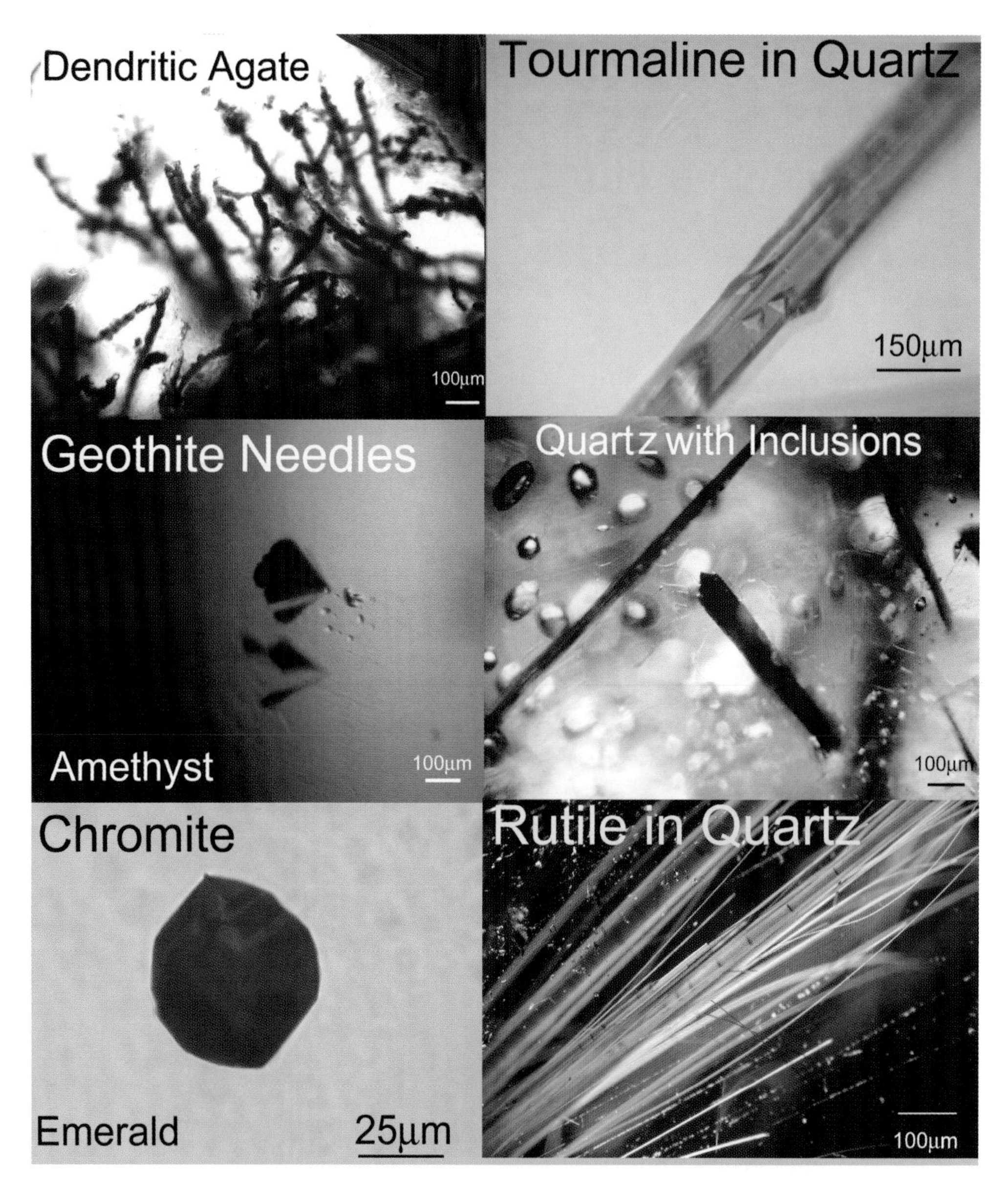

FIGURE 12.13 Common inclusions found in agate, quartz, and emerald.

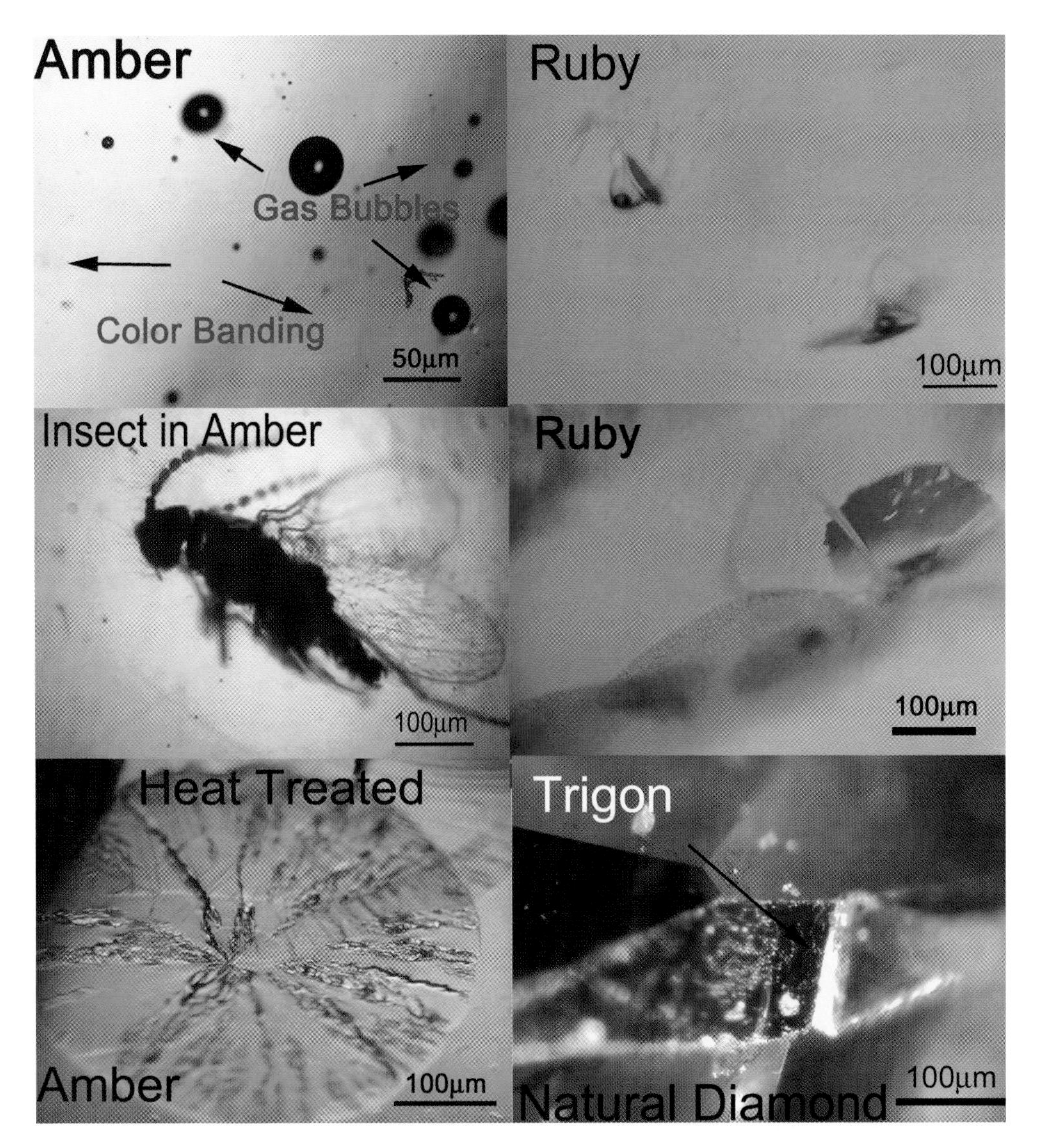

FIGURE 12.14 Common inclusions found in amber and ruby; a trigon shown on the girdle of a faceted diamond.

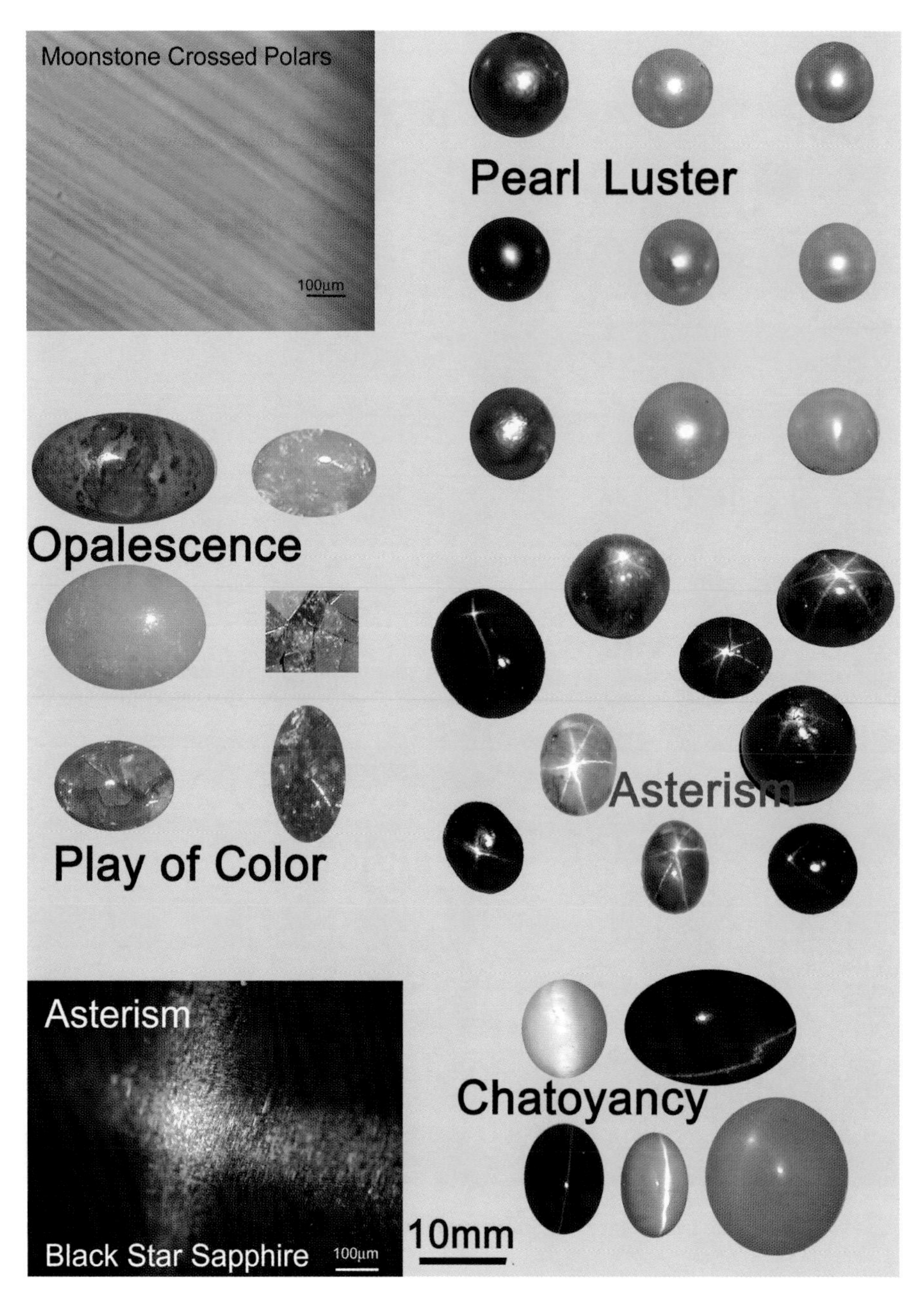

FIGURE 12.15 Examples of optical phenomena: chatoyancy, asterism, luster, and opalescence.

TABLE 12.1
Gemstone Data Reference Table[1–11]

Group	Color	RI	EBi	Pleochroism	Fluorescence	Fracture	Inclusions
Amber (organic)	Golden yellow	1.540	—	—	None to strong	Conchoidal	Air bubbles, insects, plant debris
Beryl: emerald, golden, morganite, aquamarine	Varies	1.577 to 1.583	0.005 to 0.009	Yes	None to strong	Conchoidal	Many: mica, pyrite, calcite
Corundum: ruby, sapphire	Varies; red; all others	1.762 to 1.770	0.008	Yes	None to strong	Conchoidal to granular	Many: fingerprint, silk or rutile, angular banding; synthetics; gas bubbles and curved striae
Coral (organic)	Varies; opaque	—	—	—	None	Conchoidal to granular	—
Chalcedony	Varies	1.535 to 1.540	—	No	None to medium	Conchoidal	Many
Cubic zirconium	Varies	2.150	—	—	None to medium	Conchoidal	Usually flawless
Diamond	Varies	2.417	—	—	None to strong	Perfect step break	Trigon on surface; numerous inclusions
Fluorite	Varies	1.434	—	—	None to strong	Conchoidal	Color zoning
Garnet: pyrope, rhodolite, hessonite, almandine, tsavorite	Varies	1.730 to 1.83	—	Yes	None to strong	Conchoidal	Needle inclusions; prism crystals; single and multiphase
Ivory (organic)	Yellow to white	—	—	—	Weak to strong; blue/white	Splintery; fibrous	Can be dyed
Jadeite	Varies; opaque	Spot; 1.660 to 1.680	—	—	None to strong	Granular and splintery	Greasy luster
Jet (organic)	Black; opaque	—	—	—	—	Greasy; dull; conchoidal	Lignite coal; brown streaks
Nephrite	Varies; opaque	Spot; 1.606 to 1.632	—	—	None	Greasy; dull; waxy; conchoidal to granular	Greasy; dull; waxy; some dark inclusions
Opal	Varies	1.450	—	—	None to strong; may phosphoresce	Glass-like	Possible
Pearl (organic)	Varies; opaque	—	—	—	None to strong	Uneven; step-like	Mother-of-pearl luster
Spinel	Varies	1.718	—	—	None to strong	Conchoidal	Spinel octahedral; fingerprint patterns
Tourmaline	Varies	1.624 to 1.644	0.020	Yes (strong)	None to very weak	Conchoidal	Gas bubbles, color zoning, liquid inclusions
Topaz	Varies	1.619 to 1.627	0.008	Yes	None to medium	Conchoidal to perfect	Single and multiphase inclusions
Zircon	Varies	1.985 to 25	0.04	Yes	None to medium	Conchoidal	Rutile crystals; single and multiphase

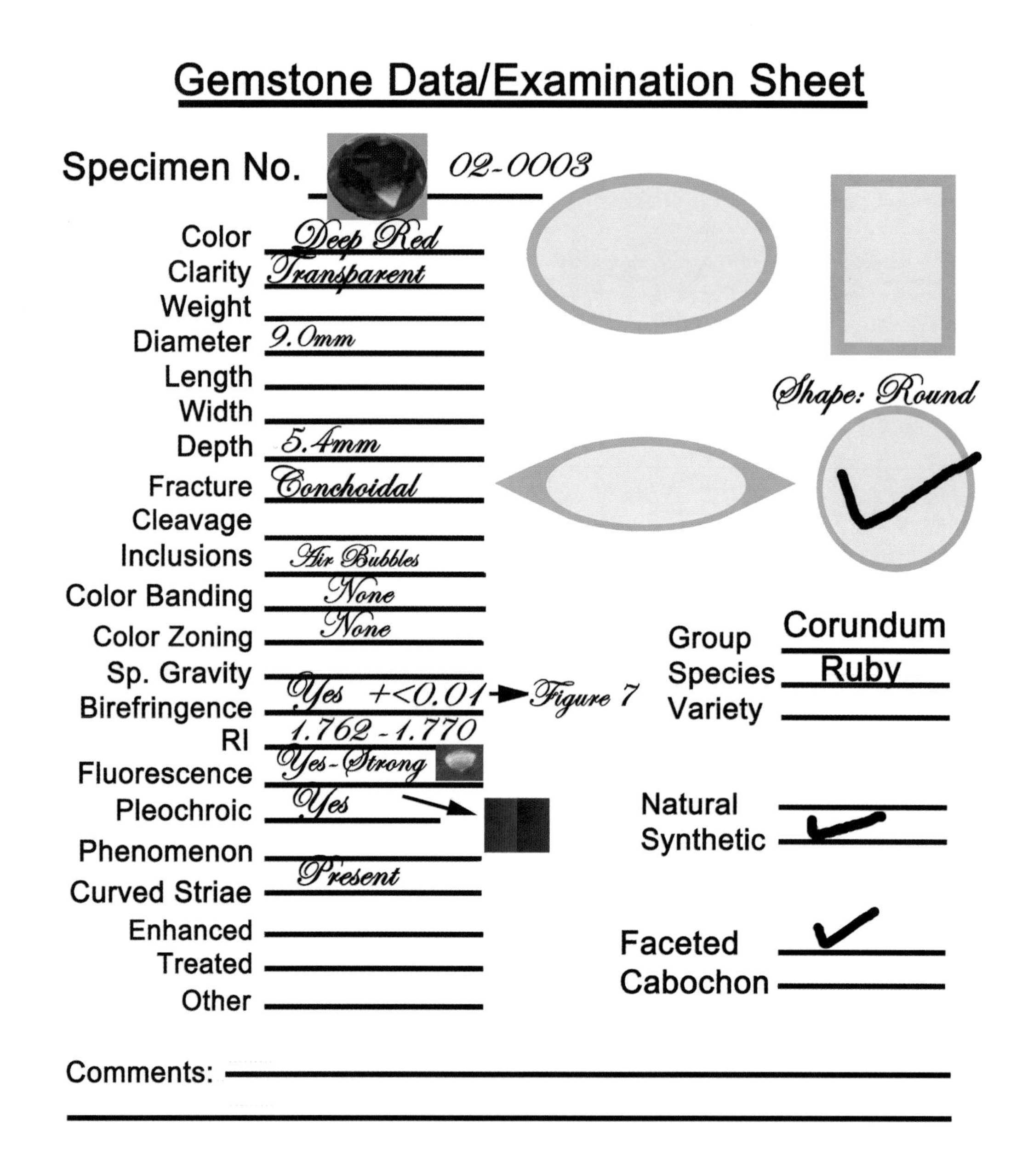

Gemstone Data/Examination Sheet

Specimen No. 02-0003

Color: Deep Red
Clarity: Transparent
Weight:
Diameter: 9.0mm
Length:
Width:
Depth: 5.4mm
Fracture: Conchoidal
Cleavage:
Inclusions: Air Bubbles
Color Banding: None
Color Zoning: None
Sp. Gravity:
Birefringence: Yes +<0.01 → Figure 7
RI: 1.762-1.770
Fluorescence: Yes-Strong
Pleochroic: Yes →
Phenomenon:
Curved Striae: Present
Enhanced:
Treated:
Other:

Shape: Round ✓

Group: Corundum
Species: Ruby
Variety:

Natural:
Synthetic: ✓

Faceted: ✓
Cabochon:

Comments:

FIGURE 12.16 Gemstone data sheet prepared for the red gemstone discussed in text and depicted in figures throughout chapter.

13 Dust Examination

Over a century ago, Hans Gross hypothesized that dust is a representation of our environment in miniature. He further proposed that recognizing the constituents of a particular dust sample would allow an analyst to determine the surroundings from which the dust originated, and thereby help solve crimes.[1] These ideas led to the development of microscopic methods and schemes to examine, identify, compare, evaluate, and employ dust analysis in solving crimes.

The primary goal of this chapter is to present a microscopic guide for the identification and characterization of the components of dust specimens mounted in a single refractive index medium, namely Cargille's Melt Mount (MM) 1.539. It is intended that the procedures discussed will provide light microscopists with a reference that will serve as an introductory guide to examining common trace materials encountered in specimens of dust during routine casework. Emphasis is placed on the use of polarized light microscopy (PLM), stereoscopic microscopy (SM), and MM 1.539 to achieve these goals.

COMPOSITION OF DUST SPECIMENS

While dust traces can encompass an infinite number of different materials, the authors have found that two primary morphological forms compose most specimens of dust: (1) fibrous materials and (2) particulate matter. The two primary morphological forms of materials composing dust specimens and other substances found in dust include:

1. Fibrous materials
   - Human hair
   - Animal hair
   - Synthetic fibers
   - Natural fibers
2. Particulate matter
   - Mineral and glass grains
   - Plaster, concrete, plastic, bricks, and paint chips
   - Metal flakes and rust fragments
   - Miscellaneous substances, i.e., vegetable matter, spices, twigs and leaf fragments, pollen grains, spores, tobacco, starch grains, feathers, wood chips, sawdust, insects and insect parts, and diatoms

PRELIMINARY EXAMINATION

All specimens of dust should be examined with a stereomicroscope for evaluation and sorting. An initial data sheet such as Table 13.1 should be prepared for each dust specimen. If a specimen appears to be homogeneous, a representative sample should be mounted on a microscope slide in Cargille's Melt Mount with a refractive index (RI) of 1.539 for the sodium D line at 25°C. The dust specimen is then covered with a No. 1½ cover glass. The Melt Mount must be heated to 60 to 70°C. It can be applied with a glass rod, eye dropper, or in stick form. The authors prefer the stick method; see Chapter 2.

Prior to mounting, a specimen may be teased with two needles to loosen the fibers and debris it contains. A representative sample of a heterogeneous dust specimen can be mounted in the same manner. However, large particles that cannot be mounted should first be sorted for separate examinations. After mounting, the specimen should be examined with a polarized light microscope for characterization and identification.

Melt Mount® 1.539 is a useful mounting medium for identifying many materials commonly found in dust. It is a stable, solvent-free, thermoplastic material. Once set, no changes in its optical properties have been observed. Its intermediate refractive index value enables a microanalyst to observe internal morphological details required for characterization and identification. When a change in relief (contrast) is required, it can be achieved simply by observing the specimen under plane-polarized light while rotating the microscope stage to change specimen orientation or by heating the specimen on a hotstage.

The degree of relief (shadowing) change between the specimen and the mounting medium depends on several important factors: (1) whether the specimen is anisotropic with respect to its optical properties; (2) the degree of birefringence of the specimen; and (3) the value of the refractive index of the mounting medium. If the substance is optically isotropic (has only one primary refractive index), no apparent change in relief will be noted when the specimen orientation is changed. If a specimen is optically anisotropic and at least one of its indices is higher or lower than that of Melt Mount 1.539, a change in relief will occur when the specimen is rotated.

In the event that a fiber or particle must be isolated or recovered from the mounted dust specimen, the Melt Mount preparation is gently heated on a hotplate, the cover glass is removed, and the item is retrieved with a forceps or fine needle while the preparation is observed with a stereomicroscope. The fiber or particle can be washed with xylene to remove excess mounting medium if the specimen is not soluble in xylene.

A data sheet (Table 13.1) should be prepared for each dust specimen. The data compiled can be used for the initial classification of the materials in the dust specimen and can guide the examiner to the appropriate identification procedure or scheme. Microscopic methods for the characterization and identification of the trace materials commonly found in dust specimens are presented in this chapter. The results obtained with these methods can be confirmed with other methods of analysis, such as micro-Fourier transform infrared (FTIR) spectroscopy, spindle-stage methods, x-ray diffraction, and scanning electron microscopy (SEM) energy dispersive x-ray analysis (EDS).

Characterization of Human Hair

Human hair can appear in dust in several forms. Complete hairs with intact proximal (root) ends, medial portions, and distal (tip) ends originating from various parts of the human body, are shed daily. They can become airborne for short periods and eventually collect in the dust of the environment. Also, hairs or hair fragments can accumulate in dust via normal grooming practices (brushing or combing) and by forcible means (pulling or cutting).

Finally, burned hair can become airborne and find its way into the dust of the environment. If such hair is severely burned, it is often unsuitable for identification or comparison purposes although it may have investigative value.

If human hair remains in dust for a prolonged period, it can become damaged by mechanical or insect action or decompose through microbial activity. Hair is composed of several layers of overlapping scales, the medulla or central canal (can appear present or absent), and the cortex or primary tissue. The cortex contains the pigment granules, cortical fusi, and other morphological features of hair (Figures 13.1 and 13.2).

Figure 13.1 also depicts a cast of the dominant scale pattern usually associated with hair of human origin. A procedure for the preparation of a temporary scale cast in Melt Mount 1.539 is covered in Chapter 2. When necessary for identification purposes, a hair can be isolated from a dust specimen as previously described, cast in Melt Mount 1.539, as detailed below, and then remounted in Melt Mount 1.539 for further study.

The specimen to be cast is placed on a microscope slide on which a thin layer of Melt Mount 1.539 has been spread over most of its top surface. The slide containing the hair specimen is then heated on a hotplate (65 to 70°C) until the solid layer of Melt Mount melts. The slide is then removed from the hotplate and allowed to cool until the Melt Mount hardens. The hair now embedded in the Melt Mount layer is peeled from the microscope slide. The resulting impression of the scale patterns can now be observed directly with a microscope at 100×.

Human hairs encountered in dust specimens usually originate from the head or pubic regions of the body. However, hairs originating from other body areas such as the face and limbs are fairly common. Therefore, examiners should become familiar with the morphologies of all types of human hairs. The morphological characteristics used to determine the somatic origins of human head and pubic hair are shown in Figure 13.3. The figures are only generalizations. Other configurations can and often do appear.

The race of the person from whom a questioned hair originated can often be important in scientific investigations. This type of determination is made on the basis of the morphological features exhibited by questioned hair specimens. Figure 13.4 depicts the primary morphological features for human head hair. Refer to Appendix A and Chapter 5 for additional information and examples.

It is important to note that the observations necessary for determinations of somatic and racial origin of questioned human hair can be made while the hair is still mounted in the matrix dust specimen, without the need for isolating, demounting, or otherwise manipulating the dust preparation. The authors have successfully compared questioned and known hair specimens while the questioned hair was still mounted in the original Melt Mount preparation.

Numerous protocols for the examination and comparison of human hair appear in the literature. Figure 13.5 depicts some of the variations seen in human head and pubic hair specimens. Refer to Chapter 5 for the human hair protocol used by the authors in their casework. A data sheet such as Table 13.2 makes the collection, tabulation, and interpretation of the data somewhat easier. The data recorded on the table can help establish the somatic and/or racial origin of a questioned hair and show similarities or differences between questioned and known specimens.

Characterization of Animal Hair

Complete animal hairs and fragments often appear in forensic dust specimens. The role of animal hair as evidence in forensic investigations has been established. Animal hair accumulates in dust in much the same way as human hair. Many domestic pets shed hair daily. Hair from grooming pets finds its way into dust. Animal hair originating from articles of clothing, other textiles, and furs can become airborne and thus be incorporated into environmental dust. One of the authors published a scheme to aid in the identification of the various species of animal hair commonly found in forensic science casework. Melt Mount 1.539 serves as the mounting medium.

Complete animal guard hairs (firm, somewhat coarse, prominent protective hairs that originate from an animal's outer coat) present in dust specimens are sorted out during initial examination. These hairs should be examined visually and with a stereomicroscope. Each hair should be sketched and measured. Its reflected light colors and color banding should be noted. All data should be compiled on a data sheet (Table 13.3). After preliminary examination, the scale pattern is cast in Melt Mount 1.539, then a wet mount of the guard hair is prepared in Melt Mount 1.539. Occasionally it becomes necessary to cross-section a guard hair for identification purposes. This can be done only when a sample is large enough. A cross section can be prepared in a few minutes with plastic microscope slides (for details, see Chapter 2). The specimen is then examined under plane-polarized light with a polarized light microscope.

The scale cast is examined first. The dominant scale pattern in the basal region (near the root) of the hair is noted. The scale patterns from root to tip are scanned and noted. Figures 13.6 through 13.8 show the six basic scale patterns. The wet mount is then examined to collect information about the transmitted light color, medullary configuration, and other specimen parameters. All observations are noted on the Table 13.3 form. Figure 13.9 depicts five primary medullary configurations.

The preliminary identification of the family or species of a questioned hair is determined by comparing the data collected on Table 13.3 with the flow chart (Figure 13.10) and with the photomicrographs and information in Appendix B. To confirm the identification, the specimen is then compared to reference standards and published literature.

Often, animal guard hairs, under (fur) hairs, and fragments can be tentatively identified as to species or family of origin on the basis of a few morphological characteristics without an elaborate identification scheme. The technique is useful when examining forensic dust specimens and requires thorough knowledge of animal hair morphology. This knowledge can be acquired by studying the morphology of hairs from known sources. Study specimens can be obtained commercially and from museum collections. An analyst should acquire this background knowledge before attempting to identify hair and hair fragments.

Characterization of Synthetic Fibers

As a result of the extensive modern production of synthetic fibers for all types of textile products, our environment is literally inundated with minute fragments of fibers. Dust specimens composed of synthetic fibers rolled into balls can be found everywhere. Dust balls are formed by the wearing down of textile materials (rugs, mats, clothing, etc.), the shedding of human and pet hair, and the deposition of natural fibers and other materials in the environment. Dust balls can be likened to soil samples and, like soil samples, often represent the environments in which they are formed. Synthetic fibers entangled in dust specimens can be identified in its original state without the need for extensive preparation.

Prior to mounting, the dust specimen should be teased with two needles to loosen the fibers, hairs, and other debris present. The dust specimen is mounted on a microscope slide in Melt Mount 1.539, as previously described, and observed under a PLM. The microscopist will observe a variety fibers in the dust specimen. He or she should single out a questioned fiber and make a number of observations.

Information concerning fiber morphology is collected first. Next, the relative refractive indices (RRIs) of the N|| and N⊥ directions of the fiber in comparison with the RRI of the mounting medium (1.539) are obtained by the Becké line (BL) method using plane-polarized light. The BL method involves aligning the elongated axis of the fiber parallel to the vibrational (preferred) direction of the polarizer. The movement of the BL is noted when the microscope focus is raised (the BL moves toward the medium of higher RI under these conditions). The elongated axis is then made perpendicular to the preferred direction of the polarizer and the movement of the BL in this orientation is noted; Figure 13.11 shows orientation and BL movement.

The questioned fiber is then observed between crossed polars. If it is optically anisotropic, the amount of retardation it exhibits is estimated using an interference chart and the appropriate compensators. The sign of elongation (SE) is determined at this point. Estimated birefringence (EBi) is computed using the collected data. Other comparative information about appearance (bright, semidull, dull, twist, crimp), degree of relief, and so on is collected and recorded in the examiner's notes or on a fiber data sheet (Table 13.4).

To determine the generic classification of an unknown fiber, the collected data in Table 13.4 is compared to the information in Table 13.5 and Figure 13.12, to known published data, and to known standards. Each type of fiber in a dust specimen is identified in the same manner. If a comparison of a fiber is desired, the questioned fiber in the matrix dust specimen and the known fiber specimens can be compared side by side on a comparison microscope (two optically bridged PLMs). Known fiber standards can be compared in the same manner. Chapter 1 of this text discusses basic PLM theory. See Chapter 7 and Appendix C for explanations of how to identify generic classifications of synthetic fibers.

Characterization of Minerals, Glass, and Related Materials

The minerals, glass, and related substances encountered in dust specimens usually originate from the soil in the surrounding region or from other sources in the environment such as vegetation, horticultural operations, animal activity, manufacturing, construction, and traffic. When a forensic dust specimen is mounted in Melt Mount 1.539 and studied with a PLM, tiny fragments of these types of materials are often observed. Just as hairs and fibers can be characterized and identified on the basis of their morphological appearances and optical properties, these materials can be identified in the same manner.

The two most common minerals and mineral-like substances found in dust specimens are grains of quartz and glass fragments. The two substances have similar morphological features, for example, conchoidal fractures and sharp edges. They appear quite similar when viewed under plane-polarized light. However, they are easily distinguishable by the appearance of interference colors in quartz grains when viewed between cross polars (see Figure 13.13).

Figure 13.14 shows other common materials. Most are identified easily on the basis of morphology and a quick determination of optical properties, i.e., degree of relief, interference colors, and so on. When a suspected mineral grain is located in a dust specimen, it is important to determine its thickness along the microscope's optic axis. An accurate thickness measurement and estimated retardation value will enable the analyst to obtain a reasonably true value of the birefringence of the grain. This in turn will allow correct identification of species. Reviews of measurement and birefringence determination can be found in Chapter 3.

Table 13.6 lists various minerals and related materials commonly found in dust specimens and their physical and optical manifestations in Melt Mount 1.539. To determine the identity of an unknown mineral or related substance, compare the data in Table 13.7 with the information in Table 13.6, a Michel–Lévy interference chart, the published data, and known standards mounted in Melt Mount 1.539. Table 13.8 lists less common components found in dust samples.

DUST EVALUATION

Once the trace contents in a questioned dust specimen have been identified, the information should be complied on a data sheet (Table 13.9). The recorded data will enable the analyst to evaluate, interpret, and compare dust specimens much more easily. The data from a tabulation sheet can be effortlessly adapted to create a computerized database that could be interpreted by the use of artificial intelligence. For that reason, a tabulation should be prepared for each dust specimen examined. Finally, a tabulation sheet can be useful when preparing or presenting poster sessions, papers, workshops, court exhibits, and sworn testimony.

REFERENCES

1. Locard, E., The analysis of dust traces. II, *Am. J. Police Sci.*, 1, 405, 1930.
2. Locard, E., The analysis of dust traces. III, *Am. J. Police Sci.*, 1, 496, 1930.
3. Longhetti, A. and Roche, G.W., Microscopic identification of man-made fibers from the criminalistics point of view, *J. Forens. Sci.*, 3, 393, 1958.
4. McCrone, W.C. and Delly, J.D., *The Particle Atlas*, 2nd ed., Ann Arbor, MI, Ann Arbor Science Publishers, 1973.
5. Petraco, N., A guide to the rapid screening, identification, and comparison of synthetic fibers in dust samples, *J. Forens. Sci.*, 32, 768, 1987.
6. Petraco, N. and De Forest, P.R., A guide to the analysis of forensic dust specimens, in *Forensic Science Handbook*, Vol. 3, Saferstein, R., Ed., Englewood Cliffs, NJ, Prentice-Hall, 1993, chap. 2.
7. Bloss, F.D., *Optical Crystallography*, Washington, D.C., Mineral Society of America, 1999, p. 109.

TABLE 13.1
Preliminary Data Sheet

Medium: Melt Mount 1.539

Case No. ______________________________

1. Primary Morphology

   Fibrous ________ Particulate __________ Both __________

2. Homogeneity and Heterogeneity

   Homogeneous? Yes _____ No _____

   Fibers: Separate _______ Clustered _________ Both _________

   Particulate: Separate _______ Clustered _________ Both _________

   Heterogeneous? Yes _____ No _____

   Aggregate of both primary forms? Yes _____No _____

   Number of possible fibrous types ______________________________

   Number of possible particulate types______________________________

3. Initial Classification

Shape______________________________

Sketch:

If fibrous:

   Hair: Human _____________ Animal _____________

   Synthetic fiber______________________________

   Vegetable fiber ______________________________

If particulate

   Mineral grain ____________ Glass chip _______________

   Other particulate(s)______________________________

Note: Used for characterization and identification of trace materials commonly found in dust. Data are obtained with unaided eye and viewed with a stereomicroscope.

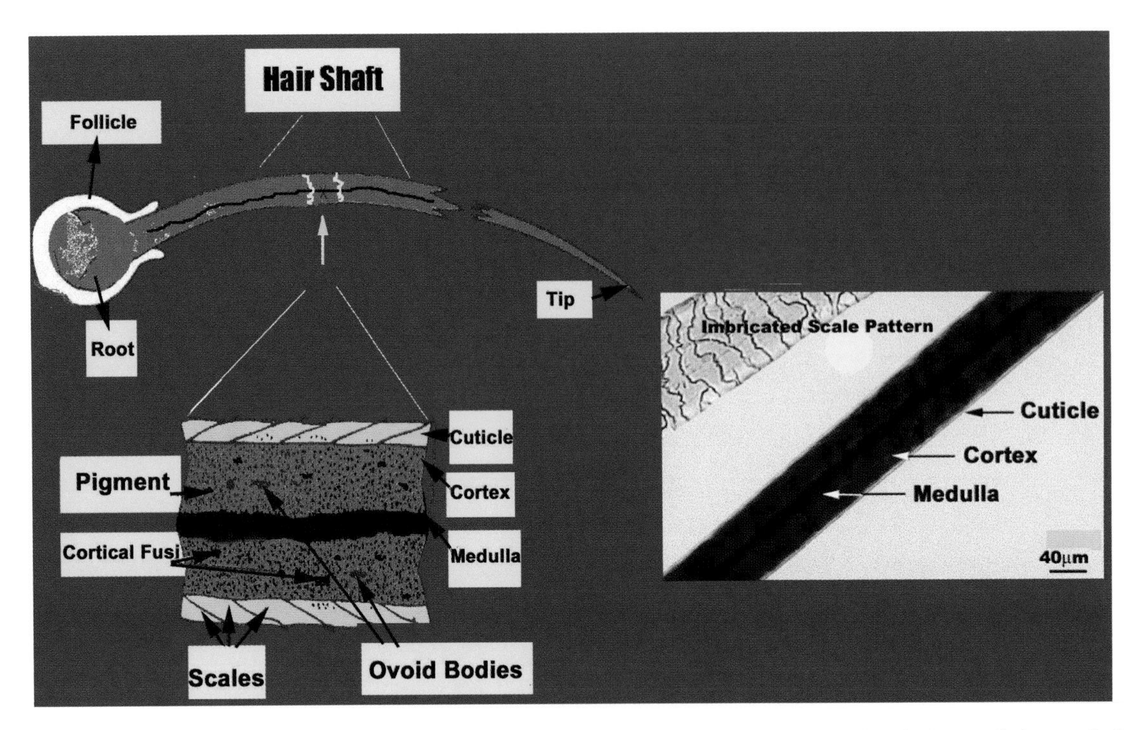

FIGURE 13.1 Right: human head hair cast and mounted in Melt Mount 1.539. Left: diagram of basic human hair morphology.

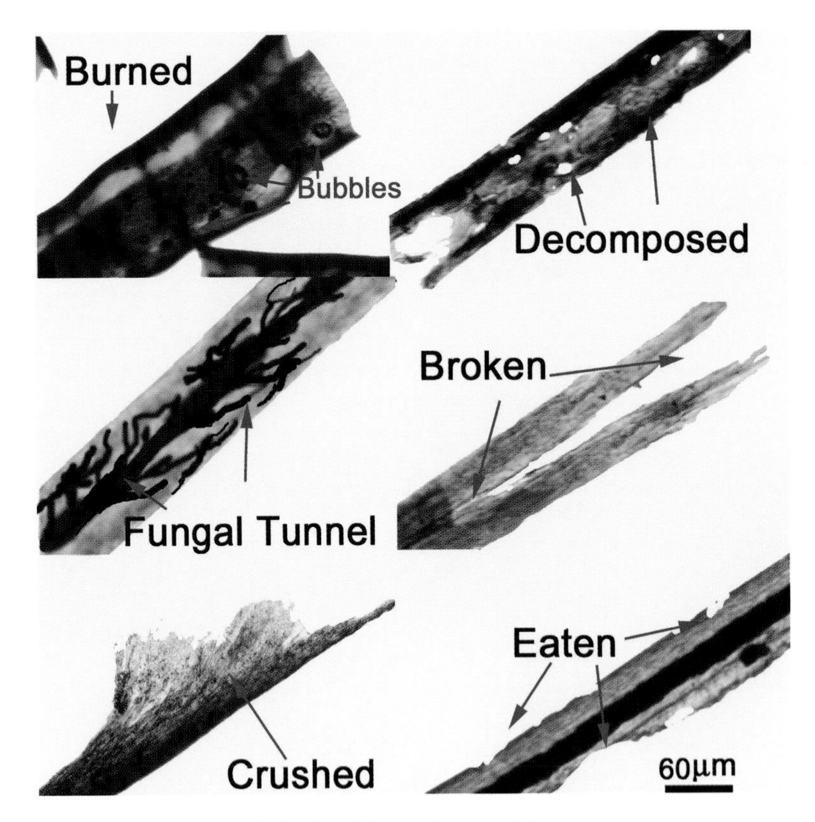

FIGURE 13.2 Several manifestations of human hairs found in specimens of dust.

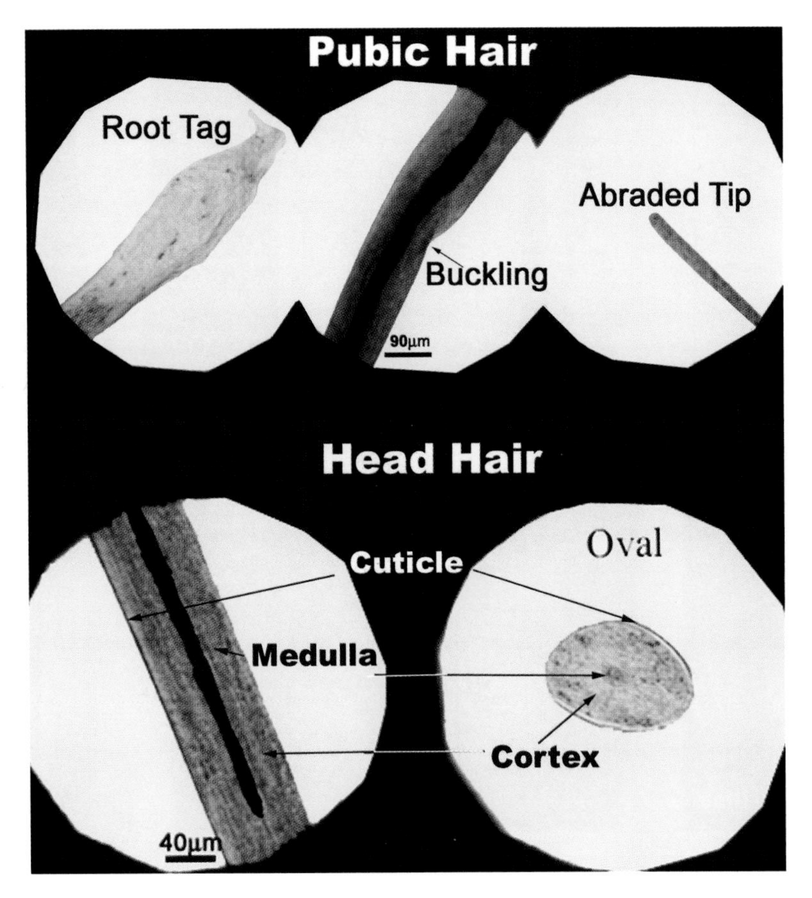

FIGURE 13.3 Morphological appearance of human head and pubic hair.

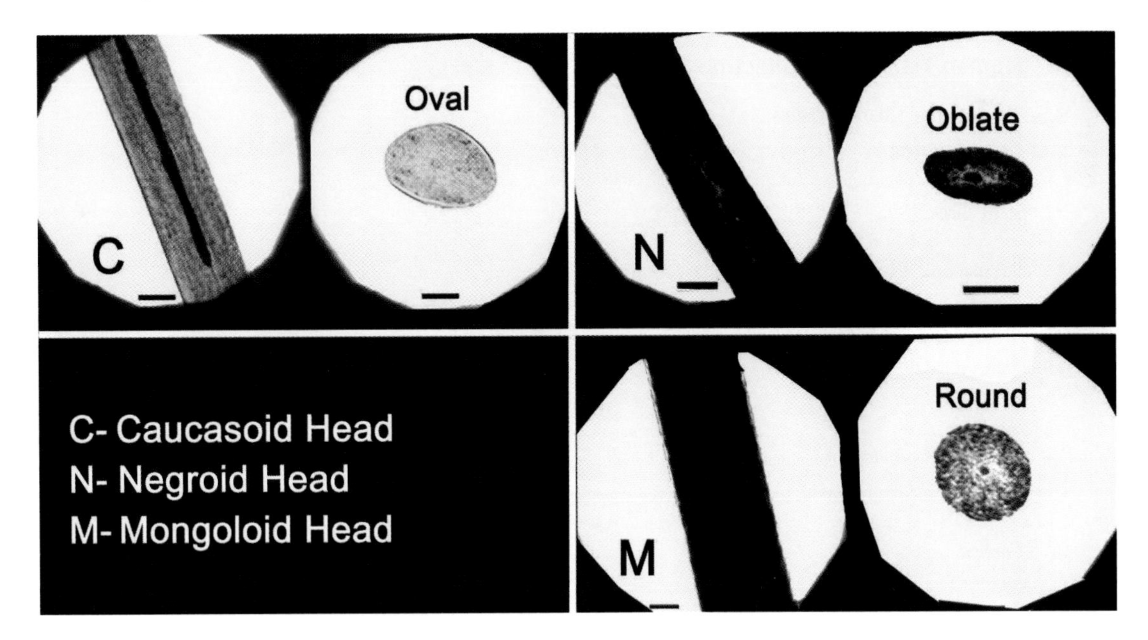

FIGURE 13.4 Three examples of human head hair. Each exhibits major morphological features associated with racial ancestry.

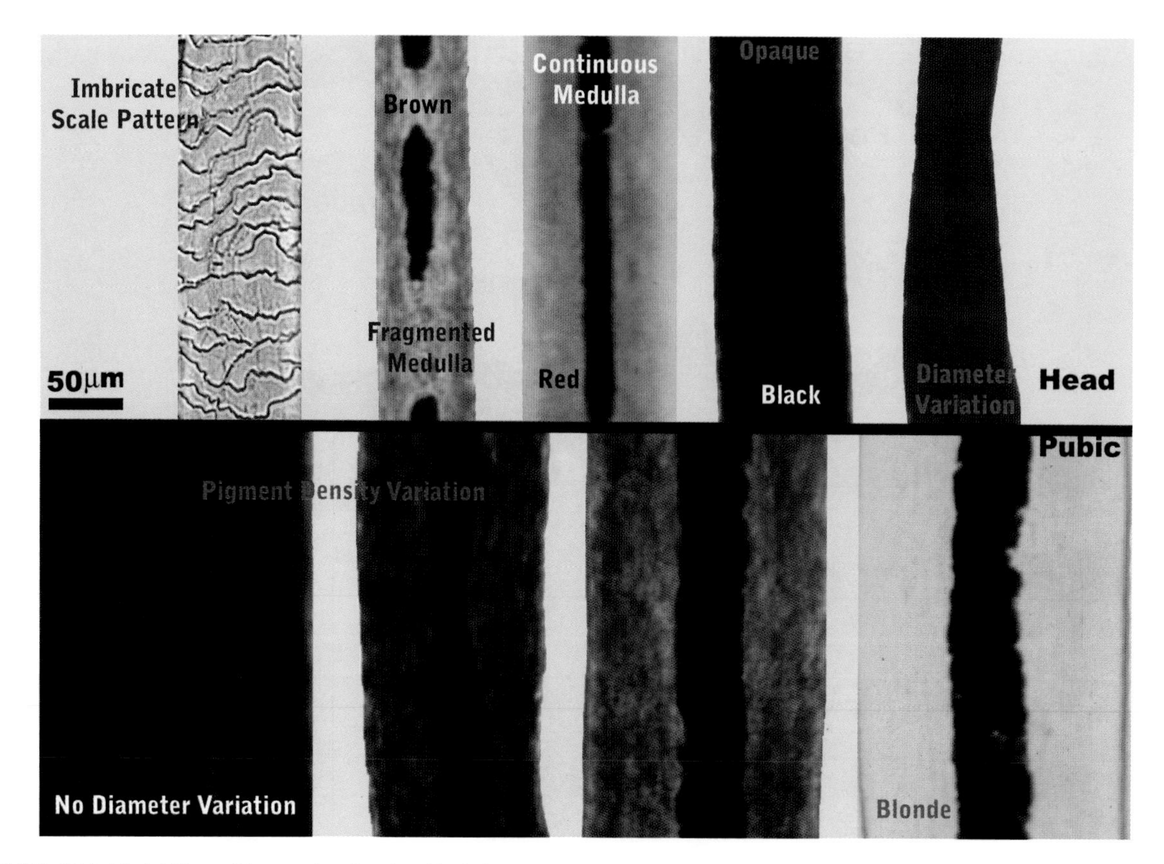

FIGURE 13.5 Variability of human head and pubic hair.

TABLE 13.2
Human Hair Data Collection Sheet

Macroscopic Characteristics

Gross features:

Length ______________________ R.L. color ______________________

Shaft shape ______________________

Microscopic Characteristics

Cuticle:

Margin ______________ Distribution ______________ Shape ______________

Color ______________ Thickness ______________

Medulla:

Medullary configuration ______________________

Diameter thickness (µm) ______________________

Cortex:

T.L. color______________X-S ______________ ______ Pigment distribution ______________

Pigment density ______________ Diameter range ______________ Cortical fusi ______________

Cortical damage ______________ Oddities ______________ Foreign debris ______________

Proximal end

Root shape ______________________

Distal end

Shape ______________________ Growth phase ______________________

TABLE 13.3
Data Sheet for Characterizing Common Animal Hair

Dominant scale pattern basal (base) region ____________

Scale pattern(s) along shaft (describe from root to tip) ____________

Appearance of cuticle (scale) margin ____________

Cortex:
Shape of hair: straight ______ curly ______ crimped ______ other ____________

Length of shaft: ______ mm

Sketch:

Color: single color ______ Bicolored ______ Multicolored ______

If single: reflected ________ transmitted ________

If bicolored or multicolored, describe from tip to root ____________

Banding reflected ____________

Banding transmitted ____________

Pigment density and distribution (e.g., heavy toward medulla) ____________

Shaft diameter (mm): range ________ average ________ maximum ________

Root shape ______

Cross-sectional shape ____________

Miscellaneous (i.e., ovoid bodies) ____________

Medullary Configuration:
Medulla: absent ________ present ____________

Primary configuration ____________

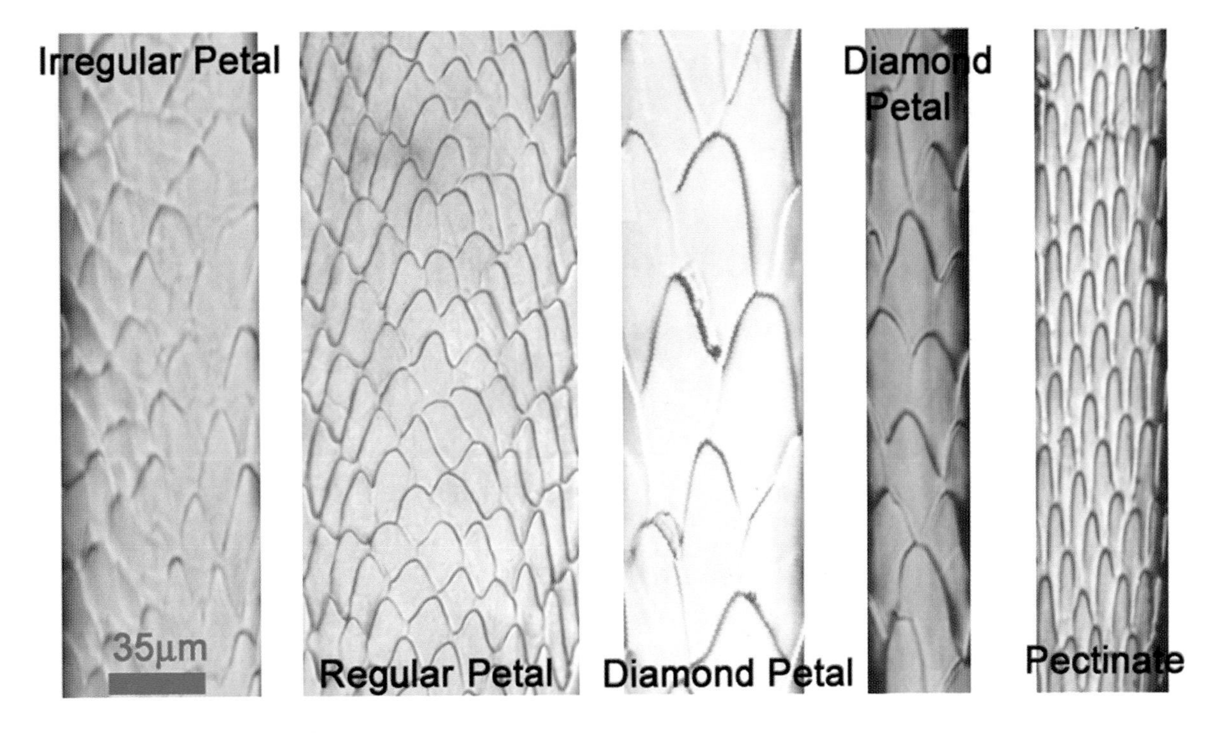

FIGURE 13.6 Petal-shaped scale patterns.

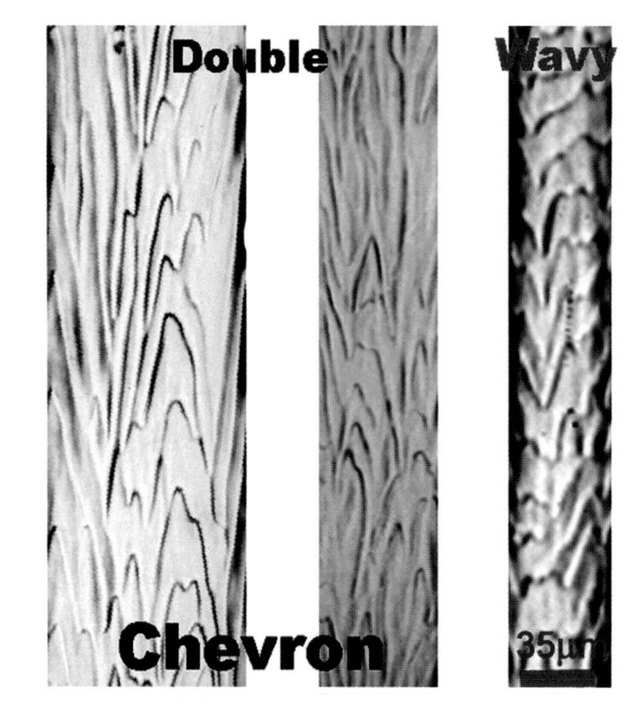

FIGURE 13.7 Basic chevron-shape scale patterns.

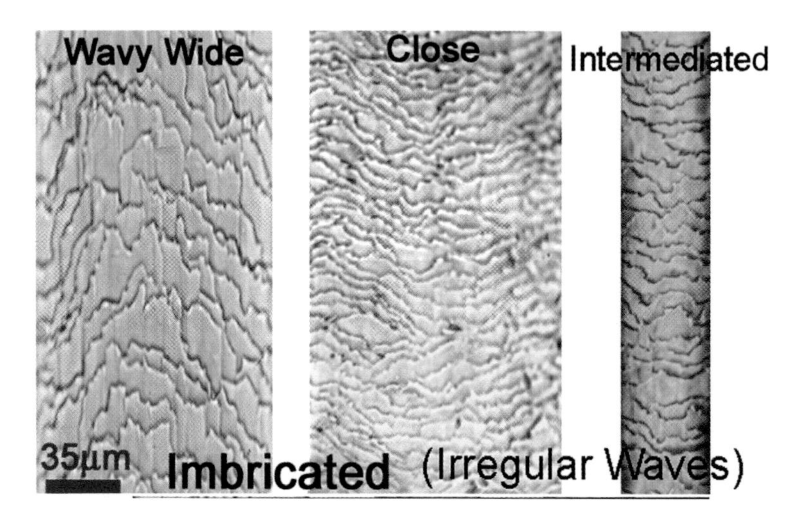

FIGURE 13.8 Basic imbricated or wavy scale patterns.

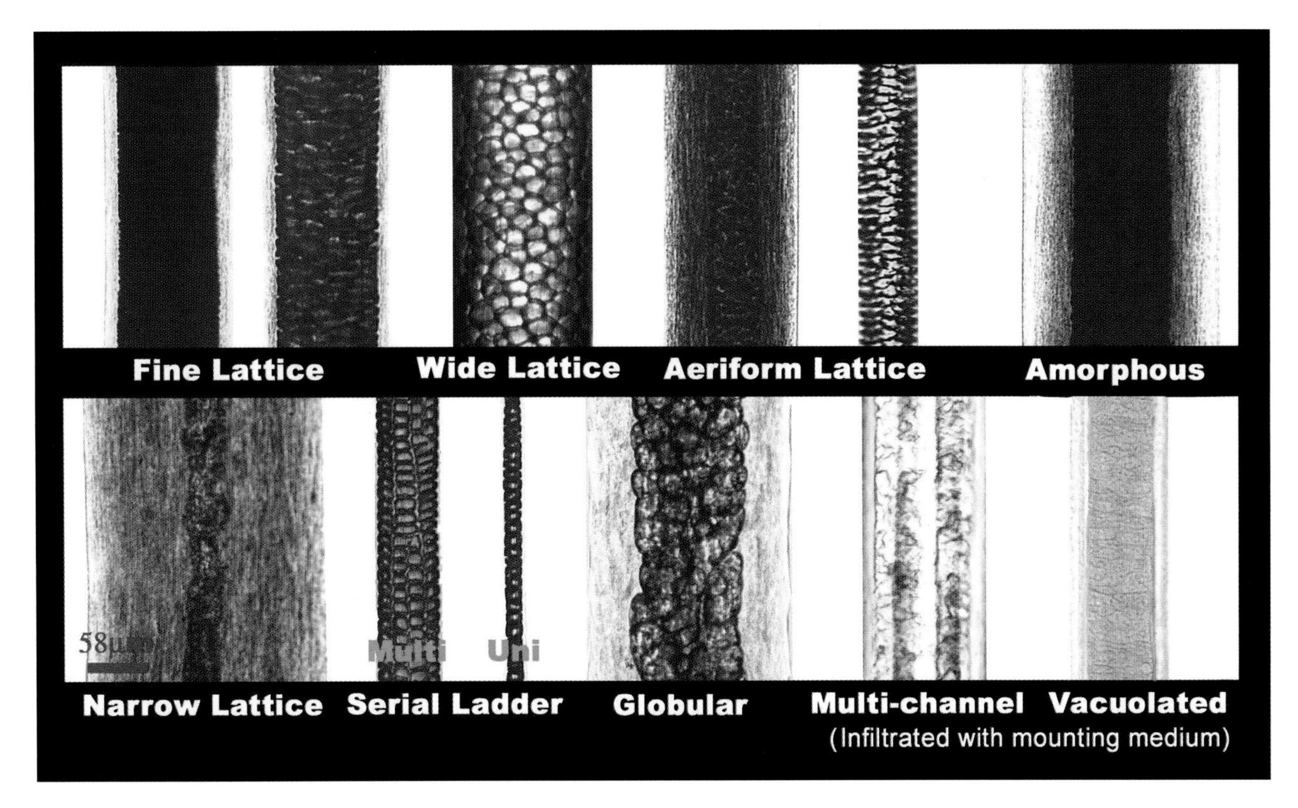

FIGURE 13.9 Basic medullary configurations.

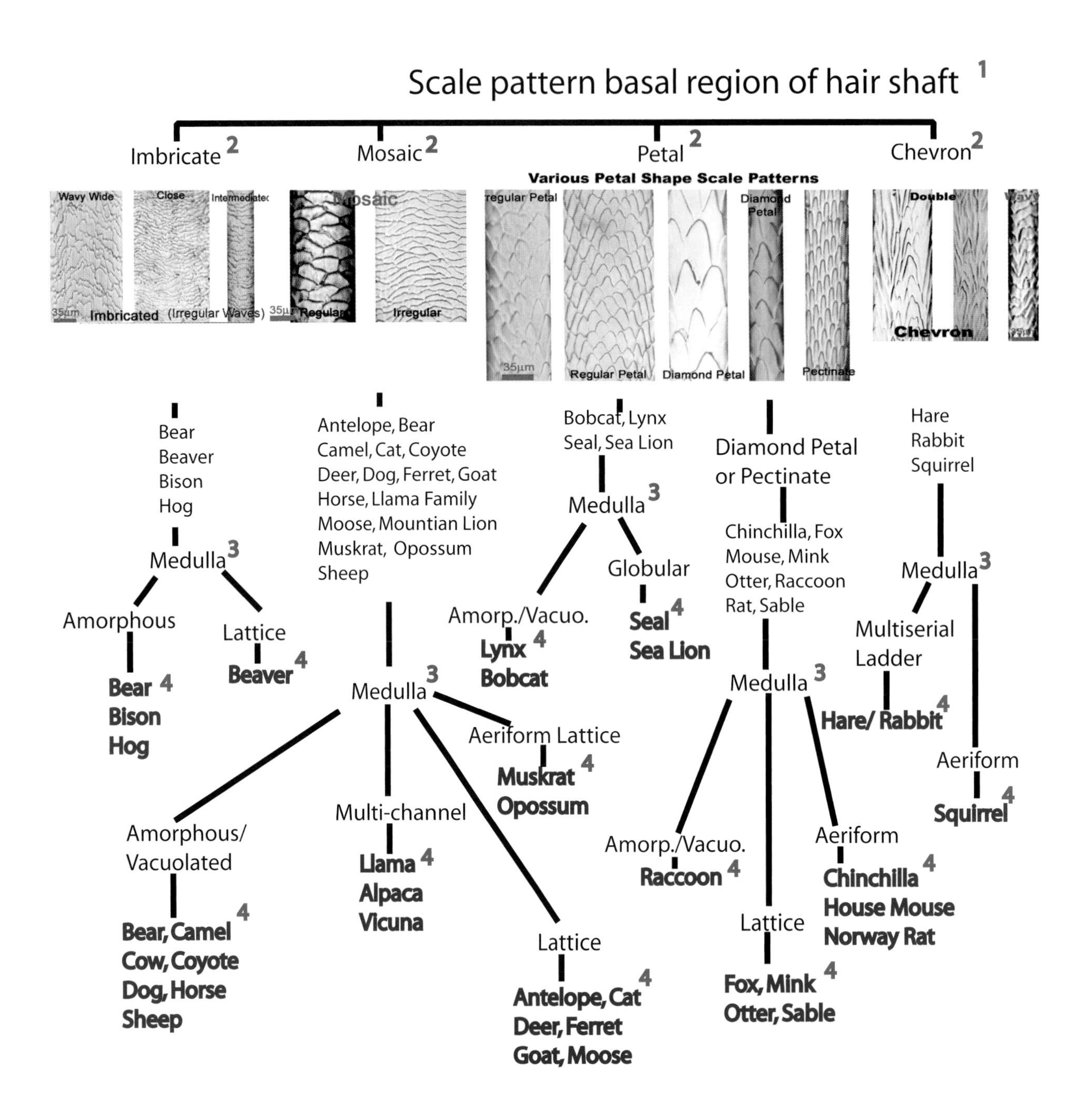

1 Principal scale pattern in the basal region of a mature guard hair.

2 Imbricated scale patterns close, intermediate, wide; mosaic scale patterns regular and irregular; Chevron scale patterns single and double; and petal scale patterns regular, irregular, diamond petal, and pectinate.

3 Primary medulla type exhibited in upper shaft or shield area of mature guard hair.

4 See Appendix B for identifying characteristics.

FIGURE 13.10 Animal hair flow chart.

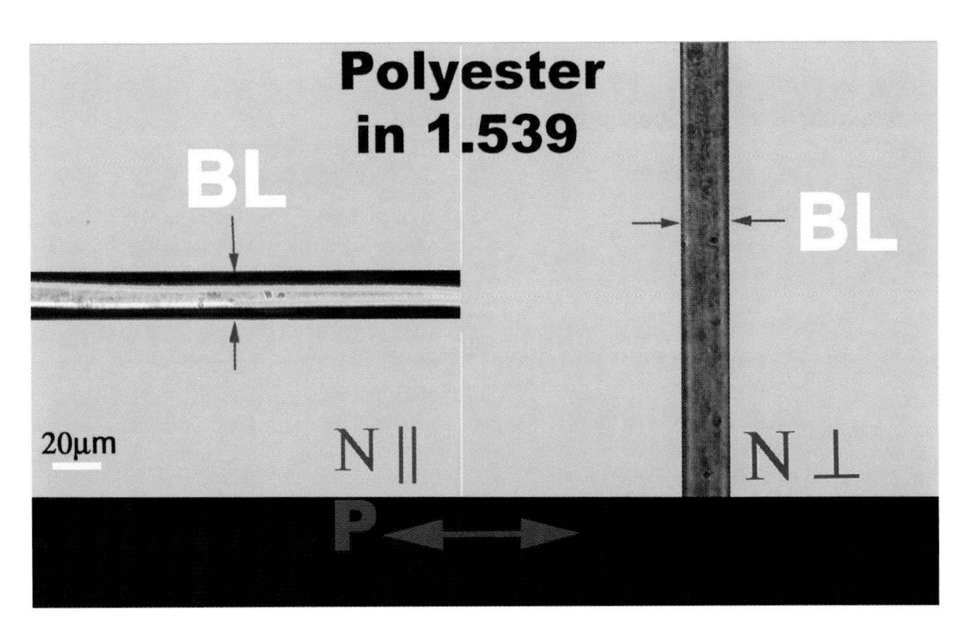

FIGURE 13.11 Fiber orientation and Becké line movement with respect to the preferred (east ↔ west) direction of the plane-polarized light provided by the condenser.

TABLE 13.4
Synthetic Fiber Data Sheet

Fiber Morphology

Longitudinal: smooth _______ striated _______ irregular _______ other_______________________________

Cross-sectional shape ____

Thickness of diameter or lobes (μm)____

Optical Data

Relative refractive indices ___________ (relative to medium = 1.539)

N parallel (n||) above _______ below _______ equal

N perpendicular (n⊥) above _______ below _______ equal

Crossed polars: isotropic _______ anisotropic _______

Estimated retardation (nm)____

Interference colors ____

Estimated birefringence ____

Sign of elongation: positive _______ negative _______

Other Comparative Information

Color: dyed _______ undyed _______

Delustering agent: bright _______ slightly dull _______ semidull _______ dull _______

Treatment: crimped _______ twisted _______ other _______

Degree of relief: low _______ medium _______ high _______

Other information ____

TABLE 13.5
Generic Classes of Synthetic Fibers Commonly Found in Dust

Generic	RRI N∥/N⊥	SE	EB_i (Range)	Relief	Cross-section
Acetate	Both <1.539	+	0.002 to 0.005	Medium to high	Serrated
Triacetate	Both <1.539	+/–	Almost 0.0; slight	Medium to high	Serrated
Acrylic	Both <1.539	–	0.001 to 0.006	Low to medium	Bean, dog bone, mushroom, round, ovoid
Modacrylic		–	0.001 to 0.003	Very low	Dog bone, multilobed
Dynel®	Both near 1.539				
SEF®	Both <1.539	–	0.001 to 0.003	Low	Irregular, ribbon
Verel®	Both <1.539	+	0.002 to 0.005	Low to medium	Irregular
Aramid	Both >1.539	+	0.120 to 0.274	Very high	Round, bean, peanut
Polyamide (nylon 6, 6.6)	N∥ >1.539; N⊥<1.539	+	0.049 to 0.063	Low to high	Round, trilobal, tetralobal
Glass and mineral wool, rock wool	Isotropic; RI 1.510 to 1.620	—	—	Low to high	Round, ovoid, irregular Fractured ends
Olefin	Both<1.539	+	0.028 to 0.034	Low to medium	Round, trilobal, delta, flat
Polyester	N∥ >1.539; N⊥ near 1.539 (> or <1.539)	+	0.098 to 0.180	Low to high	Round, ovoid Polygonal donut, trilobal, swollen ribbon
Rayon (viscose and modified)	N∥ >1.539; N⊥ ±< 1.539	+	0.020 to 0.039	Low to medium	Serrated, multilobal bean, round

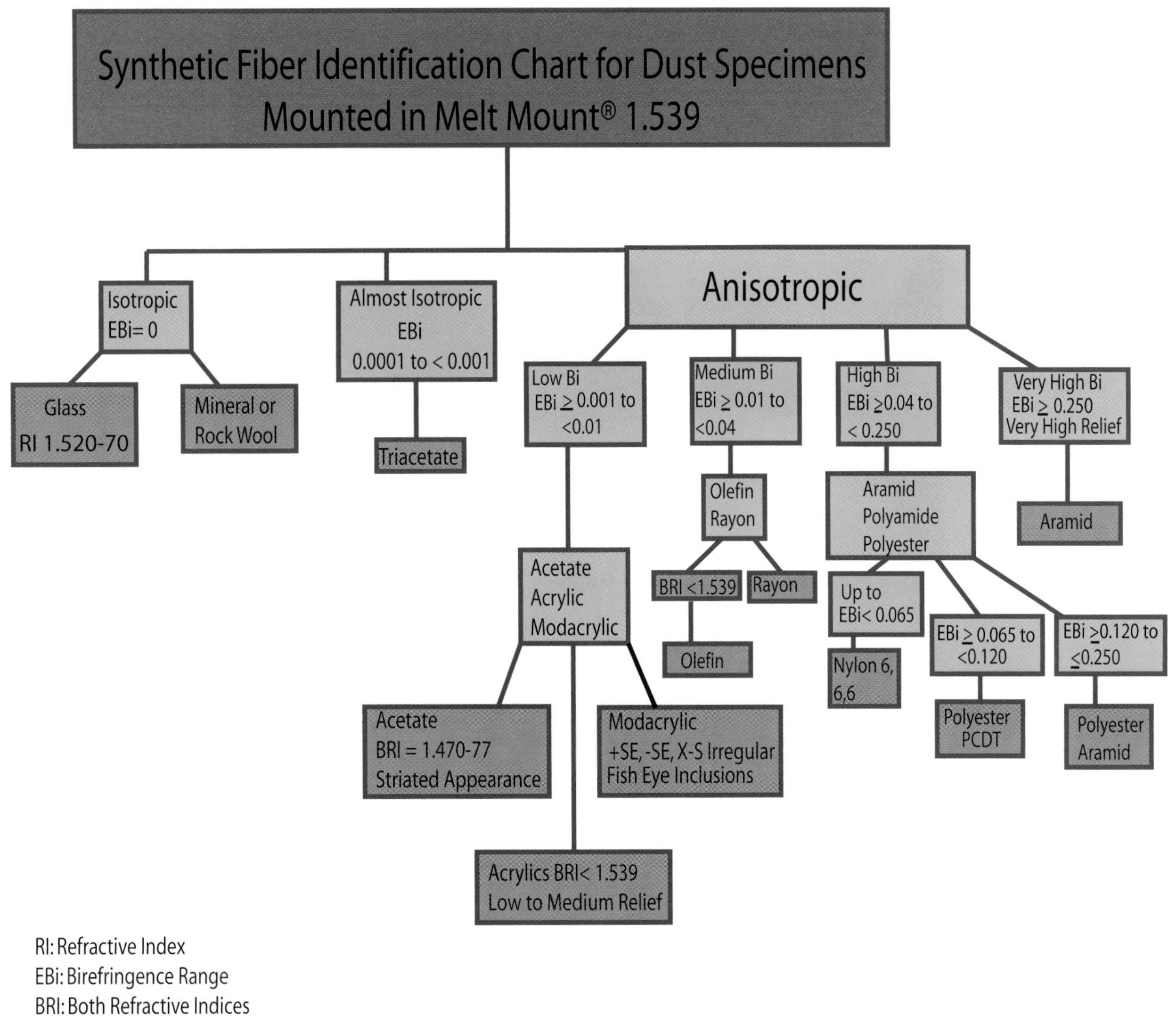

FIGURE 13.12 Flow chart for preliminary identification of synthetic fibers often found in dust specimens.

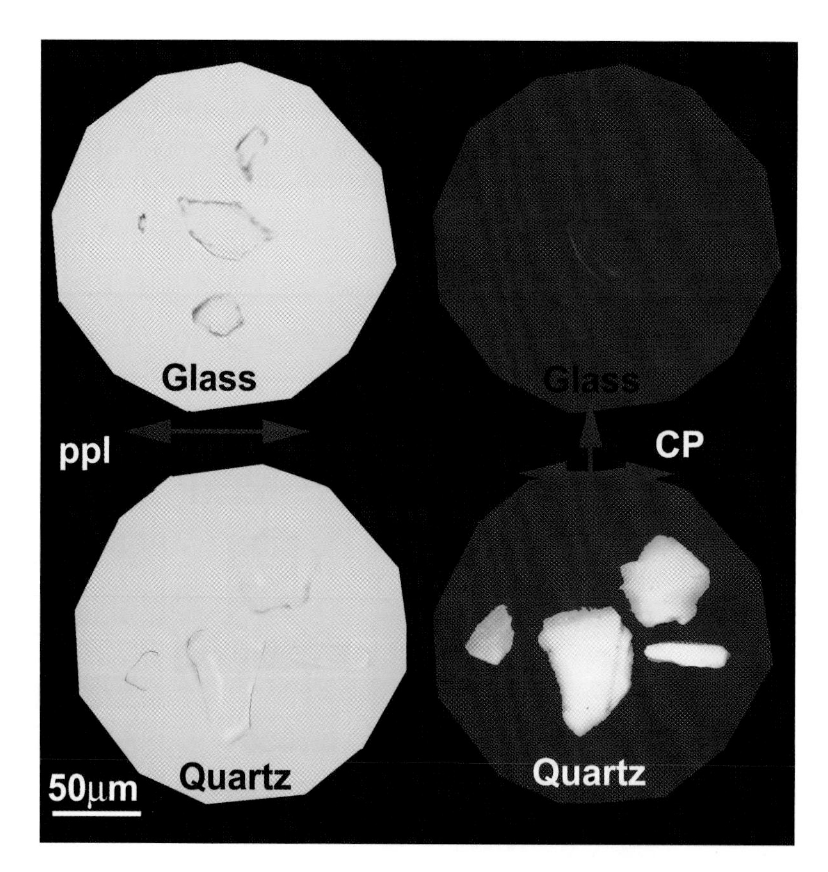

FIGURE 13.13 Top left: glass fragments. Bottom left: quartz fragments. Both are viewed with plane-polarized light. Right: same fragments viewed between crossed polars. Glass is isotropic; quartz is anisotropic.

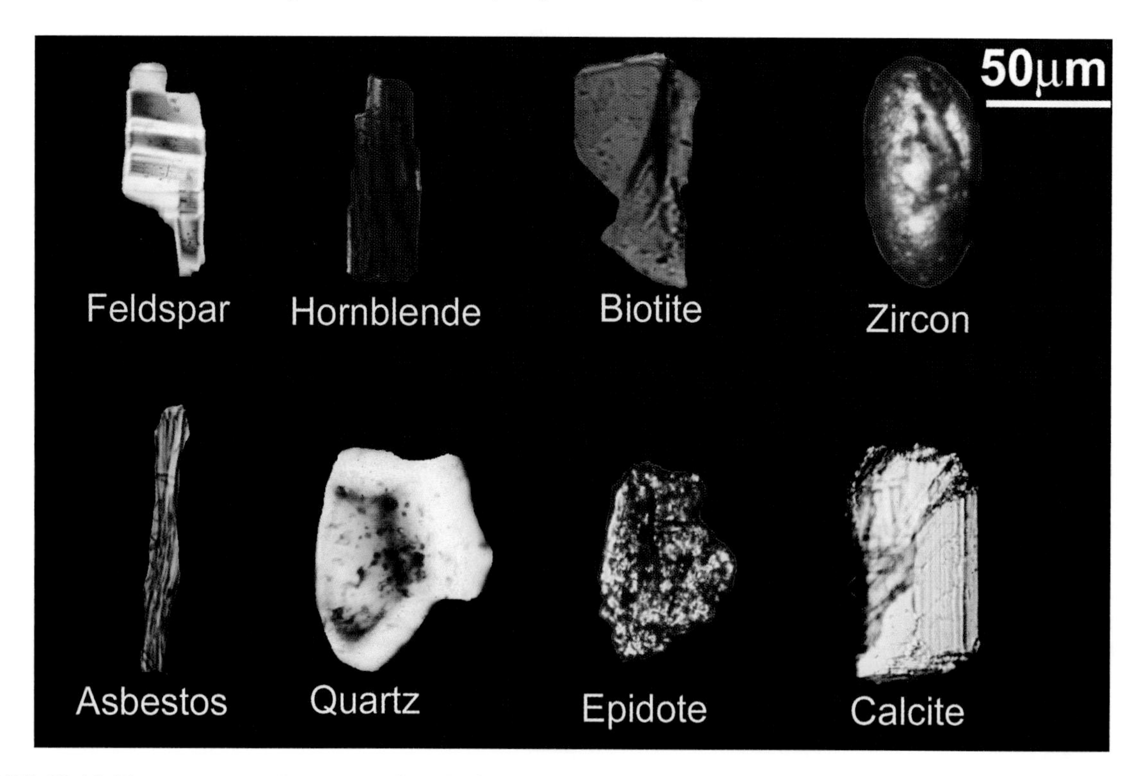

FIGURE 13.14 The appearance of common minerals viewed between crossed polars.

TABLE 13.6
Minerals and Related Materials Found in Dust Specimens

Specimen	Color	Transparency	Crystal Form	Cleavage/ Fracture	Relief	EB	Remarks
Glass	Green, amber, brown, colorless	Transparent	Amorphous	Conchoidal	Very low to high	—	Sharp edges; RI range 1.510 to 1.580; poss. strain Bi
Quartz — SiO_2	Colorless	Transparent	Hexagonal	Conchoidal	Low	+0.009	RI > MM, inclusions, uniaxial + 1 to 2° IC
Calcite — $CaCO_3$	Colorless	Transparent	Triagonal	Perfect; rhombic	Very high	+0.172	Rhombic crystals, high IC like dolomite
Gypsum — $CaSO4{\cdot}2H_2O$ (plaster of Paris)	Colorless to white	Transparent to opaque	Rhombic	Perfect	Low to medium	+0.13	RI < MM; tiny particles; surface rough
Halite — NaCl	Colorless to white	Transparent to opaque	Cubic	Perfect; cubic	Very low	—	Table salt; soluble in water; RI MM near 1.544
Feldspar group	Colorless to pink	Transparent	Monoclinic or triclinic	Perfect to good	Low	+0.005 to 0.011	Twin lamellae; albite RI ≤MM; labradorite >MM; microcline and orthoclase <MM
Mica group	Colorless, gray, green, yellowish-brown, pink	Transparent	Monoclinic	Perfect	Medium	-0.015 to 0.046	Pleochroic; multilayered; biaxial
Garnets	Colorless to pink	Transparent	Cubic	Conchoidal	High	—	Isotropic; RI >>MM; strain Bi
Zircon — $ZrSiO_4$	Colorless to amber, pink, blue, green	Transparent to opaque	Tetragonal	—	Medium to very high	Varies	RI >> MM; can appear opaque; rounded edges
Tourmaline	Colorless to yellow-brown	Transparent to opaque	Triagonal	—	Medium to high	+0.02	Very pleochroic; colors vary (red, blue, green, pink, yellow)
Hornblende (ferro)	Green to yellow-brown	Transparent	Monoclinic	Perfect	Medium	+0.023	Pleochroic; green to yellow; blue to green; RI > MM
Dolomite	Colorless; white to brown	Transparent to opaque	Triagonal	Perfect	Very High	0.184	Looks like calcite
Diatoms	Colorless	Opaque	—	—	—	—	Fine structures; varied forms

Note: Specimens described as they appear when mounted in Melt Mount 1.539.

TABLE 13.7
Mineral Data Sheet

Morphology:

Crystalline form or shape ____________________

Cleavage/fracture ____________________

Twinning/type ____________________

Thickness (μm) along microscope optic axis ____________________

Optical data (plane-polarized light):

Color ____________________

Transparency ____________________

Relief relative to Melt Mount 1.539 (plane-polarized light):

Very low _______ low _______ medium _______ high _______ very high _______

Pleochroic: yes _______ no _______ colors ____________________

Optical data (crossed polars):

Isotropic _______ anisotropic _______

Estimated retardation (nm) ____________________

Interference colors ____________________

Estimated birefringence ____________________

Sign of elongation: positive _______ negative _______

Extinction ____________________

Other Information:

Magnetic: yes _______ no _______

Other ____________________

TABLE 13.8
Substances Found Less Frequently in Casework Dust Samples

Morphology	Material
Fibrous	Asbestos, talc, fiberglass, rock wool, manila, fibers, lint, cotton fibers, plant hairs, sisal fibers, silk fibers, chicken feathers, duck feathers, pigeon feathers, insect fibers (webs), wood fibers
Particulates	Paint chips and smears, sugar, pepper, herbs, sawdust, bark, twigs, leaf fragments, spores, molds, pollen grains, concrete fragments, bricks, cinders, plaster, chalk, rust, metal flakes, explosive substances, dried blood, skin and bone fragments, other tissues, paper, insects and insect parts, small mussels, clams and shrimp, flour, starch grains, drugs, marijuana, seeds

TABLE 13.9
Dust Tabulation Sheet

Specimen no.____

Specimen source____

Human hair? Yes ______ No ______

Racial origin: Caucasoid ______ Mongoloid ______ Negroid ______ Mixed ______

Somatic origin: head ______ pubic area ______ other________________________

No. of hair types: races ______ body areas ______

Animal hair? Yes ______ No ______

Guard hair ______ tactile hair ______ fur ______ other ________________________

Species of origin: dog ______ cat ______ other________________________

No. of different species ________________________

Synthetic fibers? Yes ______ No ______

Generic classes: Acetate ______ Triacetate ______ Acrylic ______ Aramid ______ Modacrylic ______

Polyamide ______ Polyester ______ Rayon ______ Olefin ______ Glass ______ Mineral ______

Other ____

No. of different types of each generic class: Acetate ______ Acrylic ______ Aramid ______

Modacrylic ______ Polyamide ______ Polyester ______ Olefin ______ Rayon ______ Glass ______
Mineral ______ Other ________________________

Vegetable fibers? Yes ______ No ______

Type: Cotton ______ Ramie ______ Sisal ______ Flax ______ Other________________

Minerals, glass, and related materials? Yes ______ No ______

Type: Quartz ______ Glass ______ Other ______

Miscellaneous Substances? Yes ______ No ______

Type____

No. of similar materials in questioned and known dust ________________________

No. of dissimilar materials in questioned and known dust________________________

Known and questioned: similar ______ dissimilar ______ both ______

14 Case Studies

INTRODUCTION

The ability of scientists to use tiny quantities of materials to reconstruct past events, solve crimes, restore artifacts, and describe past civilizations has a long history of success. Forensic scientists frequently encounter many different types of materials during the investigation of a crime. The identification and comparison of miniscule traces of diverse materials such as synthetic fibers, natural fibers, wood fragments, paint chips, minerals, feathers, pollen grains, soil, and hair can be crucial elements in the reconstruction of an event.

Chemists are often called upon to identify tiny samples of unknown substances. Art historians must know the material from which an artifact is made before they can render an opinion as to its true origin. Art conservators repair, restore, and preserve rare paintings, ancient tapestries and textiles, statuary sculpted of a range of materials, historic buildings, etc. Before they can restore or preserve such objects, they must ascertain composition so they can properly refurbish and/or conserve the objects. Archaeologists must determine the materials used in the past before they can reconstruct their past societies and civilizations. This chapter presents casework examples that demonstrate how scientific investigators utilize minute traces of a diverse array of materials to accomplish their work.

CASE 1

On a cold winter day, a dead female was found in the alleyway of an east Harlem tenement. A California florist flower box with a plastic liner was found in close proximity to the body. The decedent was identified as a member of a well-known church and was known to have sold church literature in the buildings surrounding the alley in which her body was discovered. The detectives investigating the case forwarded the flower box, plastic liner, and the decedent's clothing to the forensic science laboratory.

Polarized light microscopy (PLM) identified tan wool fibers, red acrylic fibers, and navy blue wool fibers found on the box and liner. The three types of questioned fibers were compared microscopically with the decedent's clothing and found consistent in all respects. She had been wearing a tan wool overcoat, navy blue wool/polyester-blend slacks, and a red acrylic sweater). This finding associated the woman with the flower box and liner. In addition, light blue nylon rug fibers and several brown rabbit hairs were found on the box and liner. Similar light blue nylon rug fibers and rabbit hairs, and red nylon rug fibers, were found on the decedent's tan wool overcoat. Neither the rabbit hairs nor the nylon rug fibers could be associated with the victim's environment (her clothing or residence).

This information was conveyed to the field investigators. After further inquiries in the neighborhood, they learned of a man who sold a full-length, brown rabbit hair coat to a local man the day after the body was discovered. The investigators obtained the coat from the purchaser. The hair of the coat was compared microscopically to the questioned rabbit hairs found on the victim's wool coat and the flower box liner. The specimens of questioned rabbit hair were consistent in all physical and microscopic characteristics to the rabbit hair of the suspect's coat. This information gave the police probable cause to obtain a search warrant for the suspect's apartment.

In the suspect's apartment, two rugs were found. One was light blue and the other was red; both rugs were composed of nylon fibers. Samples of both rugs were collected by the crime scene unit and forwarded to the forensic science laboratory for comparison with the questioned rug fibers found on the victim's clothing, the flower box, and the plastic liner. Both the questioned and known rug fibers were found to be consistent in all respects. The presence of light blue nylon rug fibers, red nylon rug fibers, and brown rabbit hairs on the flower box, plastic liner, and woman's clothing enabled the author to associate the woman, the flower box, and liner found in the alleyway with the suspect and his apartment. Table 14.1 shows the various associations made possible by the trace evidence.

The investigating detective made further inquiries about the suspect. A witness stated he saw the suspect carrying a large California flower box a day or two before the body was discovered. From the evidence it was theorized that the woman was killed in the suspect's apartment, placed in the flower box, brought up to the roof of the building in which the suspect resided, and thrown off the building into the alley below. On the basis of the evidence, the suspect was arrested, indicted, and tried for murder in the second degree. Extensive testimony about the trace evidence was presented over a 3-day period. The defendant was found guilty of murder in the second degree and subsequently sentenced to life imprisonment.

This case demonstrates the kinds of associations of people, places, and things that can be made by trace evidence. Trace evidence can also reveal information about the occupations or surroundings of the principals in a case.

The next case shows how this information can be useful in another real-life situation.

CASE 2

In June 1978, the body of a woman in her 20s was found in a parking lot in midtown Manhattan. The medical examiner found black fibrous material in her hands. Microscopic examination revealed a blend of synthetic fibers consisting of 80 modacrylic fibers, 15 acrylic fibers, and 5 polyester fibers. A search of the victim's husband's van found in New Jersey produced similar looking tufts of black fibrous material. Although the van's interior had been recently cleaned and stripped of its carpeting, black fibers were found on a wooden plant holder that the husband used in his floral business and in the samples of vacuum sweepings collected from the rear of the van.

PLM analysis of the black fibers found in the van revealed a synthetic blend consisting of 80 modacrylic, 15 acrylic, and 5 polyester fibers. The comparison of the black synthetic fibers found in the victim's hands and the black synthetic fibers found in the van disclosed them to be consistent in all respects.

During the investigation, a question arose concerning the victim's husband's previous occupation. Although he was now employed in the florist industry, it was believed that he was at one time involved in the building industry as a contractor. The investigator also wanted to know whether the van had been used to transport building materials. PLM examination and analysis of the vacuum sweepings removed from the van disclosed the presence of trace materials that could be associated with the building industry (see Table 14.2).

Although the items listed on Table 14.2 do not conclusively prove the occupation of the van's owner or user, they do, at the very least, provide a strong indication. At the trial, both the black fibers and particulate matter were used to implicate the woman's husband and his accomplice in her death. On occasion, the unequivocal association of the people, places, and things involved in an event can be achieved by utilizing available trace evidence. The following case is an example.

CASE 3

In the early morning hours of April 12, 1982, atop a lonely roof garage on the west side of Manhattan, bodies of three men were found. Each man had been shot once in the back of the head. A light-colored van was seen speeding away from the scene. Hours later, two dog walkers found the body of a fully clothed woman lying face down in a secluded alley street on the lower East Side. The woman was killed in the same manner as the men on the roof garage. The condition of her body and other evidence indicated she had been shot at the garage and then transported to the alley.

An eyewitness stated that he saw a man shoot a woman and place her in a light-colored van. The gunman then chased down the three men who tried to go to the woman's aid and shot each one of them. Days later, the prime suspect in a black van was arrested in Kentucky.

More than 100 items of evidence were collected from the van and forwarded to the New York City Police Laboratory for examination. Among the items forwarded were three sets of vacuum sweepings from the van interior. An autopsy of the woman produced several items of trace evidence that were forwarded to the author for microscopic examination. The woman's clothing was also received for trace analysis.

A prime question that arose during the investigation was whether the woman's body, which was placed in a light-colored van at the garage and later left in an alley on the lower East Side, could be associated with a black van recovered more than 1000 km away from the scene. Microscopic analysis and comparison of the trace evidential materials found on the victim and inside the van made this association possible.

Table 14.3 lists trace materials the victim and the van had in common. Microscopic comparisons of the questioned human head hair present on the victim's clothing were made with known samples. Ten brown and gray Caucasian head hairs from the victim's blazer were consistent in microscopic characteristics to the defendant's known head hair sample. One chemically treated head hair found on the victim was consistent in microscopic characteristics to the known head hair sample obtained from the defendant's wife. One forcibly removed brown Caucasian head hair found on the rear door of the van interior by the Kentucky State Police laboratory personnel was found to be consistent in all microscopic characteristics with the decedent's known head hair.

A white seed recovered from the victim's mouth by the medical examiner and a white seed found in van sweepings were forwarded to an internationally known botanist for identification and comparison. At trial, the botanist testified that the two seeds were identical in all respects. Although he could not identify them, both were from the same species of plant and, if not from the same plant, probably from a rare wild flower (Figure 14.1).

Sixteen gray metallic/black paint chips found on the victim and her clothing were compared to the gray metallic/black paint removed from the van. Samples from the questioned and known sources were examined and compared by microscopic, chemical, and instrumental means. All the paint specimens from the van and from the victim were similar in all respects (Figure 14.2). The remaining items of trace evidence from the victim and the van were examined and compared microscopically and, where necessary, by chemical and instrumental methods.

The remaining types of trace evidence from the victim were found to be similar to their counterparts from the van.

Blue and black flakes of acrylic paint were found in vacuum sweepings from the van and on the suspect's sneakers. No blue and black paint flakes were found on the victim or her clothing. During a search of the defendant's residence in New Jersey, a large quantity of blue and black acrylic paint was found in the garage. It was apparent from the evidence in the defendant's garage that a large rectangular object had been painted recently with blue and black paint. The blue and black paint flakes from all the sources and the known blue (undercoat) and black (topcoat) paint from the van were compared by microscopic, chemical, and instrumental means. All the samples were consistent in every respect.

At trial, extensive testimony concerning the collection, examination, identification, and comparison of the trace evidence from the victim and the van was given over a 2-day period. When questioned about the source of the trace evidence found on the victim and her clothing, the author stated unequivocally that the trace evidence on the victim was from the defendant's van. On the basis of this evidence and circumstantial evidence, the defendant was found guilty of all charges and sentenced to 100 years in prison.

CASE 4

The complainant in a civil action alleged that a contractor dumped asphalt-contaminated soil from a highway project onto his private property. Six specimens of soil from various locations were received from the environmental laboratory investigating the case. The laboratory manager requested that they be examined to determine which one did not originate from the same location as the other five. The request was also made to carry out the examination without the examiner's knowing the origins of any of the soil specimens.

The data shown in Table 14.4 support the finding that sample S1 most likely originated from a location different from samples S2 through S6. Samples S3 and S6 were shown to have originated from similar locations, as did samples S2, S4, and S5. The findings were confirmed as correct by the requesting laboratory manager.

CASE 5

In the early spring, the body of a young adult female was seen floating down the East River near midtown Manhattan. Several joggers saw the body bobbing up and down as they ran along the boardwalk that parallels the East River. The dead woman, later identified as a missing correction officer, was pulled from the river at the lower end of the East Side. At autopsy, it was determined that the victim had been shot with a handgun. Investigation revealed that the dead woman was in the process of divorcing her estranged husband. Consequently, he became the prime suspect. A search warrant was issued for his upstate residence. While executing the warrant, the crime scene officers discovered a water-stained man's right shoe in the suspect's bedroom closet. A small quantity of sand adhered to the inside portion of its heel. The shoe was sent to the laboratory for examination.

Before removal of the sand from the heel, its presence was documented, as shown in Figure 14.3. The shoe was then examined with a stereomicroscope. The questioned grains of sand seemed to adhere to the heel and to each other. A total of 10 mg of sand was removed from the shoe. The sand specimen was not sufficient for sieving. After color analysis, the stained area of the shoe and the questioned sand specimen (QS1) were examined chemically for sodium and chloride ions. Both tested positive for both ions. Residue removed from the stained area of the shoe was examined by x-ray diffraction (XRD) analysis and emission spectrographic analysis. Samples of the river water were taken and allowed to evaporate. The known salty residue and questioned residue were examined in the same manner. Both specimens had the same trace elemental compositions, and XRD confirmed that both contained sodium chloride.

During the investigation, it was hypothesized that the suspect had taken his estranged wife to a local beach where he pulled her into the water and shot her with her own revolver. The revolver was never recovered. Known sand samples (KS1) of the local beach were collected and compared to the QS1 specimen. Table 14.5 illustrates that the known (KS1) and questioned (QS1) sand specimens were consistent in all respects.

Therefore, it was determined that the QS1 specimen may have originated from the source of KS1 (known sand specimens were collected from various locations along the East River and all were found different from KS1 and QS1). Soil-evidence testimony was presented at trial to help the jury reconstruct the event. The suspect was found guilty of second-degree murder.

CASE 6

The body of a young woman was found on a beach lying supine in the sand. She was killed by a blunt force trauma to the head. Her van was on the beach next to her. Three days after the body was found, a suspect who lived near the beach was arrested. Members of the crime scene unit searched the suspect's residence. Known sand samples (KS1) from the beach and samples from the inside of the victim's van (QS2 and QS4) were forwarded to the laboratory. Questioned sand samples from the suspect's clothing (QS1), sneakers (QS5), socks (QS6), and pant legs (QS3) were collected and forwarded for comparison with the known beach sand and questioned specimens from the van.

Every sand sample was processed through a standard sieve series (20, 40, 60, and 80 mesh). The 80-mesh fractions were used for color analysis (Figure 14.4), density gradient separation, and PLM examination. Table 14.6 contains the data used in the comparison. From Table 14.6, one can see that although numerous similarities of the known and questioned specimens are evident, significant differences made it impossible to associate the questioned specimens to the sources represented by KS1, QS2, and QS4.

CASE 7

In July 1982, a grocery store robbery turned into a homicide when the owner of the store tried to fight back and was shot with a small caliber handgun. Upon arrival at the scene, the investigating detective found a holster near the victim's hand along with a handgun. It was not known whether the holster belonged to the grocer or the robber. Later that day, a suspect was identified by an eyewitness and arrested. A search of the suspect and his residence produced a small caliber pistol. Both handguns and the holster were forwarded to the laboratory for examination.

Based on the impression found in the holster left at the scene, it was evident that the pistol obtained from the suspect's residence was most likely the weapon stored in the holster (Figure 14.5). Since the spent projectile that killed the grocer was never recovered, it became vital to further associate the pistol and the holster. On the bases of size and shape, the handgun from the scene, a small .32-caliber revolver believed to be the grocer's, was excluded as coming from the holster.

Tiny specks of dust removed from the corners of the holster and the pistol trigger area were examined and compared with PLM. A comparison of the dust from the holster to the dust from the pistol revealed 17 different fiber and hair types common to both objects. Each fiber type exhibited the same physical and optical properties and chemical composition (see Table 14.7). When confronted with this information, the suspect admitted his involvement and stated that he accidentally dropped the holster while attempting to escape. Defense counsel advised him to accept the plea bargain offered.

CASE 8

In the early morning hours of a warm day in summer of 1984, concerned neighbors discovered the bodies of an elderly couple. From the condition of the apartment and the bodies, it appeared that a simple burglary turned into a double homicide. The fire escape window of the couple's three-room Manhattan apartment was broken. Dresser drawers in the bedroom had been emptied. The couple lay in pools of blood on the living room floor. It appeared that an intruder surprised the victims and killed them with a blunt instrument. Clumps of dust were found covering the floors in each room of the apartment.

A specimen of the dust was collected by the investigator and safeguarded for future study. A few days later a suspect was arrested across town. A search of his clothing revealed a clump of dust stuck to his right shoe. The specimens were forwarded to the laboratory for examination and comparison.

Examination and comparison of the dust from the victims' apartment to the dust from the suspect's right shoe divulged that 12 materials were common to each specimen. Figure 14.6 is an overview of the materials found in the dust specimens. Table 14.8 lists the materials both specimens had in common. The results of this comparison enabled the author to report that the dust from the suspect's shoe most likely originated from the victims' apartment. When told of the results of the examination, the suspect confessed to the killing.

CASE 9

On a snowy day in March 1986, an immigrant family in New York City's Chinatown reported their 14-year-old daughter missing. As was her custom, the young girl had returned home from school in the early afternoon. The rest of the family normally arrived from work around 6 o'clock. When they arrived home, their apartment was in disarray and their daughter was missing. They immediately reported the missing daughter to the police who conducted a search of the tenement in which the family resided and the surrounding neighborhood. The girl was not found. A short time later the family received a telephone call demanding a ransom for her safe return.

The next day, the police decided to conduct a second search of the neighborhood and dogs were employed in the search. A short time later, the girl's body was found in a sleeping bag in the courtyard behind the building where she had lived. She had been strangled with a ligature. Figure 14.7 shows where the victim was found. When her body was removed from the sleeping bag, several questioned dust (QD) balls were found as shown in Figure 14.8.

The following day, a young unemployed man was arrested for kidnapping and murder. He resided across the hall from the victim in a shared three-bedroom apartment. Each occupant in the apartment had a bedroom and shared a common living room, kitchen, and bathroom. A search of the suspect's bedroom, another bedroom, the kitchen, living room, and bathroom for dust consistent with the QD proved negative. However, while searching the premises, officers noticed that the third bedroom was locked. The female occupant was out of the country. A search warrant was obtained for the locked bedroom. When the room was opened, clumps of known dust (KD) were observed in plain view on a gray-colored rug. The crime scene officers documented the location, removed the dust, packaged it, and forwarded it to the laboratory for comparison.

Examination and comparison of the QD with the KD revealed that 28 materials were common to each dust specimen. Figure 14.9 is a group of photomicrographs of six of the common materials. Table 14.9 lists all the materials found in both the QD and KD specimens.

Upon returning from her trip, the resident of the locked bedroom was questioned by detectives. She identified the sleeping bag as hers. She stored it in the closet in her room. She also told investigators she was a seamstress and did much of her work at home.

From the evidence, it was theorized that the suspect killed the victim in her apartment. He then moved the body into his neighbor's locked bedroom and placed it on the gray rug. Upon discovering the sleeping bag, he placed the victim into the bag and stored it in the unoccupied bedroom. That night he moved the sleeping bag containing the body into the courtyard. When informed of the evidence against him, he confessed, and agreed to plead guilty if his name was not released to the local Chinese community.

CASE 10

A young male was found dead in bed with an apparent bullet wound to the left side of his head. Later, a single .38-caliber deformed lead bullet was recovered from inside the pillow under the head of the deceased. At autopsy, no other spent bullets were recovered from the body. Consequently, the medical examiner wanted to confirm that the bullet recovered from the pillow caused the fatal wound. The trace evidential material present on the questioned bullet made this confirmation rather trivial and provided more information than was actually required.

As shown in Figure 14.10, the bullet (B) was covered with a large tuft of white fibers (F) and several dark hairs (H). Close scrutiny of the trace material adhering to the questioned bullet revealed four distinct layers of trace material. The outermost layer of white fiber was identified with PLM as polyester fibers of the same type of polyester used to fill the pillow. The next layer was composed of fragments of white cotton threads similar to the textile material utilized in the pillow casing. The third layer contained brown human head hair and some bone fragments. The final layer was composed of brown human head hair. Some blood and soft tissue were also mixed with the inner layers. The questioned human head hair fragments recovered from the bullet were found to be consistent in all microscopic characteristics with the known head hair specimen taken from the decedent at autopsy.

Figure 14.11 depicts the various substances found on the bullet. The trace evidence obtained from the questioned bullet made it apparent that it caused the young man's death — entering and exiting his head and then entering the pillow under his head. Confirmation by physical evidence was probably not necessary but this case illustrates how a sequence of targets through which a bullet has passed can sometimes be determined from examination of the layers of trace evidence on the targets.

CASE 11

During the commission of an apartment burglary, the resident shot the intruder once while the intruder was entering the premises through the living room window. The intruder fled on foot and sought medical treatment at a local hospital. A search of the crime scene produced one deformed .38-caliber lead bullet in the rear courtyard directly under the complainant's living room window. Several tufts of blue fibrous material were embedded on the bullet surface. The suspect was apprehended at the hospital after hospital authorities notified the police that they were treating a man with a gunshot wound.

At his arrest, the defendant, who matched the complainant's description, was wearing a blue jacket. A specimen of the jacket was forwarded with the questioned bullet for examination and comparison purposes. The questioned blue fibers from the bullet were identified by PLM as consisting of a mixture of two distinct types and colors of blue acrylic fibers. Blue polyester and acetate fibers were also found on the bullet. PLM examination of the specimen from the jacket revealed it was composed of a blend of two distinct types and colors of blue acrylic fibers. Microscopic, instrumental, and microchemical comparisons of the acrylic fibers from the bullet and the jacket revealed that they were indistinguishable. On this basis it was concluded that the questioned acrylic fibers from the bullet could have originated from the suspect's jacket. These data made possible the association of the suspect with the scene of the crime. Lengthy testimony about the fiber evidence was presented at trial and the suspect was convicted of burglary.

CASE 12

The complainant, a male Caucasian, was shot twice in the head and once in the abdomen while patronizing an uptown Manhattan social club. Three .45-caliber armament control panel (ACP) semijacketed lead bullets were recovered during a search of the premises. Two of the bullets had hollow points and one had a soft point. Two were recovered from the walls inside the premises; both were severely deformed and contained minute traces of paint and plaster embedded on their surfaces. The third (soft note) bullet was found on the floor of the club. A quantity of wood was embedded in the bullet (Figure 14.12).

One of the wooden tiles in the hardwood floor inside the club had what appeared to be a bullet hole traversing its entire thickness. The tile was removed and inexplicably forwarded directly to the property clerk. Two years later, a suspect was apprehended and charged with this crime. The gun used in the assault was never recovered. At the trial it became important to know whether the bullet found on the floor of the premises produced the hole in the tile.

The wooden tile was obtained from the property clerk's office after storage for almost 2½ years. The bullets were obtained from the ballistic section files. Examination revealed the floor tiles were constructed of slats of red oak with cushions of polystyrene foam glued to their bases. Traces of copper metal were found around the perimeter of the hole in the wooden floor tile. Microscopic examination of the wood recovered from the bullet disclosed it was mahogany. No traces of polystyrene were found adhering to the wood obtained from the hollow point bullet. However, when the top layer of wood was removed from the hollow point bullet, fragments of blue cotton denim threads were found underneath the wood. At the time of the shooting, the victim was wearing a blue denim shirt and pants outfit.

These data suggested that the questioned bullet found on the floor was not the one that made the hole in the wooden floor, but was the bullet that struck the victim's abdomen. After exiting the victim, it struck another wooden object, perhaps the bar. (No known sample of wood from the bar was available.) This information also indicated that another copper-jacketed bullet (never recovered) made the hole in the floor tile. Whether this hole was made during the commission of this crime or another event remained unknown. This information was reported to the district attorney for use at the trial.

CASE 13

During the execution of a search warrant at a suspected illicit narcotics-processing location, an officer taking part in the raid was struck and killed by a projectile that passed through the door leading to the only bedroom in the apartment. A search of the crime scene produced five deformed .38-caliber lead bullets. One of the questioned bullets (Q1) was found on the living room floor at the end of a hallway leading from the bedroom. Three of the questioned bullets (Q2 through Q4) were found inside the bedroom. The fifth questioned bullet (Q5) was discovered in the bedroom closet. A sixth bullet (Q6) was removed from the officer at autopsy.

Stereomicroscopic examination of the deformed lead bullets revealed the presence of trace evidential material embedded on all their surfaces. During the investigation, it became crucial to know the trajectories of the questioned projectiles. The trace material present on each bullet made these determinations possible.

Analysis of the paint found on bullets Q1 and Q6 and the known paint specimens removed as standards from the bedroom door (which had two apparent bullet holes) revealed numerous colors of paint that were consistent in physical and chemical properties. The data in Table 14.10 indicates that bullets Q1 and Q6 did in fact go through the bedroom door. Figure 14.13 depicts the various colors of paint found on the surfaces of projectiles Q1 and Q6 as they compare to the layers seen in the cross-section of a known paint chip removed from the bedroom door.

Stereomicroscopic examination, PLM, microchemical tests, and instrumental analysis of the various colors of paint found on questioned bullets Q2 and Q5, and the known paint specimens removed from the bedroom walls helped to prove that Q2 and Q5 struck the walls. Q2 then ricocheted to the floor where it was found. Q5 passed through the bedroom wall into the closet where it was found (see Figure 14.14).

Microscopic, microchemical, and instrumental analysis and comparison of the trace evidential material present on the deformed Q4 lead bullet and the covering obtained from the bedroom floor made it obvious that Q4 was fired directly into the bedroom floor (see Table 14.11).

Finally, PLM and x-ray diffraction analysis of the trace evidential material present on Q3, and the known material forming the bedroom ceiling, made it evident that the deformed Q3 lead bullet struck the bedroom ceiling, ricocheted, and fell to the bedroom floor where it was recovered by the crime scene officer.

CASE 14

Two police officers responded to a radio call citing "shots fired." Upon arriving at the scene, they observed a male Caucasian acting irrationally. He approached the officers with a shiny object in one of his hands. One of the officers ordered him to stop but he continued to approach. One of the officers drew his revolver and fired one shot in the direction of the subject who fell to the ground. The object in the man's right hand was a knife. He had apparently suffered a gunshot wound and was removed to the local hospital where he was pronounced dead.

The police officer who fired in the direction of the subject did not believe that the shot he discharged struck the subject. During the subsequent investigation, the crime scene unit found bloodstains inside the reported location. This and other circumstantial information indicated that the subject was shot inside the building by a female occupant. The crime scene officers searched the location for the spent bullet fired by the officer. A deformed lead bullet was found in the alleyway adjacent to the building. A bullet impact mark (BIM) was found on the side of the same building. A portion of the brick that contained the BIM was removed and the questioned bullet and brick were forwarded to the laboratory for examination.

The spent bullet and brick fragment containing the BIM were examined with a stereomicroscope. No traces of clothing fibers, blood, or other tissues were found on the brick. Part of the brick contained a mark that was determined to be metallic lead. The mark had the same shape as the nose of the deformed spent projectile. A chevron-shaped piece of cinder was found embedded in the nose of the bullet. A piece of cinder with an identically shaped cross-section was found on the exposed surface of the brick in the same area as the BIM as shown in Figure 14.15. The two pieces of cinder constituted a jigsaw match, thus proving conclusively that the officer did not shoot the suspect.

CASE 15

Some of the enjoyment obtained from restoring historic homes and structures derives from trying to identify the myriad materials used in their construction. Mortars, in particular, although usually similar in mineral composition, are often reinforced with whatever fibrous material happened to be available when the mortar was mixed. In the restorations of many of the notable homes and buildings around New York City, the authors have identified some unusual reinforcement materials (see Figure 14.16). Table 14.12 lists fibrous materials found in mortars from historic structures in New York City.

CASE 16

Paints and protective coatings are arguably the most studied materials in the disciplines of fine art and architectural restoration as well as in archeological conservation. Whether in the restoration of historically important structures or artifacts or in the cleaning, protection, and refurbishment of artistically important paintings, the paint used to decorate, protect, or prepare the work is always studied in great detail.

In the early 1980s, the people of the U.S. undertook a restoration of perhaps one of their most important symbols. The Statue of Liberty showed many signs of deterioration and aging and was in danger of structural failure. A massive restoration was undertaken for the statue's 100th birthday. Every part of the superstructure was dissected and studied in great detail. The chemical compositions of the paints and coatings used to protect the copper metal skin from the effects of erosion and the harsh weather of New York Harbor were studied and scrutinized.

In particular, the interior coatings at the base of the extended arm were analyzed to determine the composition of the waterproofing used over the last century (see Figure 14.17). Figure 14.18 shows a cross-section of the paint removed from the base of the extended arm. Figure 14.19 shows the micro-FTIR spectra of the top blue-green coating and the bottom black coating. The information obtained from the analysis of the paint chip layers helped experts restore the statue to its original condition.

CASE 17

Natural stones have been used throughout the ages to build homes, churches, and many other structures; to mark graves; to carve statues; and to construct monuments. Consequently, fine arts and architectural conservators often encounter these materials in their casework.

In the mid 1980s, a restoration of an ancient cemetery in Hartford, CT was undertaken. The source of the sandstone used to make many of the headstones had to be determined before recommendations for headstone restoration could be proposed. Figure 14.20 illustrates the condition of the old cemetery at the beginning of the restoration. After comparing many specimens of sandstone from the tri-state area to the sandstone composing the decaying headstones, it was determined from their mineral composition that the original sandstone was quarried locally in Connecticut.

CASE 18

In the early 1980s, during restoration of a major park in New York City, it became important to determine whether the statue of Atlas holding up the world (Figure 14.21) was made of marble or limestone. The PLM examination of tiny specks of stone removed from the statue revealed that it was composed primarily of limestone.

CASE 19

In the late 1980s, the Metropolitan Museum of Art in New York presented an exhibition on Napoleon Bonaparte. During preparation of the exhibit, dust was discovered in Napoleon's uniform pockets. Although some materials found in the dust were consistent with the textiles of the uniform (primarily dyed wool and fine silk; Figure 14.22), some debris appeared to be vegetable matter.

Speculation about the identity of this mysterious material initiated theories that it might be incense or licorice root. A specimen was prepared and examined with a PLM. The morphological features revealed the material was coarse tobacco dust (Figure 14.23). Confirmation of the molecular structure of the vegetable debris was made with micro-FTIR (Figure 14.24) and the results were reported to the museum's textile conservator.

CASE 20

During recent restoration of a very expensive antique gilded frame, a specimen was forwarded for analysis to one of the authors. Apparently, much of the old gold gilding was peeling away from the wooden surface of the frame and the gilder was having a hard time restoring the frame. A microscopic examination of the gesso chips revealed a green mold growing on the underside of the gilding material (Figure 14.25).

A recommendation was made to clean the surface of the wooden frame with a 10% solution of bleach mixed with distilled, de-ionized water to remove mold spores, fungus, and mildew. Another recommendation was to experiment with gesso and various fungicides to determine whether growth of these organisms could be retarded.

CASE 21

Recently, one of the authors was asked to identify a green corrosive material present on an early American large-cent coin. The coin dated from 1850s (Figure 14.26). Two minute specimens of the green material were removed from the coin surface. One was mounted in Melt Mount 1.539 for PLM examination, and one was used for microchemical analysis. PLM examination revealed that the green material was a copper salt known as verdigris. Confirmation of the presence of copper was made with squaric acid (Figure 14.27).

CASE 22

A client who purchased a white pendant from a dealer wanted to know whether the pendant (Figure 14.28) was made of real ivory or an imitation. Two tiny specimens of the material were removed from the pendant. One was mounted in Melt Mount 1.539 for PLM examination and the other was used for IMS analysis. Both specimens were compared to known elephant ivory, and found to be the same in all physical and chemical properties (Figures 14.29 and 14.30). After review of the data, the client was told that she purchased a real ivory pendant.

CASE 23

On the morning of September 11, 2001, the most horrific attack on U.S. soil occurred. In less than an hour, three hijacked commercial airplanes were deliberately crashed into buildings that represented America's greatness. After the initial shock and horror of the event, the people of the U.S. banded together and started one of the largest humanitarian and recovery efforts in history.

In New York City, the two towers of the World Trade Center (WTC) financial complex were reduced to a fine powdery dust by two airplanes. The recovery started almost immediately. Men and women from all parts of the country dug with their hands and simple tools to search for hundreds of survivors trapped in the dangerous piles of rubble. Unfortunately, many of the rescuers and residents of the areas surrounding the WTC almost immediately started to show signs of respiratory health problems in the form of dry, relentless coughs. Many colleagues of one of the authors requested analysis of dust specimens gathered from Ground Zero and the surrounding areas.

Since the destruction of the WTC was a unique, cataclysmic event, an analytical method to study the dust specimens had to be developed. Macroscopically, each bulk specimen appeared somewhat like recently erupted volcanic ash. A few early specimens were examined with a stereomicroscope, then tiny aliquots of a few bulk specimens were studied with a PLM and revealed that each bulk sample was composed of myriad materials. All the building materials and contents were literally pulverized by the collapse. Figures 14.31A through C show some of the materials analyzed. Dust data count sheets were prepared from these initial observations (Figure 14.32).

Next, a random sample of each specimen was taken. Each bulk specimen was thoroughly loosened and mixed gently using an agate mortar and pestle and equally divided into eight aliquots. Each aliquot was divided into eight equal portions. Each portion was placed on a microscope slide, covered with a No. $1^1/_2$, 22-mm round cover glass, dispersed evenly in Melt Mount 1.539, and labeled for identification.

Finally, a quantitative particle count of each specimen was carried out with a PLM fitted with a Chaulkly point-count reticle. At least 1000 particles were counted from the microscope slide preparation made from each bulk specimen. The results were recorded on the dust data sheets and the data used to compute the percent of each material present in the average specimen. Table 14.13 lists the percentages.

In the final analysis, one can only speculate whether the dust generated by the collapse of the buildings will exert current or long-term effects on the health and lives of the people of New York City and those who helped them.

CONCLUSION

It is the authors' sincere hope that these case studies illustrate to the reader how powerful and versatile the polarized light microscope and the methods of chemical microscopy are as analytical tools for characterizing minute quantities of a diverse array of materials.

TABLE 14.1
Major Associations of People, Places, and Things (Case 1)

Association	People		Places		Things	
	Victim	Suspect	Alley	Apartment	Liner	Box
Textile fibers from victim's clothing	—	×	—	×	×	×
Suspect's rabbit hair coat	×	—	×	—	×	—
Nylon rugs in suspect's apartment	×	—	×	—	×	×

TABLE 14.2
Trace Materials Associated with Building Industry (Case 2)

Building Materials
Dried adhesive compound
Fiberglass insulation and resin
Gypsum fragments (plaster)
Iron shavings and filings
Mica chips
Plate glass fragments
Red brick fragments
Wood chips and shavings (sawdust)

TABLE 14.3
Trace Evidence Recovered from Victim and Van Interior (Case 3)

Trace Evidence	Victim	Van and Vacuum Sweepings
Brown and gray human head hair	Head, wool blazer	Back door of van, hair brush, sweepings, miscellaneous items
White seeds	Mouth	Sweepings
Gray and metallic black paint chips	Hair and wool blazer	Sweepings and floor
Clear, green, and amber glass fragments	Wool blazer and sheet	Sweepings and miscellaneous items
Cellophane	Wool blazer	Floor
Urethane foam	Wool blazer	Sweepings, miscellaneous items, foam mattress
Sawdust	Hair, wool blazer, and sheet	Sweepings and miscellaneous items
Black olefin plastic	Skirt	Floor
White and brown/white dog hair	Wool blazer	Sweepings and miscellaneous items

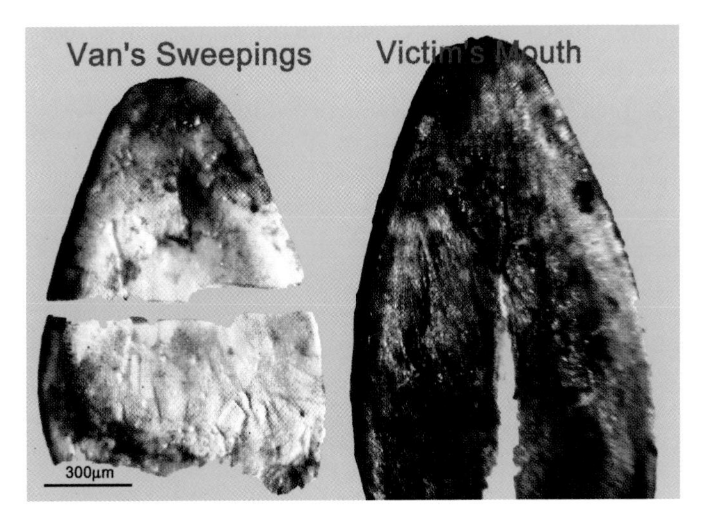

FIGURE 14.1 Right: seeds from victim's mouth. Left: sweepings from van floor.

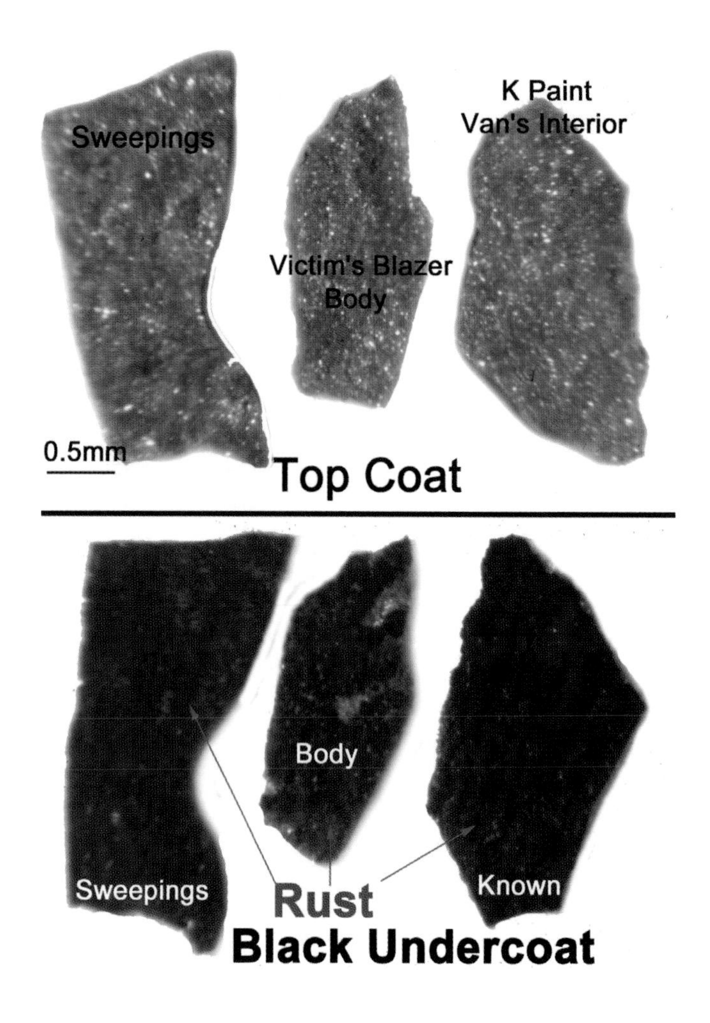

FIGURE 14.2 Left: paint chip found in vacuum sweepings from the van floor. Center: paint chip removed from the victim's wool blazer. Right: paint chip (known specimen) removed from the van interior.

TABLE 14.4
Relevant Data Taken from Soil Data Sheets (Case 4)

Specimen	MSC Soil Color	Construction Debris	Vegetation	Asphalt
S1	7.5 YR 5/4	None	(–)	(–)
S2	10 YR 5/2	Plaster, concrete, architectural glass	(+)	(+)
S3	2.5 Y 5/4	Plaster, concrete	(–)	(++)
S4	10 YR 5/2	Plaster, concrete	(+)	(+)
S5	10 YR 4/4	Plaster, brick, concrete	(+)	(+)
S6	2.5 Y 5/2	Plaster, concrete	(–)	(+++)

Asphalt contents: (+) light; (++) medium; (+++) heavy.

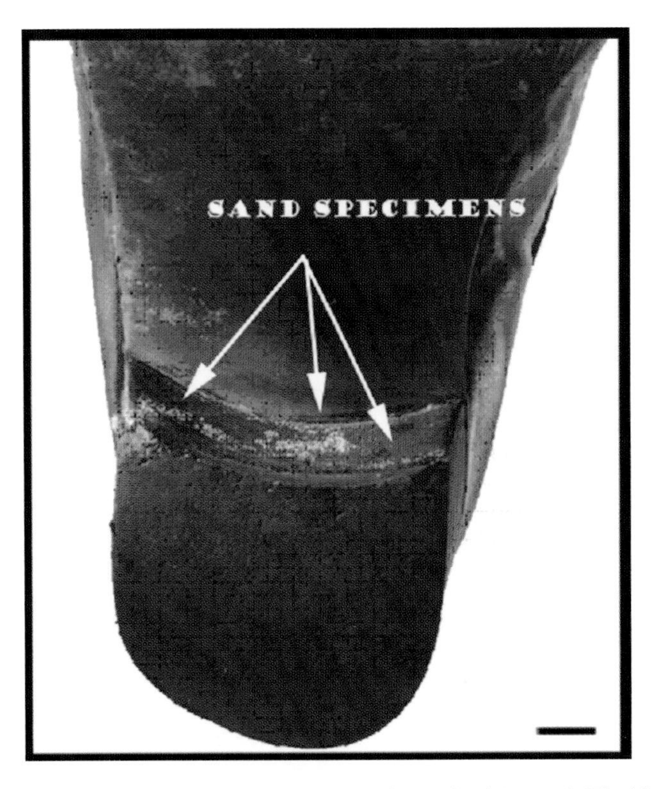

FIGURE 14.3 Right shoe with questioned sand specimens (QS1) adhering to heel (arrows). The black bar is equal to 12 mm.

TABLE 14.5
Associative Data from Soil Data Sheets (Case 5)

Specimen	Color	Minerals Present	% Minerals	Ions Present	Morphology
KS1/QS1	5Y 7/2	Quartz	88	Na+, Cl−	Rounded
		Hornblende	2		
		Garnet	5		
		Hematite	Trace[a]		
		Magnetite	1		
		Limonite	Trace		
		Tourmaline	1		
		Shell parts	2		

[a] Trace = less than 1%.

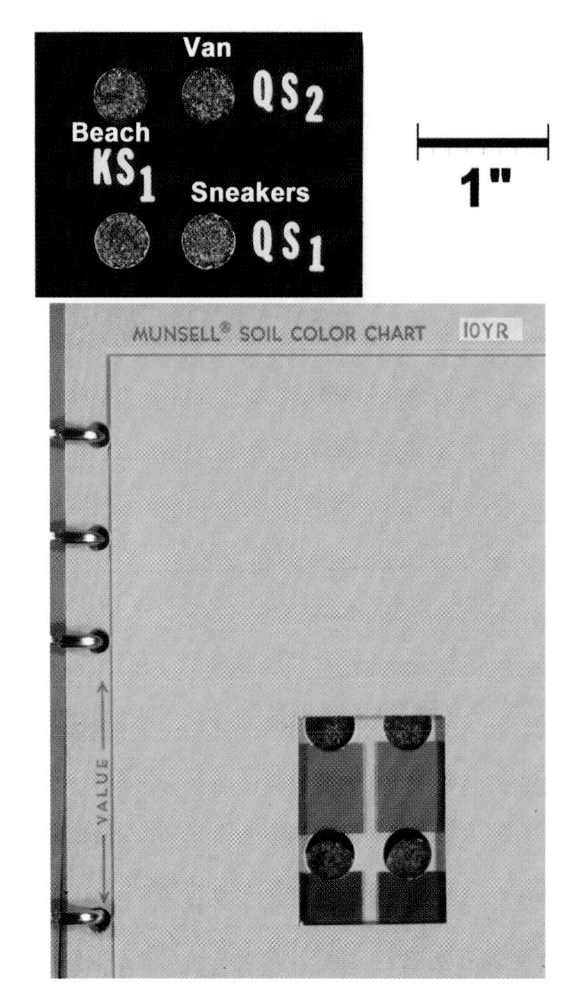

FIGURE 14.4 Munsell soil color chart system used to determine the color of the sand samples.

TABLE 14.6
Associative Data from Soil Data Sheets (Case 6)

Specimen	Location	Color	Mineral	%	Flora Debris	Asphalt
KS1[a]	Beach	2.5Y 8/2	Quartz	83	(–)	
QS2 and QS4	Victim's van	2.5Y 8/2	Magnetite	10.5	(–)	
			Hematite	4.5		
			Garnet	1.3		
			Feldspar	Trace[b]		
			Hornblende	Trace		
			Tourmaline	Trace		
QS1 and QS5	Suspect's sneakers	2.5 Y 7/2	Quartz	89.1	Trace	+
			Magnetite	5.4		
			Hematite	4.9		
			Feldspar	Trace		
			Hornblende	Trace		
QS3	Suspect's pants	2.5 Y 7/2	S/A[c]	S/A	Trace	
QS6	Suspect's socks	2.5 Y 7/2	S/A	S/A	Trace	+

[a] All three specimens are the same.
[b] Trace = less than 1%.
[c] S/A = same as above.

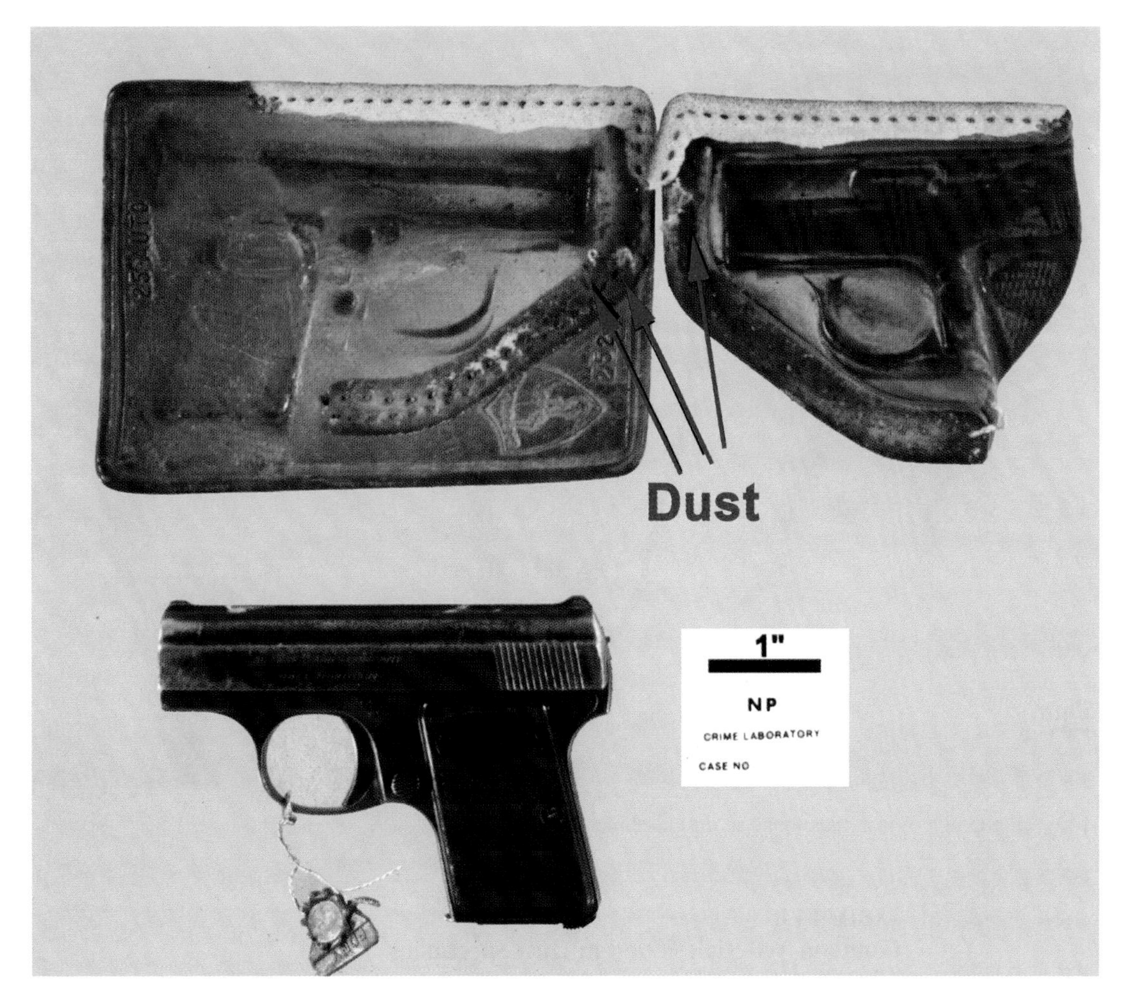

FIGURE 14.5 Holster obtained from the crime scene and pistol recovered from the suspect's residence. The dust specimens were obtained from areas indicated by the arrows.

TABLE 14.7
Trace Evidence in Common from Holster and Pistol Trigger (Case 7)

Fibers	Hairs
Cotton: colorless, blue, red, white/black	Human head: brown
Acrylic: navy blue, red, green	Negroid
Nylon: red, blue, green, pink, brown, black	Animal: dog, cat (tan)
Polyester: light blue, gray	

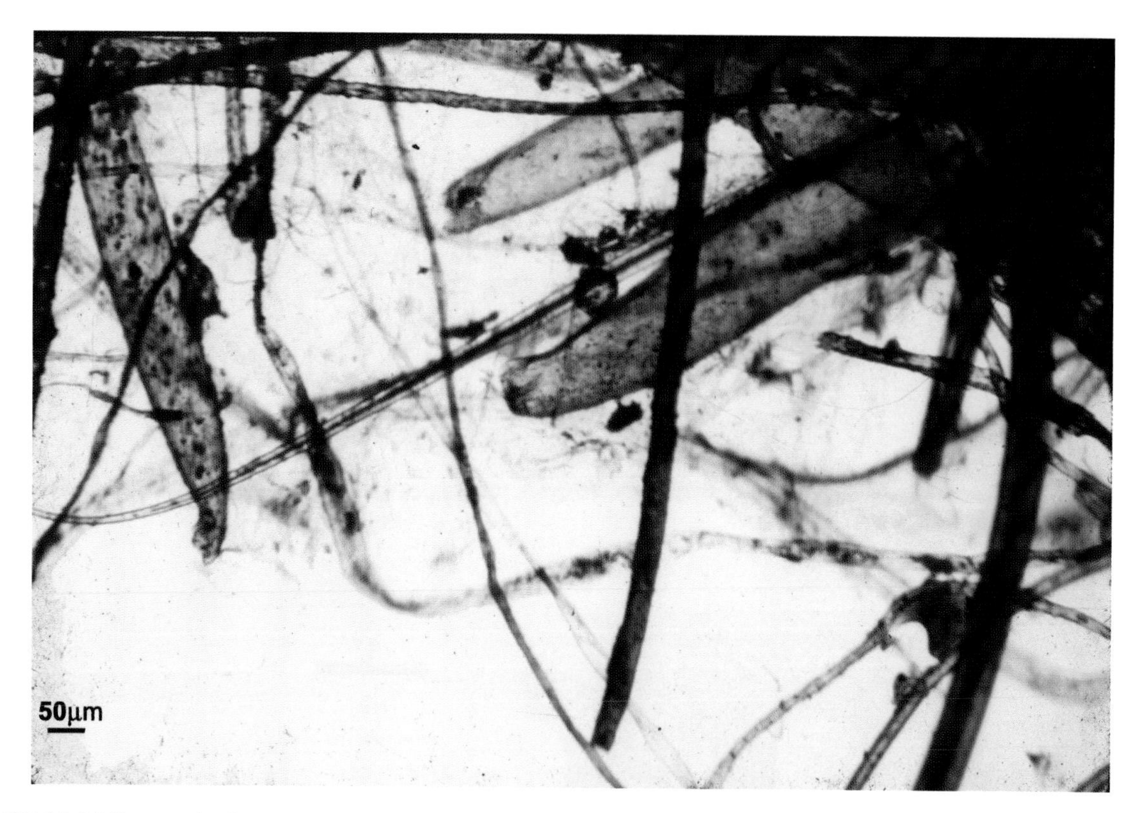

FIGURE 14.6 Micrograph of materials found in dust specimens.

TABLE 14.8
Common Materials Found in Dust Specimens from Victim's Apartment and Suspect's Right Shoe (Case 8)

Material
Partially eaten Caucasoid head hair
Human head hair
Duck down
Cockroach parts
Dried human blood
Spider silk
Quartz fragments
Feldspar fragments
Mica fragments
Manila fibers
Cotton fibers
Red nylon rug fibers

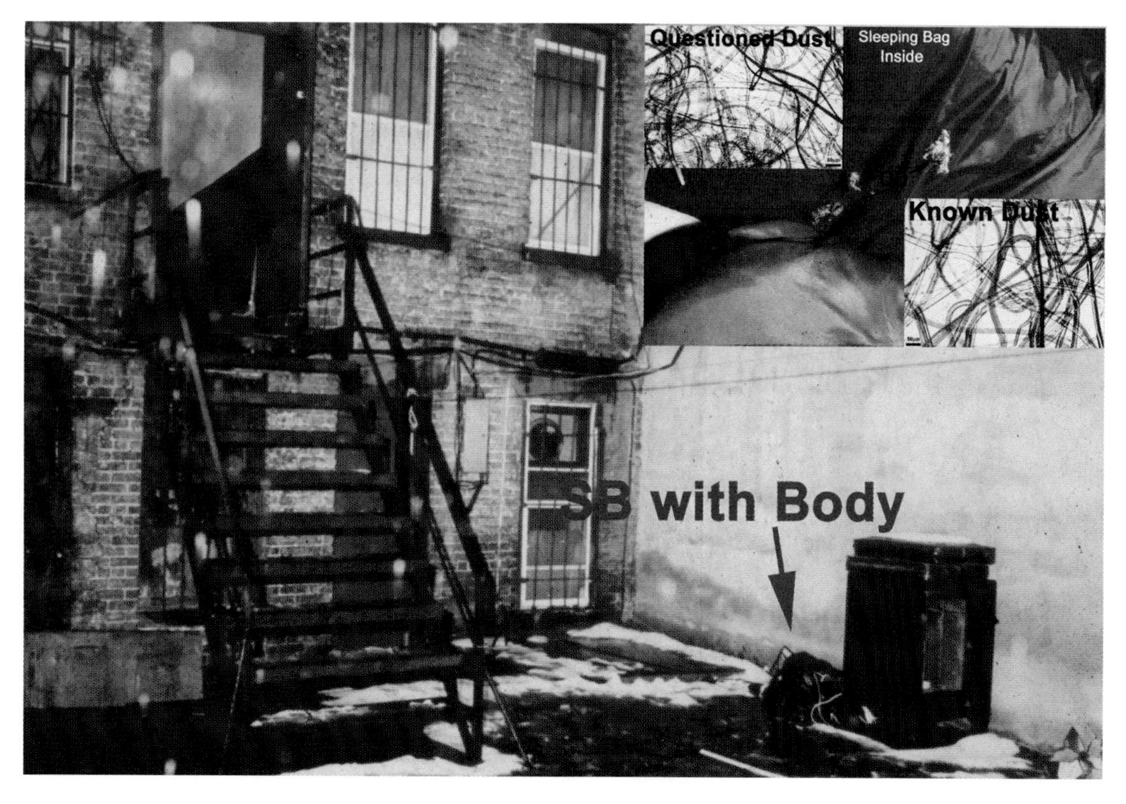

FIGURE 14.7 Victim found behind building in a sleeping bag next to a dumpster.

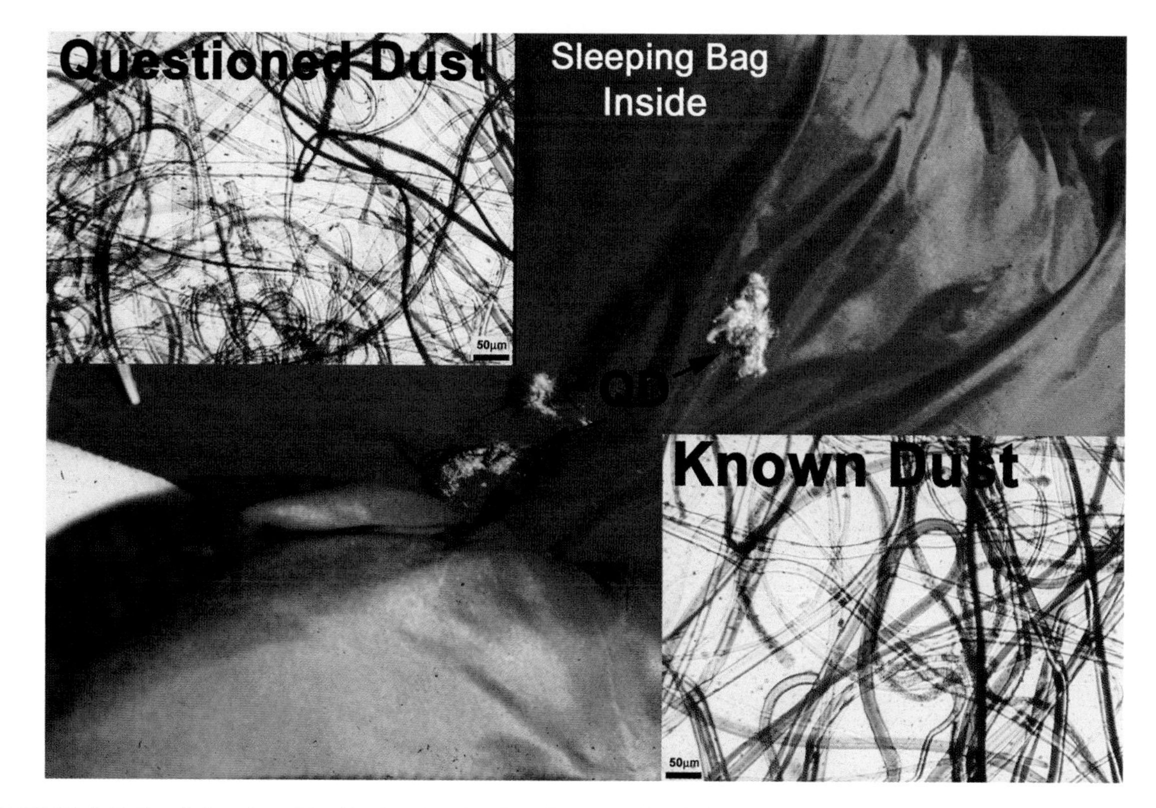

FIGURE 14.8 Balls of dust found inside the sleeping bag (arrows). Top left: photomicrograph of questioned dust found in the sleeping bag. Bottom right: photomicrograph of known dust found in the locked bedroom.

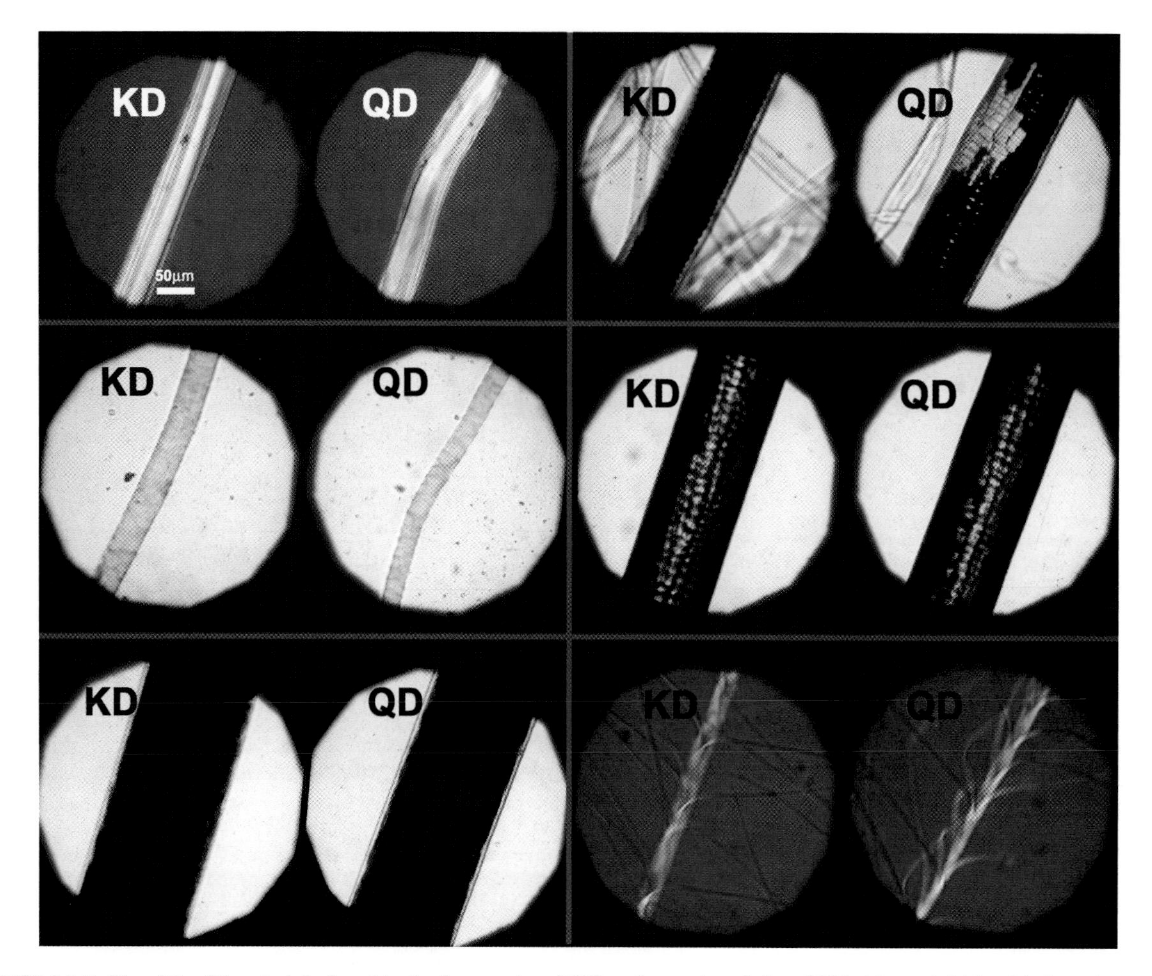

FIGURE 14.9 Six of the 27 materials found in the known dust (KD) and questioned dust (QD): Mongoloid head hair, polyester rug fibers, duck down, aqua-colored rabbit hair, aqua-colored wool, and brown-colored rabbit hair.

TABLE 14.9
Materials Found in Questioned and Known Dust[a] (Case 9)

Material
Cotton: white, pink, green, blue, yellow
Acrylic: white, pink, red, purple, blue, light gray
Acetate: orange
Modacrylic: black
Olefin: blue
Nylon: red, blue, orange/brown
Polyester: light gray (substrate from rug at crime scene)
Wool: black, orange, blue, aqua
Mongoloid head hair (consistent with occupant)
White dog hair
Rabbit hair: natural brown, aqua
Duck down
Manila fiber fragments

[a] Questioned dust from sleeping bag under body. Known dust from gray rug at crime scene. All materials had the same physical and optical properties and chemical compositions.

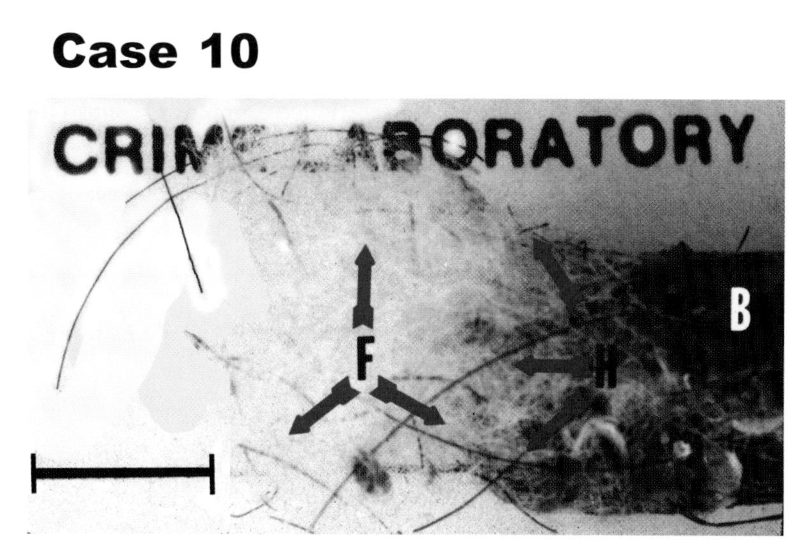

FIGURE 14.10 Deformed lead bullet with embedded tufts of white polyester fibers and brown hair fibers. The black bar equals 55 mm.

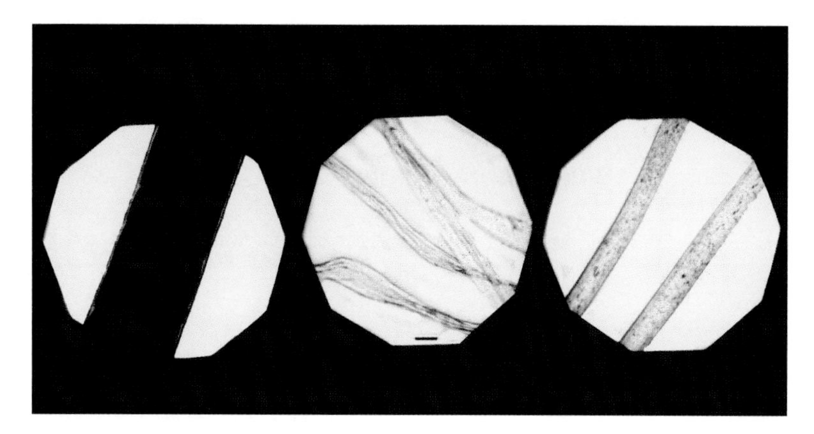

FIGURE 14.11 Right: micrograph showing outermost layer of polyester fibers. Center: cotton fiber. Left: questioned head hairs on surface of bullet. The black bar equals 20μm.

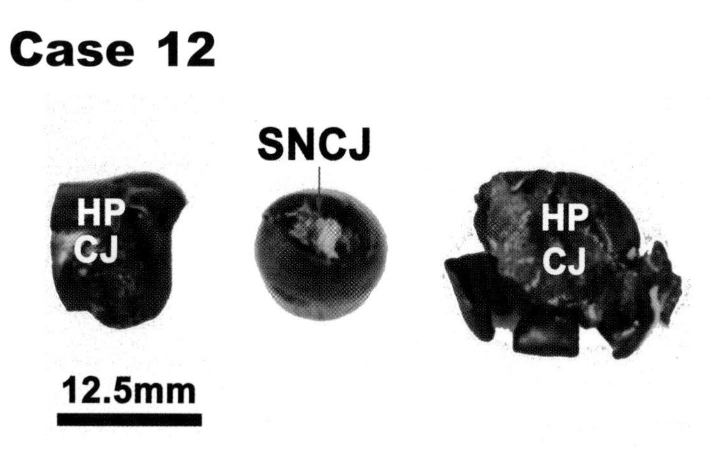

FIGURE 14.12 Three copper-jacketed spent projectiles. The center projectile contained the questioned wooden debris. Hollow point copper jacketed (HPCJ); soft nose copper jacketed (SNCJ).

TABLE 14.10
Various Colors of Paint Found on Questioned Samples Q1 and Q6 and on Known Specimen Removed from Bedroom Door (Case 13)

Known Paint (Layered)	Q12 Paint (Smeared)	Q6 Paint (Smeared)
White (top)	—	—
Beige	—	+
Medium blue	+	+
Light blue	+	+
Blue-green	—	+
White	+	+
White	+	+
White	+	+
Beige	+	+
Base wood (bottom)	+	—

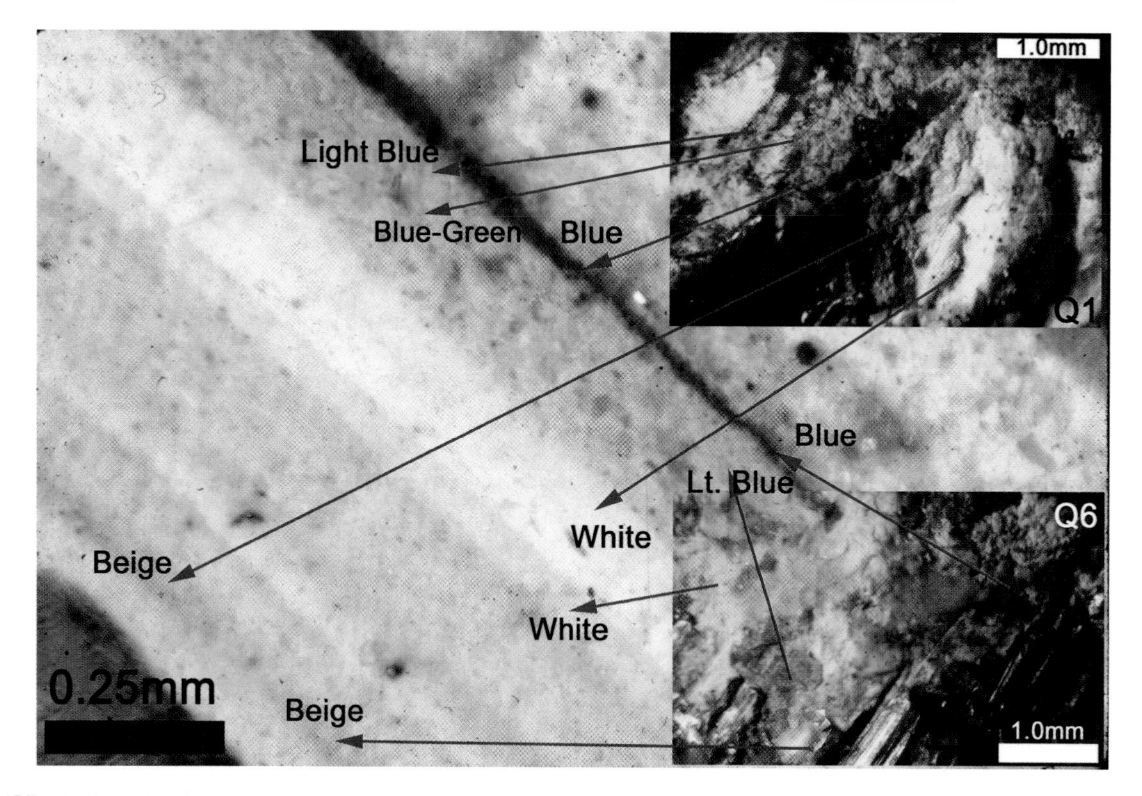

FIGURE 14.13 Various colors of paint found on the surfaces of Q1 and Q6, and corresponding colors in the layers of the known paint standard from the bedroom door. The presence of various colors of paint on the projectiles established that both bullets were shot through the bedroom door.

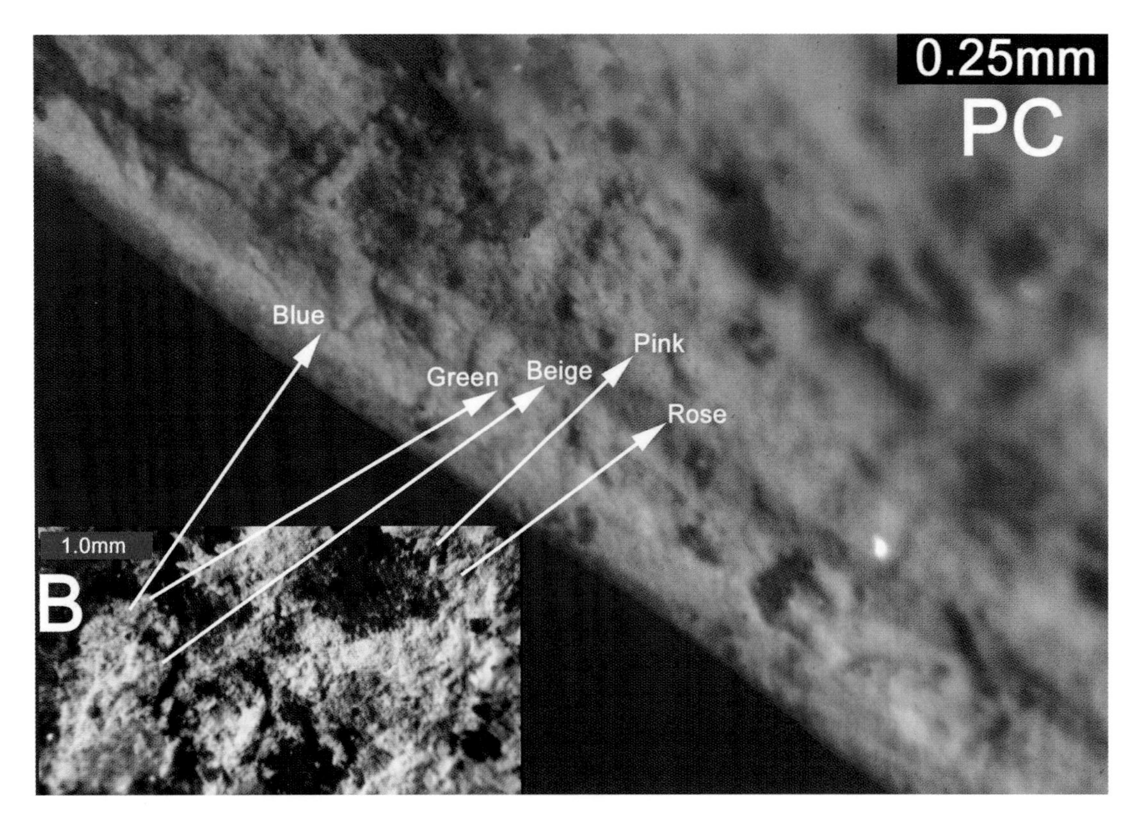

FIGURE 14.14 Colors of paint found on the surfaces of Q2 and Q5 and corresponding colors in the layers of the known paint standard from the bedroom interior walls.

TABLE 14.11
Common Trace Materials from Known Bedroom Floor Covering and Questioned Specimen 4 (Case 13)

Known Floor Covering	Q4
White/yellow vinyl floor covering	+
Metallic flakes	+
Wooden subflooring	+

FIGURE 14.15 Left: portion of brick bearing a bullet impact mark. Right: deformed lead bullet with imbedded chevron-shaped cinder.

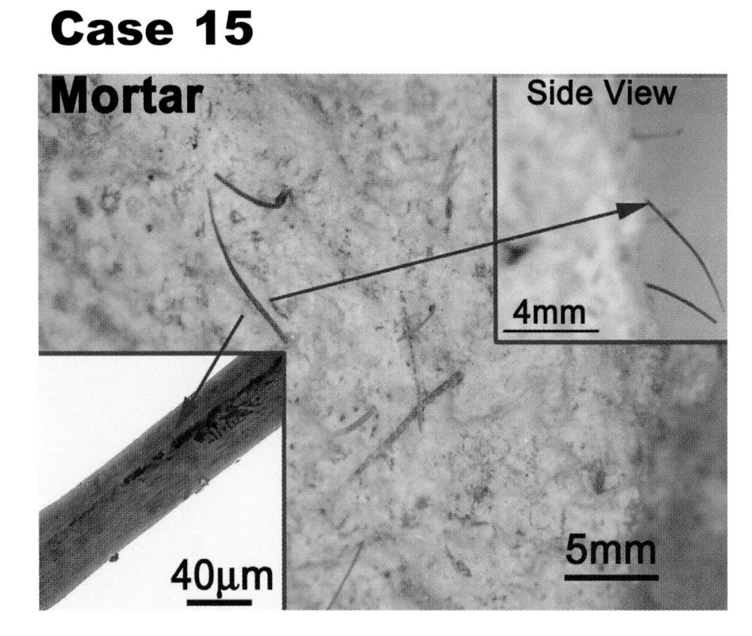

FIGURE 14.16 Specimen of mortar from the Dyckman house in New York City. Note the brown-colored Caucasoid head hairs used in the interior mortar.

TABLE 14.12
Fibrous Materials from Mortars in Buildings in New York City (Case 15)

Fibrous Material
Human head hair
Horse hair
Cow hair
Deer hair
Straw
Vegetable fibers

Case 16

FIGURE 14.17 Interior of extended arm of the Statue of Liberty prior to restoration. Note red circle around an area where one of many coating specimens was collected.

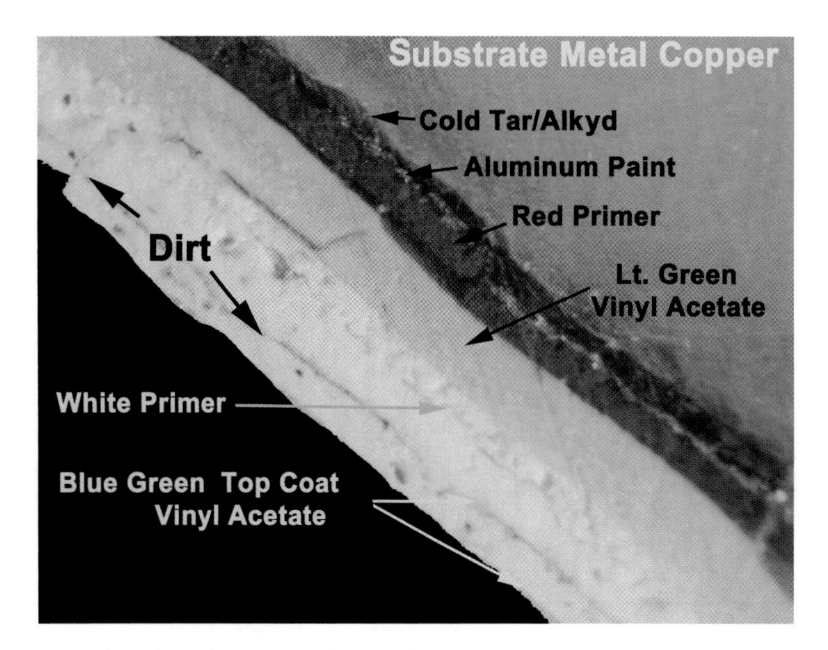

FIGURE 14.18 Cross-section of a paint chip removed from the interior base area of the extended arm of the Statue of Liberty.

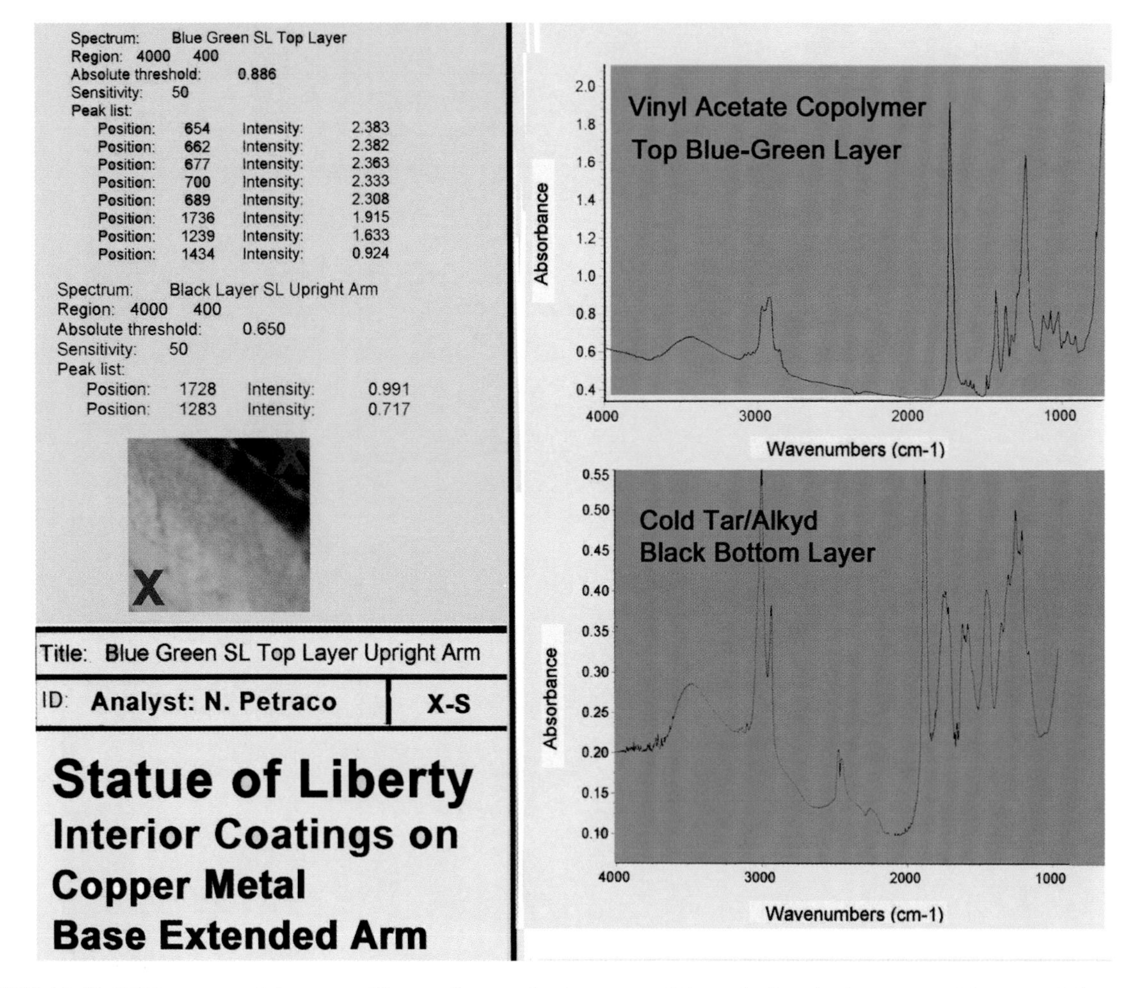

FIGURE 14.19 FTIR spectra of the top and bottom layers of paint removed from the interior base area of the extended arm of the Statue of Liberty.

Case 17

FIGURE 14.20 Ancient cemetery in Hartford, Connecticut. Inset: photomicrograph of sandstone specimen removed from an old tombstone.

Case 18

FIGURE 14.21 Statue of Atlas holding up the world. Inset: photomicrograph of limestone specimen removed from the statue.

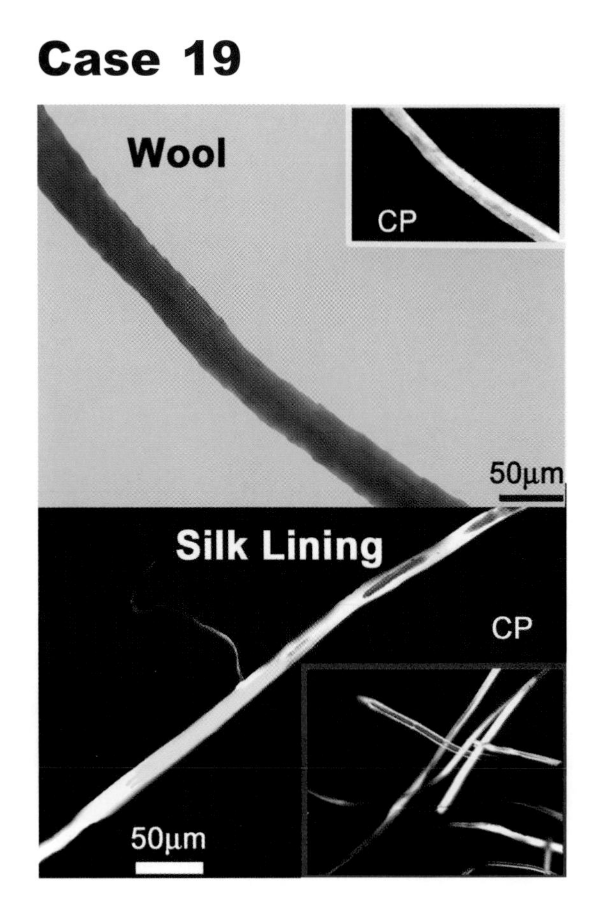

FIGURE 14.22 Wool and silk samples from Napoleon's uniform.

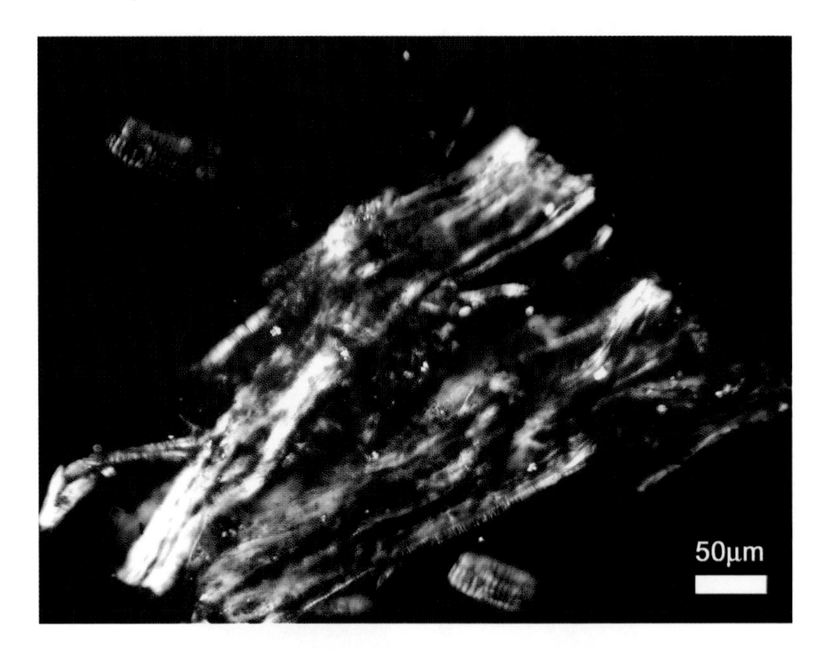

FIGURE 14.23 Fragment of tobacco dust found in a pocket of Napoleon's uniform.

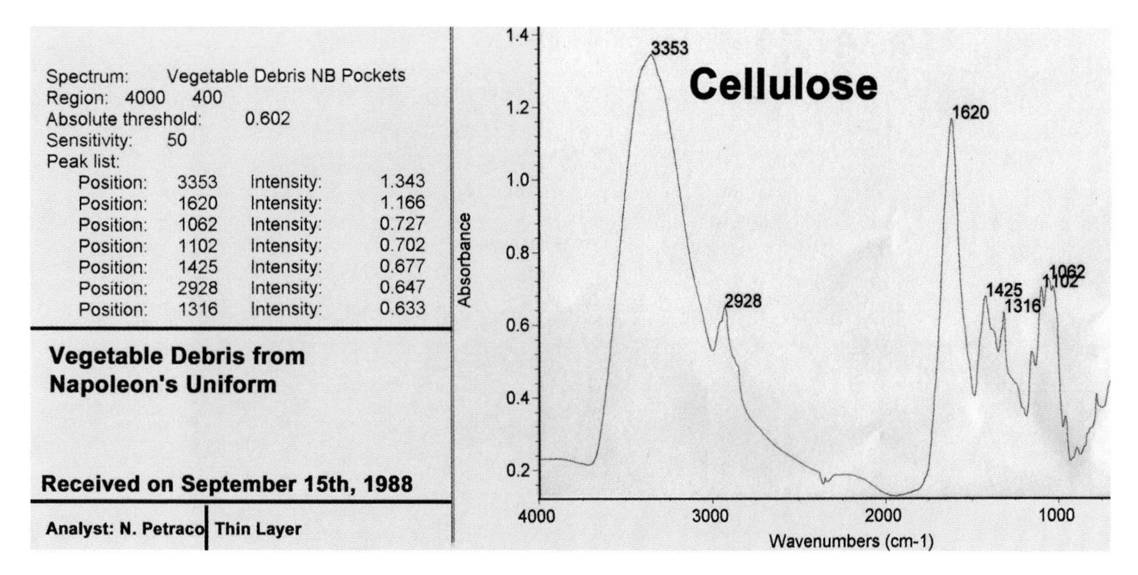

FIGURE 14.24 FTIR spectrum of vegetable debris from Napoleon's uniform pockets.

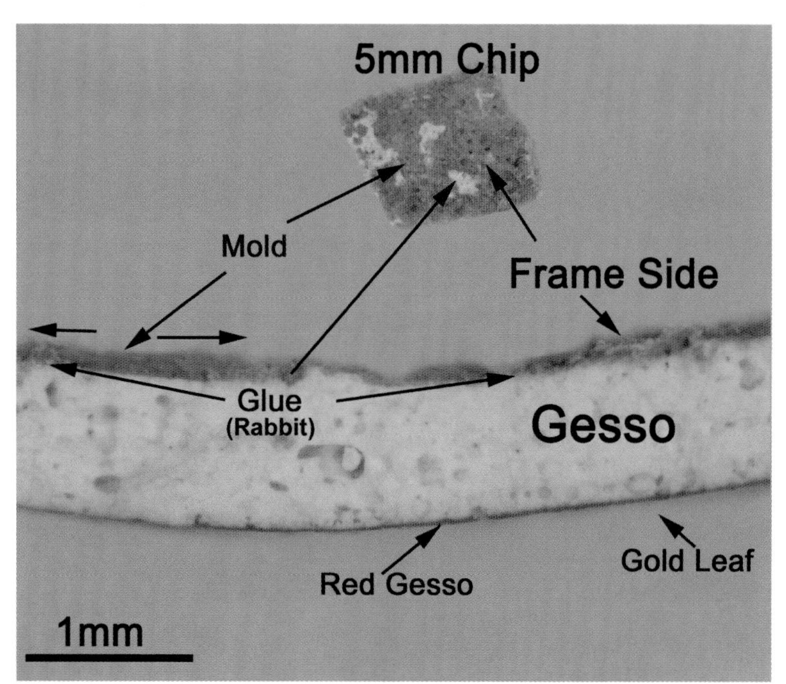

FIGURE 14.25 Green mold on underside of gesso chip.

Case 21

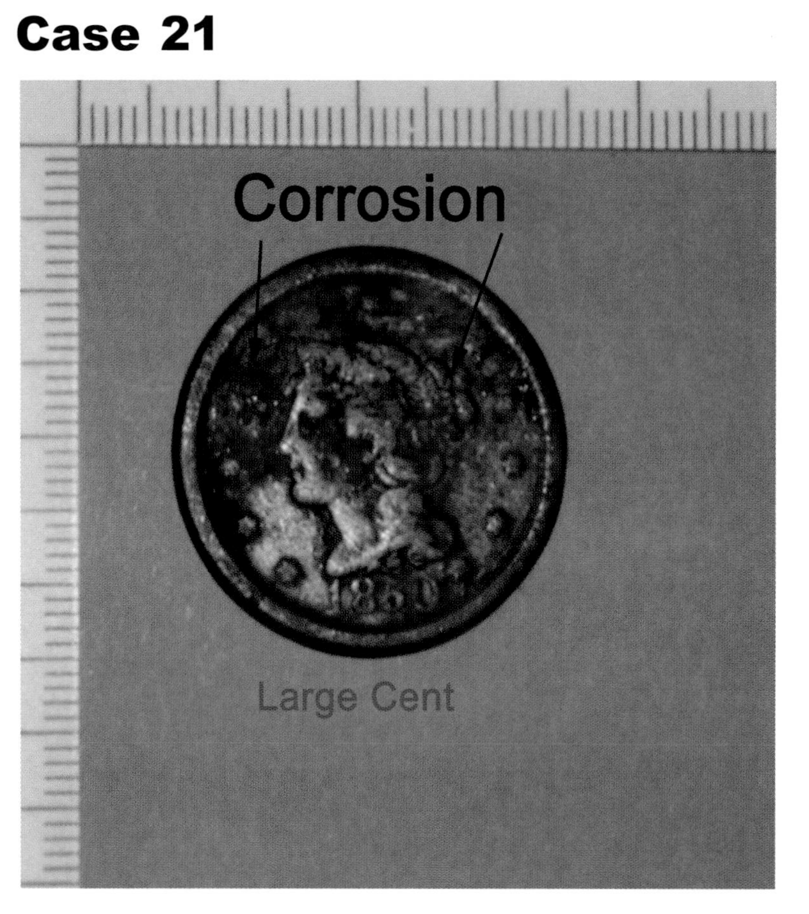

FIGURE 14.26 Large U.S. cent coin minted in 1850. Note green corrosion product.

FIGURE 14.27 Specimen of verdigris removed from the coin and copper squarate crystals formed from copper ions in solution with the squaric acid reagent.

Case 22

FIGURE 14.28 Questioned carved ivory pendant.

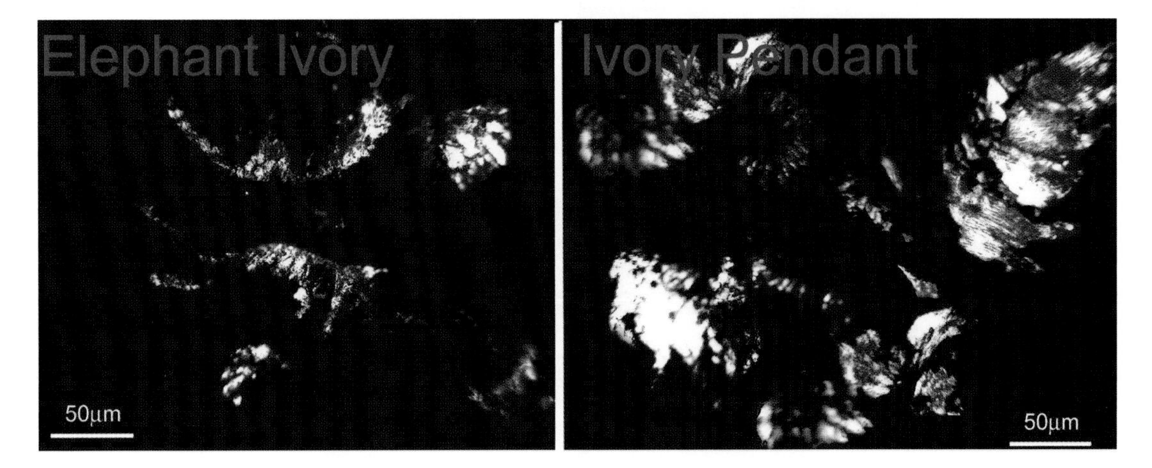

FIGURE 14.29 Photomicrographs of ivory dust from known elephant tusk and from pendant. Note low birefringence and fibrous texture.

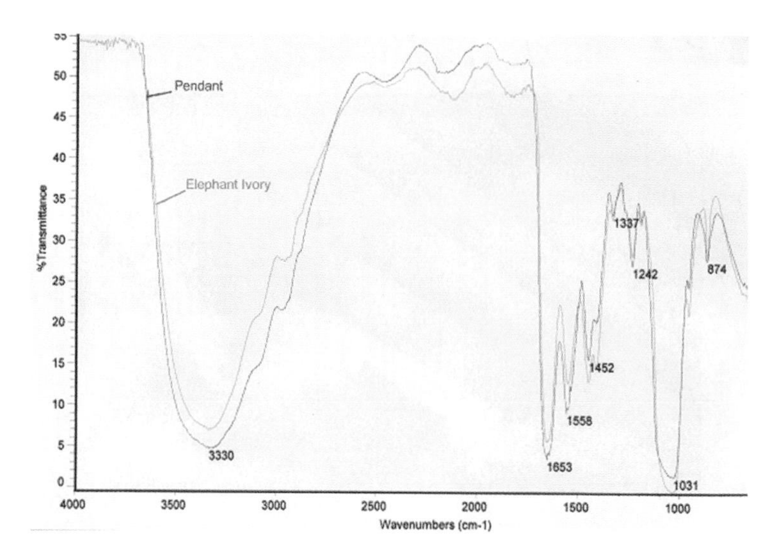

FIGURE 14.30 IMS spectra of known elephant ivory and questioned pendant.

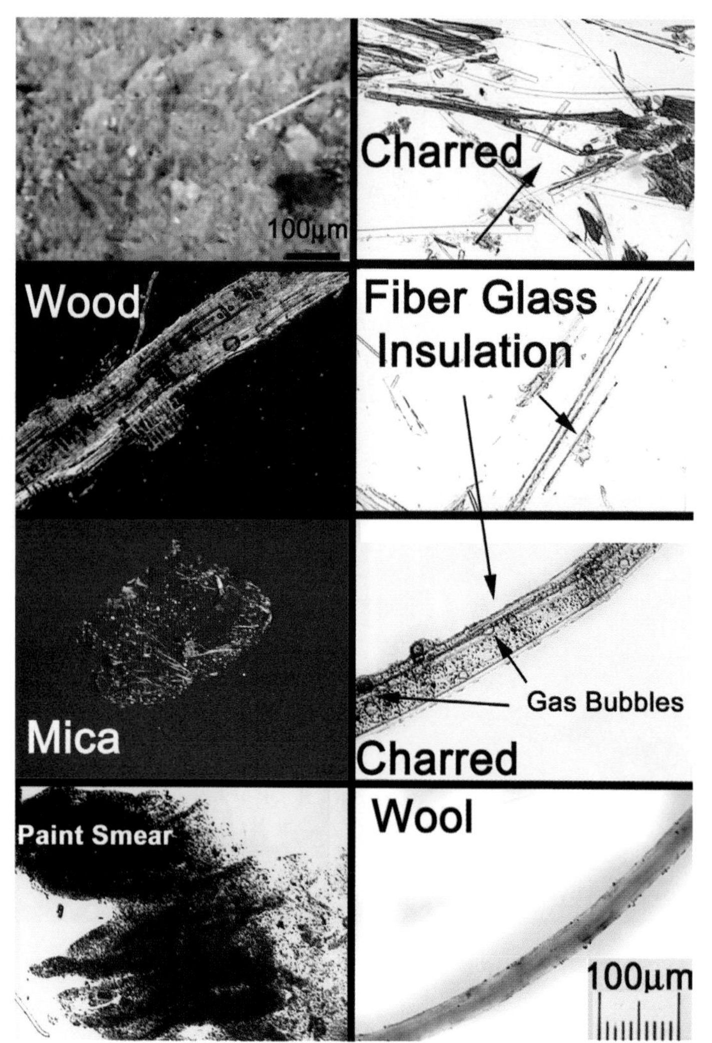

FIGURE 14.31A Top left: photomicrograph of a bulk specimen. The remaining photomicrographs are of materials found in bulk sample.

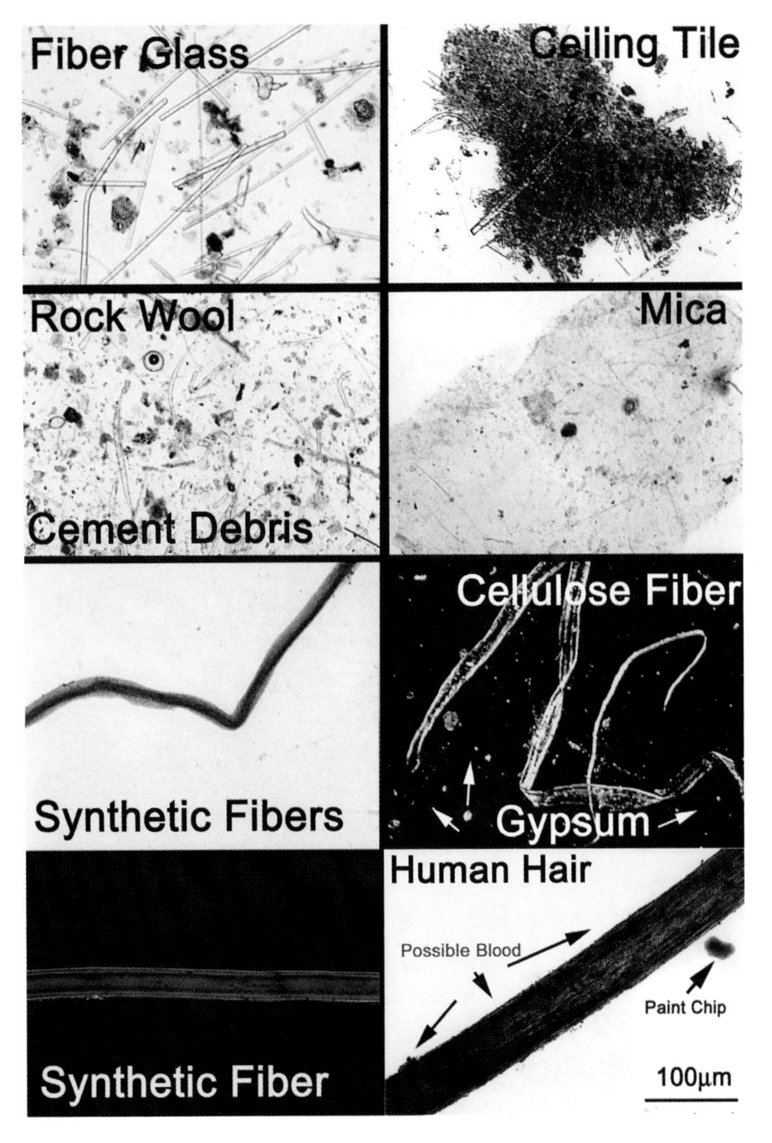

FIGURE 14.31B Particulates found in bulk sample.

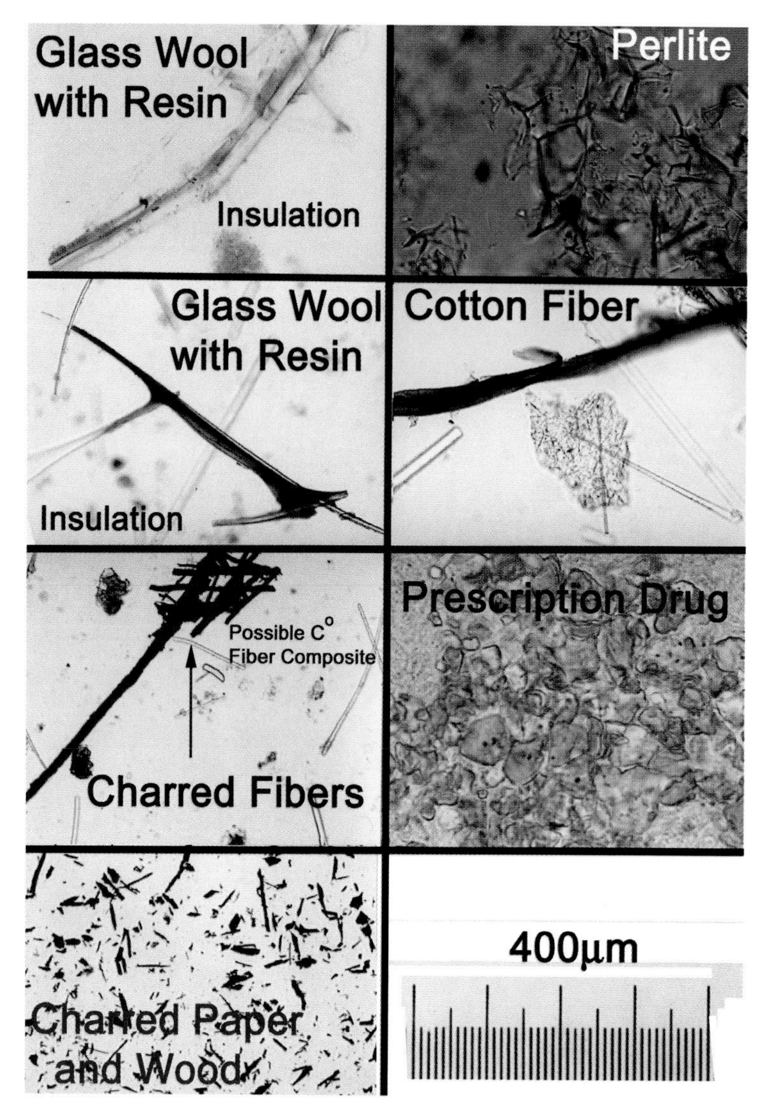

FIGURE 14.31C Particulates found in bulk sample.

Location Obtained____________________
Sample No.__________, Date __________, Time__________
Collector__________, Analyst__________

Material	Counts						Totals	%
Fiberglass with Resin								
Colorless								
Red/Pink								
Yellow								
Blue/Black								
Orange								
Rock Wool								
Asbestos								
Synthetic Fibers								
Human Hair								
Animal Hair								
Natural Fibers								
Paper Fibers								
Ceiling Tile Debris								
P. Human Remains								
Mica Flakes								
Plaster of Paris								
Concrete Debris								
Paint Smears								
Al° Fragments								
Wood Fragments								
Foam Fragments								
Charred Wood								
Charred Debris								
Perlite								
Drugs								
Plastic Fragments								
Glass Fragments								
Miscellaneous								

M.M. 1.539 Cargille®
© Taka - N. Petraco July 17th, 2002, Copyright WTC Dust Form ©

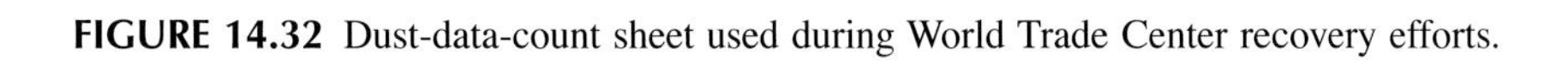

FIGURE 14.32 Dust-data-count sheet used during World Trade Center recovery efforts.

TABLE 14.13
Tabulation of Data Obtained from World Trade Center Dust Specimens (Case 23)

Materials	Count	Percent of Total
Fiberglass and rock wool	1615	45.1
Asbestos	5	Trace[a]
Synthetic fibers	72	2.0
Human remains	47	1.3
Natural fibers	49	1.4
Paper fibers	74	2.1
Ceiling tiles	73	2.0
Mica flakes	76	2.1
Plaster and concrete	1138	31.8
Paint smears and chips	18	Trace
Metal flakes	19	Trace
Wood fragments	20	Trace
Foam fragments	6	Trace
Charred wood and debris	257	7.1
Plastic fragments	5	Trace
Perlite	8	Trace
Drug fragments	12	Trace
Glass fragments	50	1.4
Unknowns	40	1.1
Totals	3584	100%

[a] Trace = less than 1%.

APPENDIX A

Human Hair Atlas

This appendix was designed to provide a compilation of photomicrographs illustrating the many characteristics used to identify, classify, and compare human hairs. It supplements the material presented in Tables 5.1 and 5.2 in Chapter 5. The photomicrographs were collected from numerous forensic hair examiners as the best available examples of hair characteristics listed in the SWGMAT guidelines. They can be used as teaching aids and as references.

Each illustration includes a short description of the trait depicted and a bar scale. Sources are indicated by the initials of contributors. The authors provided illustrations that do not include source lines. All the specimens included in this appendix originated from casework situations. The scale, unless otherwise noted, is equal to 40 µm. Anyone wishing to reproduce any of the illustrations in the atlas should contact the source of the original image for permission.

CORTEX

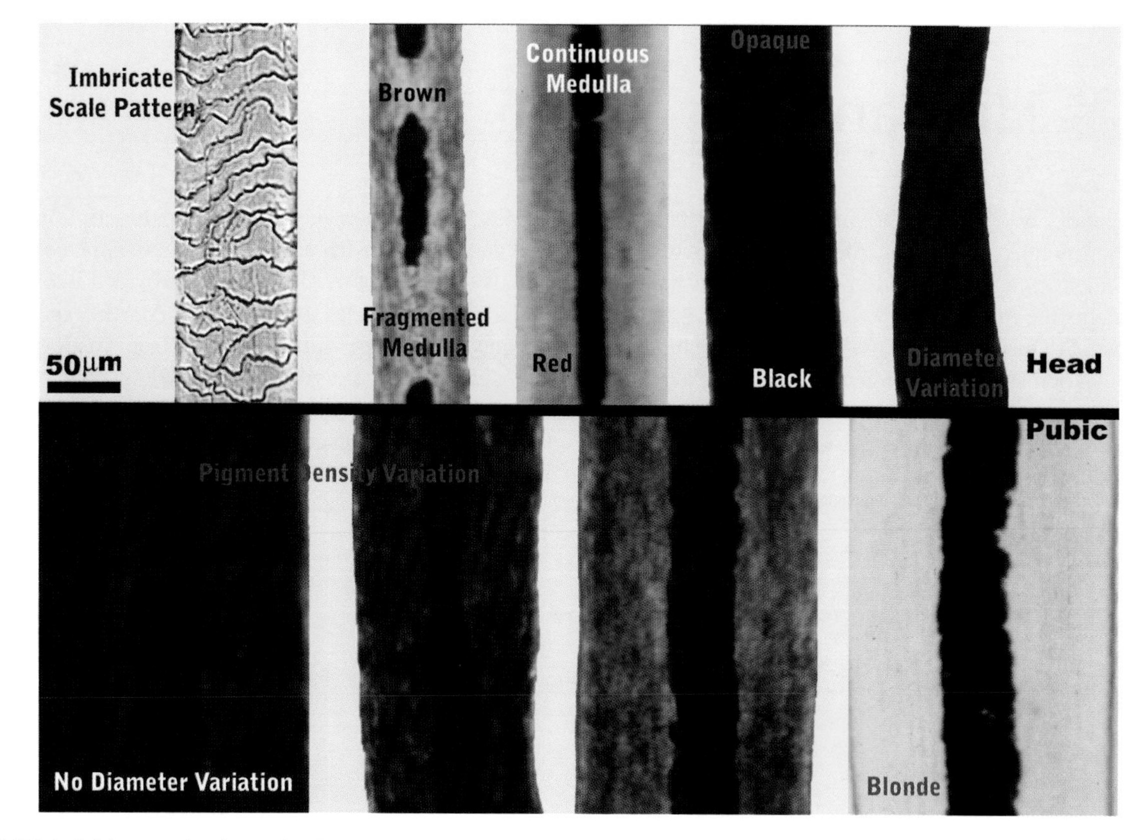

FIGURE A.1 Pigment density and color.

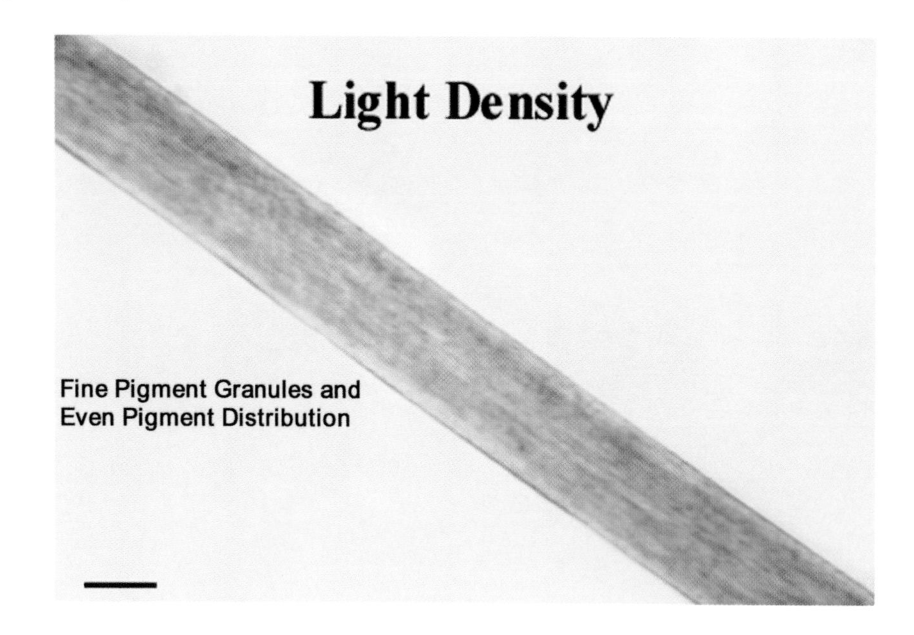

FIGURE A.2

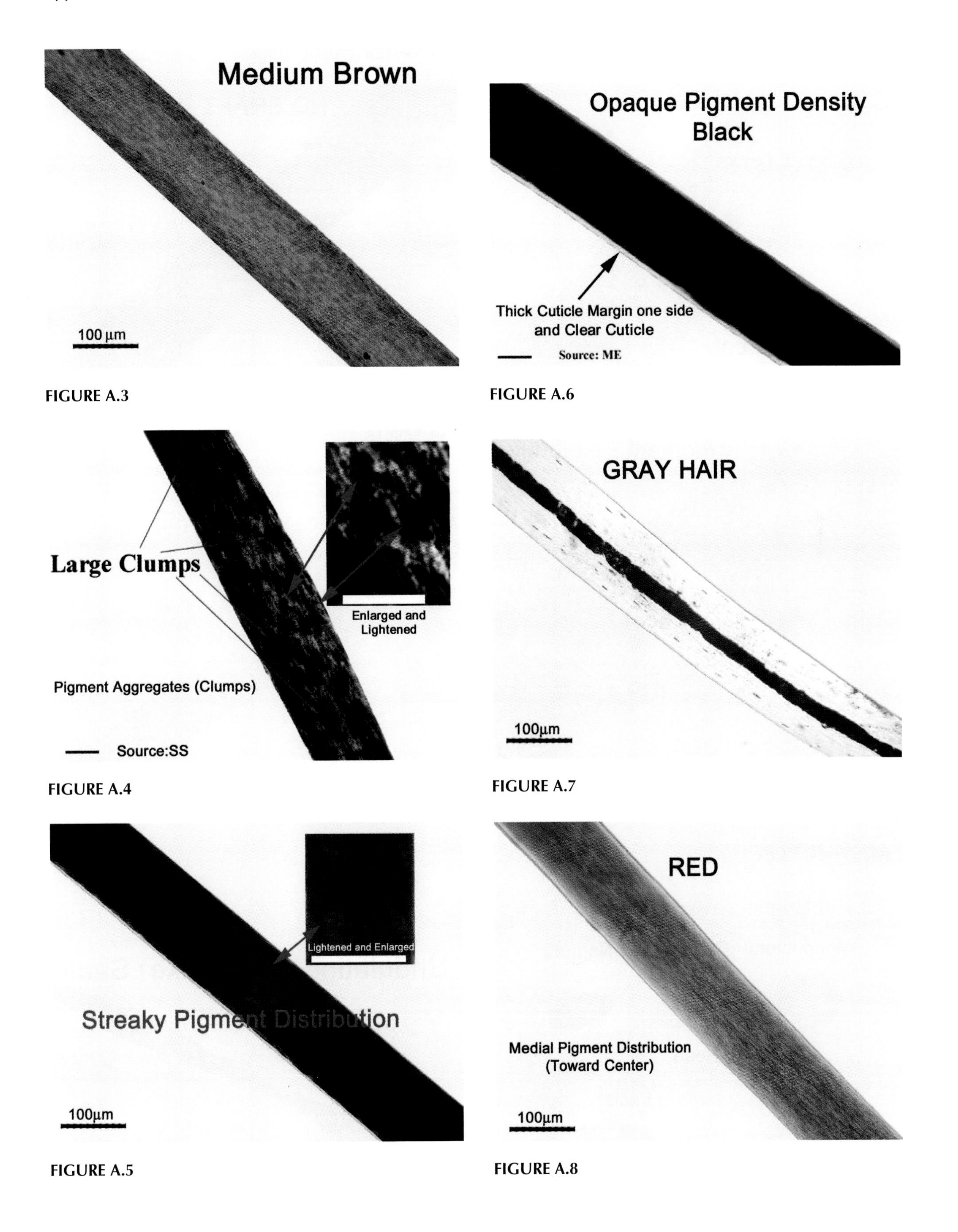

FIGURE A.3

FIGURE A.6

FIGURE A.4

FIGURE A.7

FIGURE A.5

FIGURE A.8

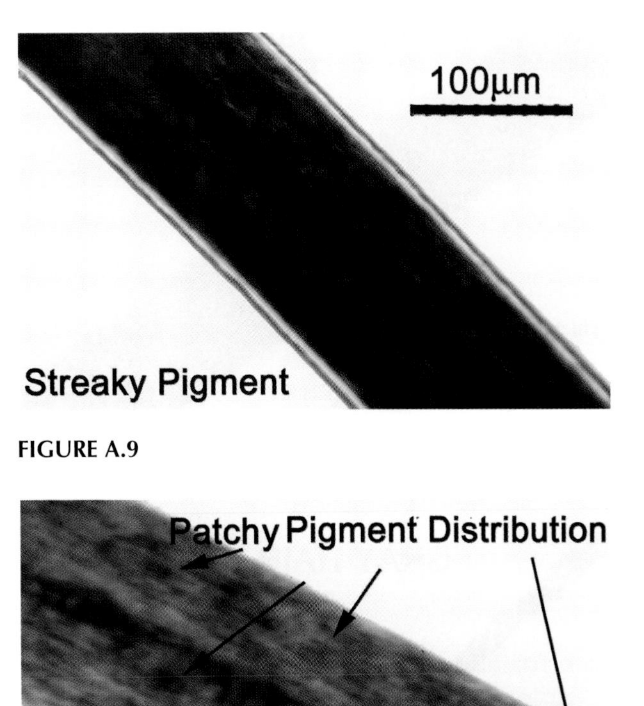

FIGURE A.9

FIGURE A.10

CROSS SECTION

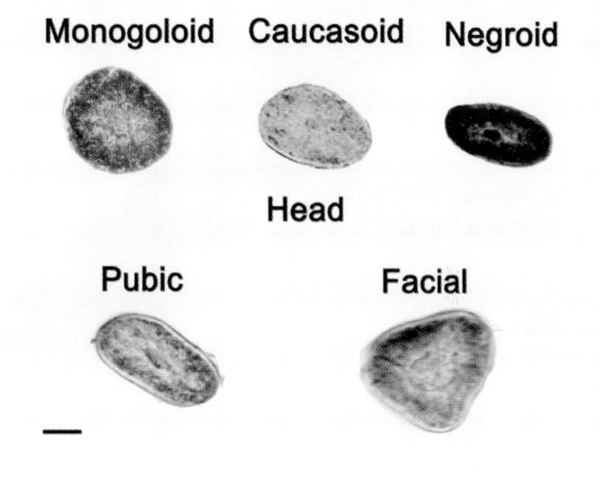

FIGURE A.11

DIAMETER VARIATION

FIGURE A.12

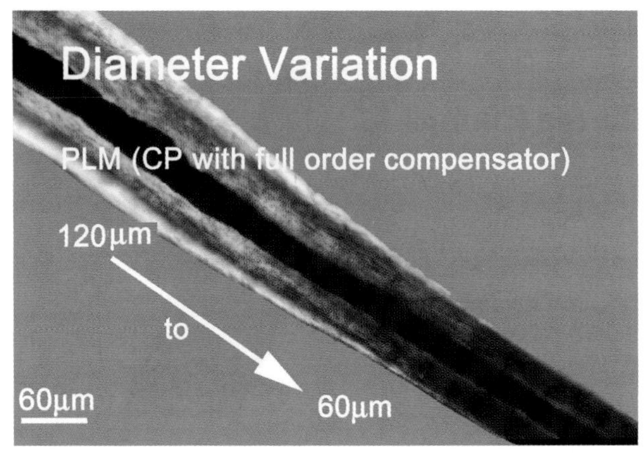

FIGURE A.13

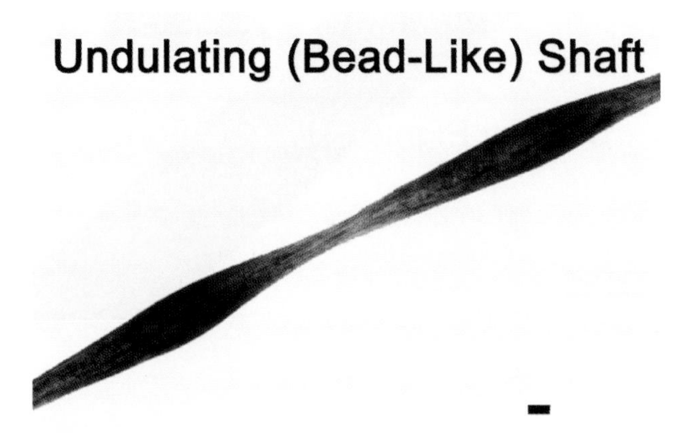

FIGURE A.14

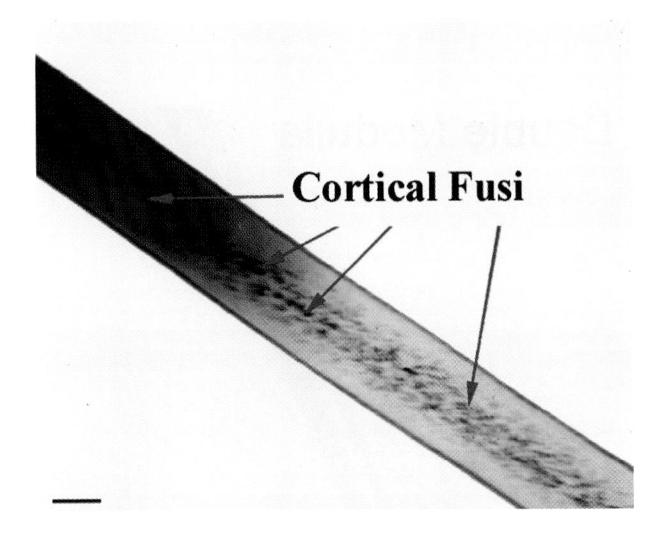

FIGURE A.15

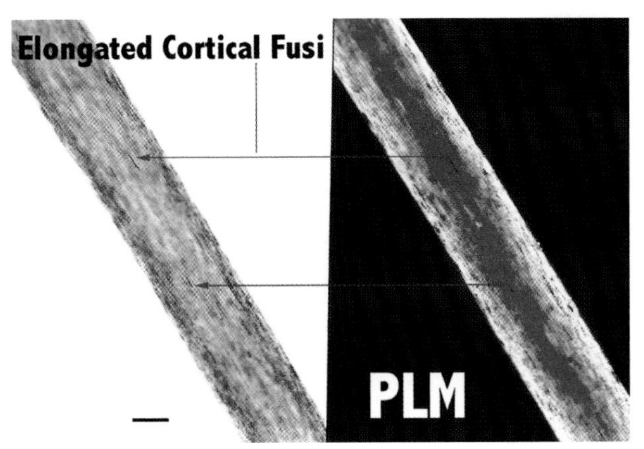

FIGURE A.17

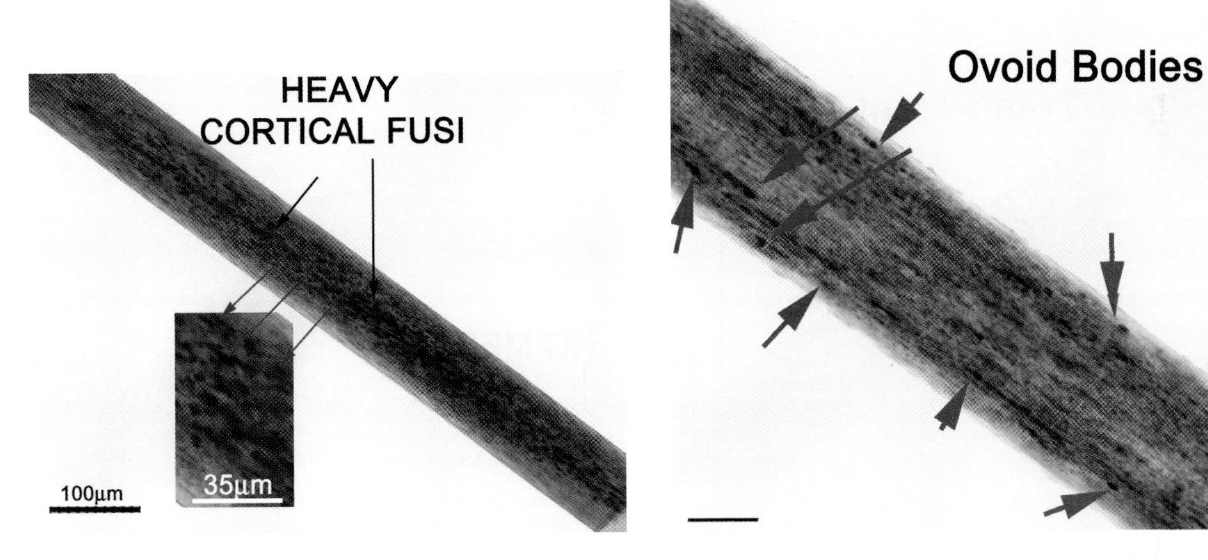

FIGURE A.16

FIGURE A.18

MEDULLA

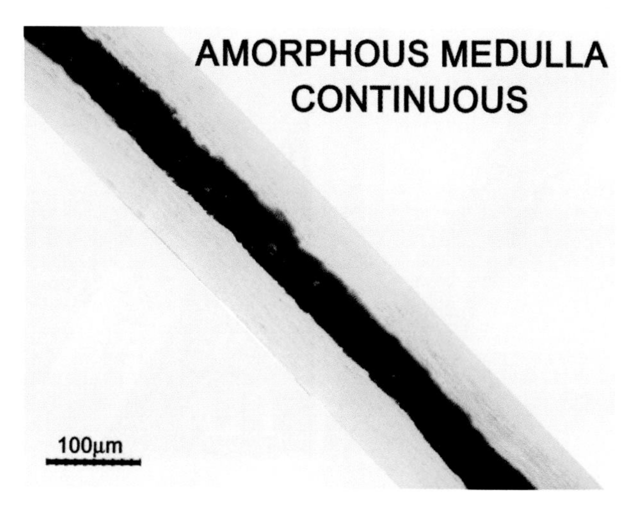

FIGURE A.19

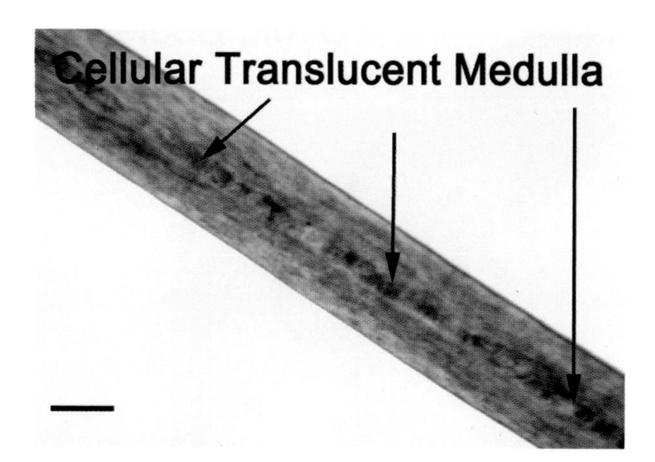

FIGURE A.20

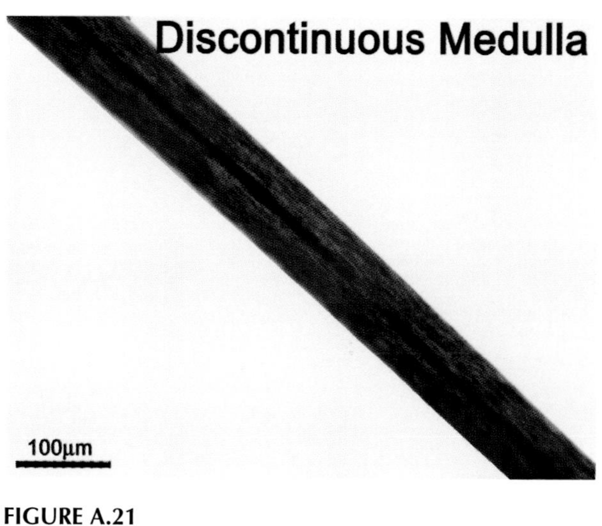

FIGURE A.21

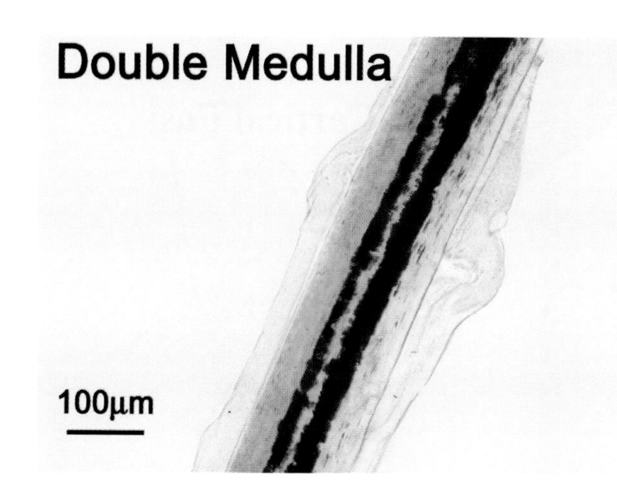

FIGURE A.22

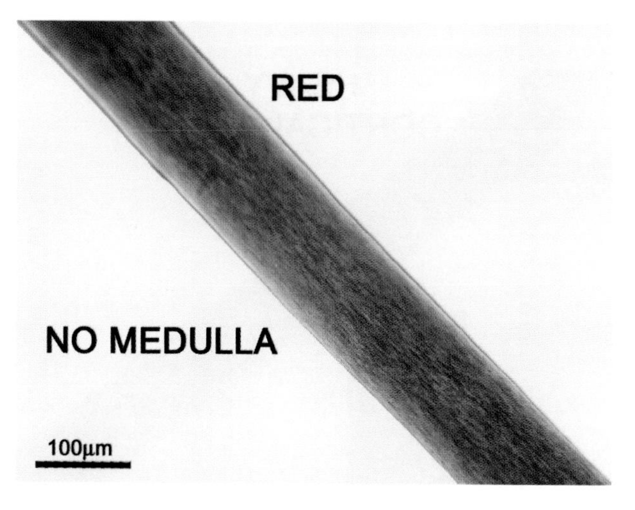

FIGURE A.23

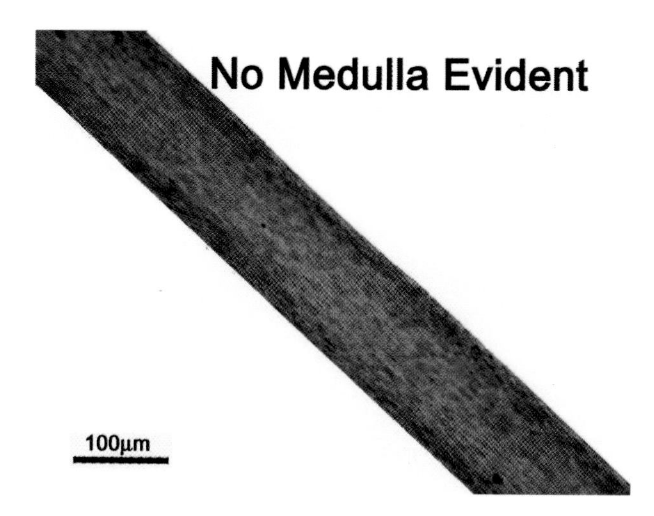

FIGURE A.24

CUTICLE

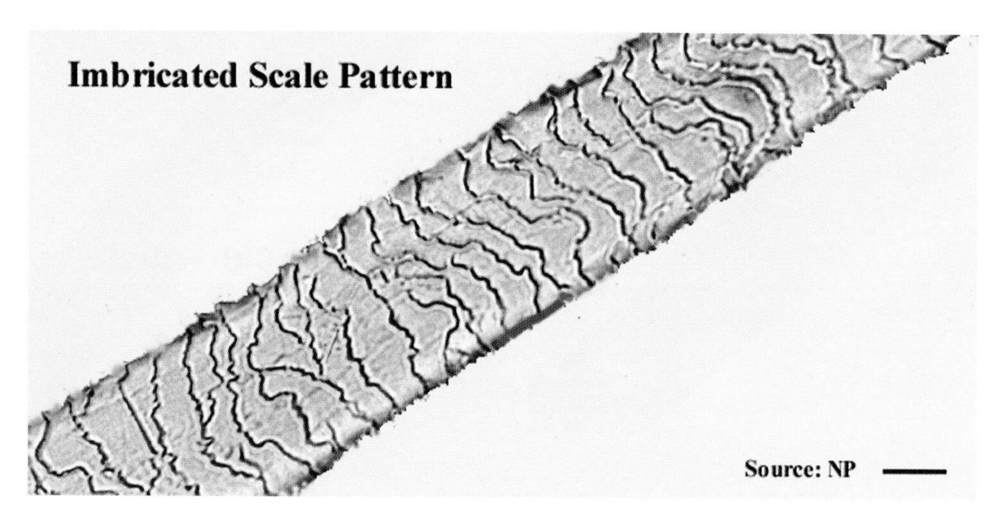

FIGURE A.25

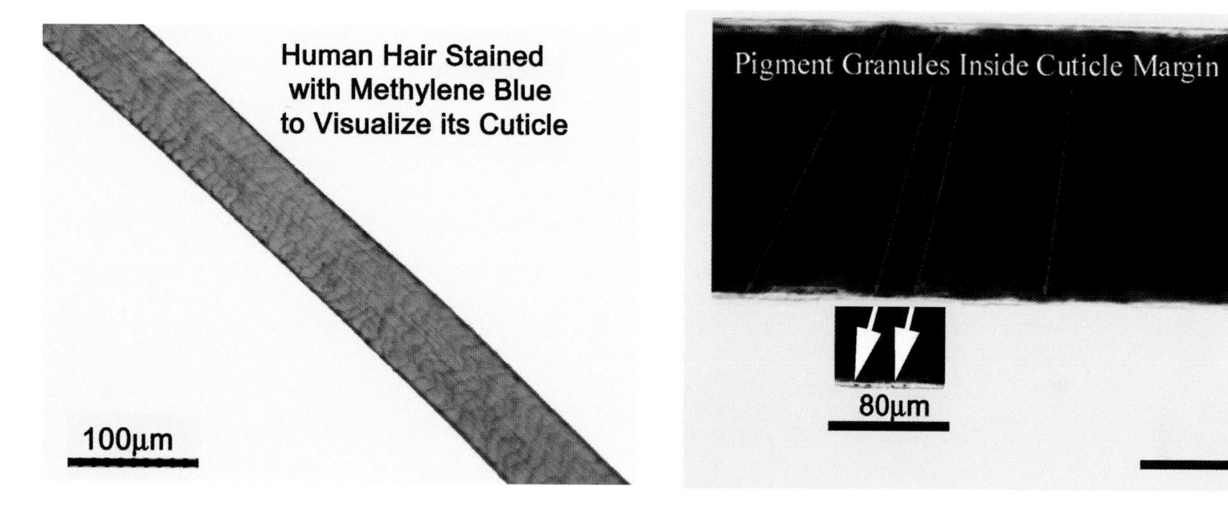

FIGURE A.26

FIGURE A.28

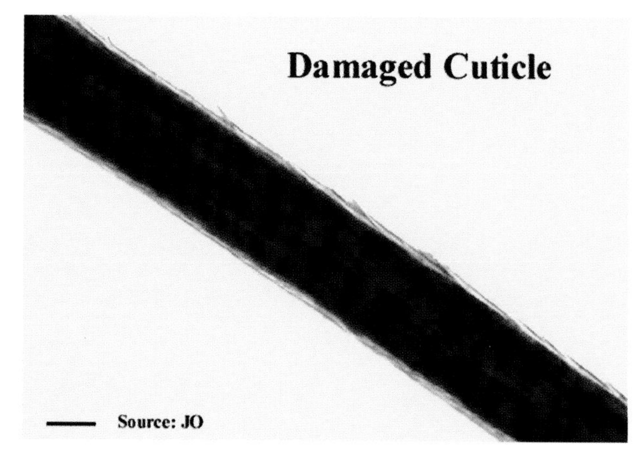

FIGURE A.27

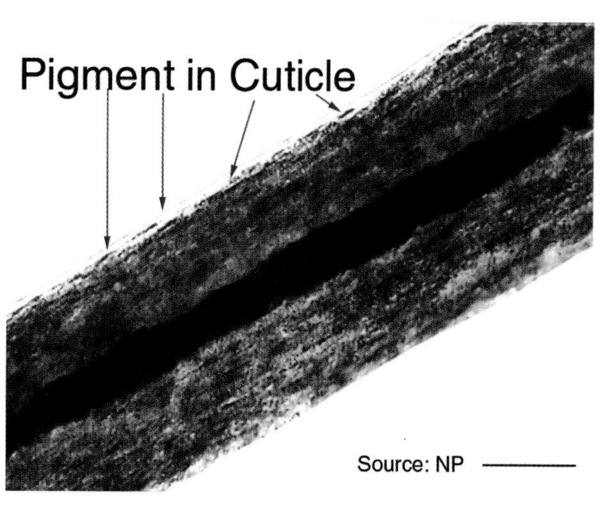

FIGURE A.29

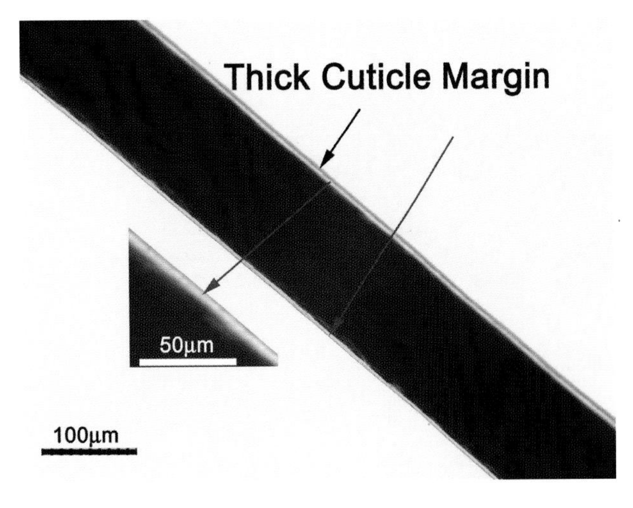

FIGURE A.30

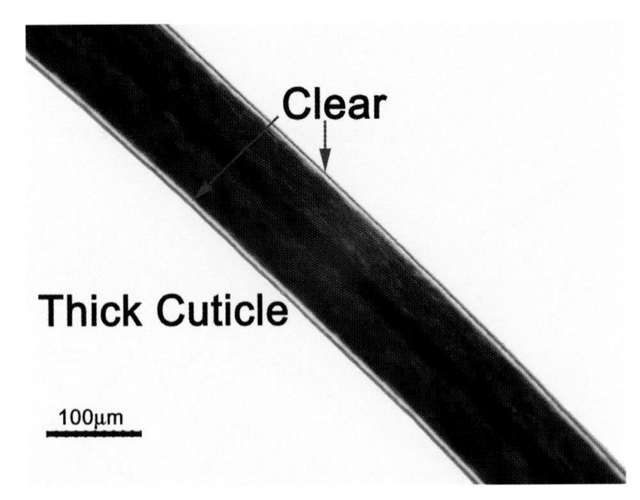

FIGURE A.31

ROOTS AND TIPS

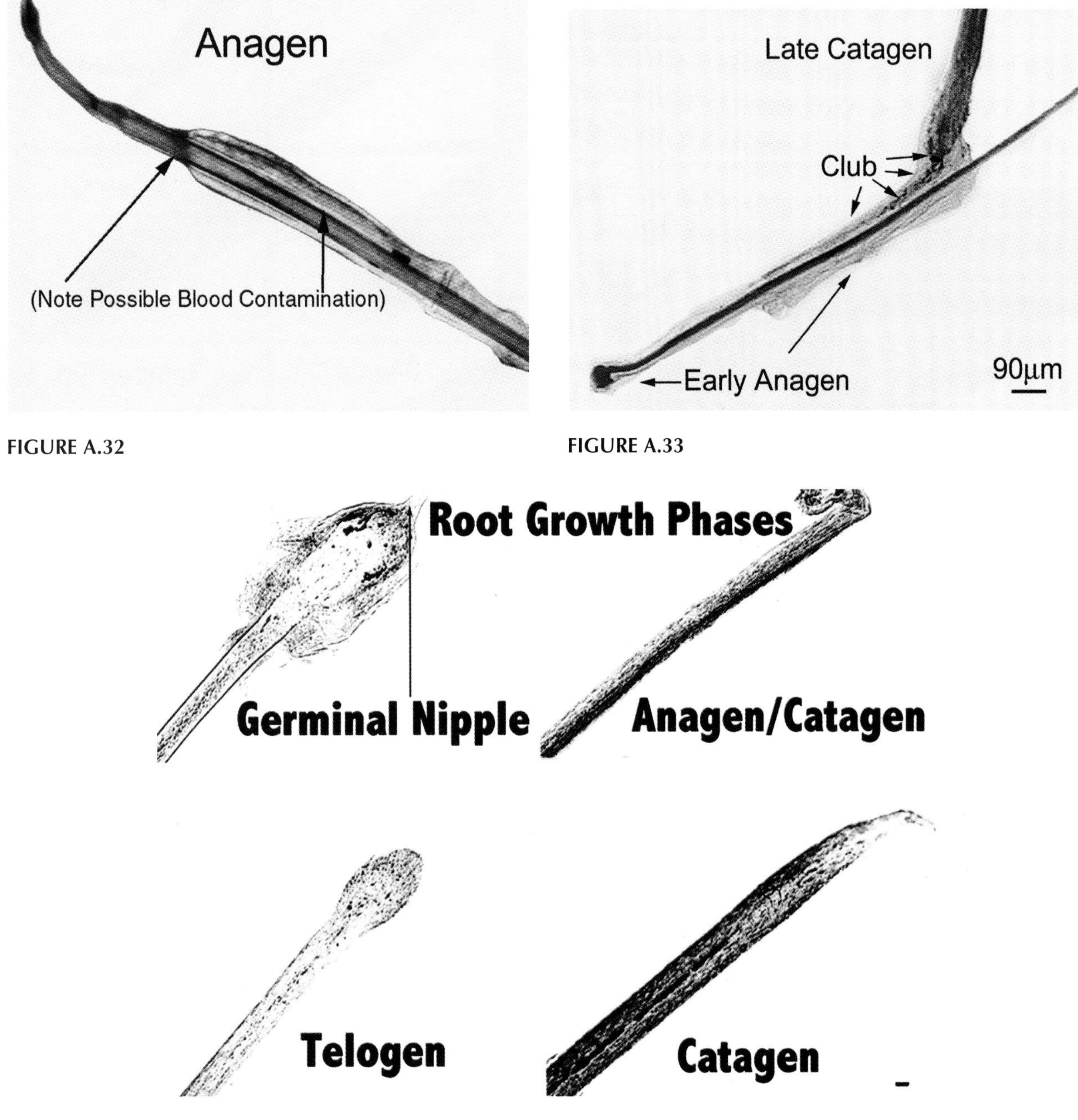

FIGURE A.32

FIGURE A.33

FIGURE A.34

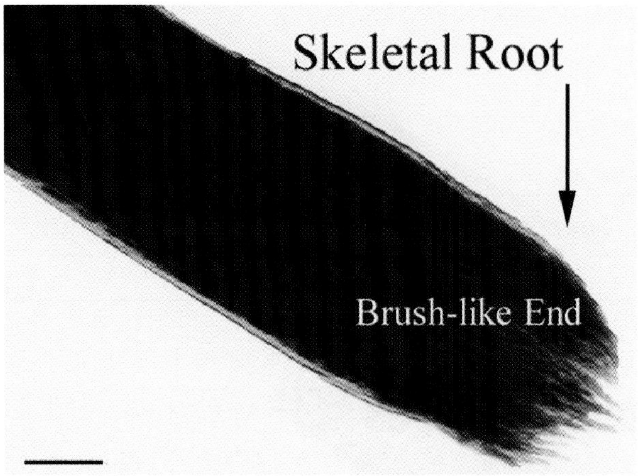

FIGURE A.35

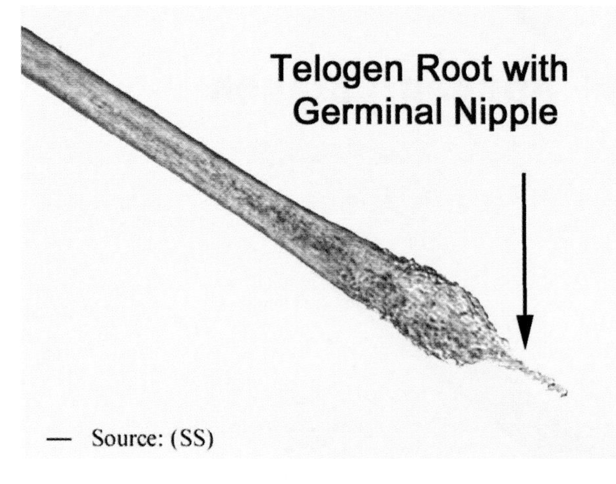

FIGURE A.36

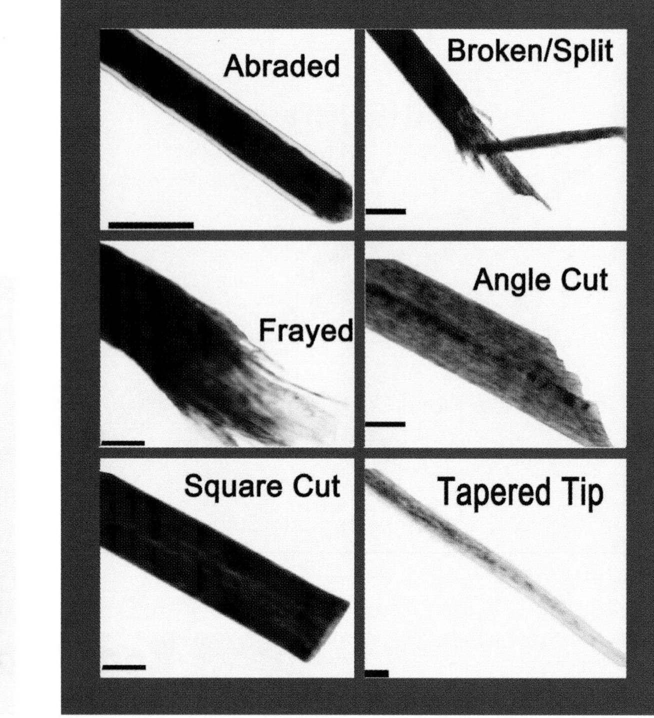

FIGURE A.37

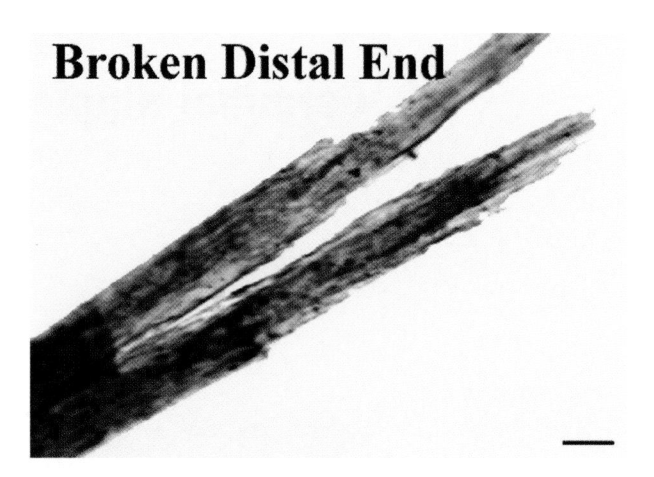

FIGURE A.38

SOMATIC ORIGIN

HEAD

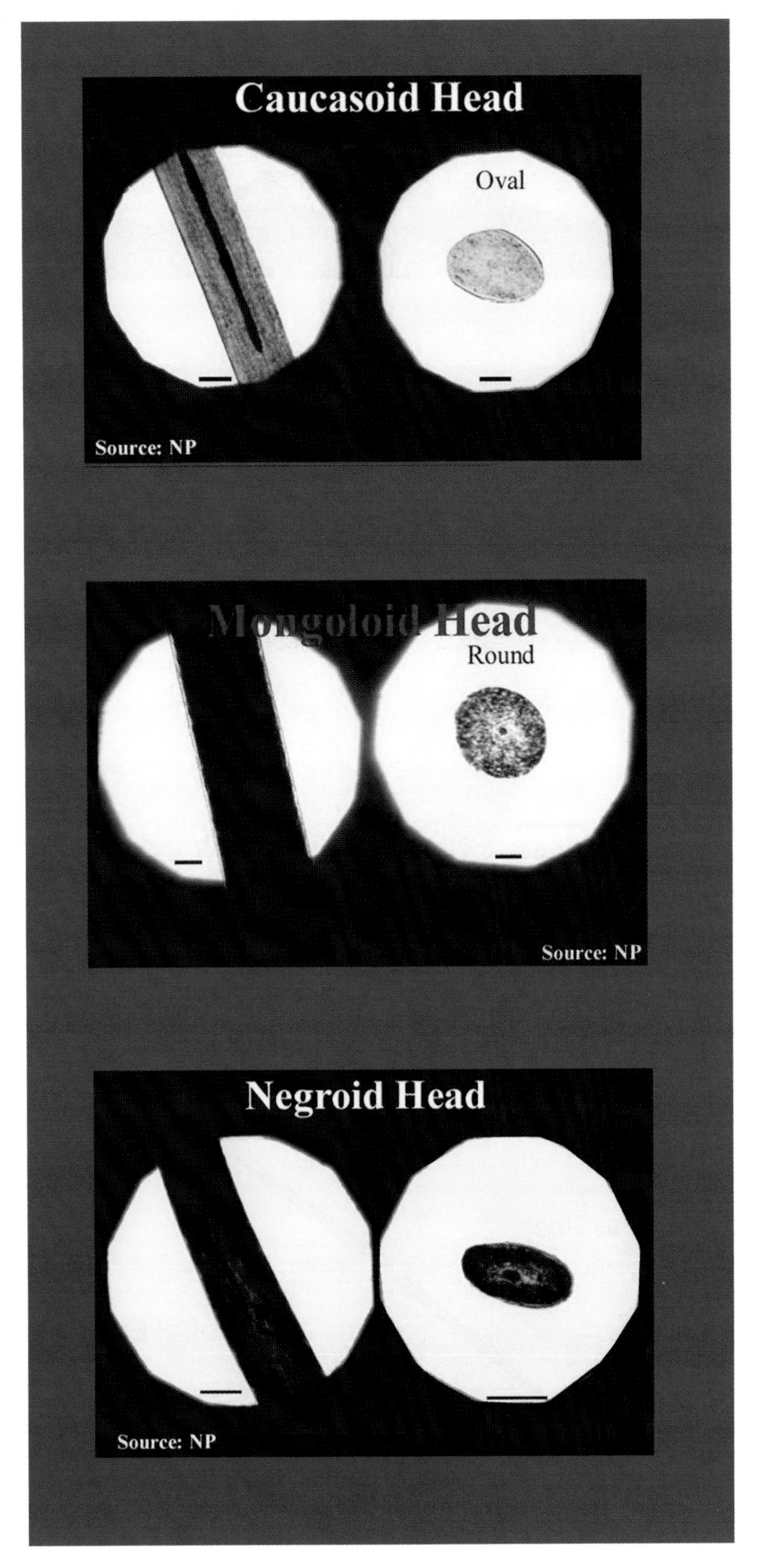

FIGURE A.39

FIGURE A.40

FIGURE A.41

FIGURE A.42

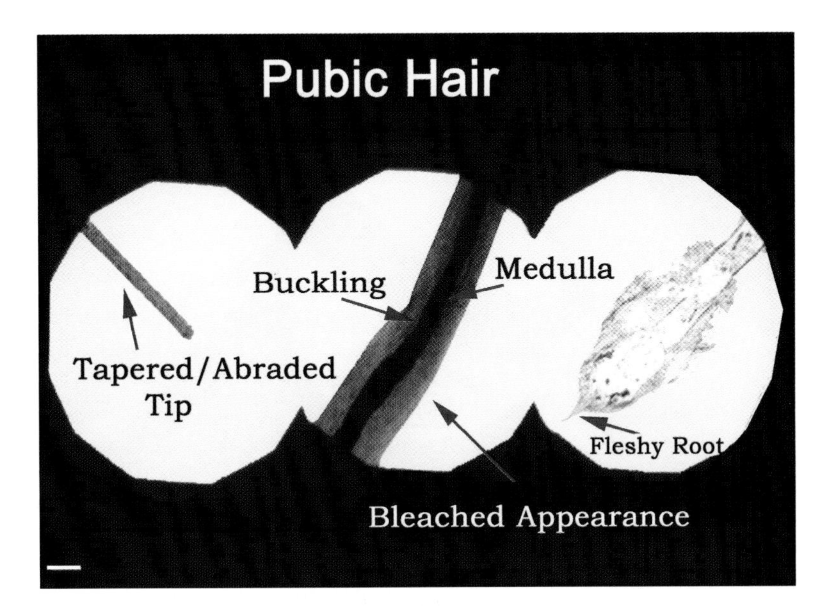

FIGURE A.43

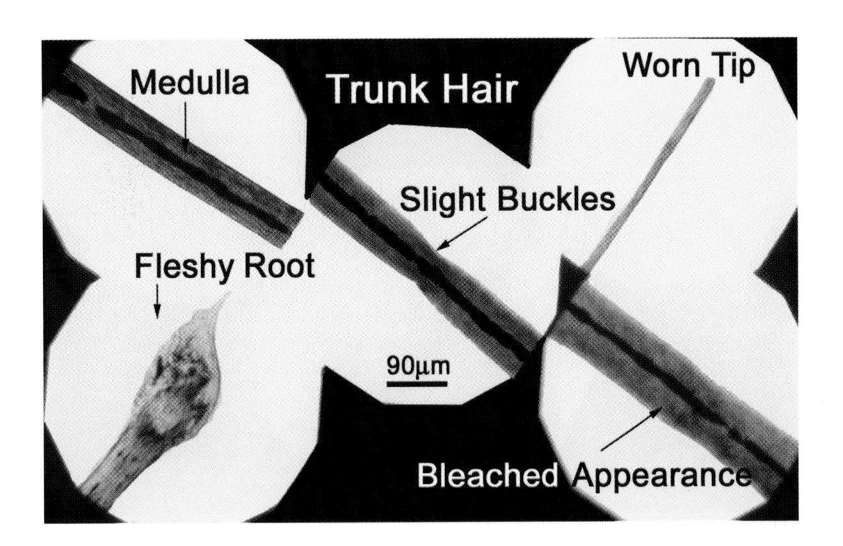

FIGURE A.44

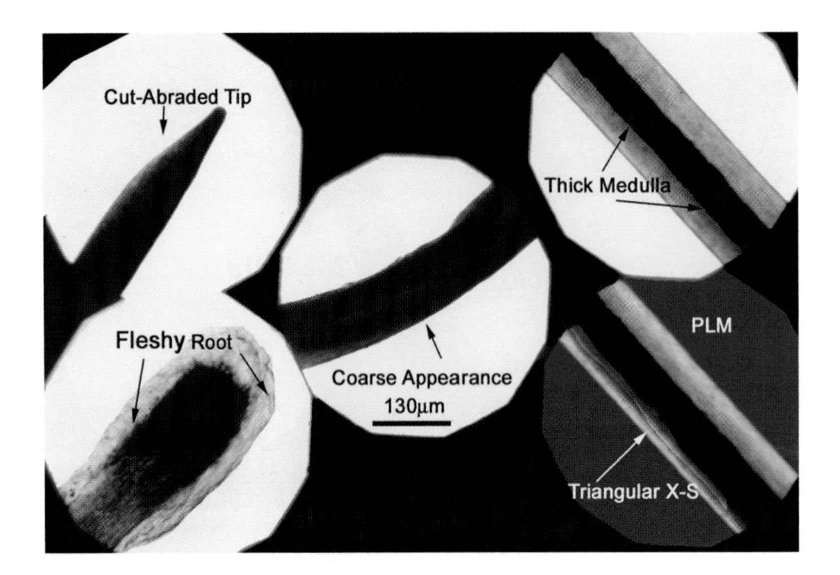

FIGURE A.45

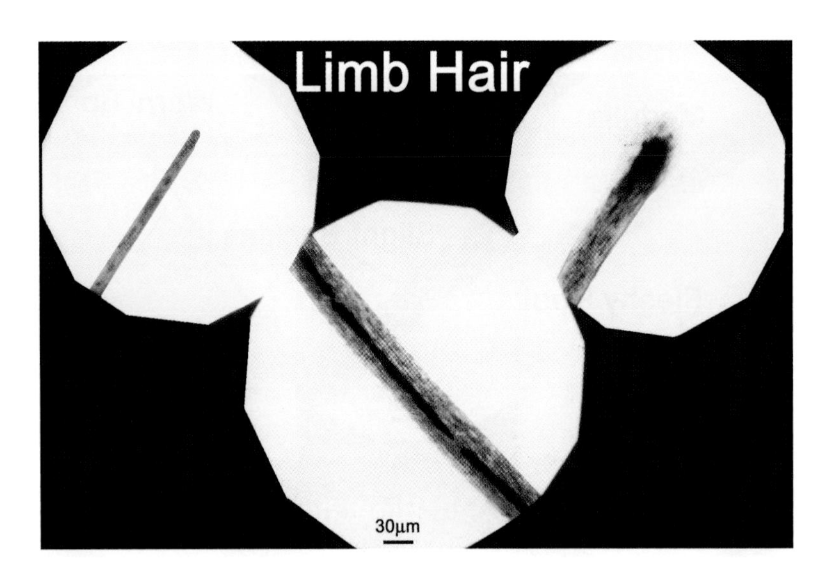

FIGURE A.46

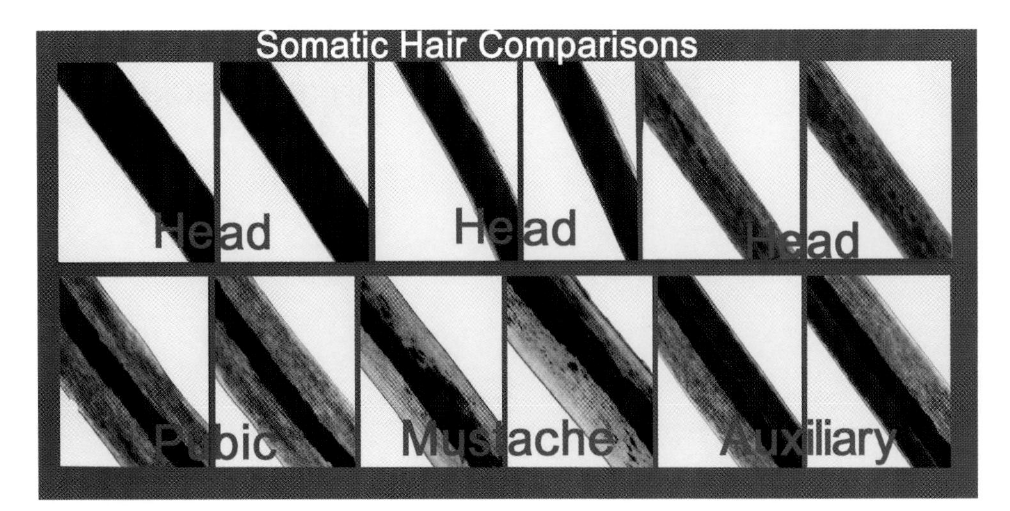

FIGURE A.47

REGION OF ORIGIN

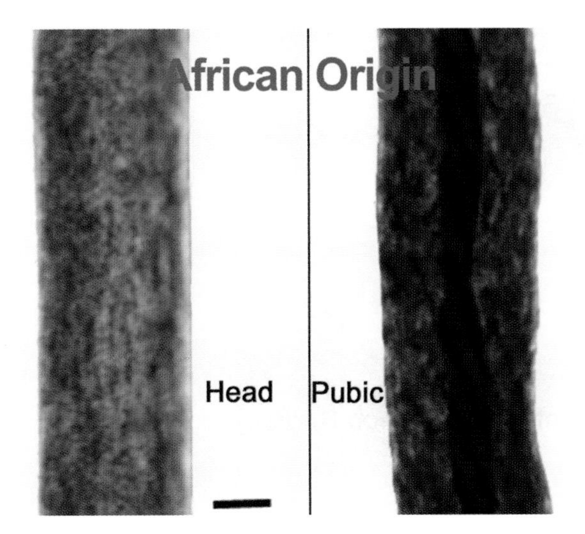

FIGURE A.48

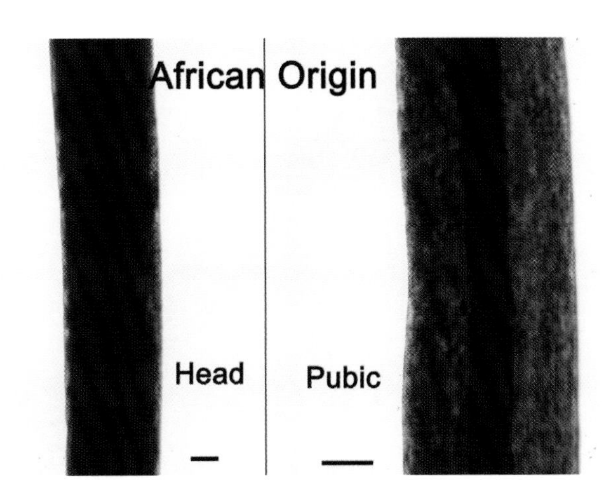

FIGURE A.49

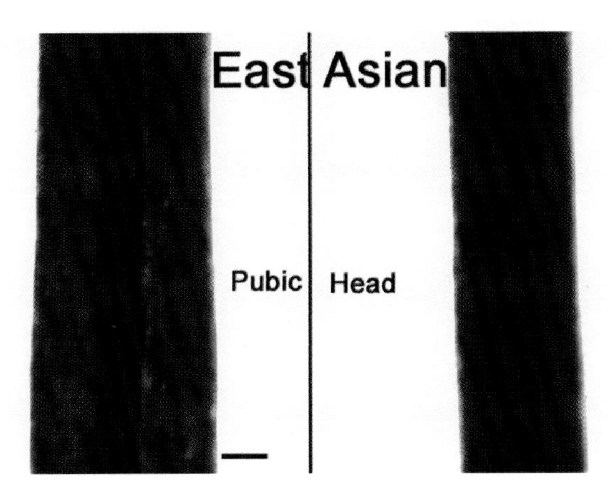

FIGURE A.50

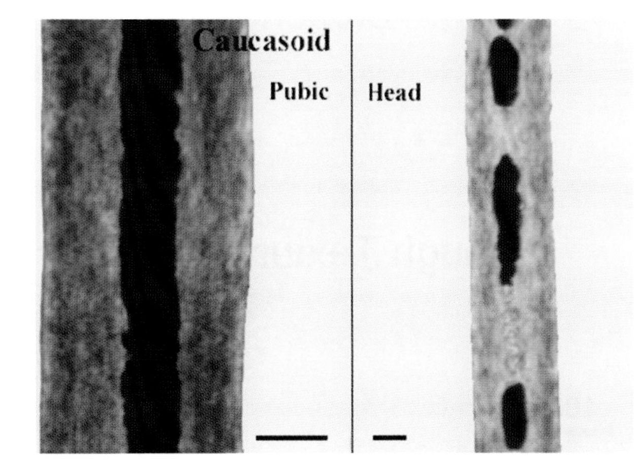

FIGURE A.51

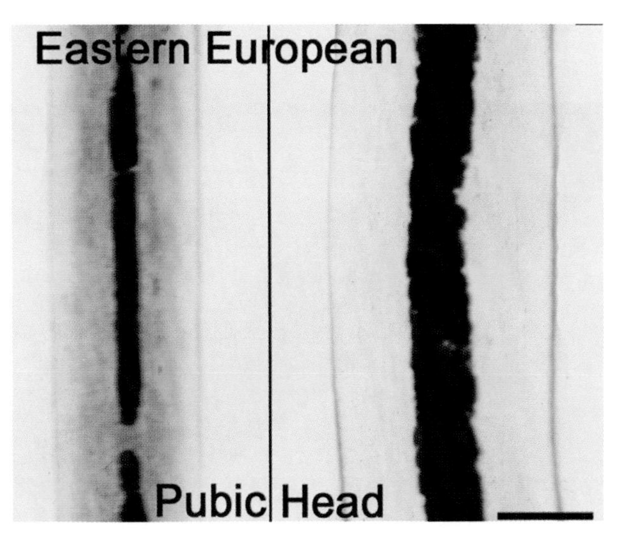

FIGURE A.52

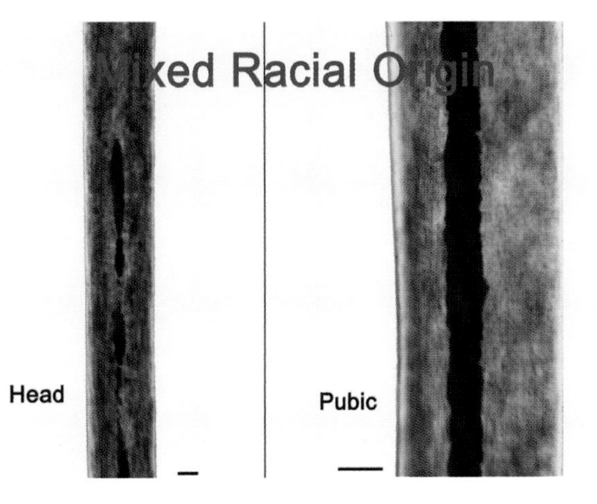

FIGURE A.53

TREATED HAIR

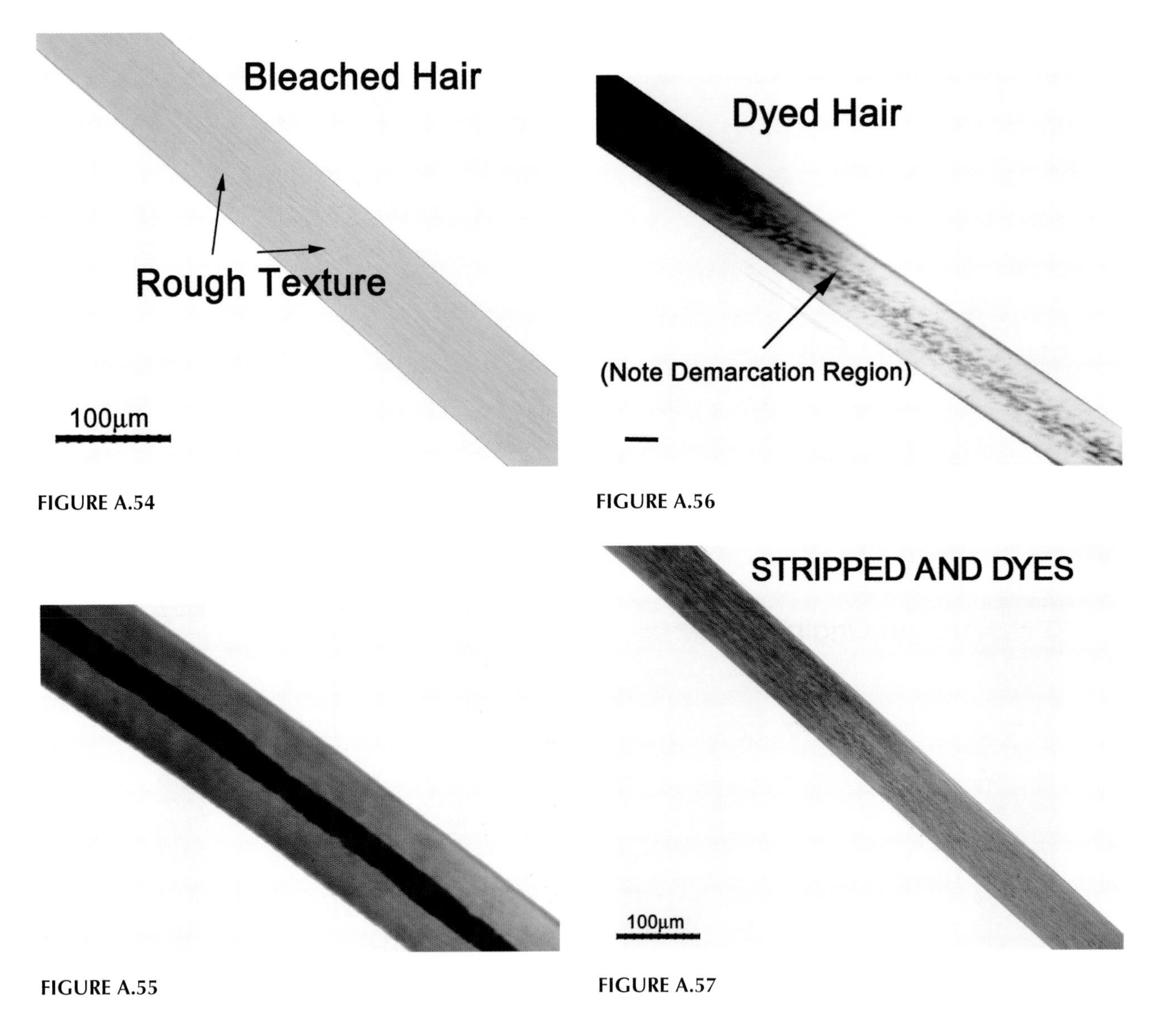

FIGURE A.54

FIGURE A.56

FIGURE A.55

FIGURE A.57

DAMAGE AND ARTIFACTS

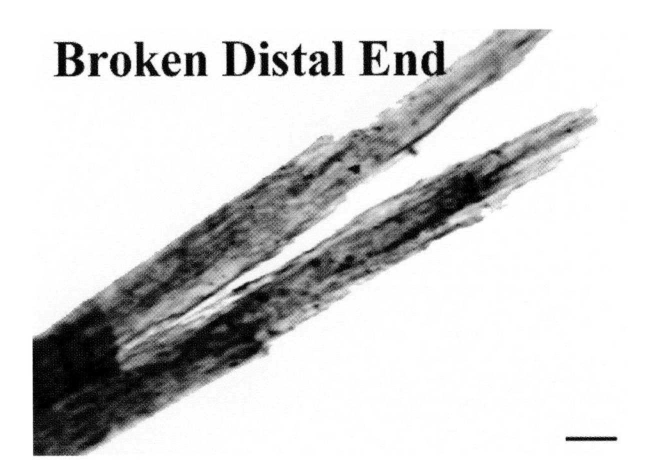

FIGURE A.58

FIGURE A.59

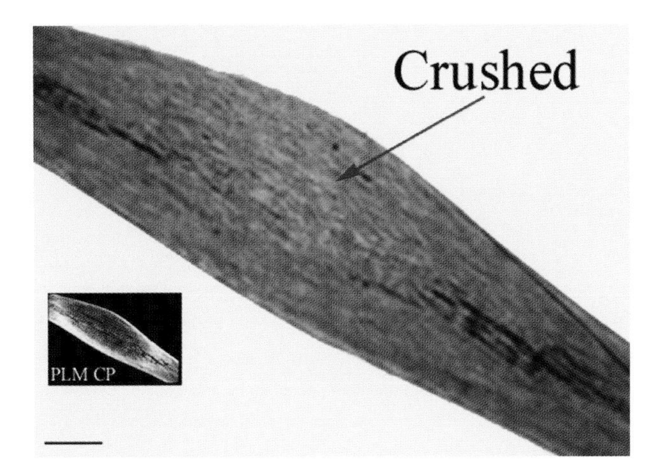

FIGURE A.60

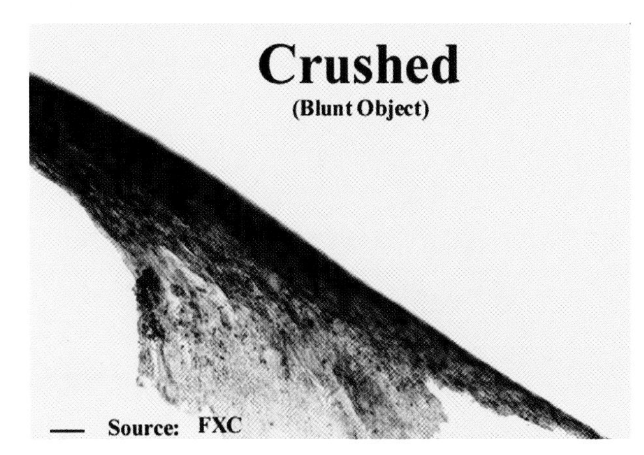

FIGURE A.61

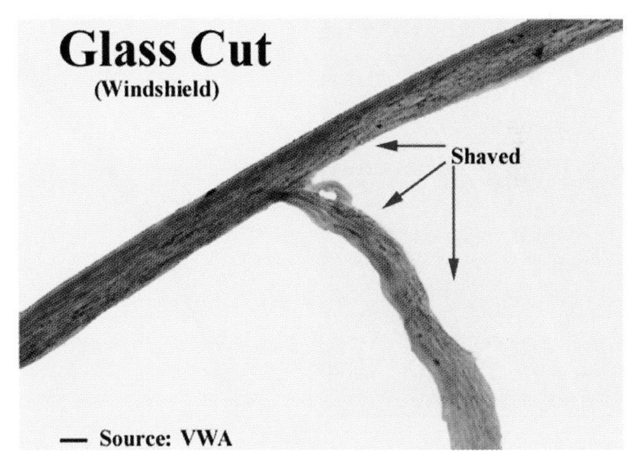

FIGURE A.62

FIGURE A.63

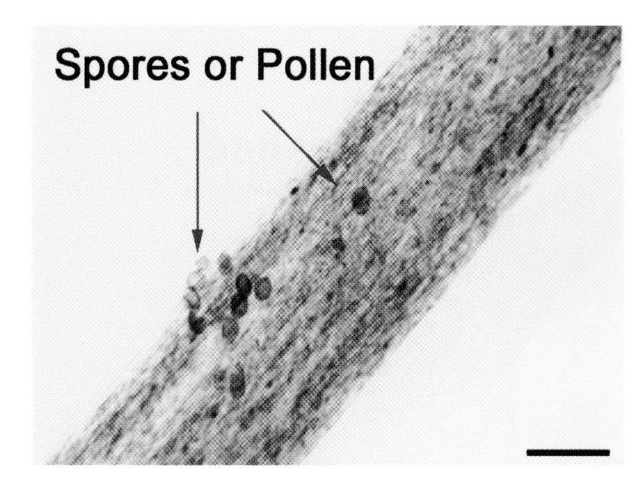

FIGURE A.64

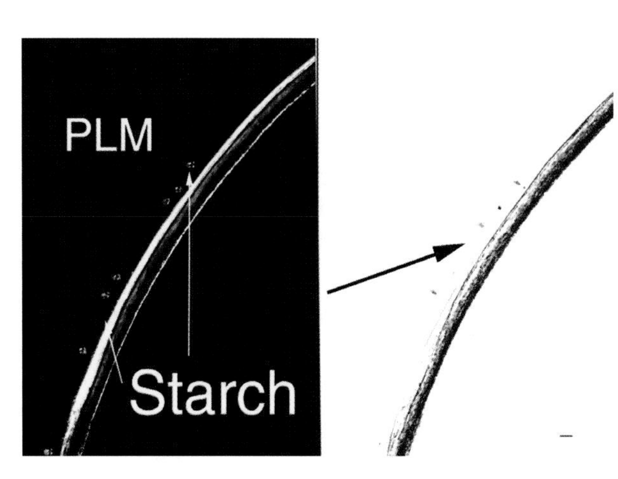

FIGURE A.65

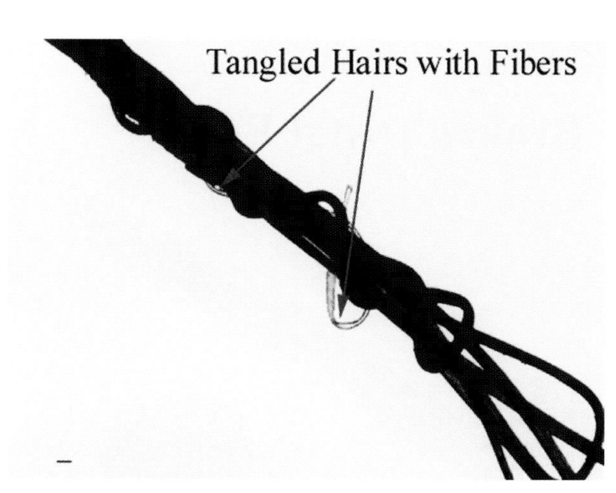

FIGURE A.66

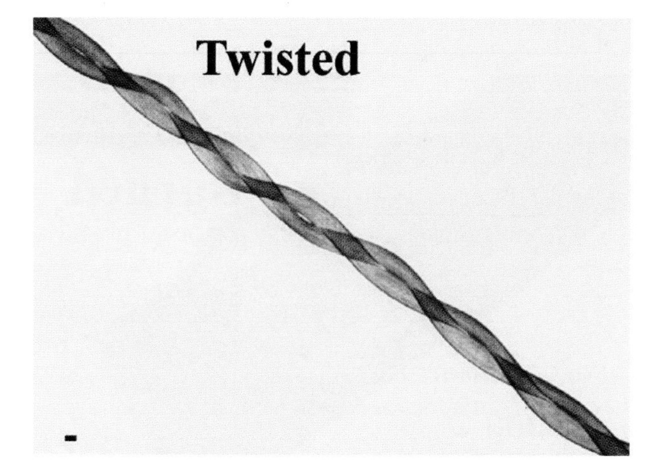

FIGURE A.67

DISEASES AND CONDITIONS

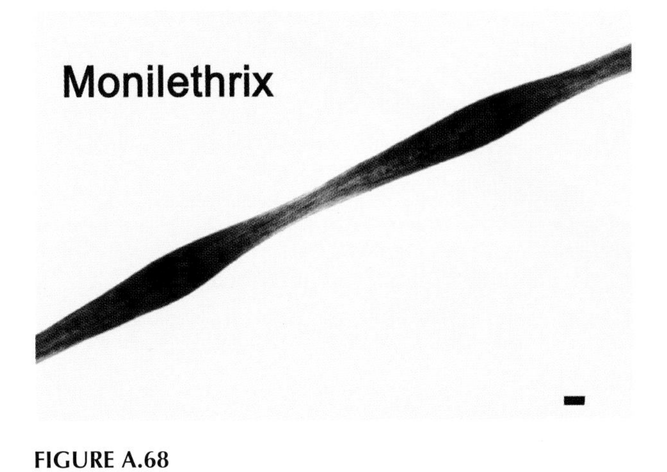

FIGURE A.68

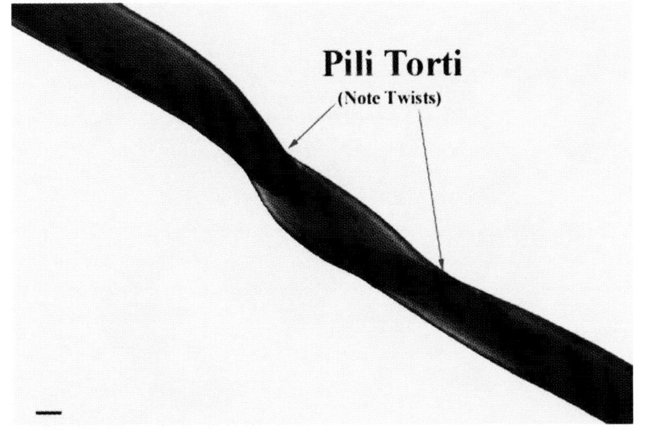

FIGURE A.69

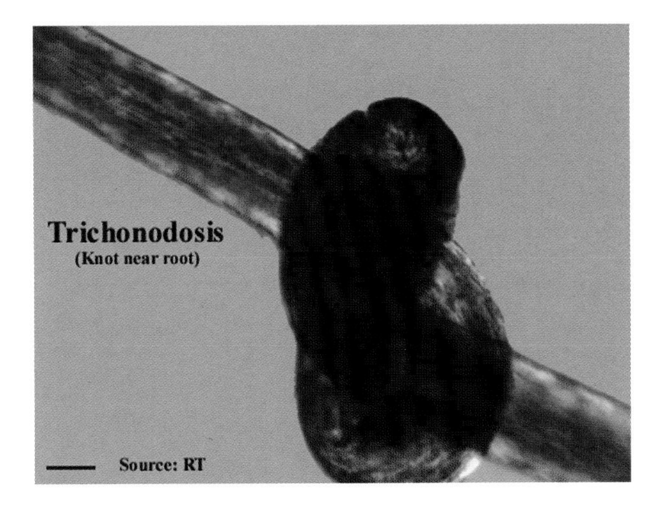

FIGURE A.70

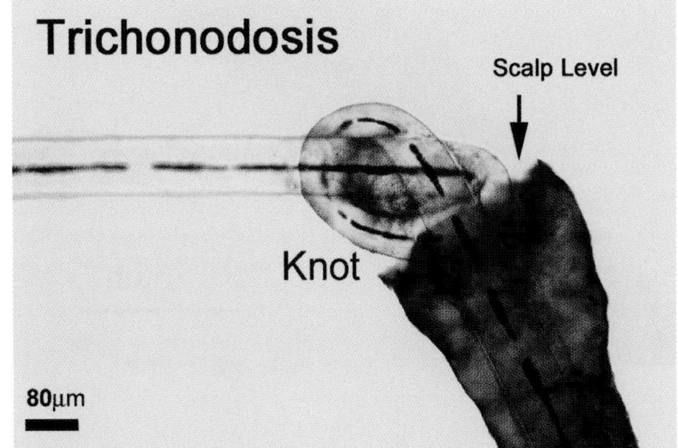

FIGURE A.71

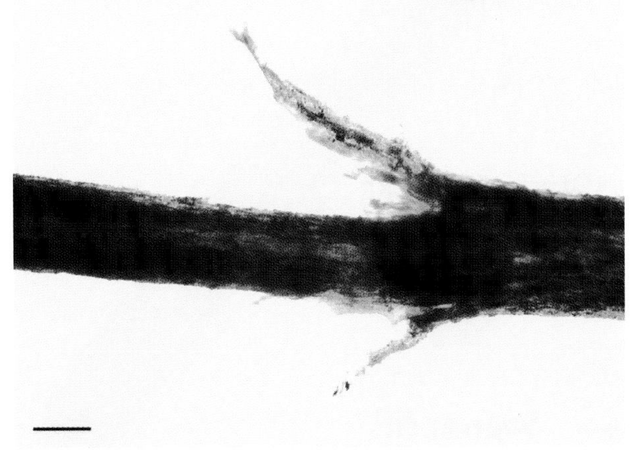

FIGURE A.72

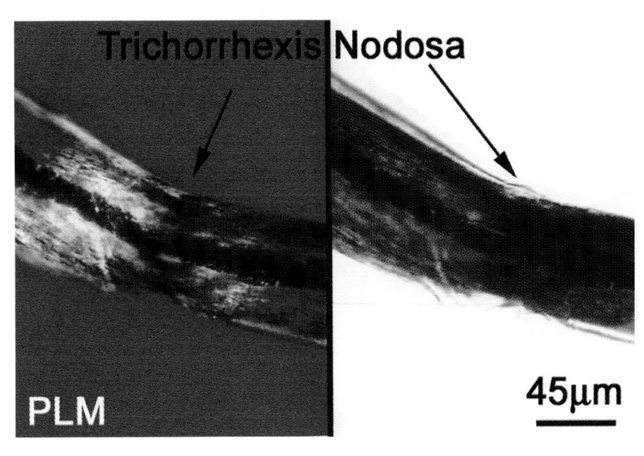

FIGURE A.73

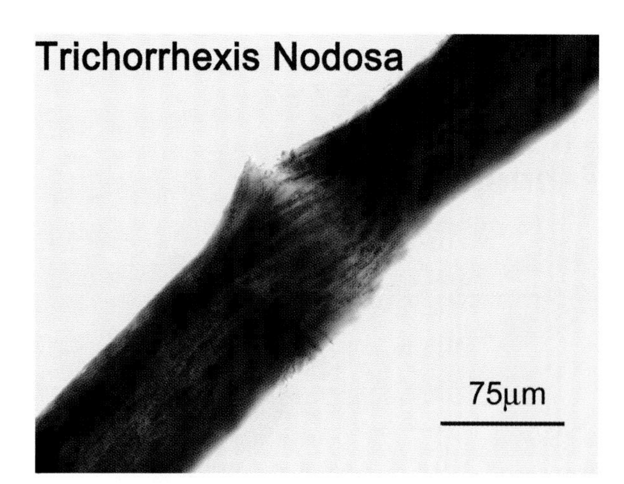

FIGURE A.74

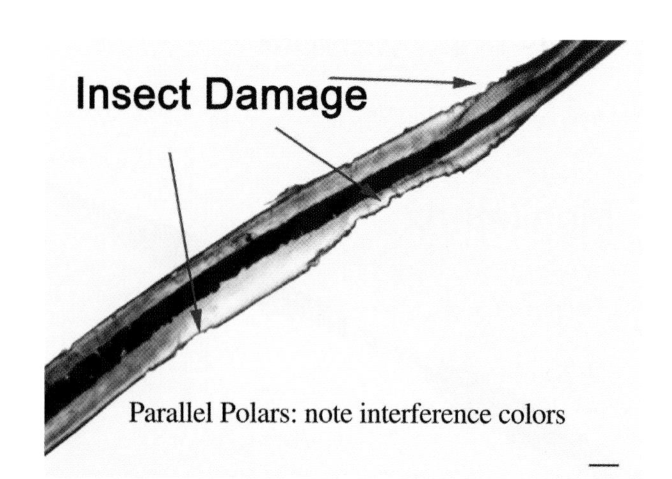

FIGURE A.77

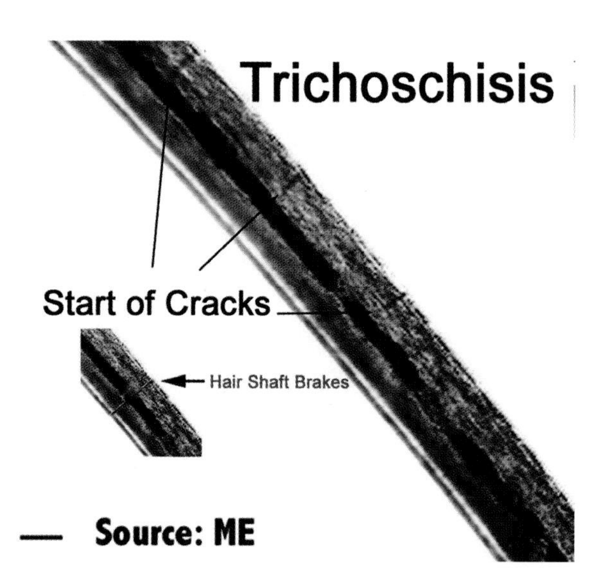

FIGURE A.75

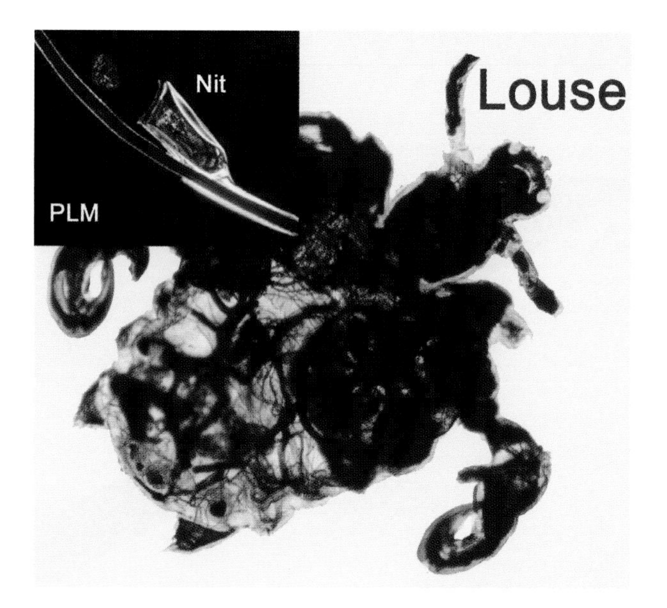

FIGURE A.78

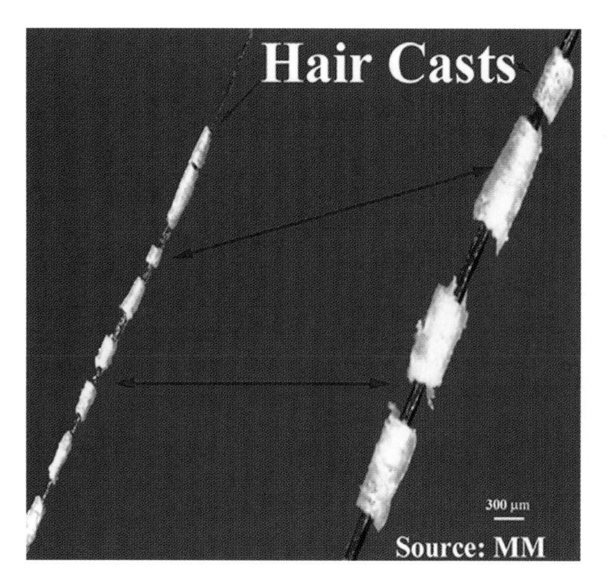

FIGURE A.76

CONTRIBUTORS TO HUMAN HAIR ATLAS

Valerie Wade Allison (VWA)
Forensic Laboratory
New York City Police Department

Francis X. Callery (FXC)
Crime Scene Unit
New York City Police Department

Mary Eng (ME)
Forensic Laboratory
New York City Police Department

Lisa Faber (LF)
Forensic Laboratory
New York City Police Department

Melanie McMillin (MM)
Forensic Laboratory
New York City Police Department

Judy O'Connor (JO)
Forensic Laboratory
New York City Police Department

Susan M. Shankles (SS)
Police Department Crime Laboratory
Tucson, AZ

Roscida Tordesillas (RT)
Forensic Laboratory
New York City Police Department

APPENDIX B

Animal Hair Atlas

This appendix is intended for the preliminary identification of commonly encountered mammalian guard hairs. Each entry contains one or more photographs of the hair of the subject's family, genus, and/or species. A short verbal description of identifying features of guard hairs is presented for each genus. Scale pattern configurations (SPCs) for each guard hair are given from root (basal) end to tip end. Only the primary medullary classification (PMC) is provided for each guard hair. The cited literature is the primary source for the information and data found in this appendix. Readers are referred to the literature for positive identification of questioned guard hair specimens. The value of the scale in each photomicrograph is as noted. Anyone wishing to reproduce the illustrations should contact the authors.[1–16]

REFERENCES

1. Glaister, J., *A Study of Hairs and Wools Belonging to the Mammalian Group of Animals Including a Special Study of Human Hair, Considered from Medico-Legal Aspects*, Cairo, MISR Press, 1931.
2. Hausman, L.A., Mammal fur under the microscope, *J. Am. Museum Nat. History*, 20, 434, 1920.
3. Hausman, L.H., Structural characteristics of the hair of mammals, *Am. Naturalist*, 54, 496, 1920.
4. Hausman, L.A., Further studies of the relationships of the structural characteristics of mammalian hair, *Am. Naturalist*, 58, 544, 1924.
5. Hausman, L.A., Recent studies of hair structure relationships, *Sci. Monthly*, 30, 258, 1930.
6. Mathiak, H.A., A key to hairs of the mammals of southern michigan, *J. Wildlife Mgt.*, 2, 251, 1938.
7. Brown, F.M., The microscopy of mammalian hair for anthropologists, *Proc. Amer. Philos. Soc.*, 85, 1942.
8. Mayer, W.V., The hair of California mammals with keys to the dorsal guard hairs of California mammals, reprinted from *Am. Midland Nat.*, 28, 480, 1952.
9. Wildman, A.B., *Microscopy of Animal Textile Fibres*, Leeds, WIRA, 1954.
10. Adoryan, A.S. and Kolenosky, G.B., *A Manual for the Identification of Hairs of Selected Ontario Mammals,* Wildlife Research Report 90, Ottawa, Department of Lands and Forests, 1969.
11. Moore, T.D., Spence, L.E., Dugnolle, C.E., and Hepworth, W.G., *Identification of the Dorsal Guard Hairs of Some Mammals of Wyoming,* Cheyenne, Wyoming Fish and Game, 1974.
12. Brunner, H. and Conan, B.J., *The Identification of Mammalian Hair*, Melbourne, Iukata Press, 1974.
13. Appleyard, H.M., *Guide to the Identification of Animal Fibres*, 2nd ed., Leeds, WIRA, 1978.
14. Moore, T.D., Spence, L.E., Dugnolle, C.E., and Hepworth, W.G., *Identification of the Dorsal Guard Hairs of Some Mammals of Wyoming,* Wyoming Fish and Game, Cheyenne, 1974.
15. Hicks, J.W., Microscopy of Hair, Issue 2, Washington, D.C., U.S. Government Printing Office, 1977.
16. Petraco, N.A., Microscopical method to aid in the identification of animal hair, *Microscope*, 35, 83, 1987.

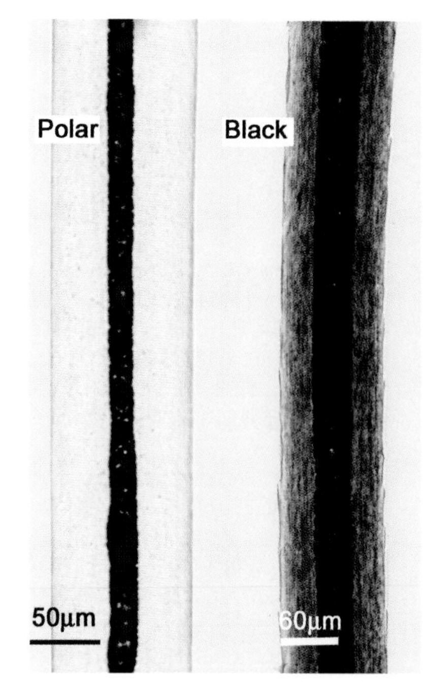

Family Ursidae
Genus *Urus*
Bears
PMC: Amorphous cellular
SPC: Imbricate or
mosaic to imbricate
MI < 0.5
Pigment distribution even
Unbanded color varies:
Polar (yellow/white)
Grizzly (dark brown to light brown)
Black (black, dark brown)

FIGURE B-1 Bear.

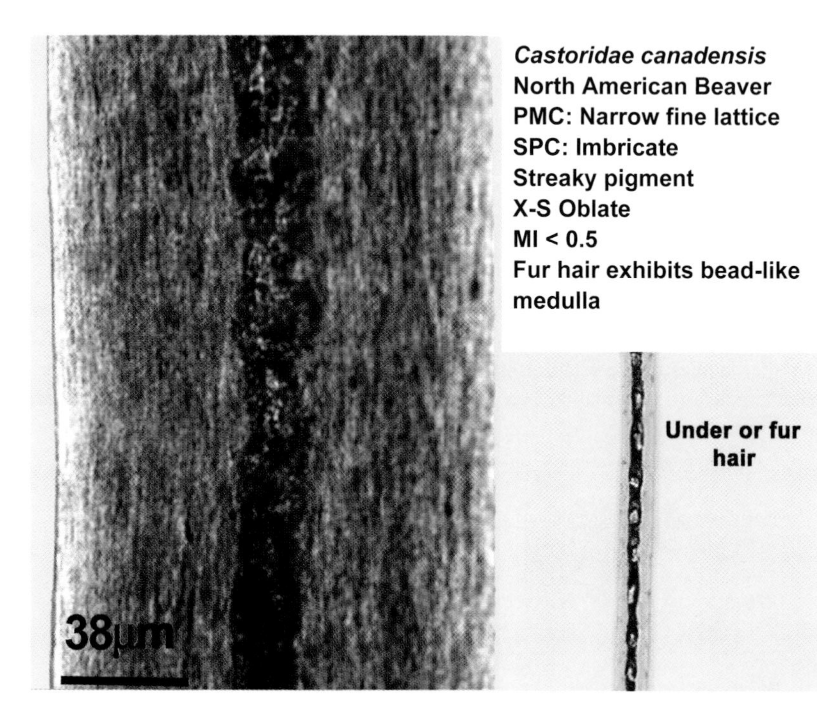

FIGURE B-2 Beaver.

Family Bovidae
Bison bison
PMC: Amorphous
SPC: Imbricated
Medulla off center
Pigment streaky
Coarse unbanded hair

FIGURE B-3 Bison.

Camelus bactrianus
Camel
PMC: Continuous amorphous
SPC: Mosaic
Pigment fine, dense, streaky
Underhair cashmere, very fine wool
MI approx. 0.5
Color tan to brown

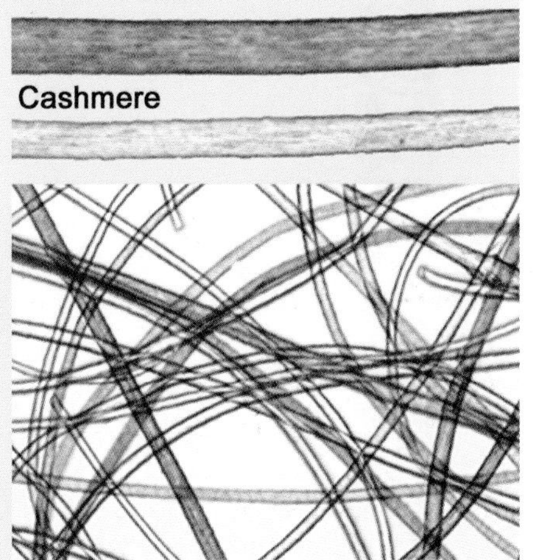

FIGURE B-4 Camel.

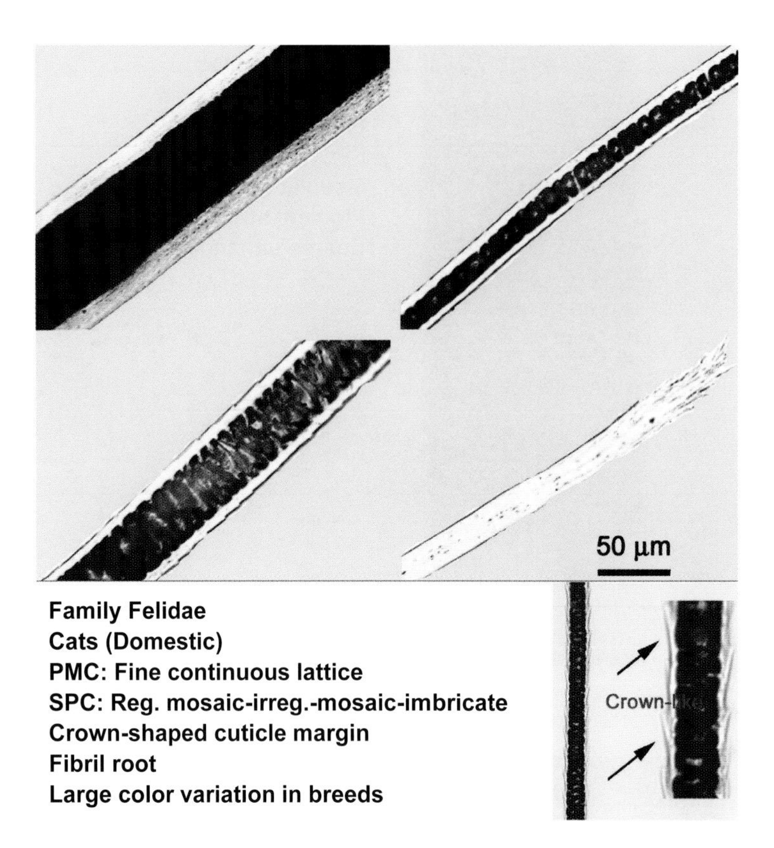

FIGURE B-5 Domestic cat.

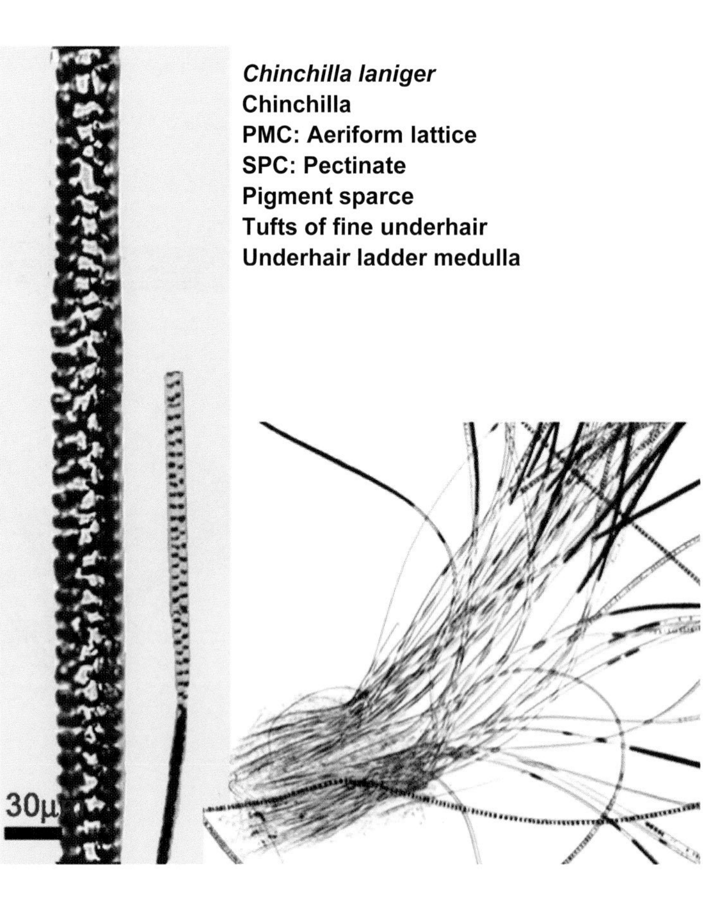

FIGURE B-6 Chinchilla.

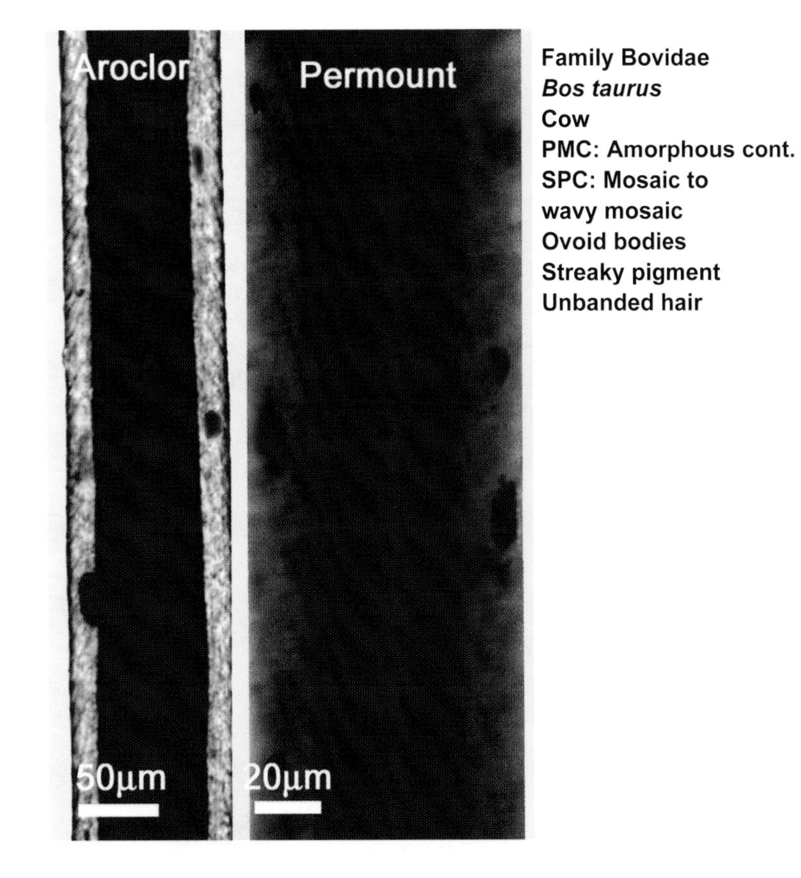

FIGURE B-7 Cow.

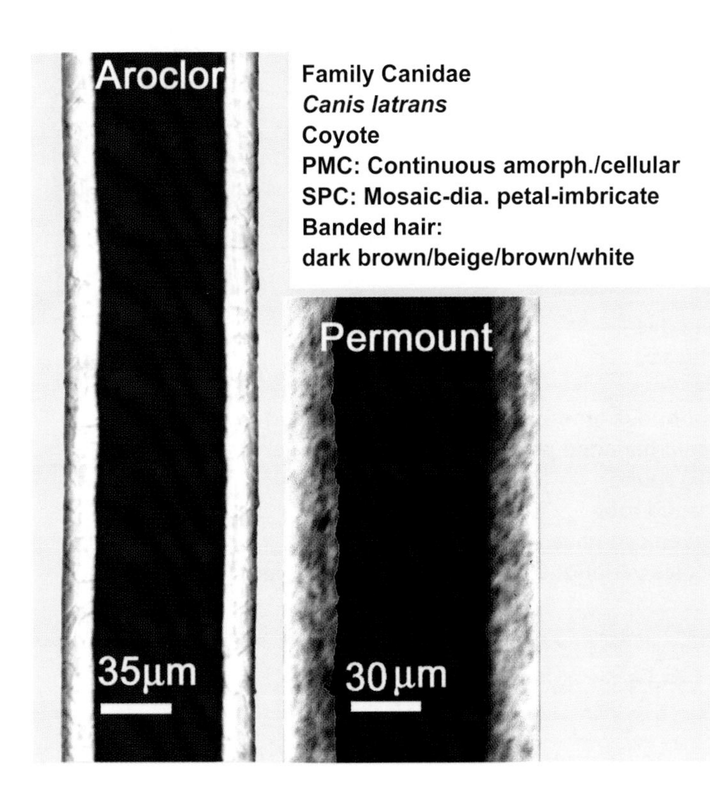

FIGURE B-8 Coyote.

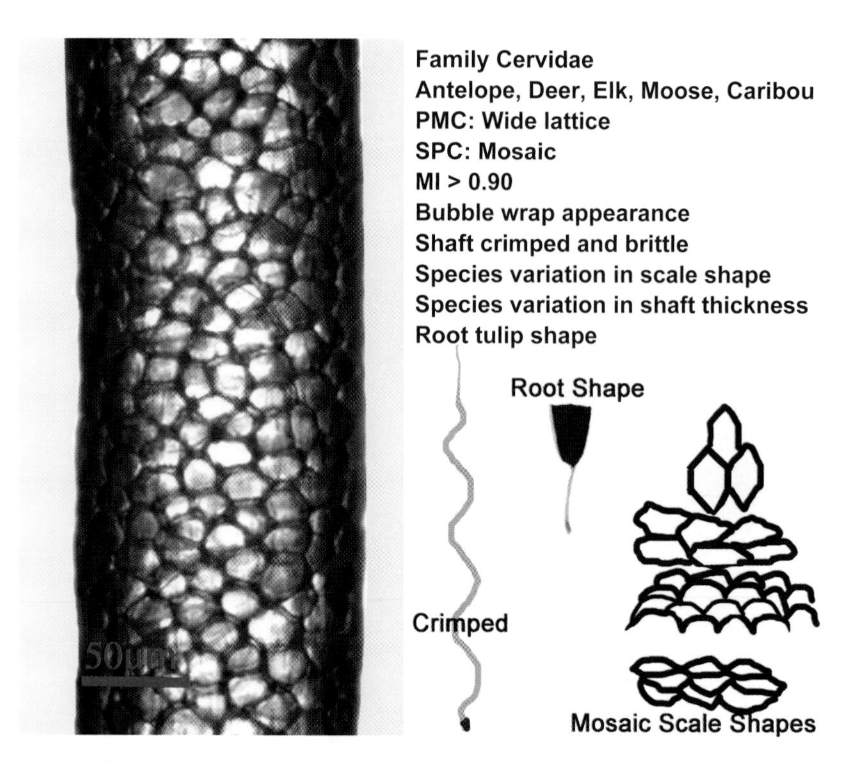

FIGURE B-9 Deer, antelope, elk, moose, caribou.

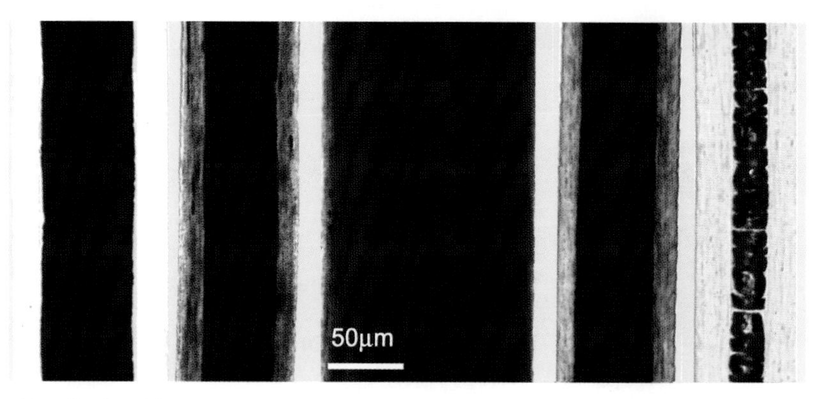

Family Canidae
Canis familiaris
Dog
PMC: Continuous amorphous, can be vacuolated
SPC: Mosaic-diamond petal-imbricate
X-S usually round
Shade-shaped root
Pigment even can have large aggregates
Large species variations in color, length, and texture

FIGURE B-10 Dog.

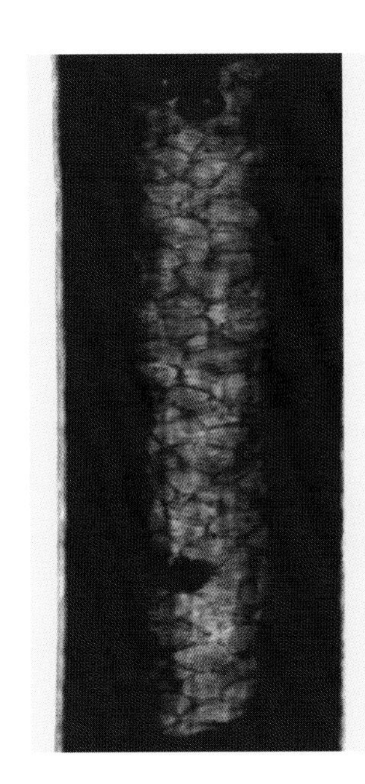

Genus *Capra*
Goat
PMC: Wide lattice
SPC: Mosaic-wavy mosaic-imbricate
Scallop-shaped cells
Pigment sparse to opaque
X-S elongated
Color varies: black, brown, white
Underhair often used as wool

FIGURE B-11 Goat.

Sus scrofa
Domestic Hog
PMC: Amorphous continuous
and discontinuous, can be absent
SPC: Imbricate
Medulla often absent
Diameter straight and very thick up to 200 μm
Shaft length 4 to 8 cm
Color white, can vary brown to black
Tips often frayed or split
Cuticle margin thick

FIGURE B-12 Hog.

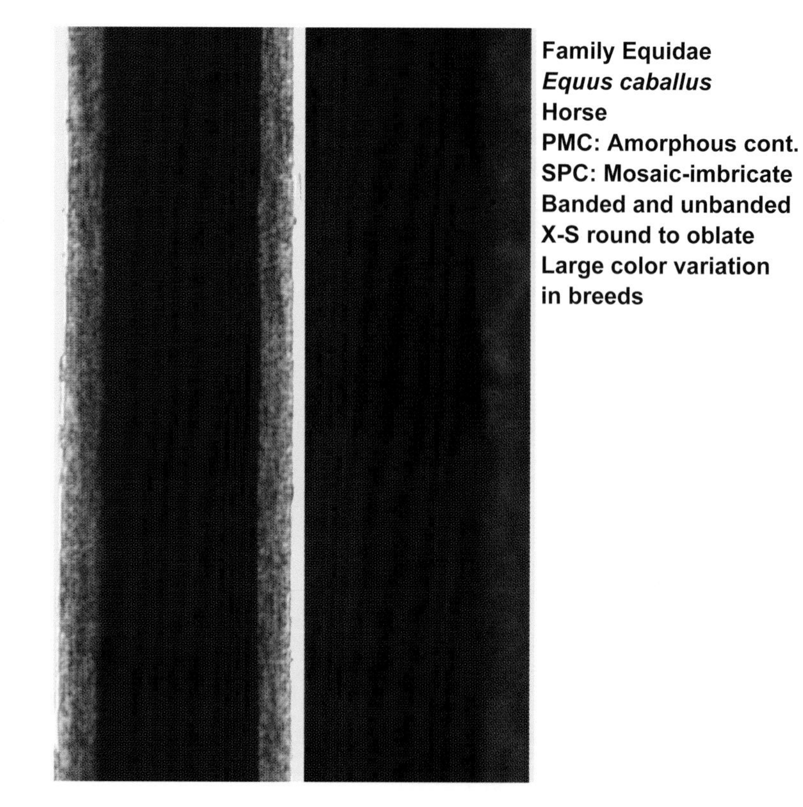

FIGURE B-13 Horse.

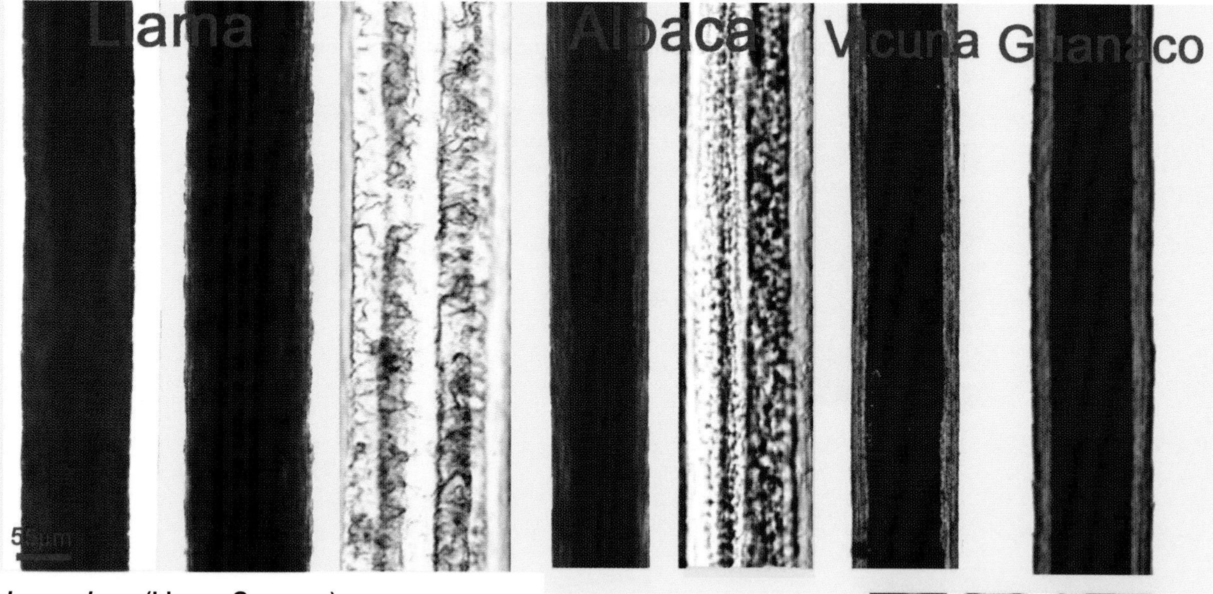

Lama glama **(Llama, Guanaco)**
Lama pacos **(Alpaca)**
Lama vicugna **(Vicuna)**
PMC: Continuous often with multichannels
SPC: Irreg. mosaic-irreg.-irreg. imbricated (wavy)
Thick cuticle margin
Pigment often streaky appearance
X-S often exhibits multiple channels

X-S

FIGURE B-14 Llama, alpaca, vicuña, guanaco.

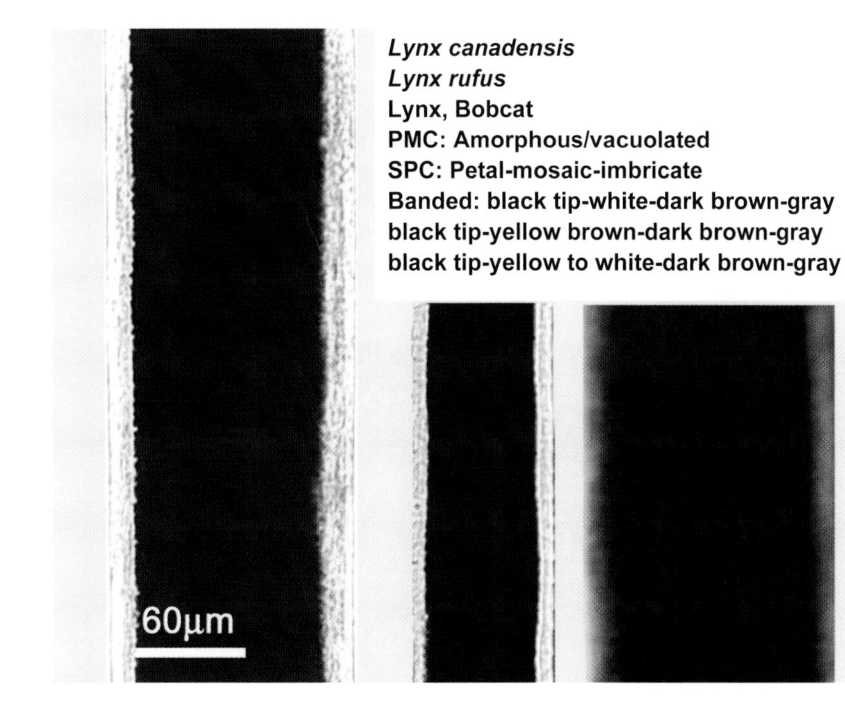

FIGURE B-15 Lynx.

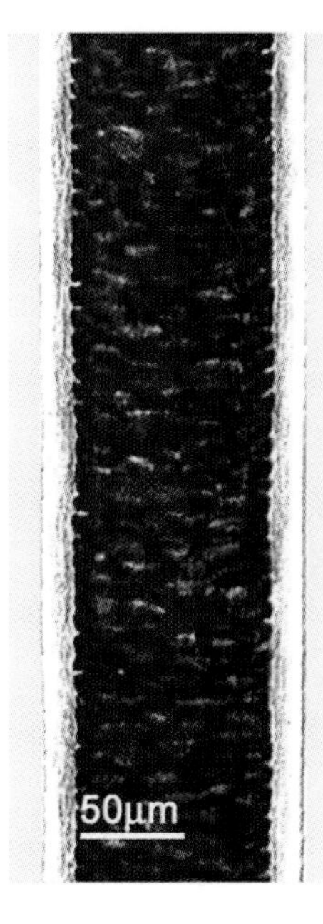

Family Mustelidae
Marten americana
Marten
PMC: Aeriform lattice
SPC: Diamond petal-petal-imbricate
Unbanded tricolor: Black-reddish brown-gray
Diameter up to 100 µm
(Close to Mink, important color differences and shaft thickness differences)

FIGURE B-16 Marten.

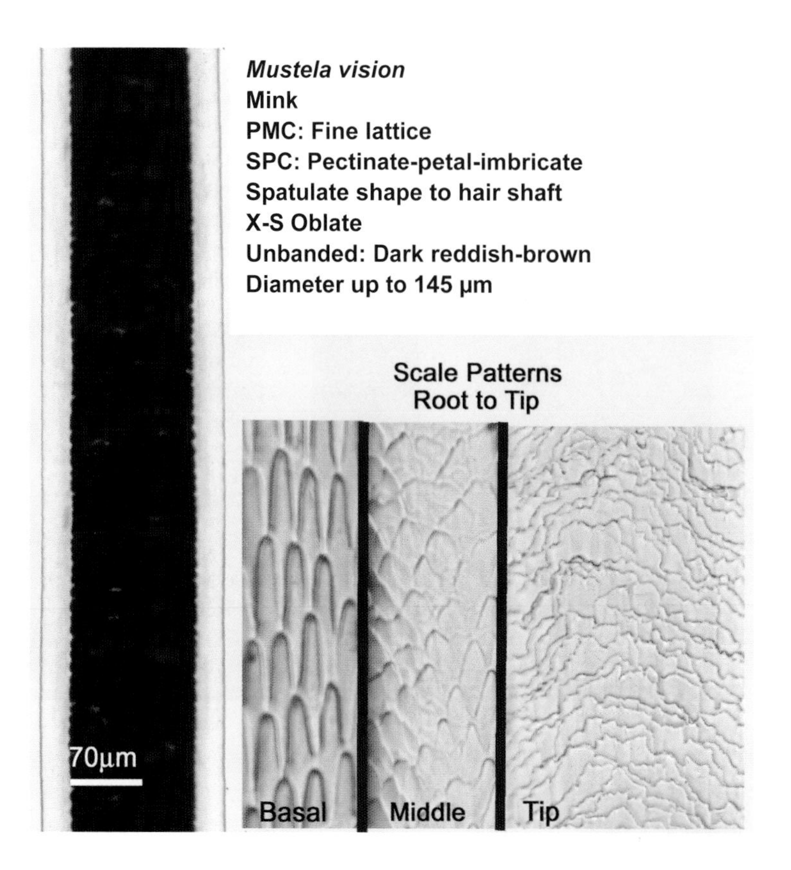

FIGURE B-17 Mink.

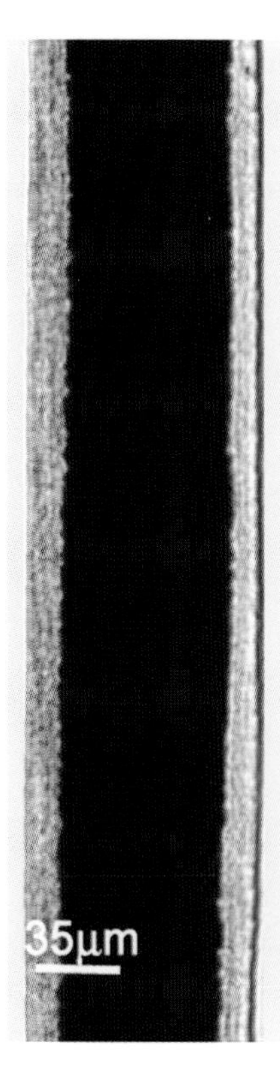

Family Felidae
Felis concolor
Mountain Lion
PMC: Continuous amorphous,
can be vacuolated if infiltrated with MM
SPC: Mosaic to imbricate
Banded: Black-lt. brown-dark brown or black
Diameter up to 115 μm

FIGURE B-18 Mountain lion.

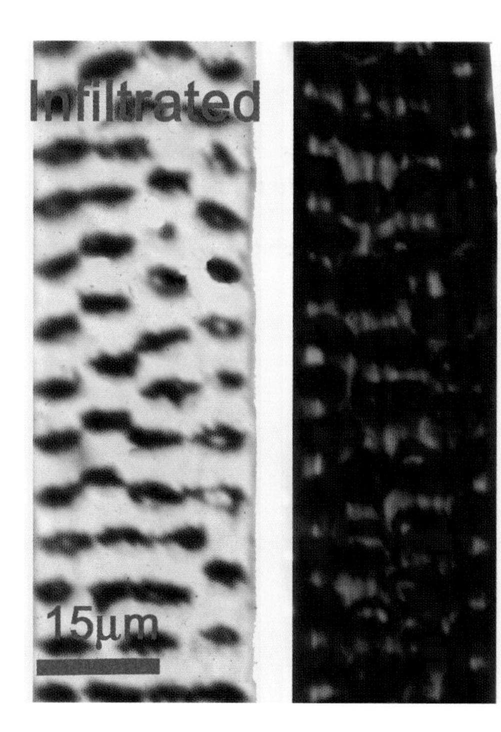

Family Muridae
Mus musculus domesticus
House mouse
PMC: Aeriform lattice
SPC: Diamond petal,
wavy mosaic
Shaft length aver. 6.5 mm
Shaft diameter aver. 30 µm
Banded: Black-yellow-gray
Unbanded: Black-gray

FIGURE B-19 Mouse.

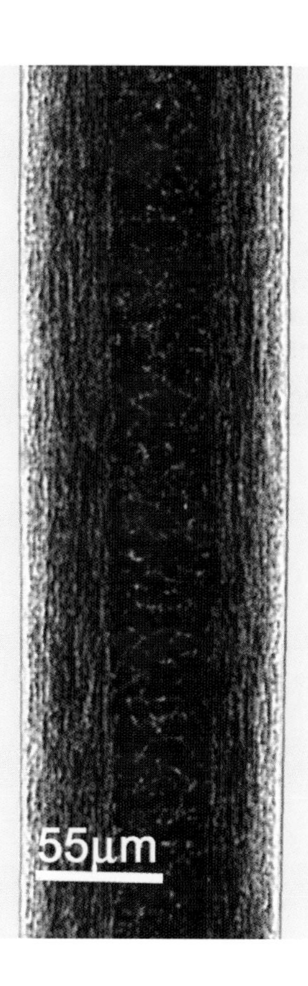

Ondatra zibethicus
Muskrat
PMC: Aeriform lattice
SPC: Mosaic-chevron-imbricate
Pigment distribution streaky
Unbanded:
Color reddish brown,
some bicolored—
red/brown to dark brown

FIGURE B-20 Muskrat.

Family Didelphidae
Didelphis marsupialis
Opossum
PMC: Aeriform lattice
SPC: Mosaic-diamond petal-wavy mosaic
Bicolored brown to white
Length 35-50 mm
Pigment streaky

FIGURE B-21 Opossum.

Lutra lutra
Otter
PMC: Wide lattice
SPC: Pectinate/diamond petal—irregular imbricate
Cigar-shaped hair shaft
Unbanded dark brown
Pigment dist. even with streaks
Shield diameter wide, up to 185 µm

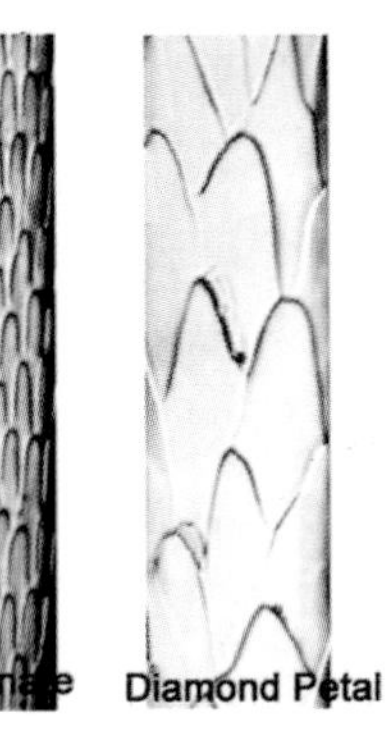

FIGURE B-22 Otter.

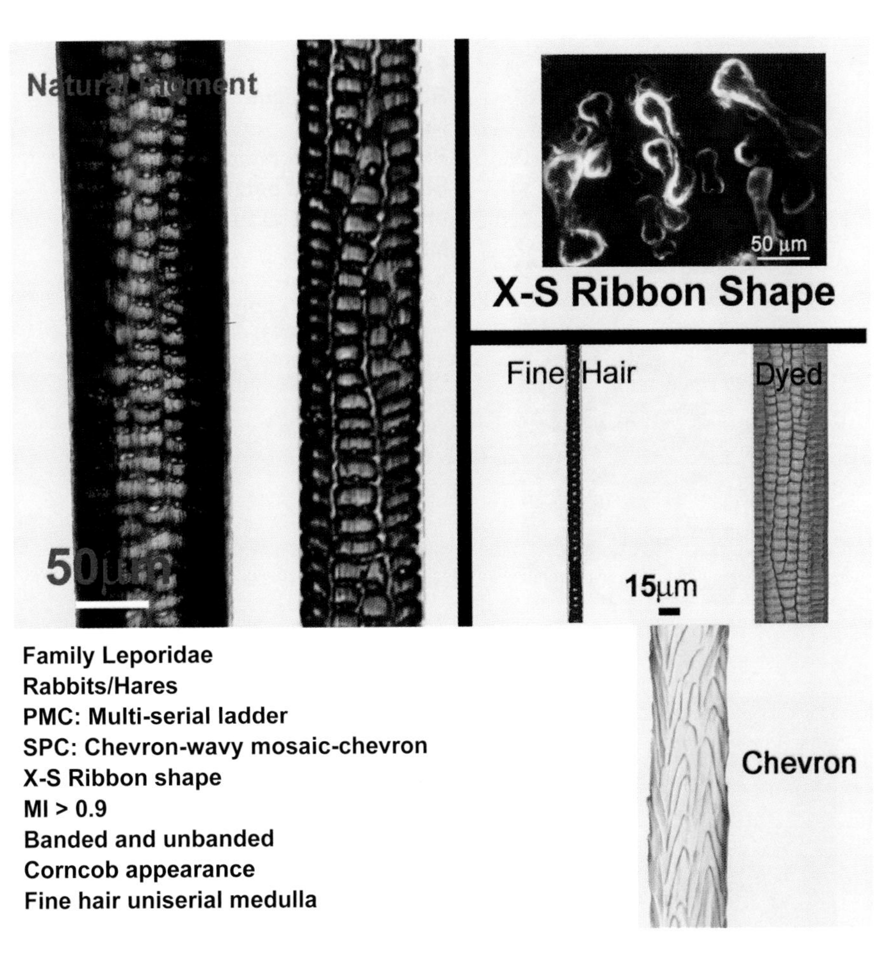

FIGURE B-23 Rabbit, hare.

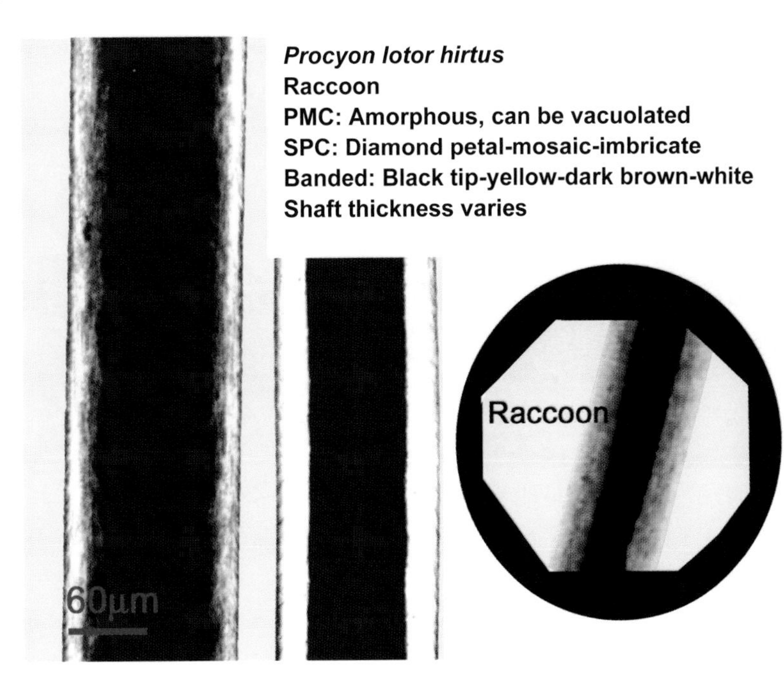

FIGURE B-24 Raccoon.

Family Murida
Rattus norvegicus
Norway Rat
PMC: Aeriform lattice
SPC: Diamond petal-imbricate
or diamond petal to mosaic
MI > 0.90
X-S Oblate
Hair length 12 to 20 mm

FIGURE B-25 Rat.

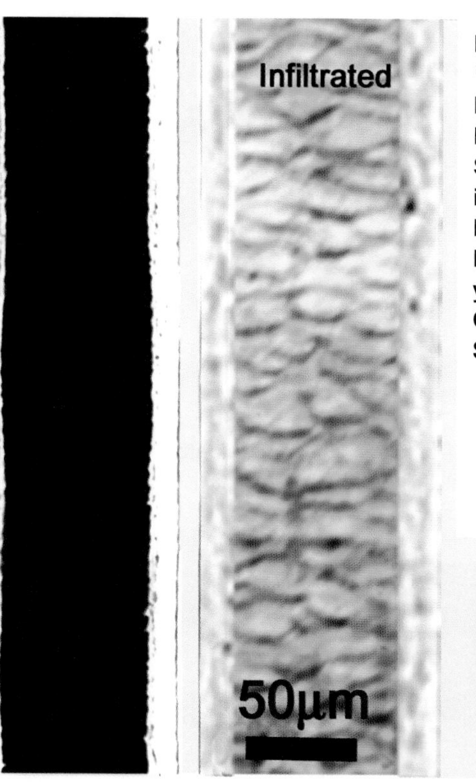

Family Canidae
Vulpes fulva
Red Fox
PMC: Fine lattice
SPC: Diamond petal-mosaic-imbricate
Pigment sparse
Banded: Reddish brown-yellow-brown gray
Cuticle thin
Shaft length up to 56 mm

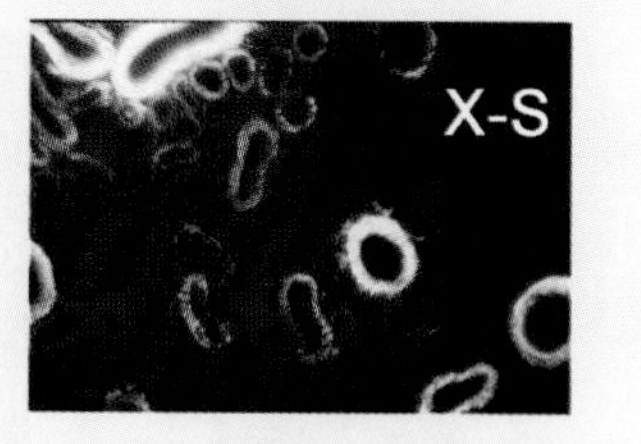

FIGURE B-26 Red fox.

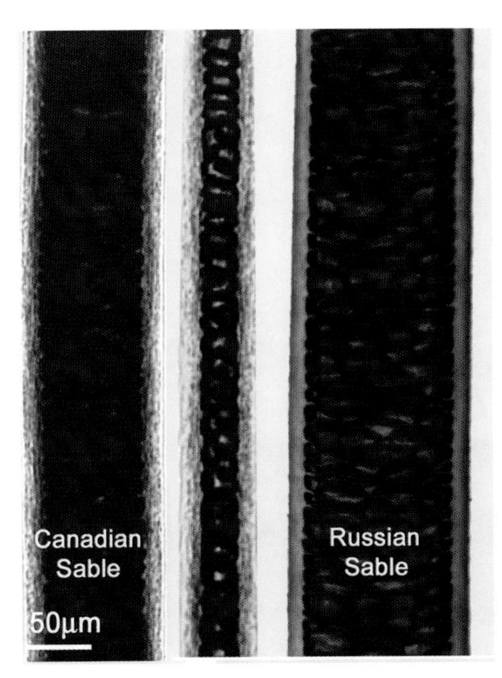

Martes zibellina
Sable
PMC: Wide lattice
SPC: Diamond petal-mosaic-imbricate
MI > 0.8
Pigment sparse and even

FIGURE B-27 Sable.

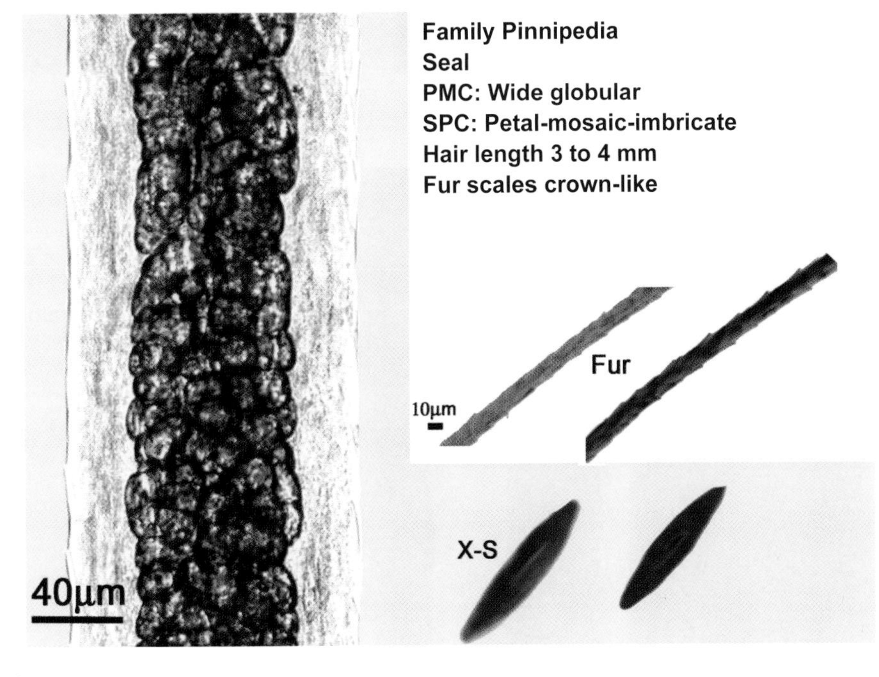

FIGURE B-28 Seal.

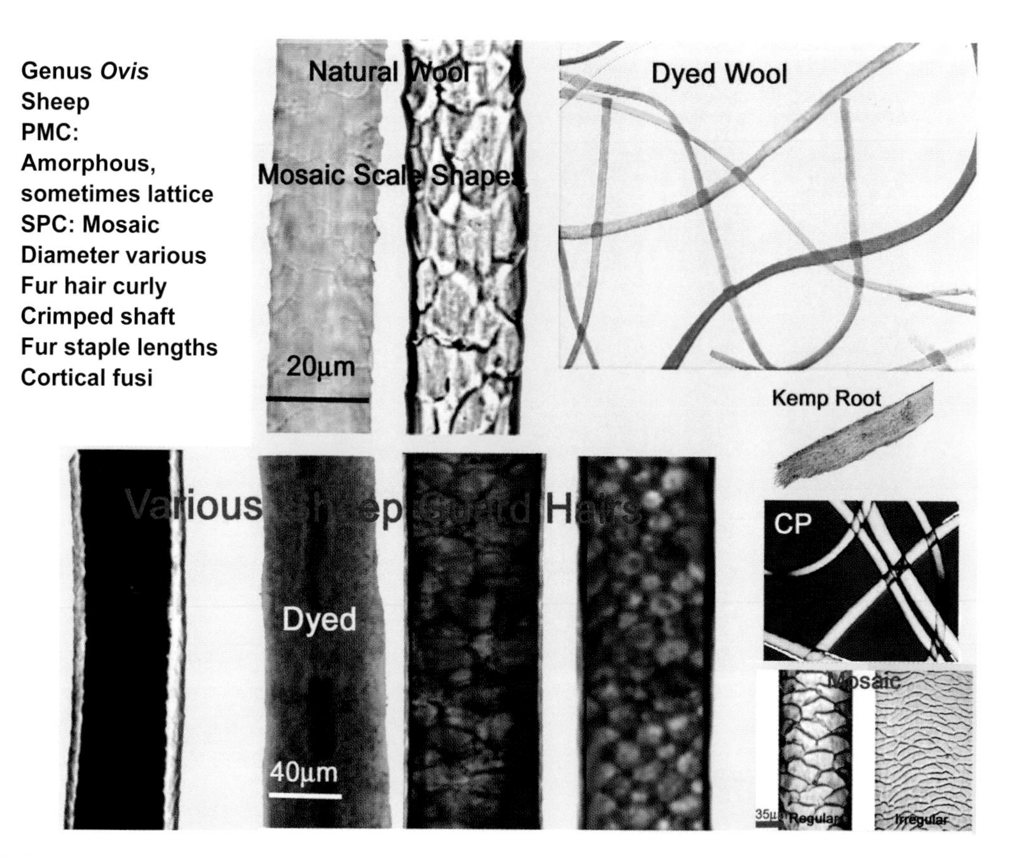

FIGURE B-29 Sheep.

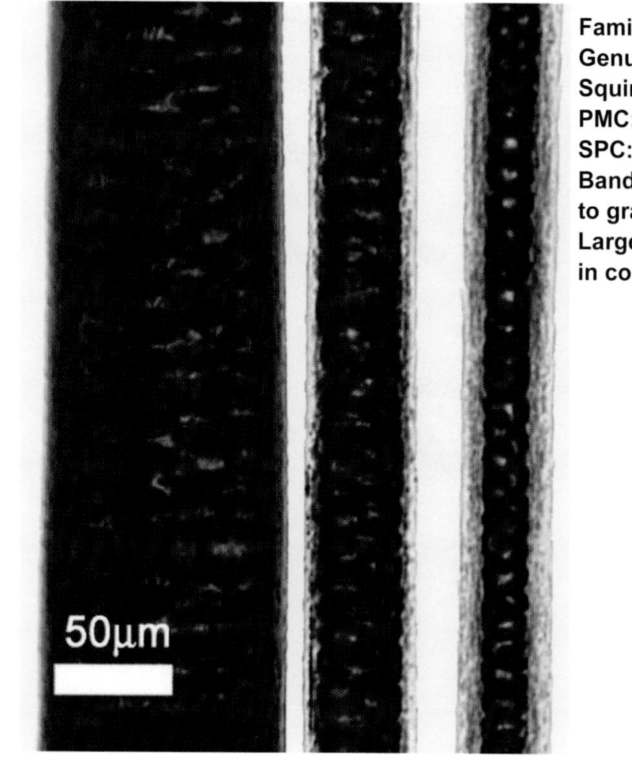

Family Sciuridae
Genus *Scirurs*
Squirrel
PMC: Aeriform lattice
SPC: Chevron-wavy mosaic
Banded: Black-brown
to gray
Large species variation
in color

FIGURE B-30 Squirrel.

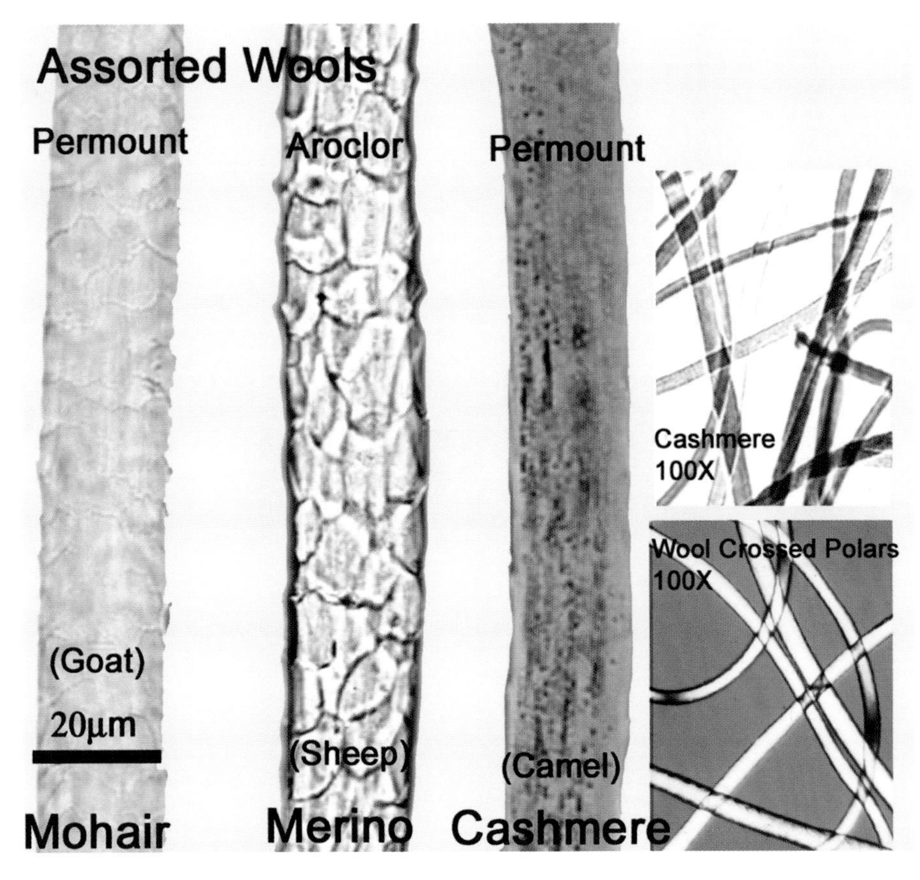

FIGURE B-31 Mohair, merino, and cashmere wools.

APPENDIX C

Synthetic Fibers

This appendix was designed to provide a compilation of photomicrographs illustrating many of the physical characteristics and optical properties used to identify, classify, and compare synthetic fibers. It illustrates the features and properties cited in Tables 7.1 and 7.2, Figures 7.2 through 7.8, and Figure 7.10 in Chapter 7. This appendix is divided into seven categories:

1. Cross-section and longitudinal views
2. PLM appearance utilizing crossed polars, plane-polarized light, and interference colors
3. Becké line movement
4. Fiber treatment
5. Fluorescence
6. Pleochroic fibers
7. Miscellaneous fibers

Each illustration includes a short description of the specimen and features demonstrated. The vibration direction of the plane-polarized light is indicated where necessary. A scale in the form of a red bar is included in each photomicrograph.

The photomicrographs were taken by the authors in the course of their casework and are intended to illustrate many of the characteristics and concepts discussed in Chapter 7. They can be used as both teaching aids and references. Anyone wishing to reproduce any illustrations included in this appendix should contact the authors and/or the publisher for permission.

CATEGORY 1: CROSS-SECTIONAL AND LONGITUDINAL VIEWS

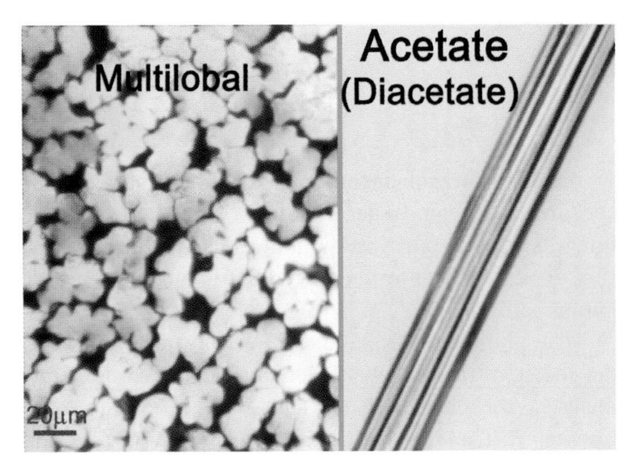

FIGURE C.1

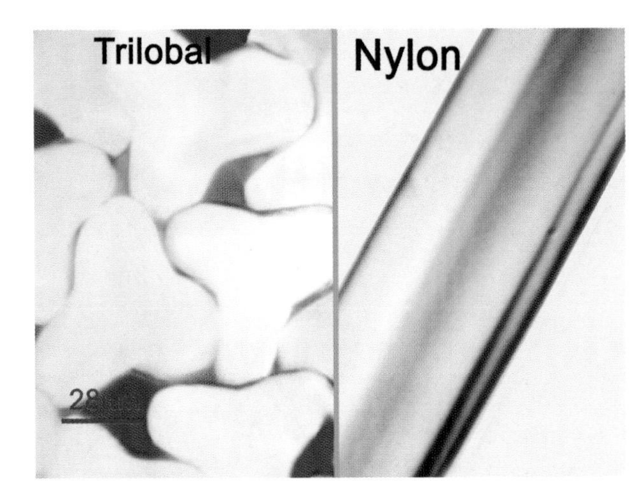

FIGURE C.2

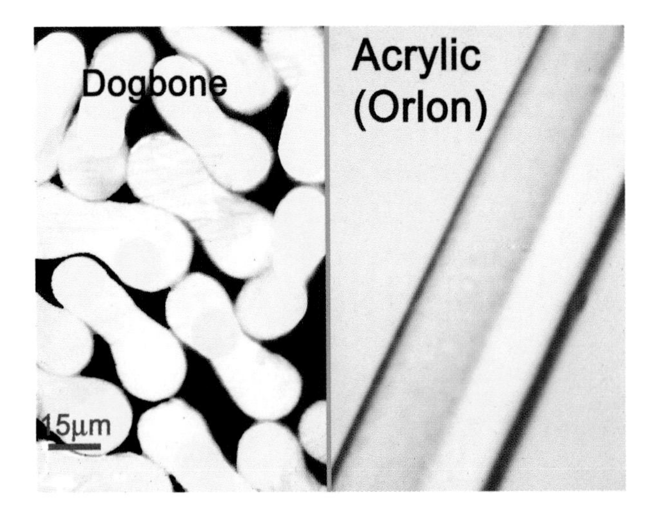

FIGURE C.3

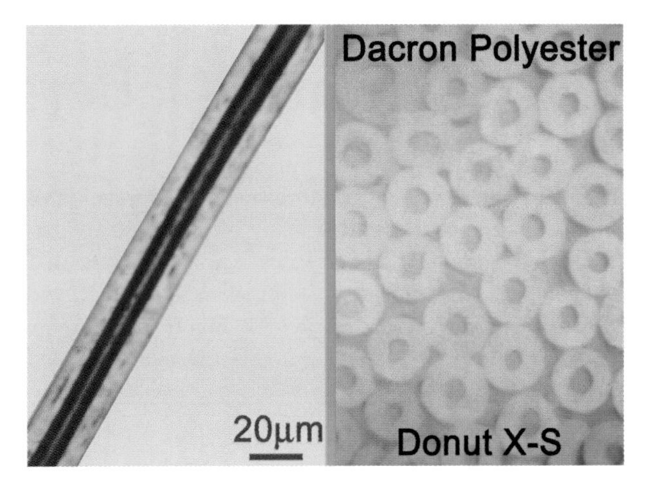

FIGURE C.4

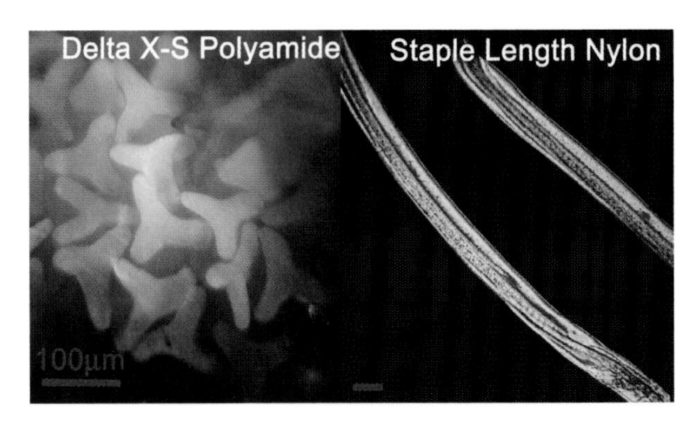

FIGURE C.5

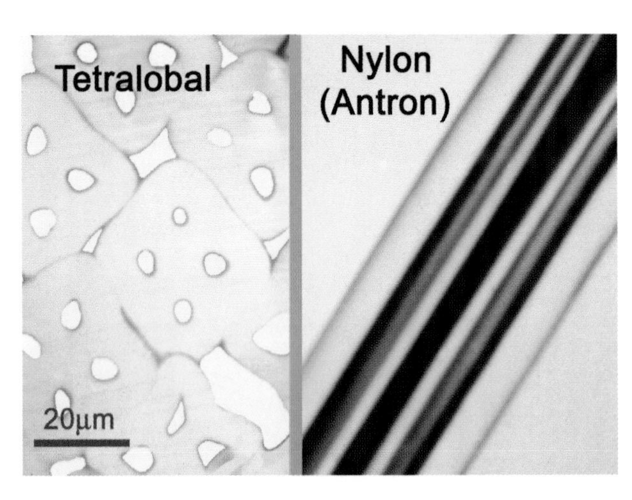

FIGURE C.6

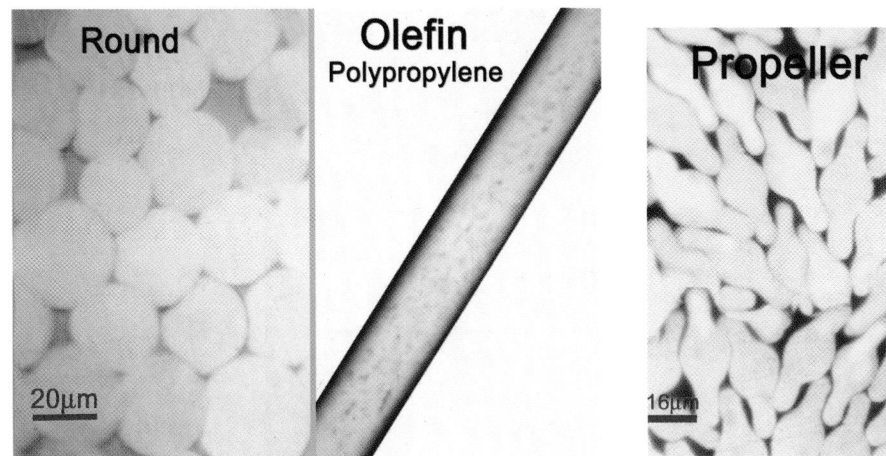

FIGURE C.7

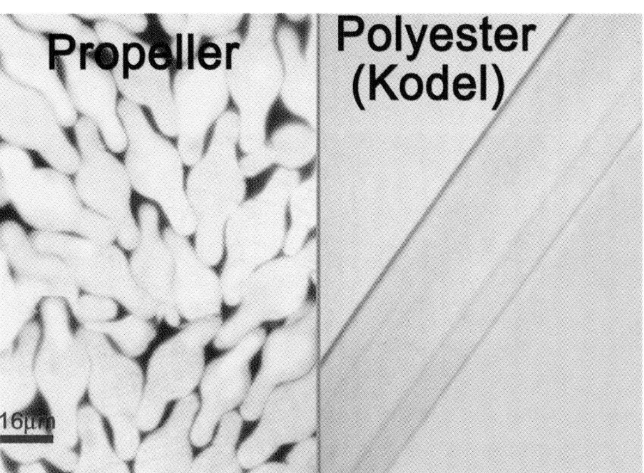

FIGURE C.10

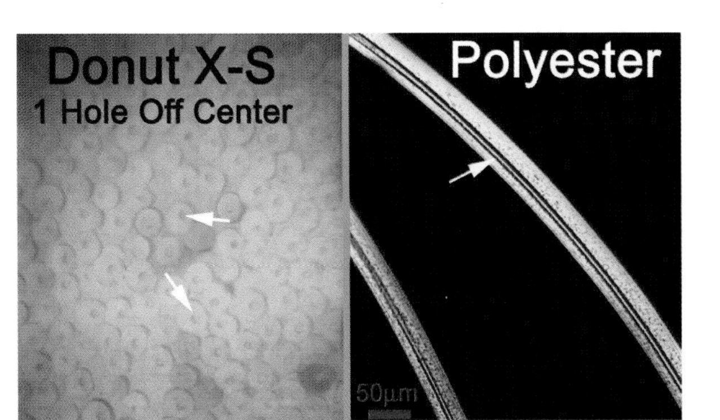

FIGURE C.8

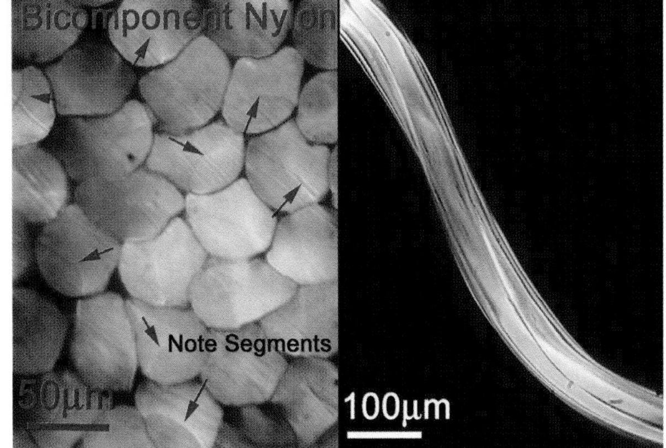

FIGURE C.11

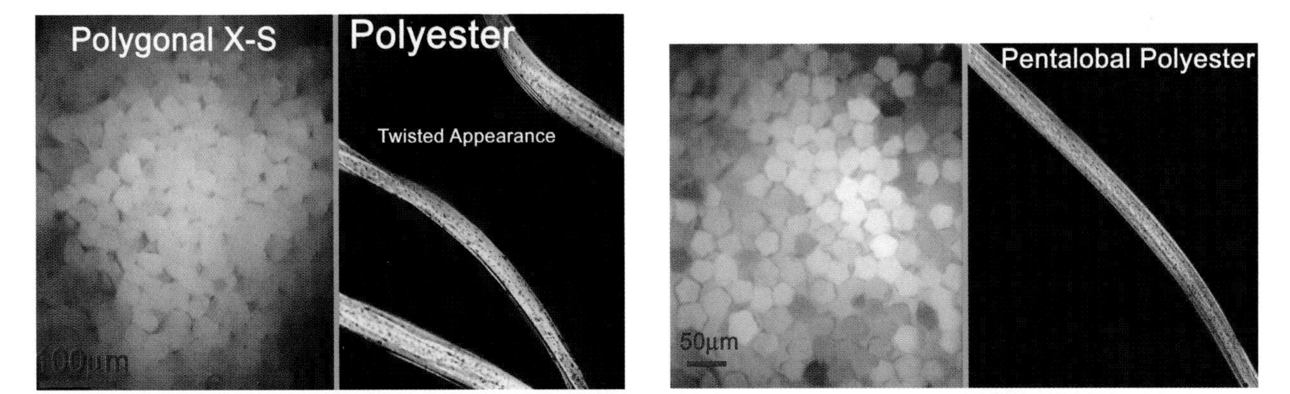

FIGURE C.9

FIGURE C.12

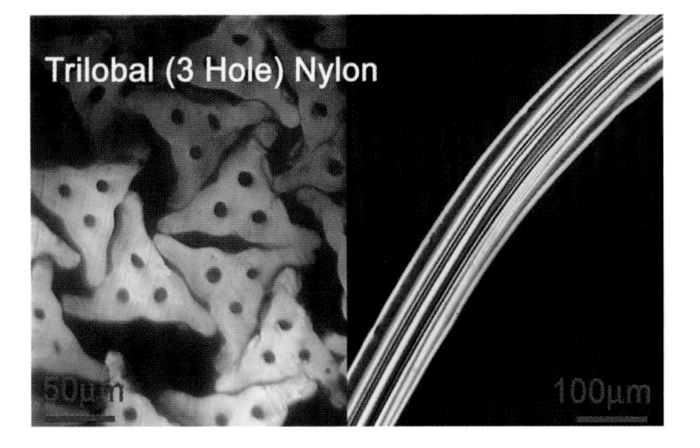

FIGURE C.13

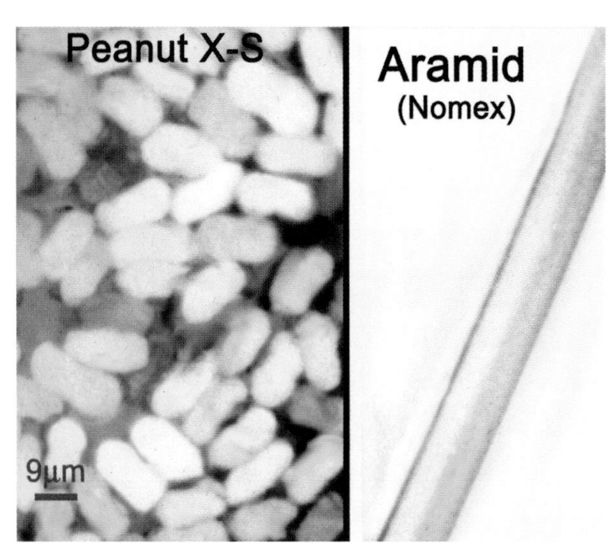

FIGURE C.15

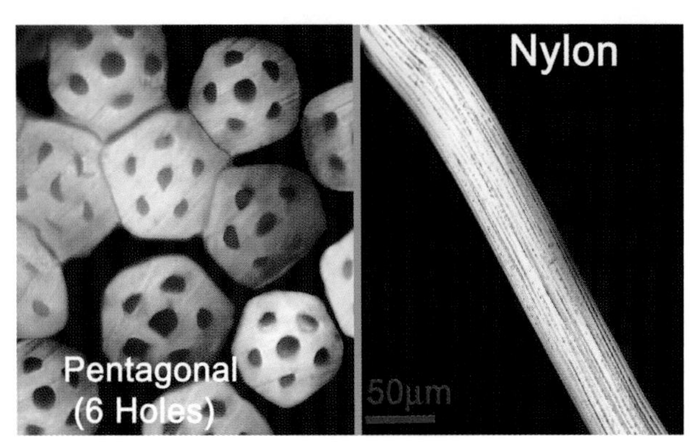

FIGURE C.14

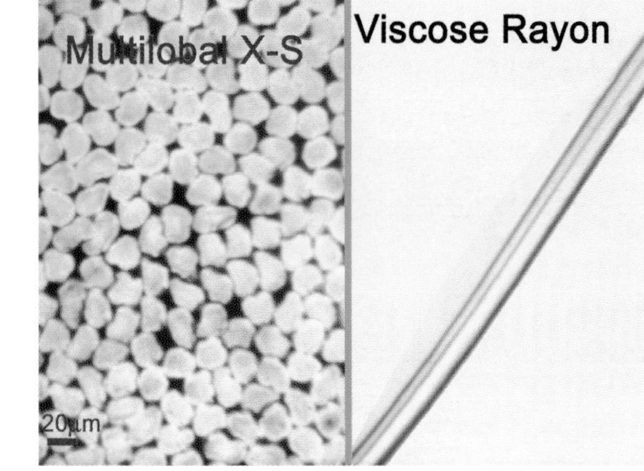

FIGURE C.16

CATEGORY 2: PLM APPEARANCE

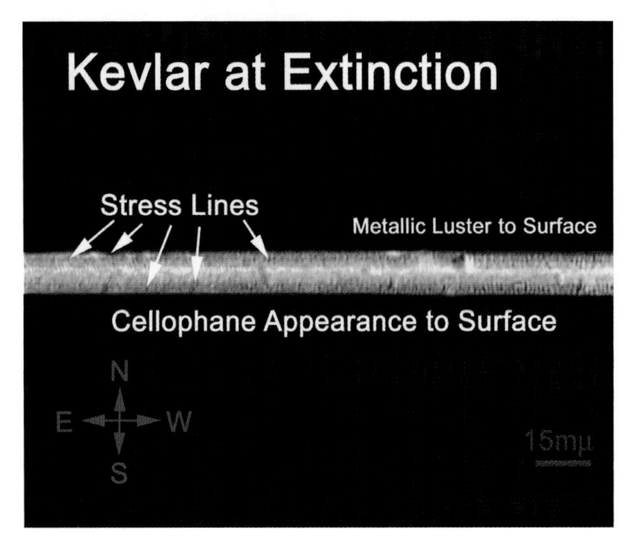

FIGURE C.17

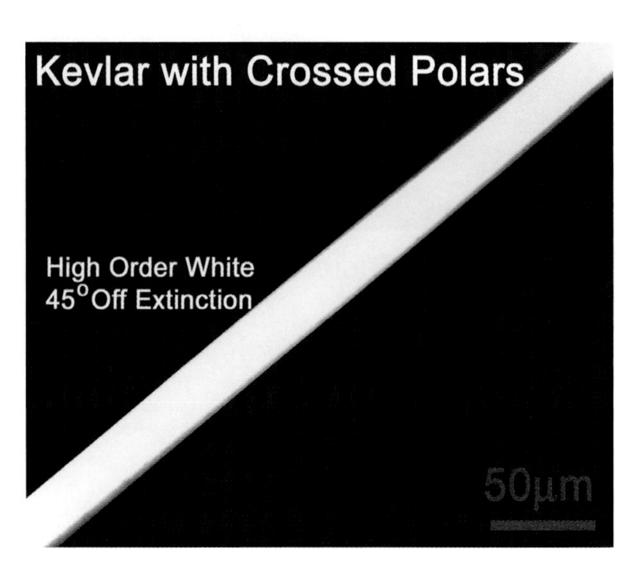

FIGURE C.18

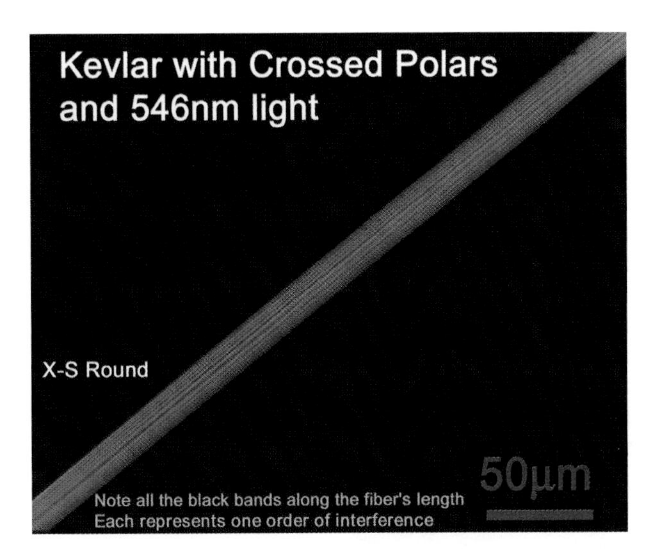

FIGURE C.19

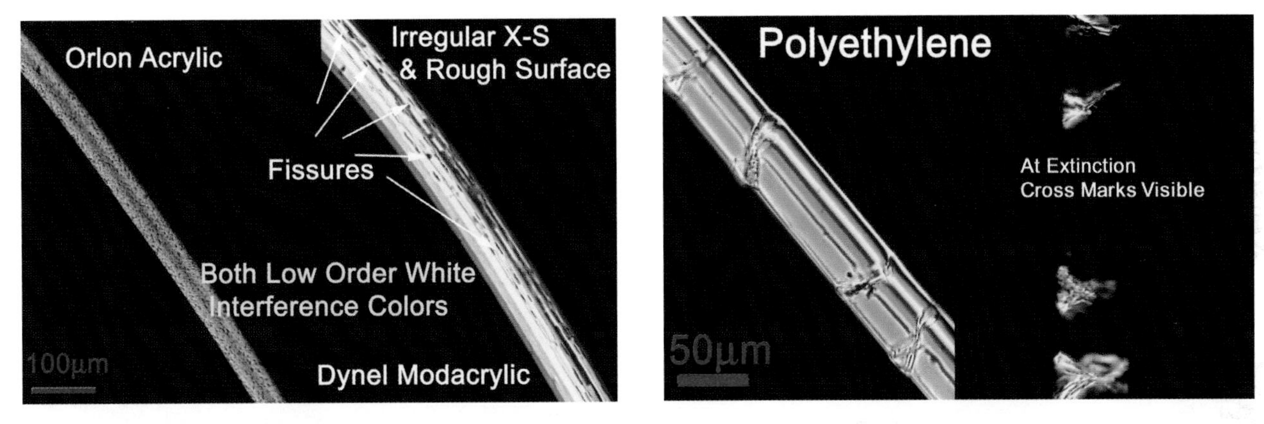

FIGURE C.20

FIGURE C.22

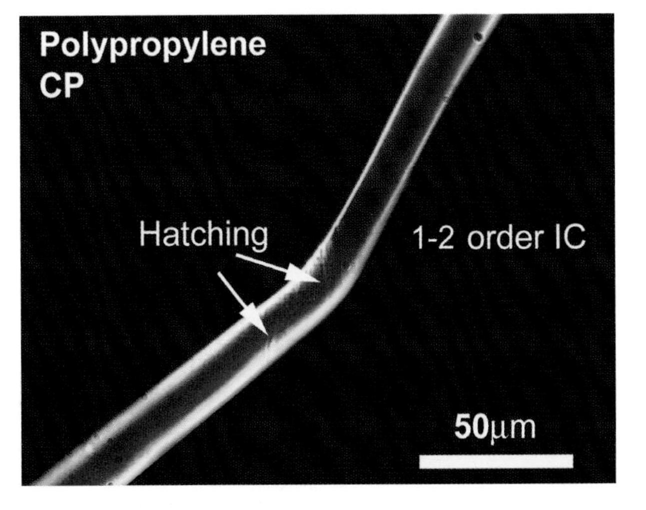

FIGURE C.21

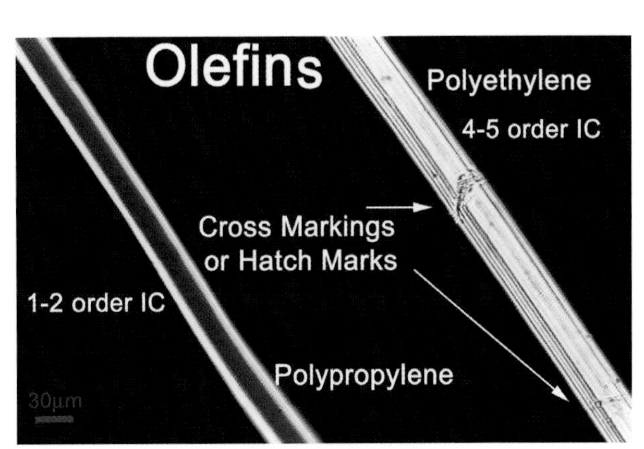

FIGURE C.23

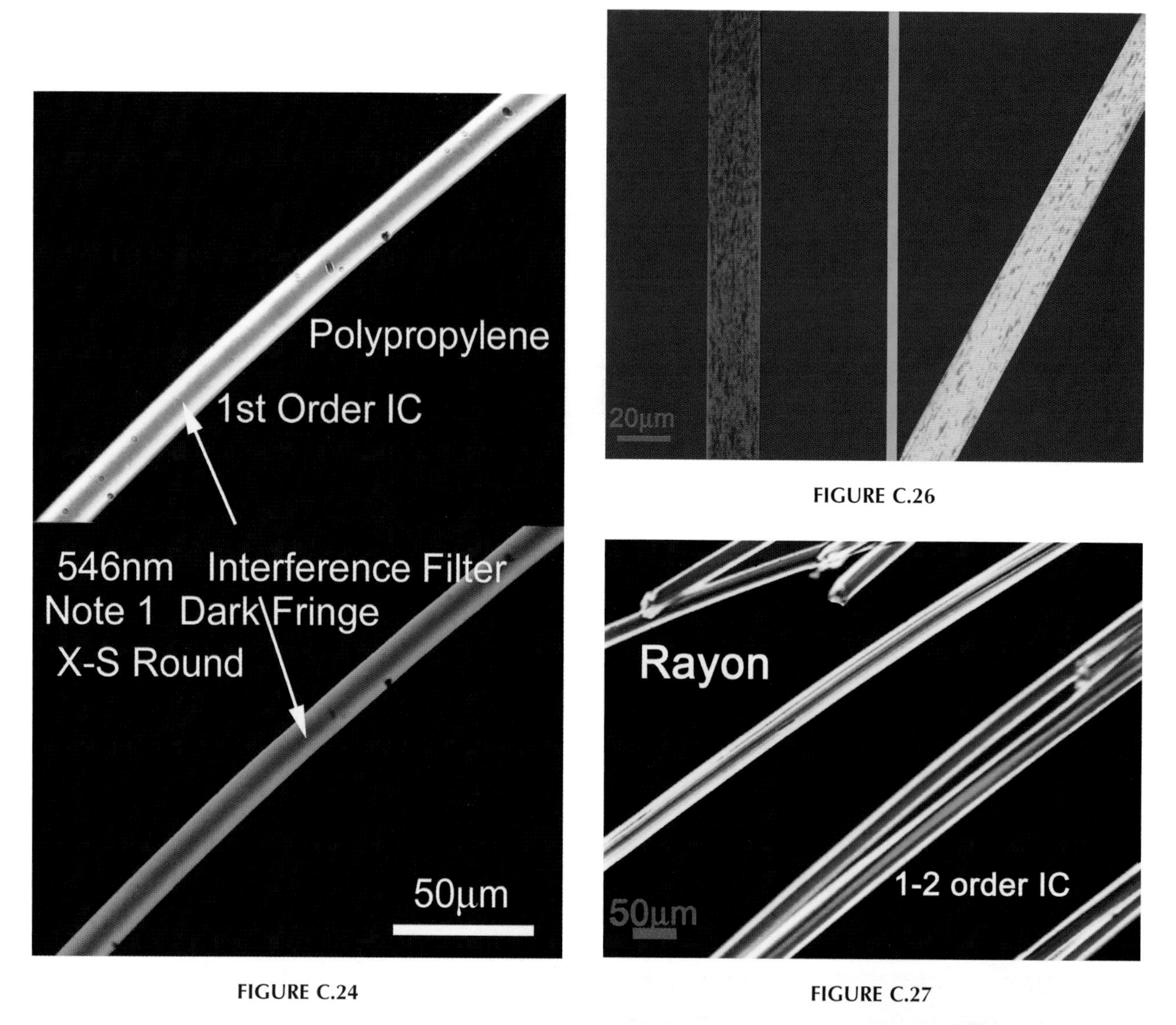

FIGURE C.24

FIGURE C.26

FIGURE C.27

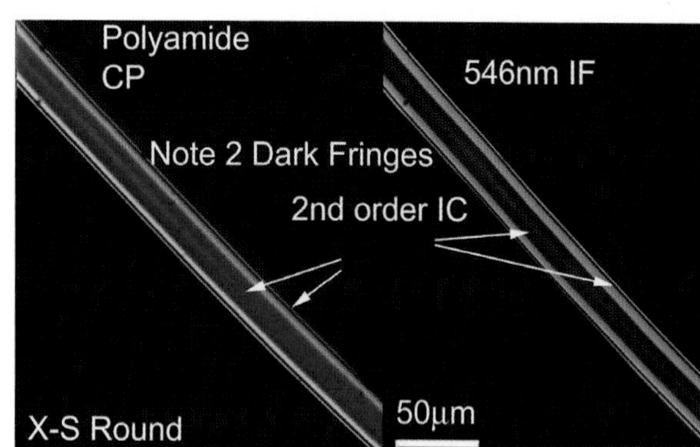

FIGURE C.25

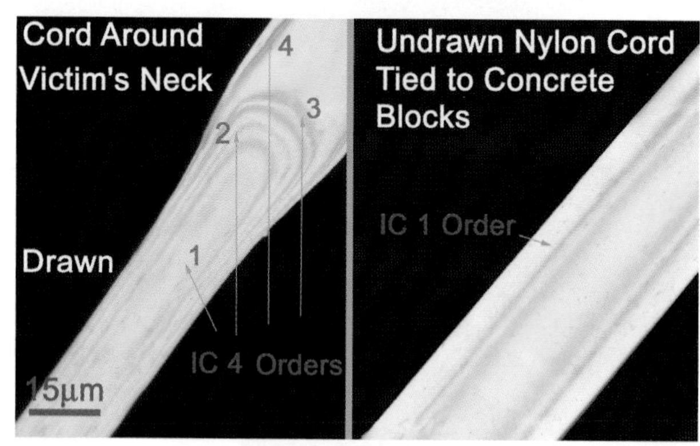

FIGURE C.28

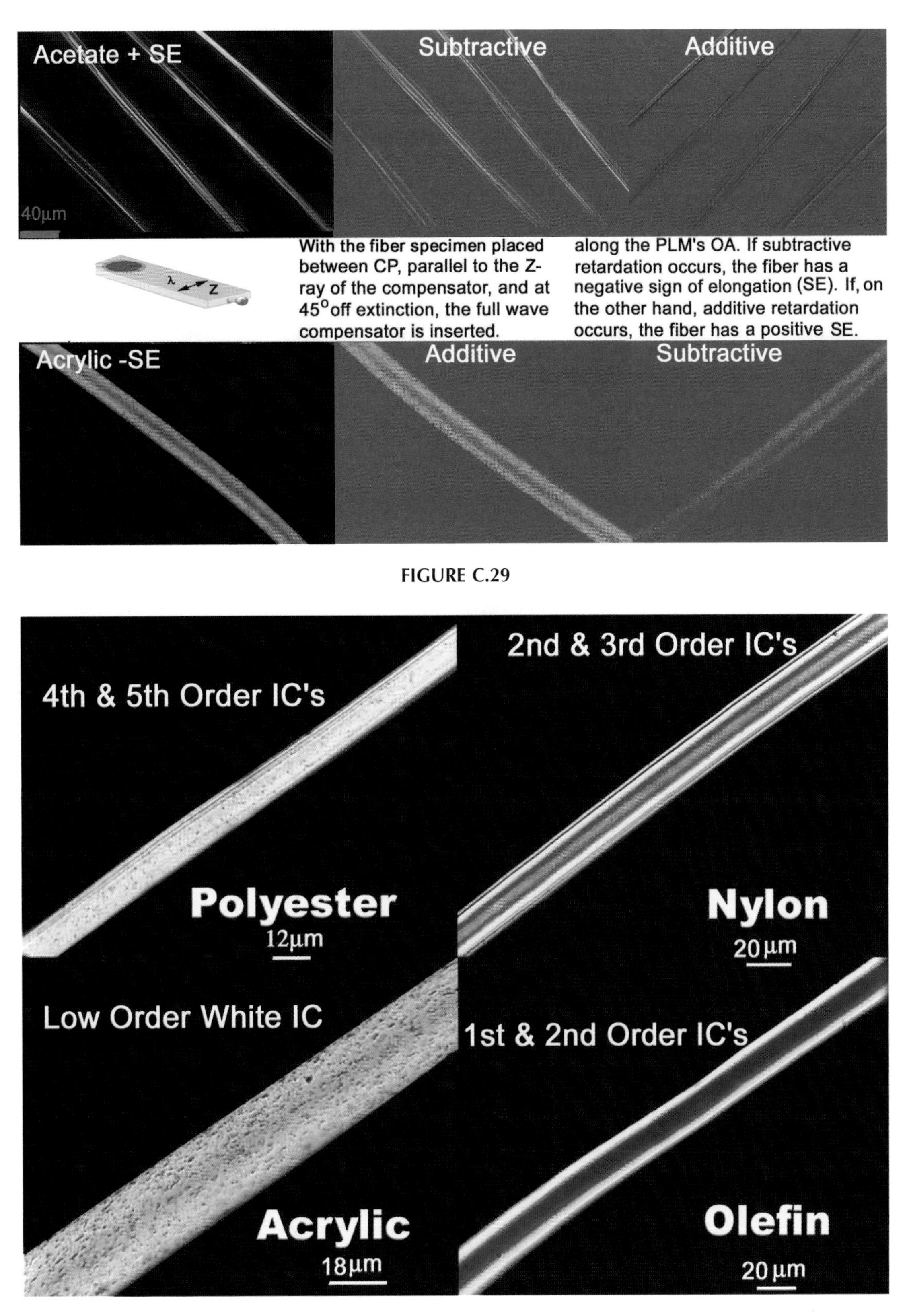

FIGURE C.29

FIGURE C.30

CATEGORY 3: BECKÉ LINE MOVEMENT

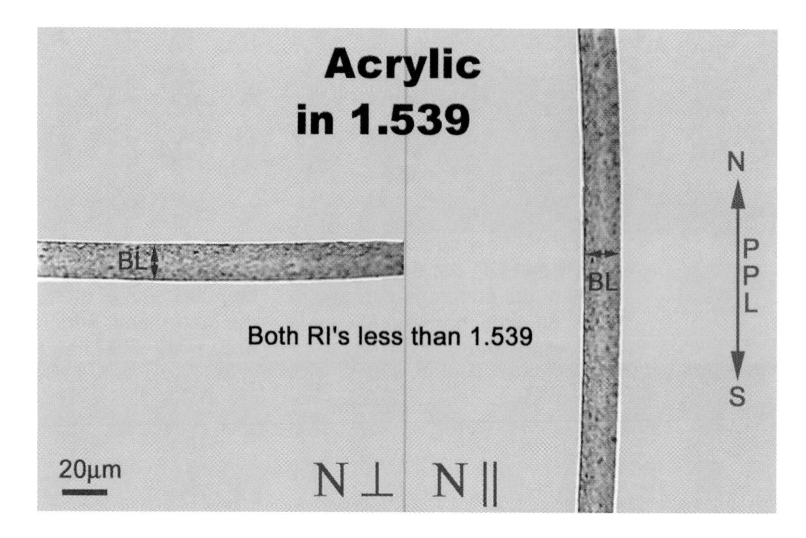

FIGURE C.31

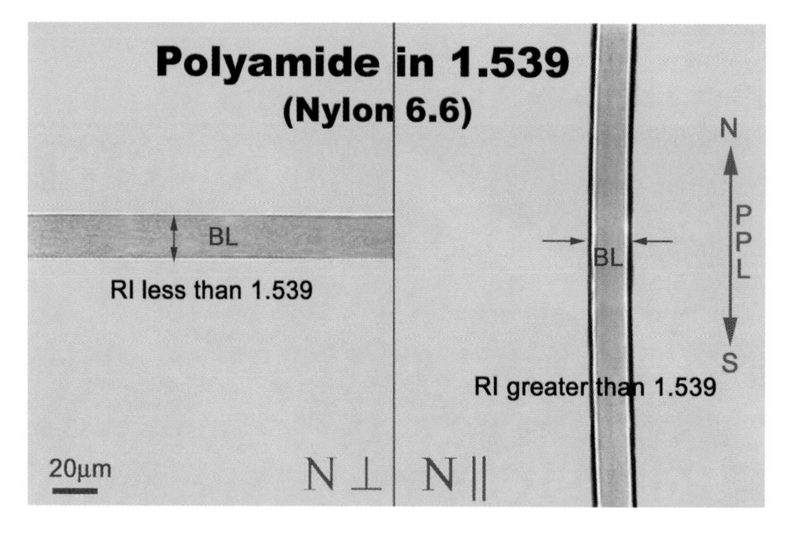

FIGURE C.32

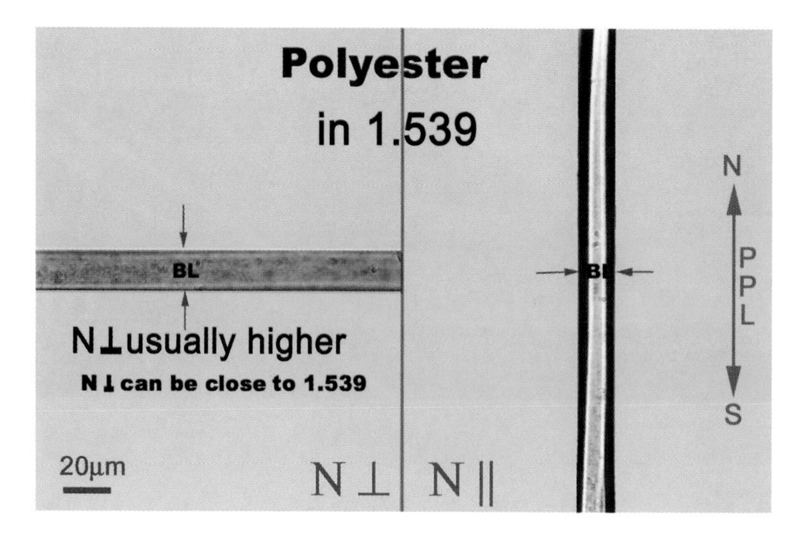

FIGURE C.33

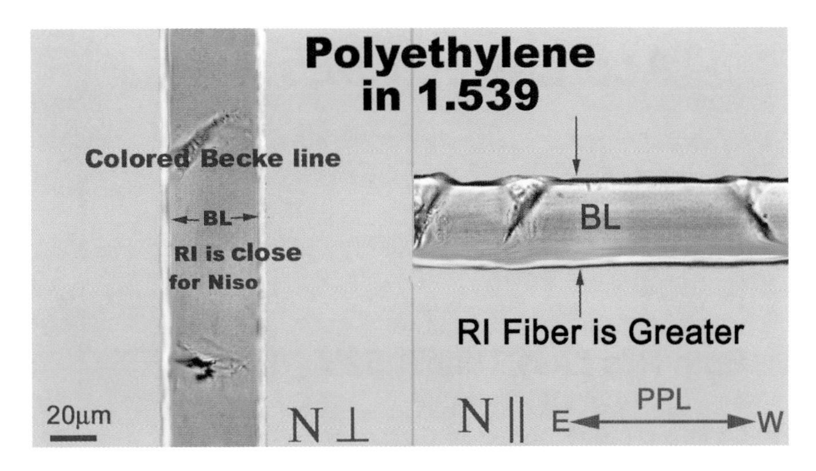

FIGURE C.34

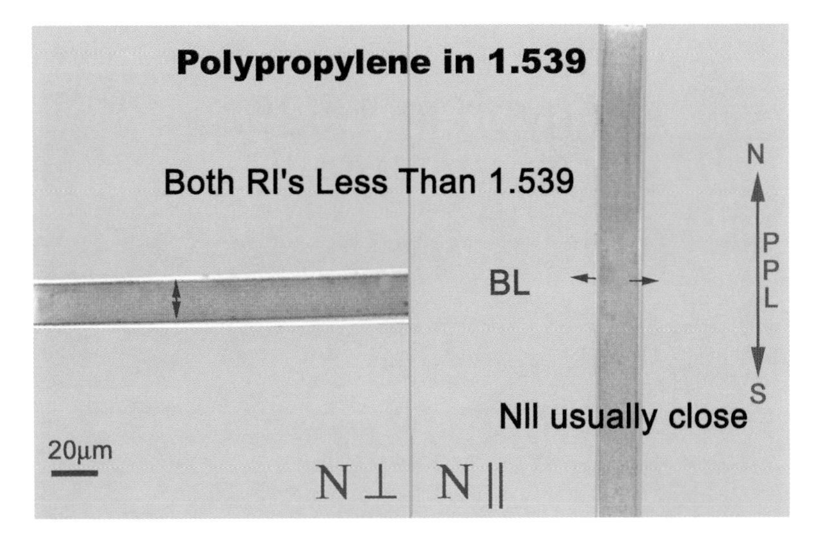

FIGURE C.35

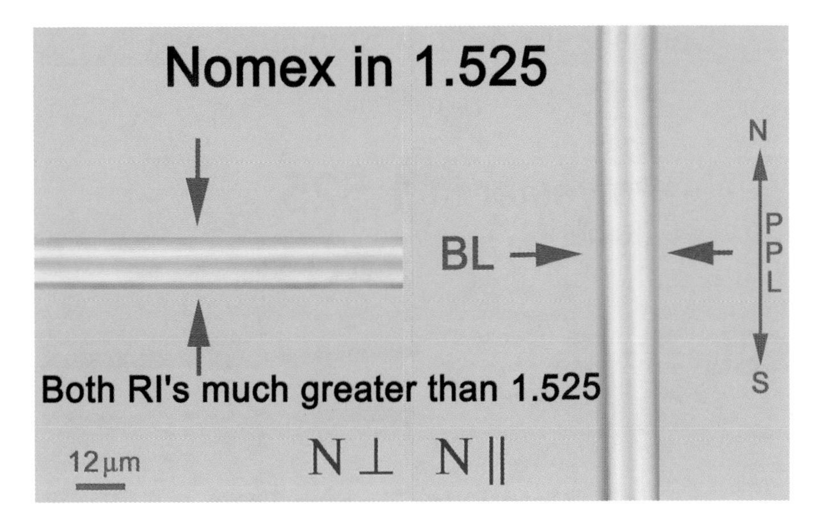

FIGURE C.36

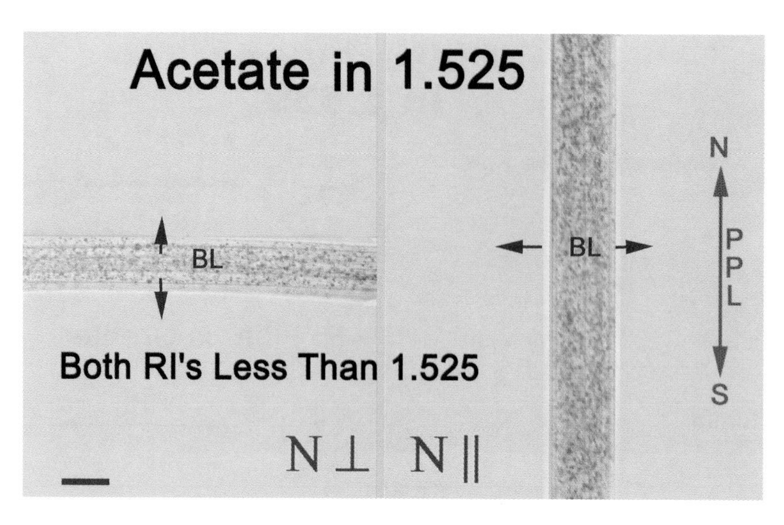

FIGURE C.37

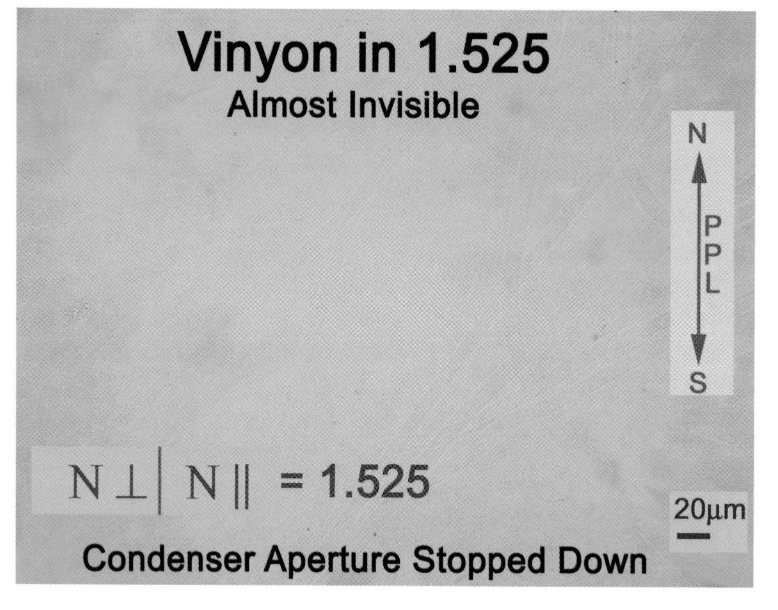

FIGURE C.38

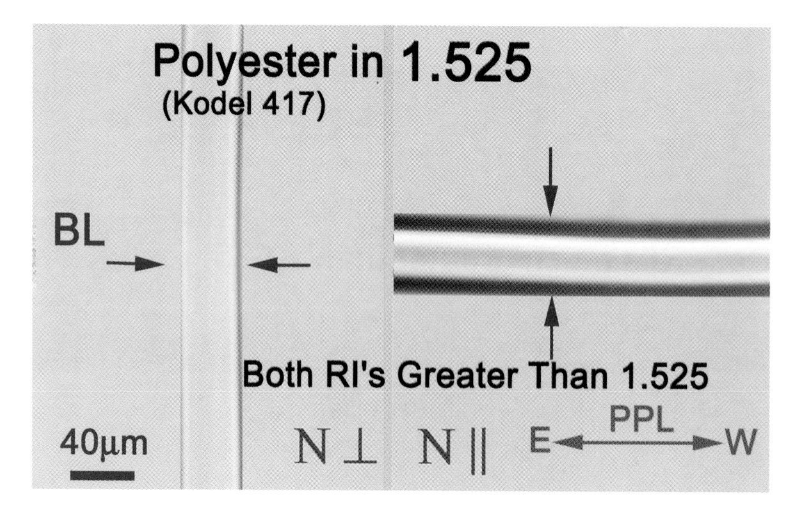

FIGURE C.39

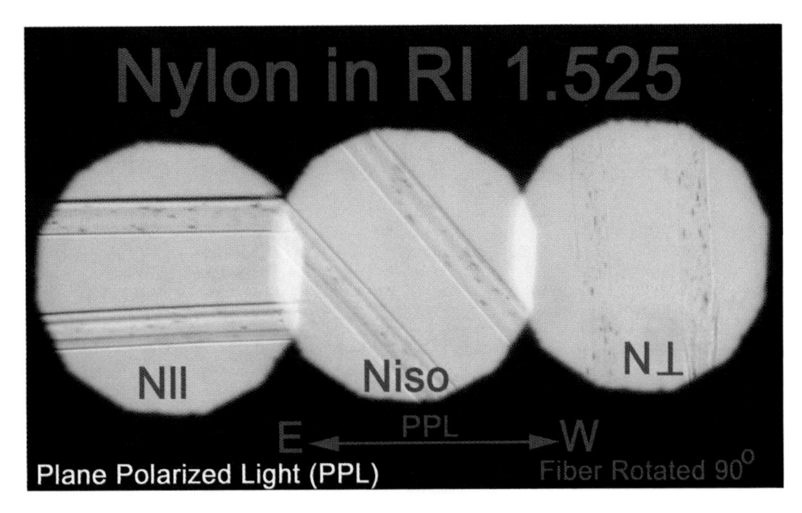

FIGURE C.40

CATEGORY 4: FIBER TREATMENT

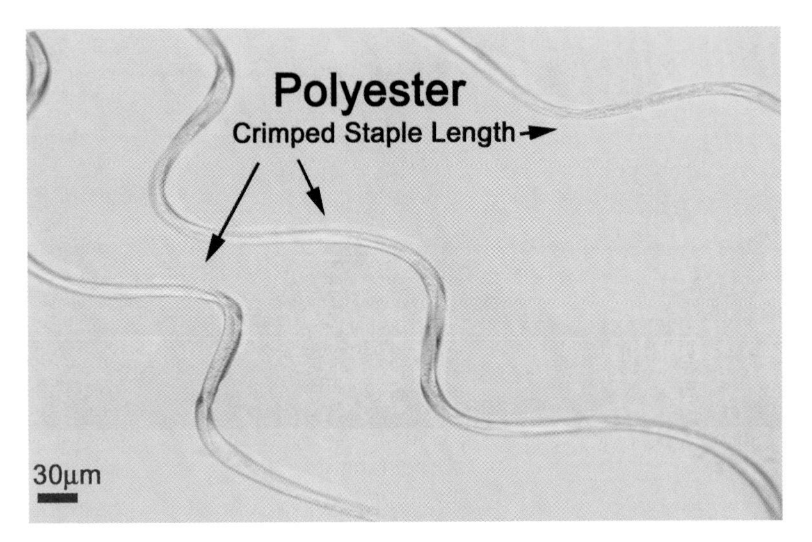

FIGURE C.41

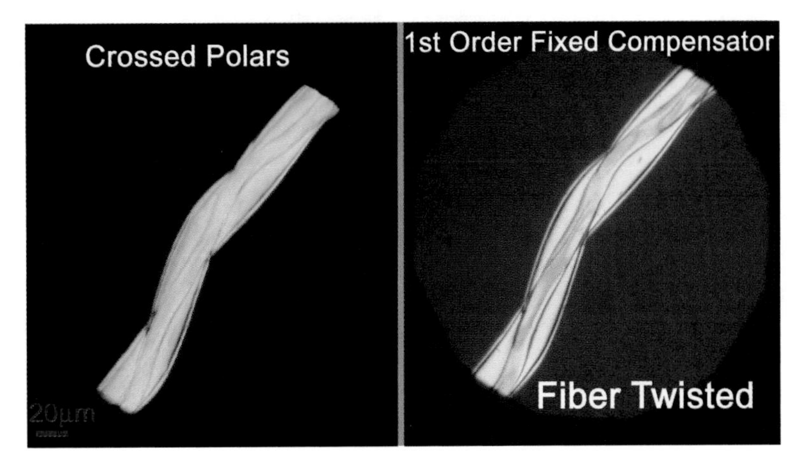

FIGURE C.42

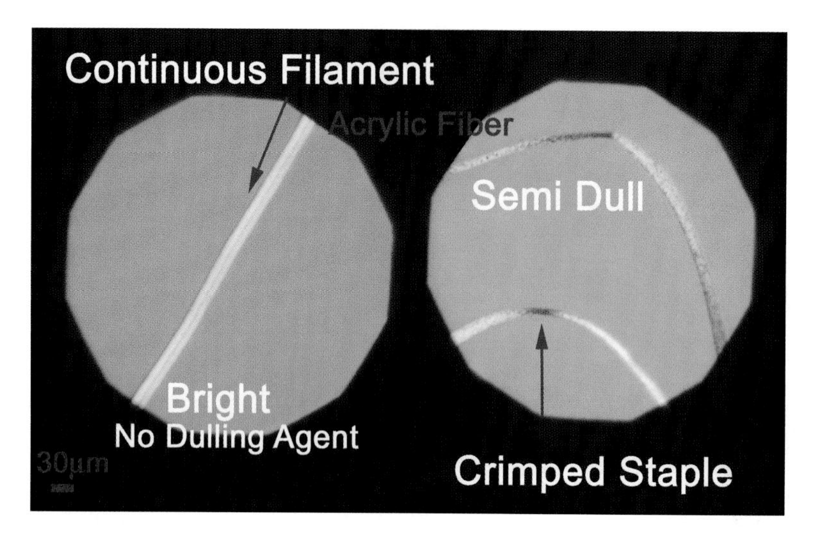

FIGURE C.43

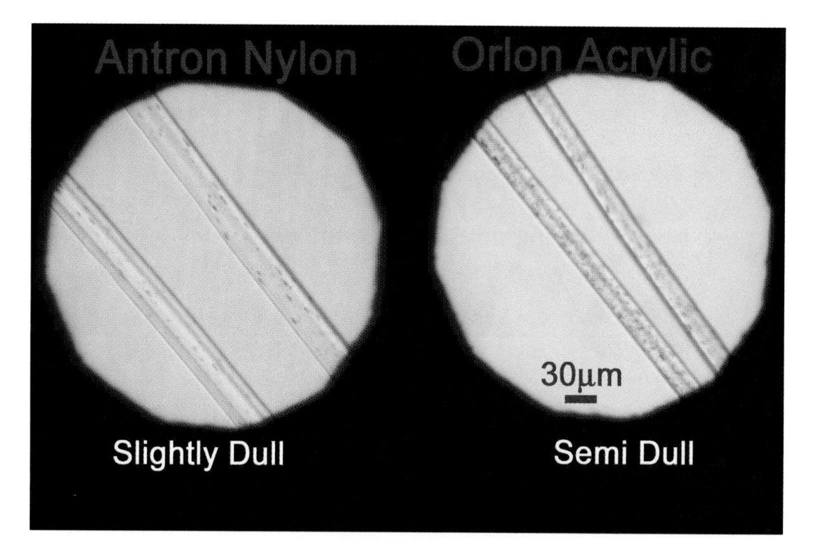

FIGURE C.44

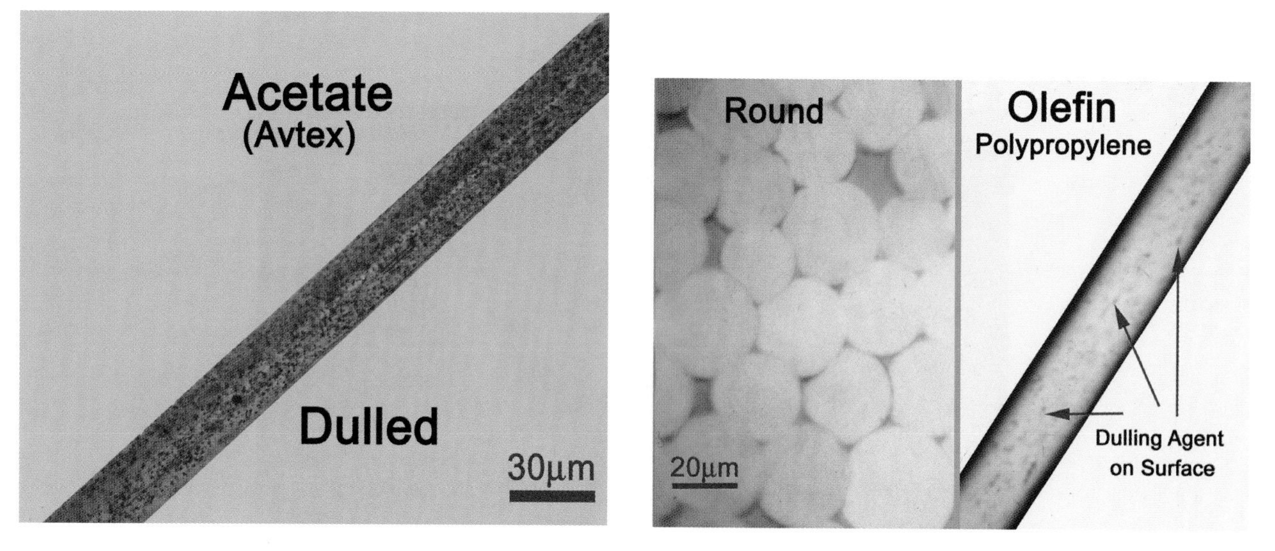

FIGURE C.45

FIGURE C.46

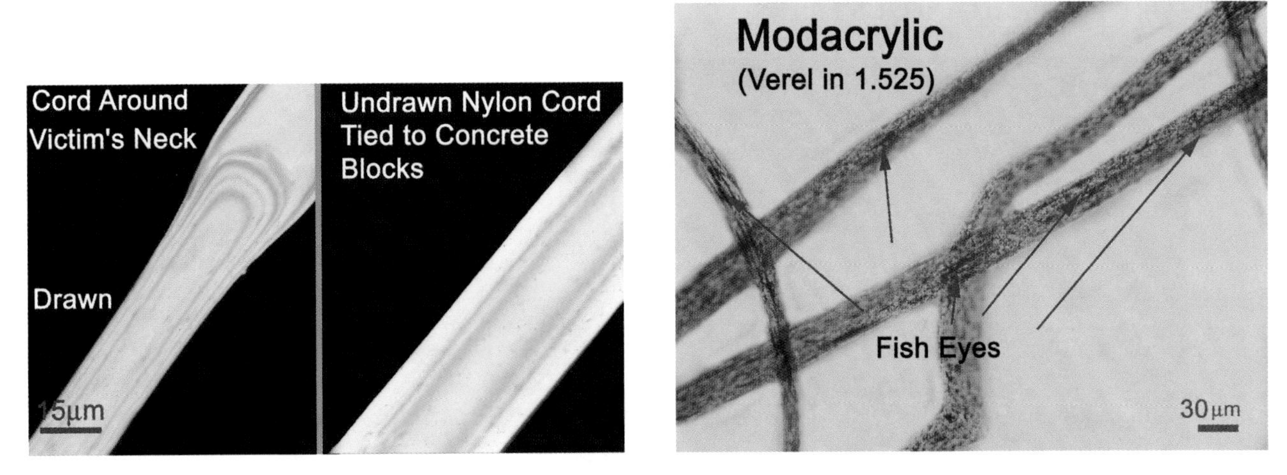

FIGURE C.47

FIGURE C.48

CATEGORY 5: FLUORESCENCE

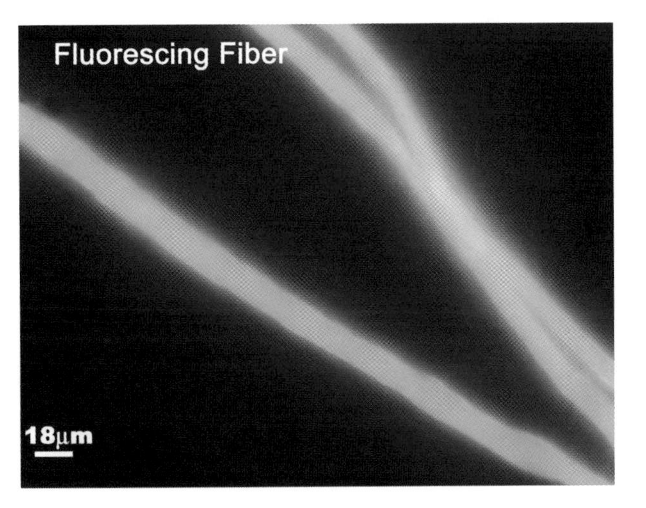

FIGURE C.49

CATEGORY 6: PLEOCHROISM

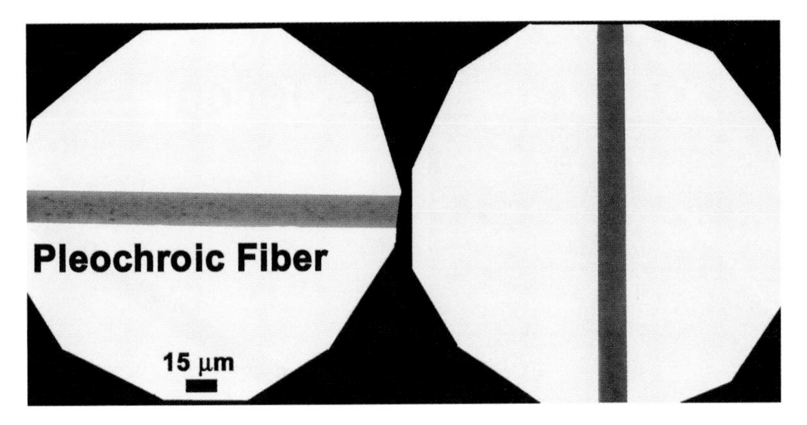

FIGURE C.50

CATEGORY 7: MISCELLANEOUS FIBERS

FIGURE C.51

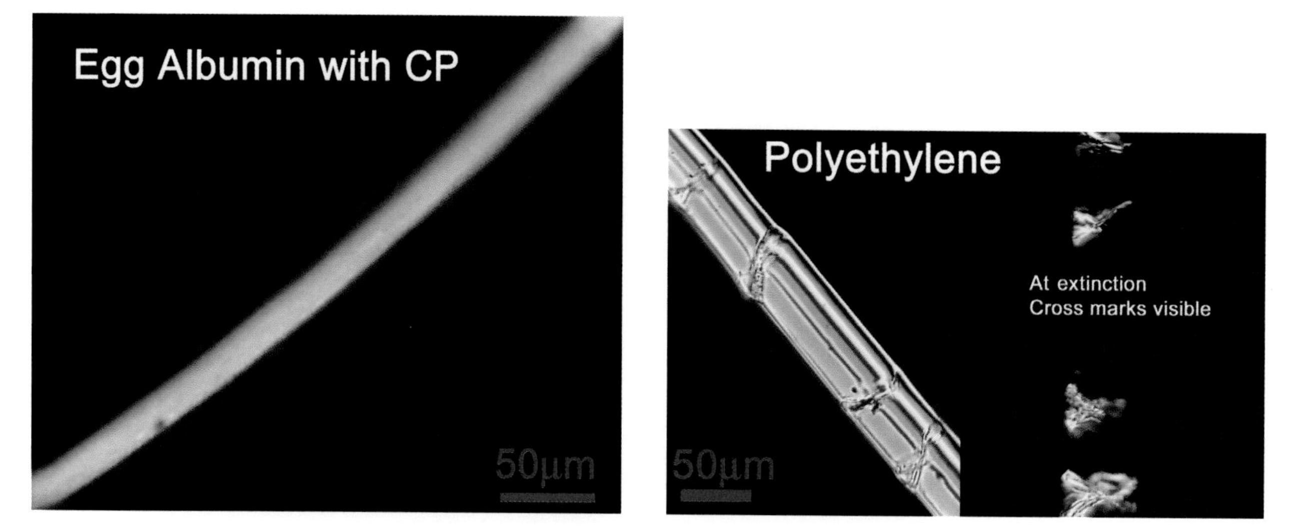

FIGURE C.52

FIGURE C.54

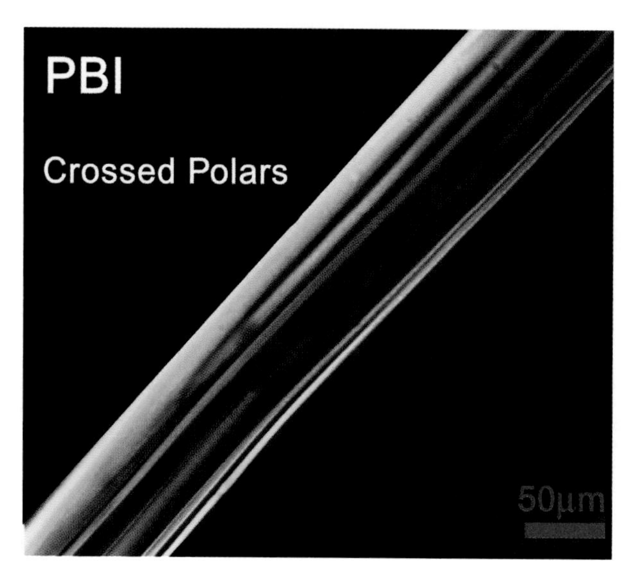

FIGURE C.53

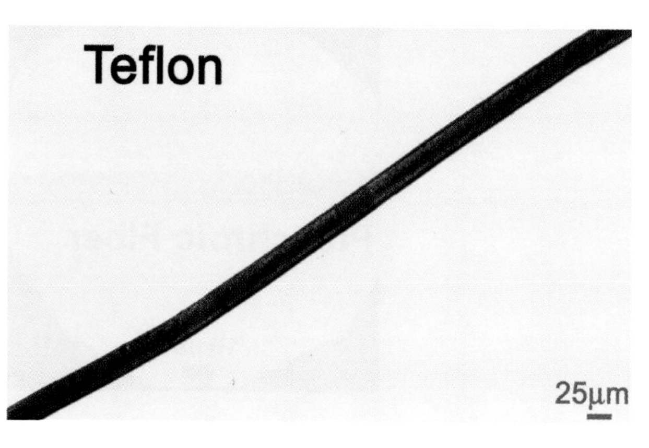

FIGURE C.55

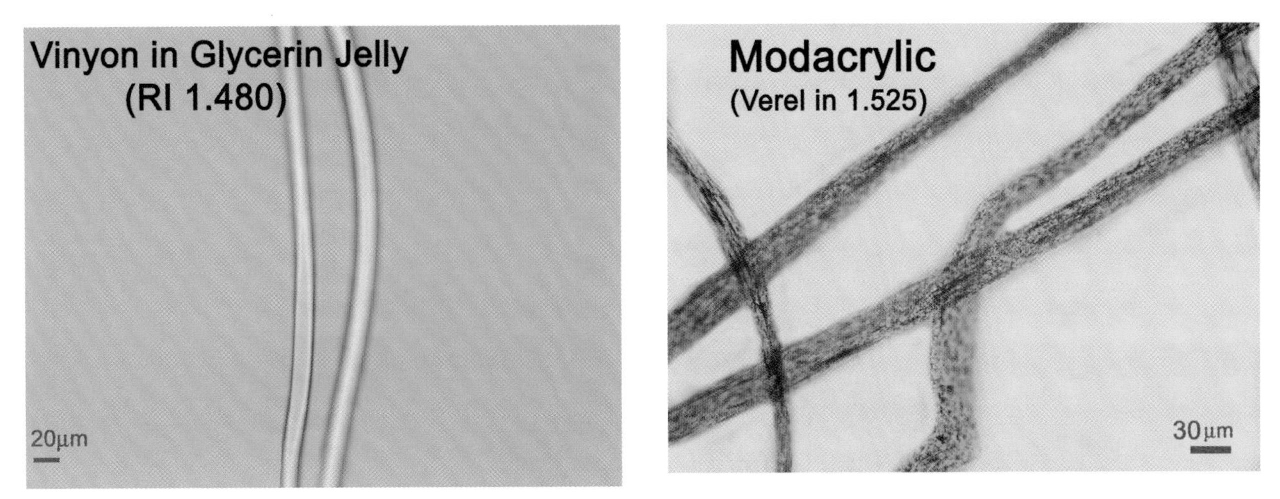

FIGURE C.56

FIGURE C.57

APPENDIX D

Paints and Pigments

This appendix was designed as a compendium of photomicrographs illustrating common materials used as pigments, extenders, fillers, and binders in paints for art and architectural purposes throughout the ages. Most materials are shown as they appear mounted in Melt Mount 1.539. Substances mounted in other media are clearly noted. Photomicrographs of many crystals forming and chemical reactions occurring during microchemical testing are included. The scale in each photomicrograph is equal to 40 μm unless otherwise noted. The materials are for the most part arranged in the following order:

1. Binders
2. Pigments, extenders, and fillers
3. Pigment color order:
   a. Whites
   b. Blacks
   c. Browns
   d. Blues and greens
   e. Purples and violets
   f. Yellows, oranges, and reds

ADDITIONAL SAMPLE PREPARATION PROCEDURE

Sample preparation for hard-to-dissolve specimens:

1. With a fine tungsten needle, remove a tiny fragment of pigment, extender, or paint flake from a specimen cross-section.
2. Place the test substance on a microscope slide and add a small droplet of freely prepared aqua regia (1:1 HCl and HNO_3) to the substance.
3. Gently heat the preparation on a hotplate under a hood or at least in a very well-ventilated area. The residue can be tested as-is or dissolved in deionized distilled water or any other aqueous acid, base, or organic solvent. If the specimen fails to dissolve, it should be fused as described in Chapter 4.

TESTS FOR BINDERS

WAXES

A tiny fragment of the specimen is placed onto a microscope slide. Small drops of methylene chloride and xylene are added to the specimen. The preparation is gently heated with a small flame. The solvent dissolves the wax; upon evaporation, the wax will be deposited in concentric rings about the pigment. The wax is isolated and melted on a hot stage.

Wax	Melting Range (°C)	Color/Characteristics
Beeswax	62 to 65	Yellow to brown
Carnauba	84 to 86	Light yellow with green lumps, brittle
Candelilla	68 to 70	Brown to yellow-brown, brittle
Montan	80 to 86	Dark brown

DRYING OILS

Over the course of time, drying oils are converted to oxynes. One can identify fats and fatty acids that are esters of glycerol by the conversion of glycerol or glycerides to acrolein. There is no way to differentiate the various kinds of oils by microchemical tests. Linseed oil has a distinct odor.

ANIMAL FATS

Animal fats alter considerably over time and may leave residues consisting largely of free fatty acids.

OILS AND FATS

Oils and fats can be visualized by staining with Nile blue or Sudan black.

RESINS

Resins respond to a few color reactions; they turn green with copper acetate and red with alum tincture. They also respond to the Storch–Morawski test. On a microscope slide, dissolve a few minute grains of the specimen in a drop of acetic anhydride and heat gently. Add a drop of 24N H_2SO_4. A blue-violet to red color is produced if rosin or rosin oil is present.

Another method is to place a tiny fragment of the resinous test material in a drop of anhydrous ethyl alcohol on a microscope slide. Note whether the material is soluble, partially soluble, or insoluble in ethyl alcohol. Evaporate the alcohol and, with a hot bar or hotstage, determine the melting point or melting range of the specimen (see table below).

Resin	Melting Point/ Melting Range (°C)	Color/Characteristics
Copal (hard)	Varies, 180 to 340	Insoluble in alcohol
Copal (soft)	Varies, 150 to180	Partially soluble in alcohol
Damar (dammar)	120	Soluble in alcohol
Mastic	95	Very soluble in alcohol
Rosin	Varies, 100 to 130	Pale yellow to amber, very soluble in alcohol
Sandarac	135 to 145	Dull red, very soluble in alcohol

Proteins

Millon's reagent method: add reagent to a tiny fragment of the specimen and heat gently. A dark red color indicates protein.

Biuret reagent for peptides, proteins, amino acids and albumin: place a flake of the specimen on a microscope slide and add a drop of Biuret's reagent. The formation of a pink-violet color is a positive result for an amide group.

Casein differs from egg white in that it contains phosphorus. Casein can be detected by extraction from the color layer with alkali, precipitating it with dilute acid, and subsequently determining the presence of nitrogen, sulfur, and phosphorus.

Casein test for amino acids (to distinguish casein compounds from egg-based tempera paints): add a drop of ninhydrine reagent in ethanol or acetone to a test fragment of suspected casein. Formation of a deep blue-purple color indicates an amino acid. Ninhydrine reacts with casein (amino acid) and not egg-based tempera paints (whole proteins).

Feigl's test for native protein: add a drop of reagent (0.1% solution of the potassium salt of the ethyl ester of tetrabromophenolphthalein in ethanol) to a particle of test substance. Wait 1 minute, then add a drop of 0.2 *N* acetic acid. A blue solid in a yellow solution indicates a positive reaction.

Gum Arabic

Gum arabic can be detected by adding a drop of a 4% solution of orcinol which, after adding concentrated hydrochloric acid and heating, causes the formation of a red-violet color complex that turns into a blue precipitate soluble in alcohol. Gum arabic is soluble in cold water.

Collagen and Gelatin

Add a drop of reagent made from equal volumes of dilute aqueous solutions of fuchsin and picric acid to a tiny fragment of the test substance and allow the mixture to soak 1 to 2 minutes. Remove excess stain by decantation and absorption of the stain with wedges of filter paper. Wash the specimen several times with water to remove all traces of stain. A blood-red color develops if the test substance contains a collagenous material, i.e., leather.

Starches

Starch grains and starch paste can be recognized by the iodine reaction Gram stain. Herzberg stain yields a dark blue color reaction with carbohydrates.

Cellulose

Cellulose can be detected by adding a particle of test substance to a drop of Herzberg stain. A red-blue color indicates cellulose.

Clays

Clays can be distinguished by staining with malachite green dissolved in nitrobenzene after heating in a drop of dilute hydrochloric acid.

REFERENCES

1. Chamot, E.M. and Mason, C.W., *Handbook of Chemical Microscopy,* Vol. 1, New York, John Wiley & Sons, 1930.
2. Chamot, E.M. and Mason, C.W., *Handbook of Chemical Microscopy,* Vol. 2, New York, John Wiley & Sons, 1931.
3. Eibner, A., Microchemical examination of media (abstract), *Mouseion*, 20, 5, 1932.
4. Faust, G.T., *Staining of Clay Minerals as a Rapid Means of Identification in Natural and Beneficiated Products*, Washington, D.C., U.S. Bureau of Mines, 1940, p. 3522.
5. Vesce, V.C., *Classification and Microscopic Identification of Organic Pigments*, Mattiello, J.J., Ed., New York, John Wiley & Sons, 1942.
6. Feigl, F., *Qualitative Analysis by Spot Tests,* 3rd ed., New York, Elsevier, 1946.
7. *Studies in Conservation*, 1, 1, 1952 (now published quarterly).
8. Schaefer, H.F., *Microscopy for Chemists*, New York, Van Nostrand, 1953.

9. Kirk, P.L., *Crime Investigation*, New York, Interscience, 1953, p. 257.
10. McCrone, W.C., *Fusion Methods in Chemical Microscopy,* New York, Interscience, 1957.
11. Bloss, F.D., *An Introduction to the Methods of Optical Crystallography*, New York, Holt, Rinehart & Winston, 1961.
12. Schneider, F.L., *Qualitative Organic Microanalysis,* New York, Academic Press, 1964.
13. Gettens, R.J. and Stout, G.L., *Painting Materials: A Short Encyclopedia*, New York, Dover Publications, 1966.
14. Crown, D.A., *The Forensic Examination of Paints and Pigments*, Springfield, IL, Charles C Thomas, 1968.
15. Fulton, C.C., *Modern Microcrystal Tests for Drugs*, New York, Interscience, 1969.
16. Adrosko, R.J., *Natural Dyes and Home Dyeing*, New York, Dover Publications, 1971.
17. Fiegl, F., *Spot Tests in Inorganic Analysis*, 6th ed., Amsterdam, Elsevier, 1972.
18. McCrone, W.C., Delly, J.G., and Palenik, S.J., Eds., *The Particle Atlas*, Vols. 2 and 5, Ann Arbor, MI, Ann Arbor Science Publishers, 1973 and 1979.
19. Stevens, R.E., Squaric acid: a novel reagent in chemical microscopy, *Microscope*, 22, 63, 1974.
20. Wills, W.F., Jr. and Whitman, V.L., Extended use of squaric acid as a reagent in chemical microscopy, *Microscope*, 25, 1, 1977.
21. Thornton, J.I., Forensic paint examination, in *Forensic Science Handbook*, Saferstein, R., Ed., Englewood Cliffs, NJ, Prentice-Hall, 1982, chap. 10.
22. Fleischer, M., Wilcox, R.E., and Matzko, J.J., *Microscopic Determination of the Nonopaque Minerals*, Washington, D.C., U.S. Geological Survey, 1984.
23. Delly, J.G., Microchemical tests for selected cations, *Microscope*, 37, 139, 1989.
24. Wills, W.F., Jr., Squaric acid revisited, *Microscope*, 38, 169, 1990.
25. Mayer, R., *The Artist's Handbook of Materials and Techniques*, 5th ed., New York, Viking, 1991.
26. Koleske, J.V., Ed., *Paint and Coatings Testing Manual of the Gardner–Sward Handbook*, 14th ed., Philadelphia, American Society for Testing and Materials, 1995.
27. American Society for Testing Materials, Standards E 1610 through 1695, West Conshohocken, PA.
28. Ryland, S.G., Infrared microspectroscopy of forensic paint evidence, in *Practical Guide to Infrared Microspectroscopy*, Humecki, H.J., Ed., New York, Marcel Dekker, 1995, chap. 6.
29. Jungreis, E., *Spot Test Analysis: Clinical, Environmental, Forensic, and Geochemical Applications*, 2nd ed., New York, John Wiley & Sons, 1997.
30. SWGMAT, Forensic Paint Analysis and Comparison Guidelines, May 2000 revision, Washington, D.C., U.S. Department of Justice.

MEDIUMS

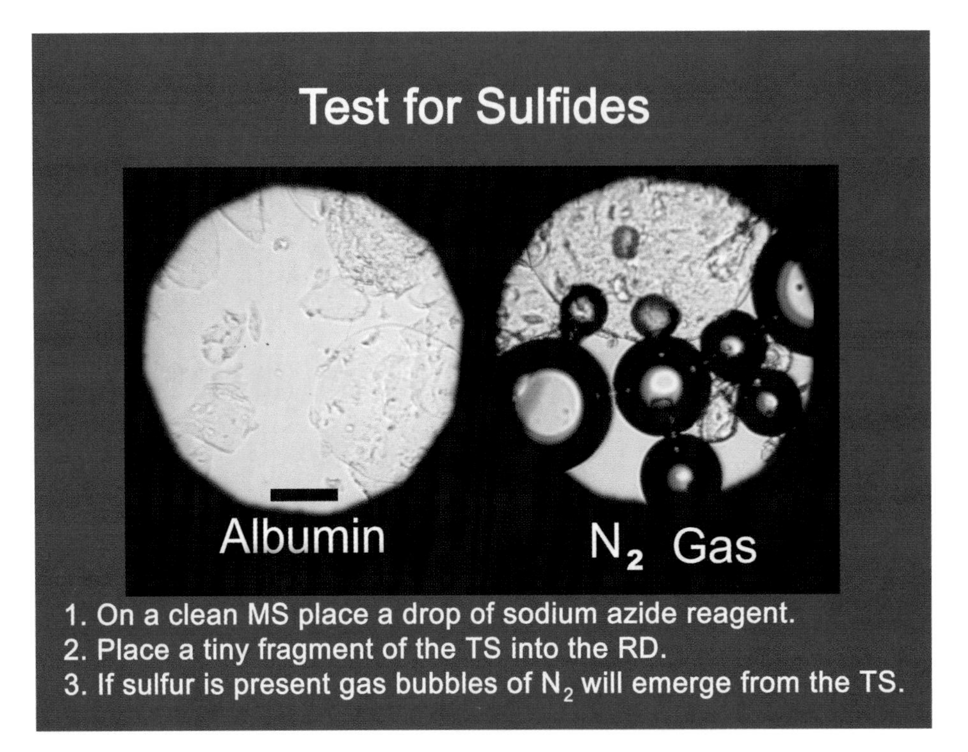

FIGURE D.1

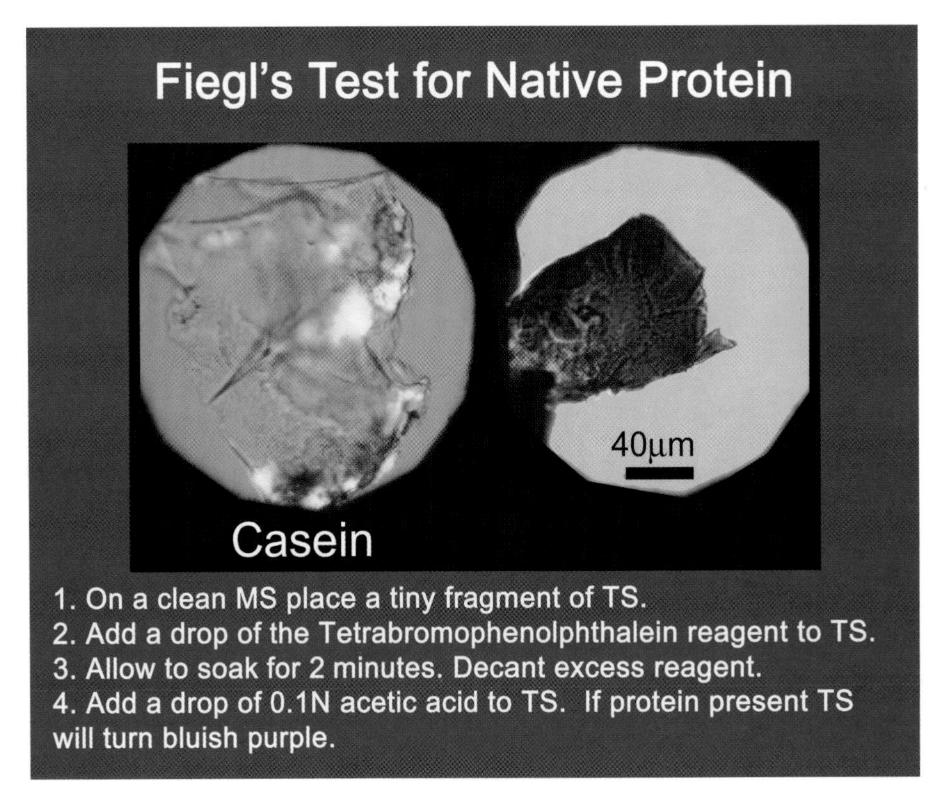

FIGURE D.2

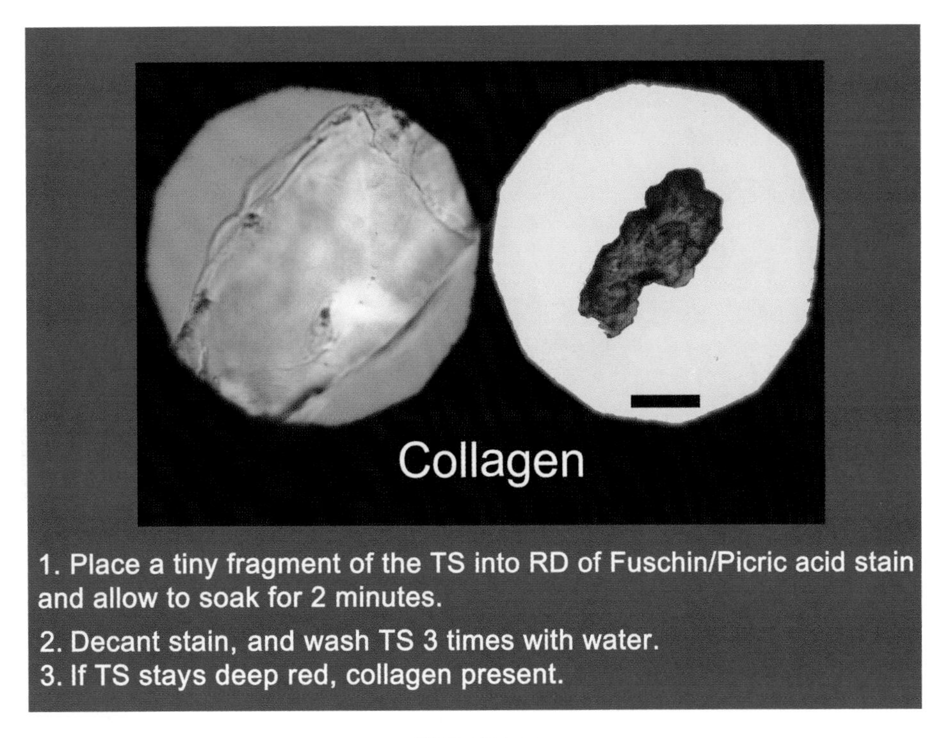

FIGURE D.3

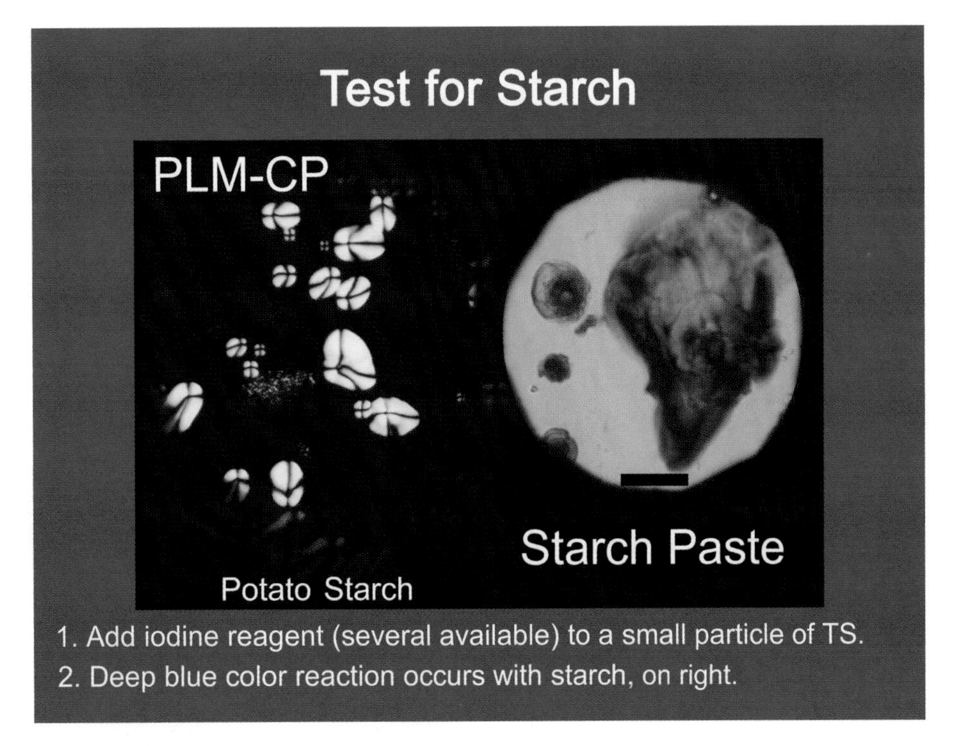

FIGURE D.4

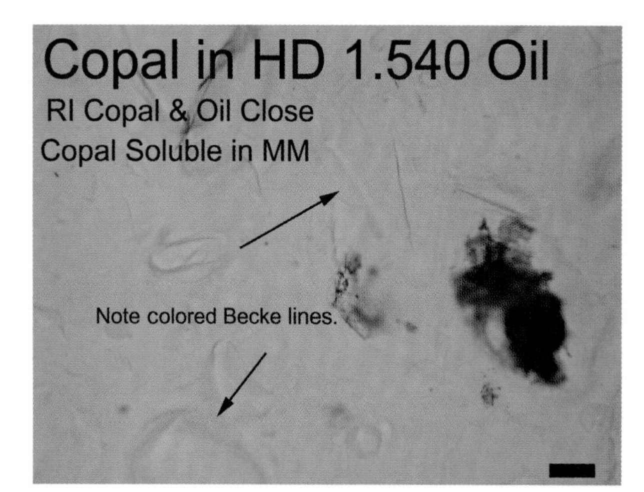

FIGURE D.5

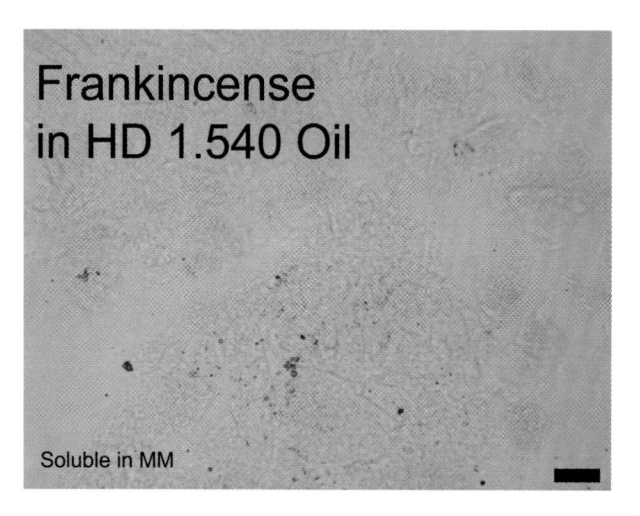

FIGURE D.6

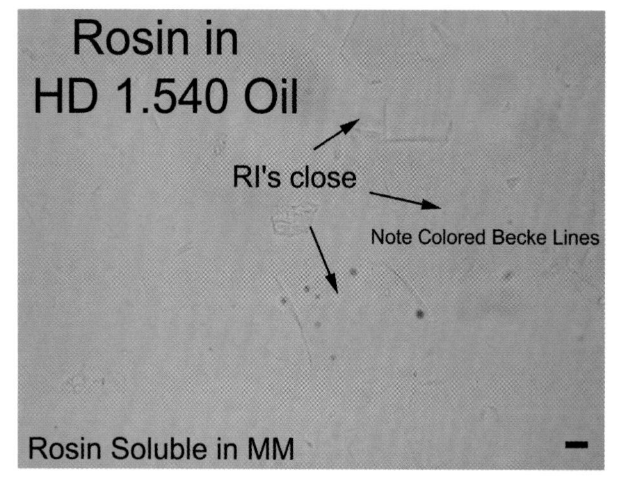

FIGURE D.7

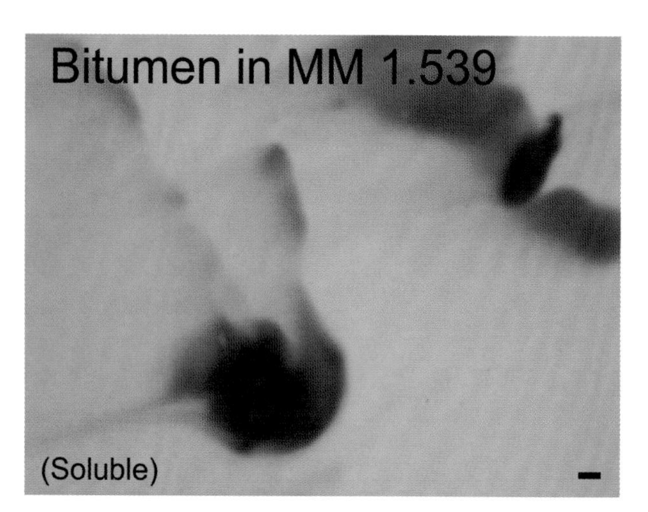

FIGURE D.8

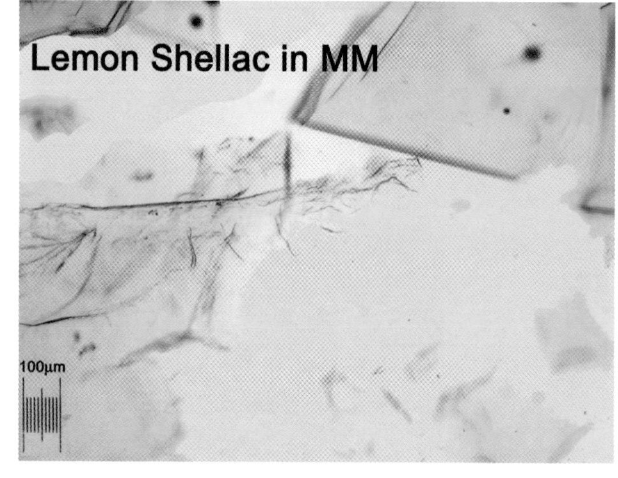

FIGURE D.9

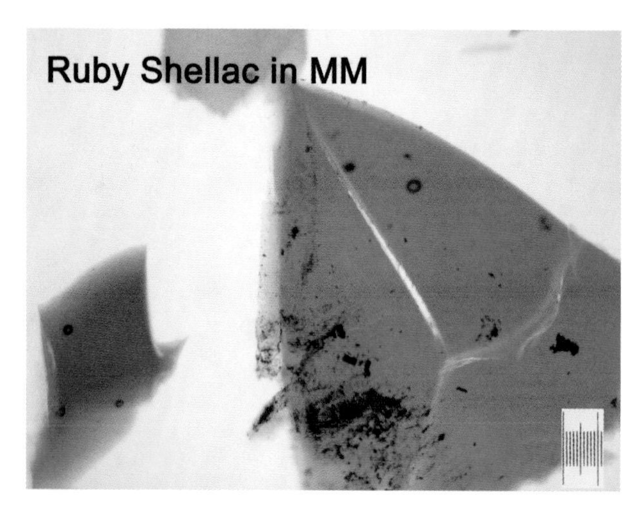

FIGURE D.10

FILLER

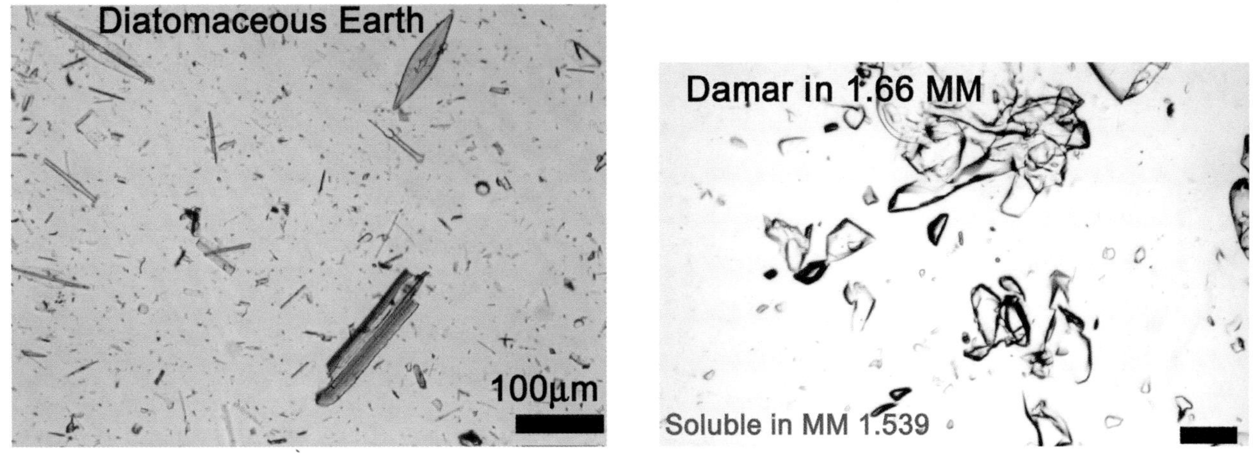

FIGURE D.11

FIGURE D.12

PIGMENT: BLACKS

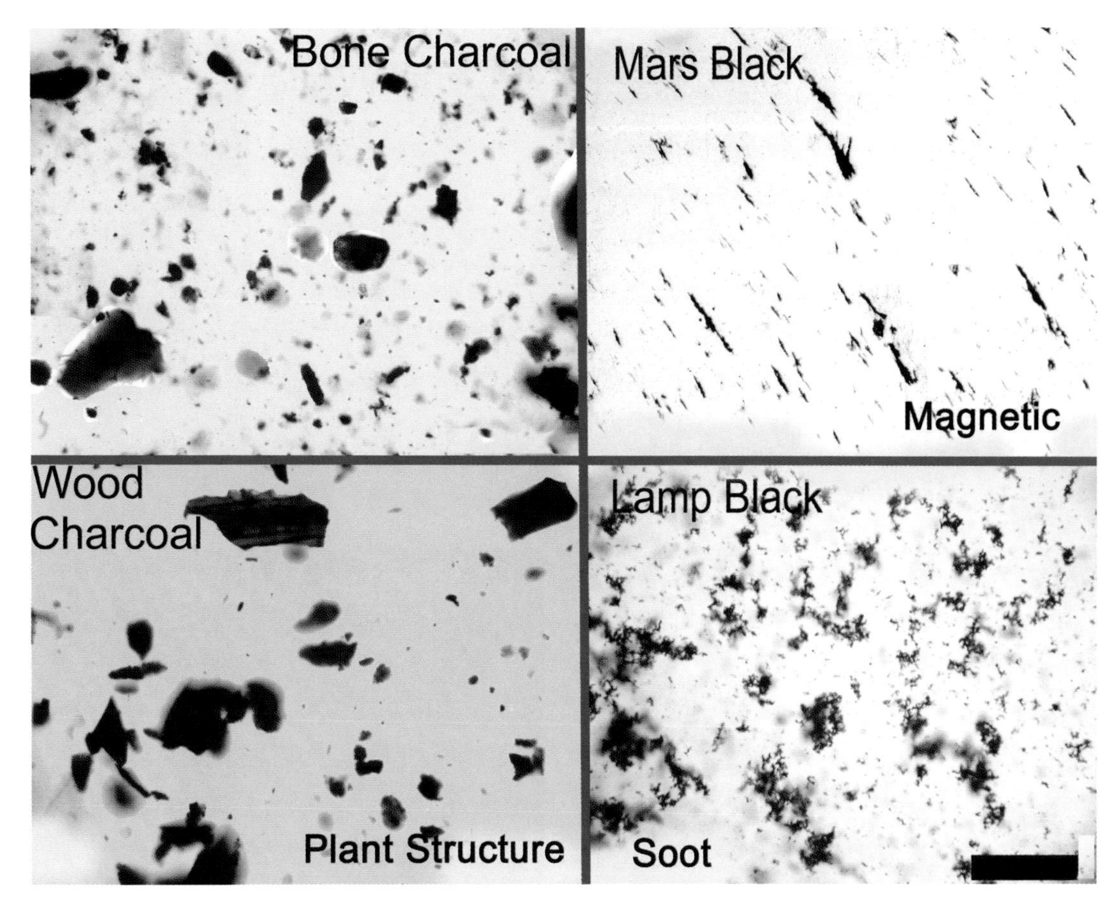

FIGURE D.13

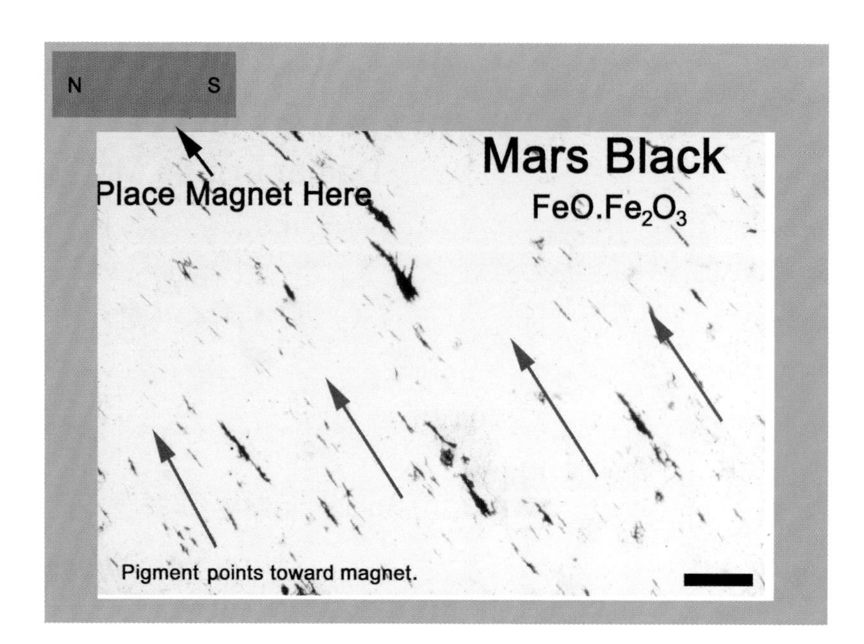

FIGURE D.14

PIGMENT: WHITES

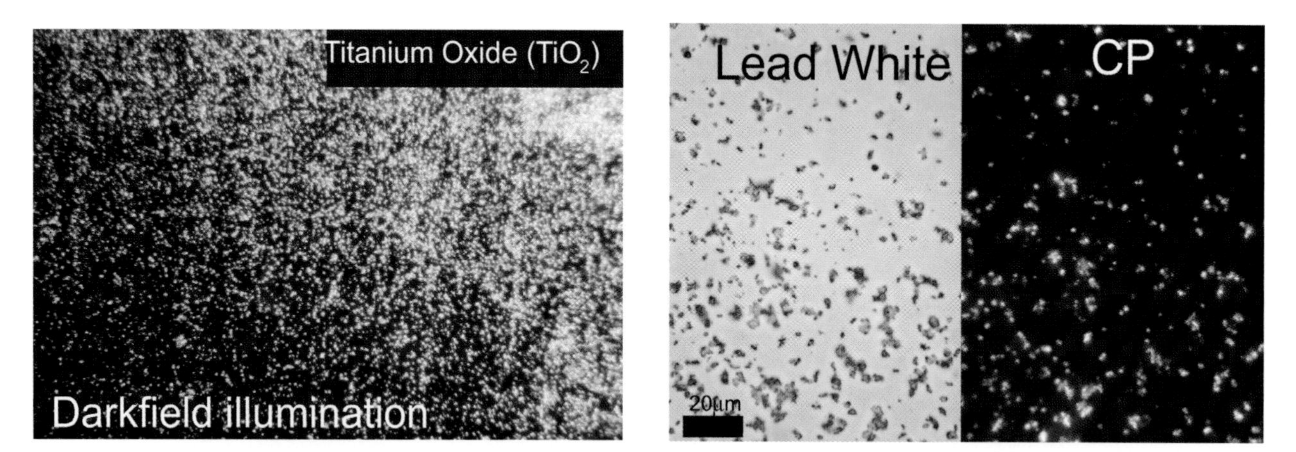

FIGURE D.15

FIGURE D.17

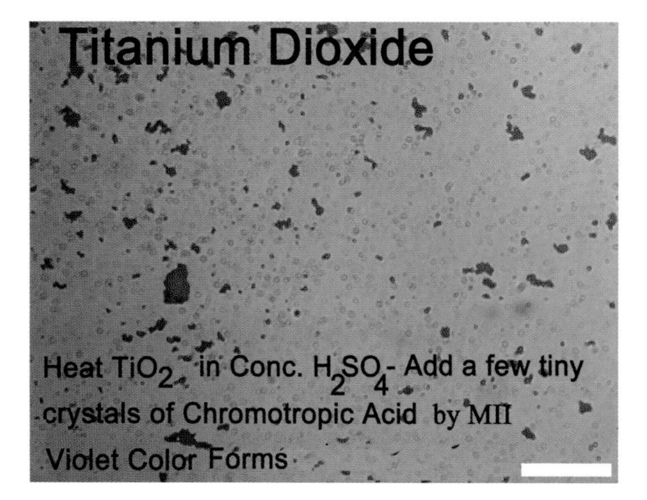

FIGURE D.16

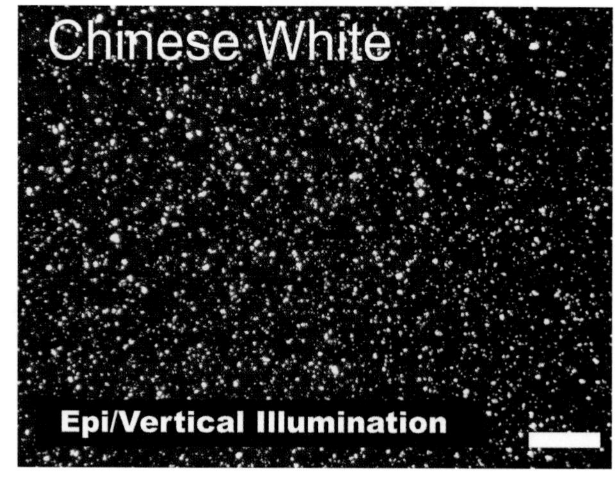

FIGURE D.18

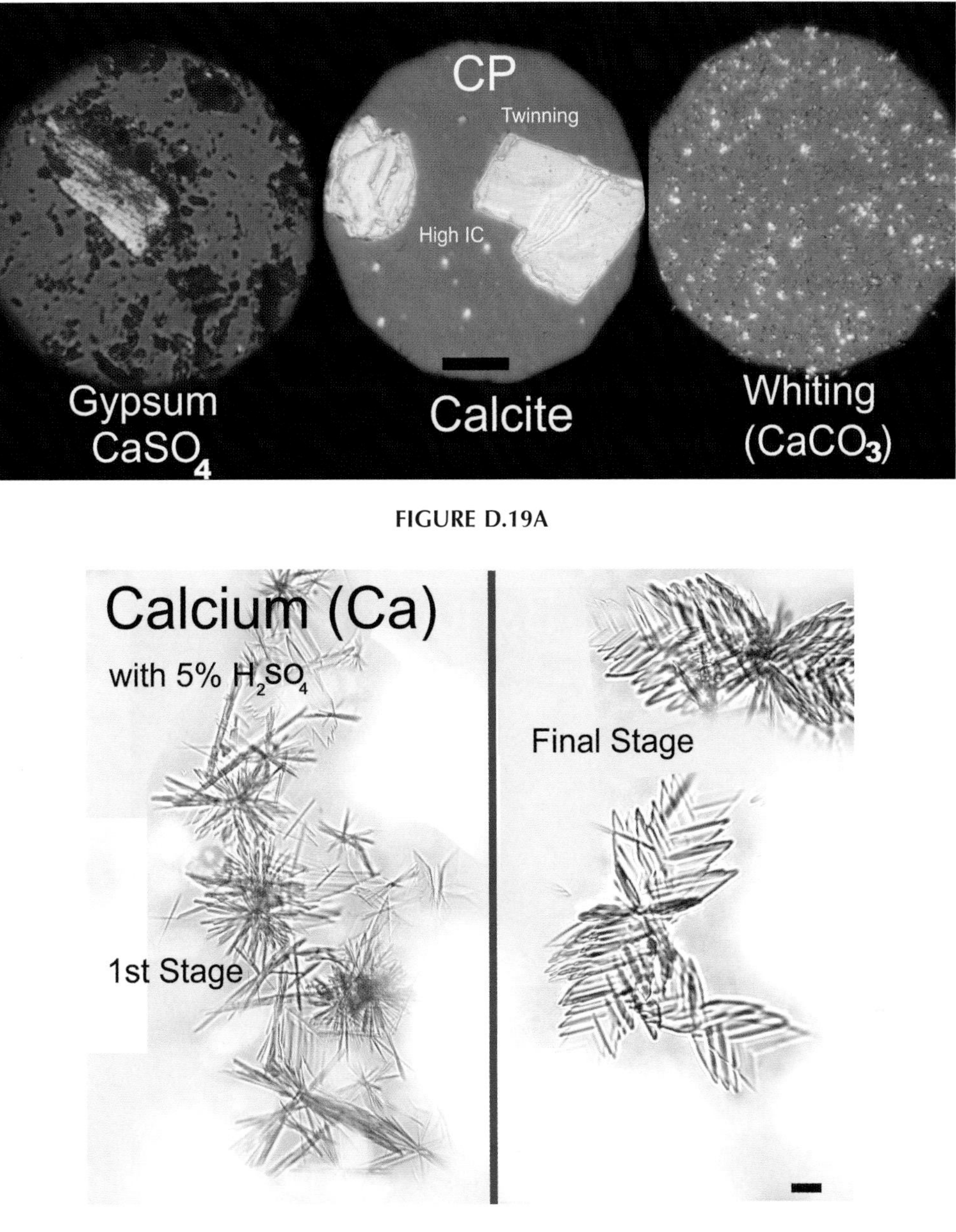

FIGURE D.19A

FIGURE D.19B

EXTENDERS

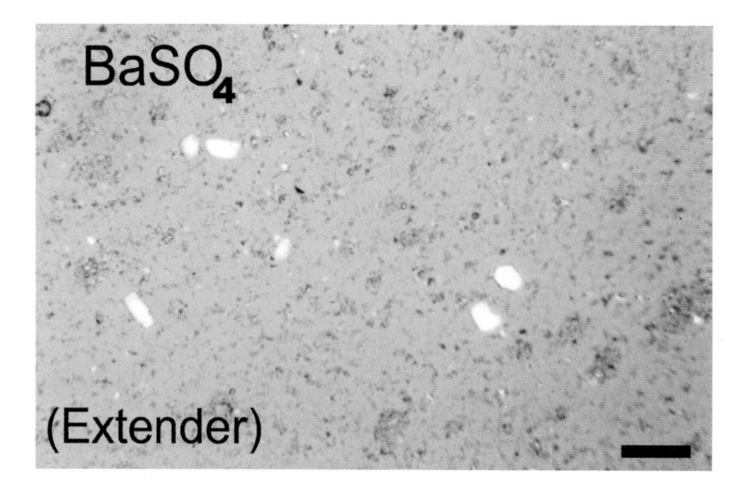

FIGURE D.20A

FIGURE D.20B

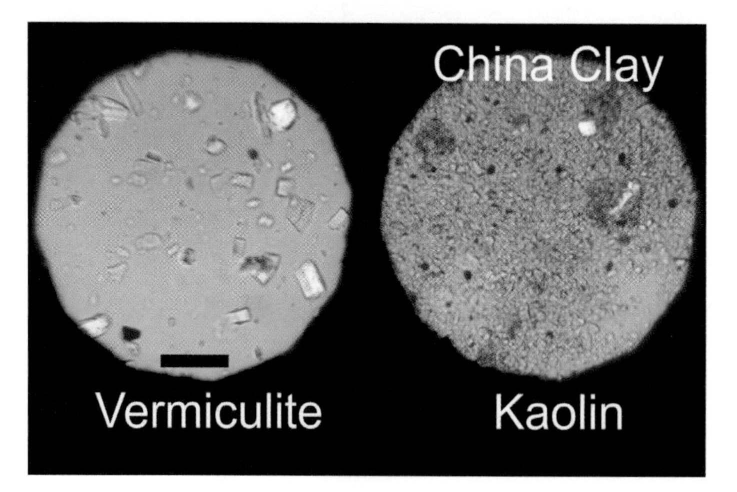

FIGURE D.21

PIGMENTS (continued)

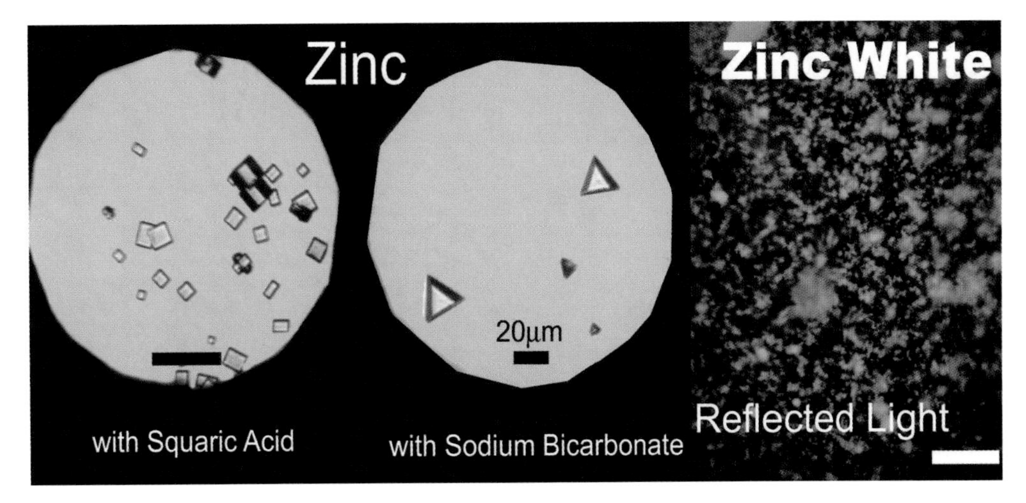

FIGURE D.22

PIGMENT: BLUES

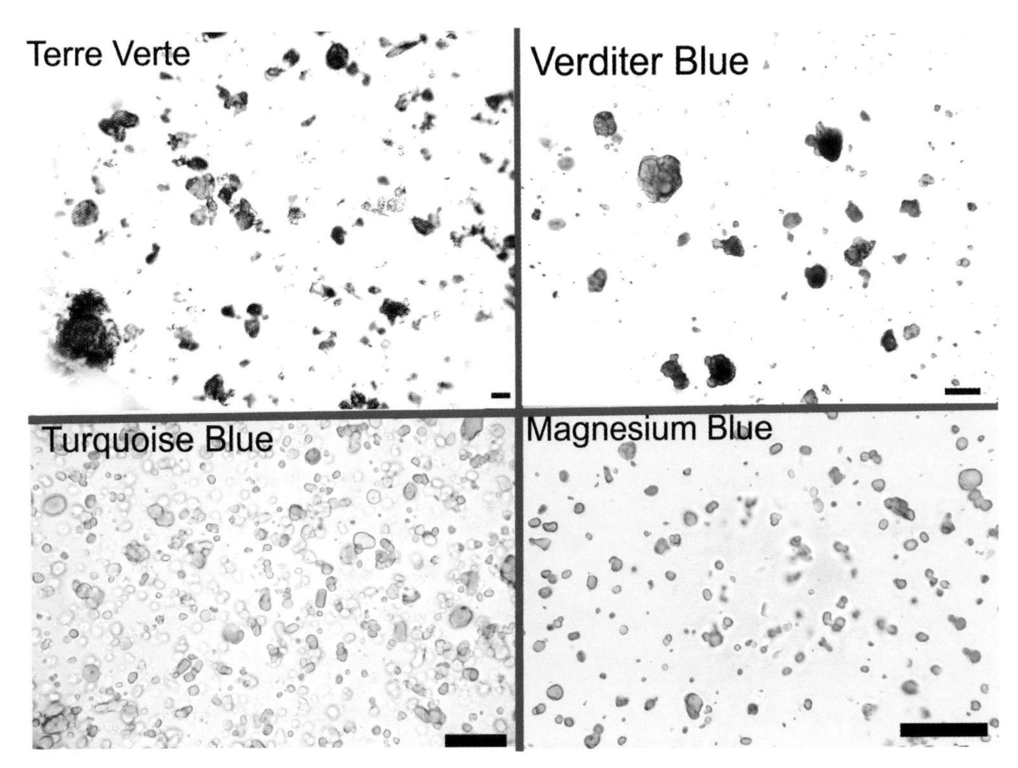

FIGURE D.23

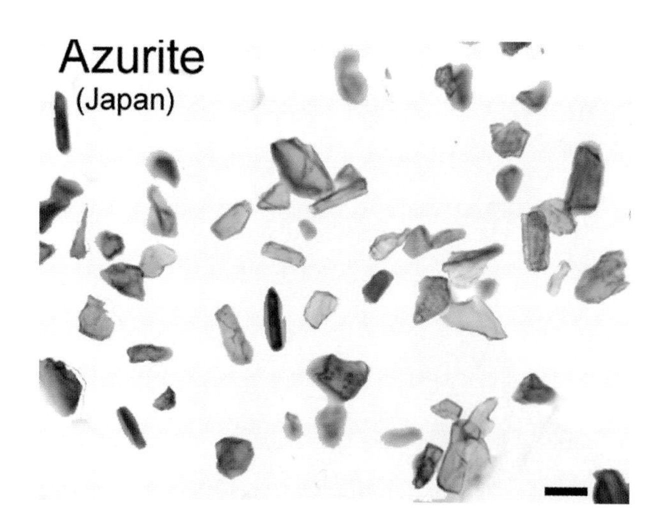

FIGURE D.24A

FIGURE D.24B

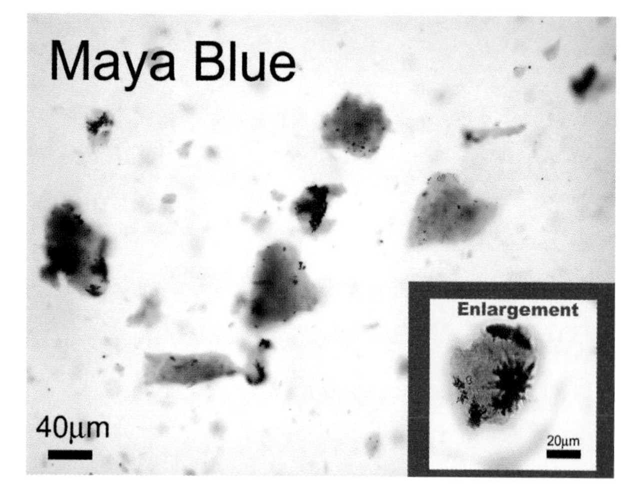

FIGURE D.25

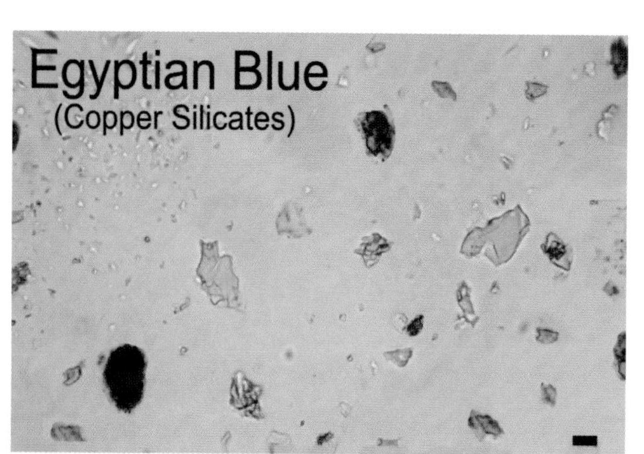

FIGURE D.26

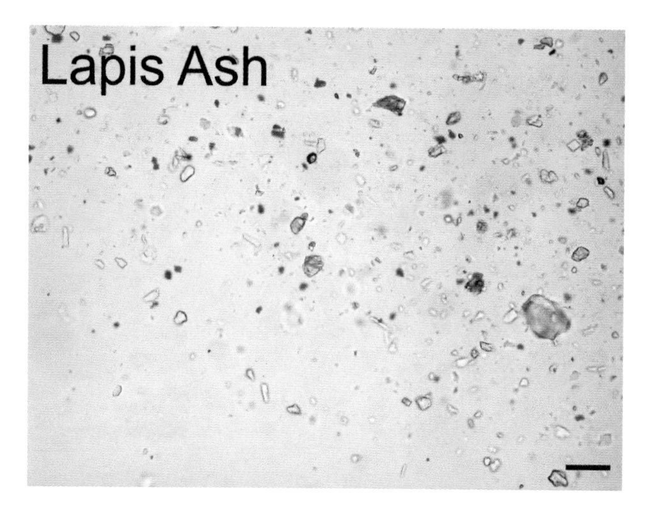

FIGURE D.27A

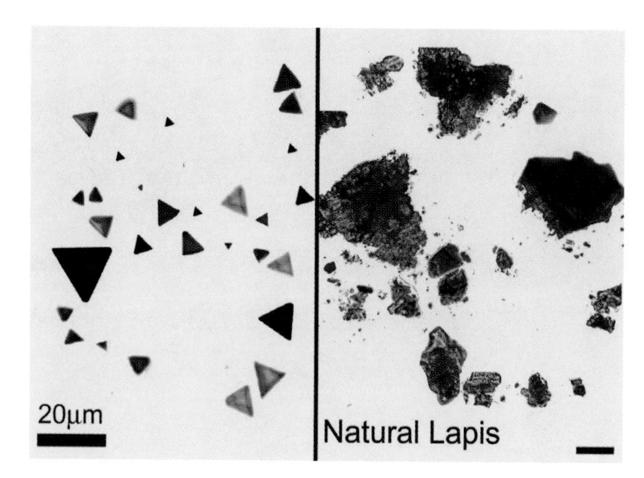

FIGURE D.27B

FIGURE D.28

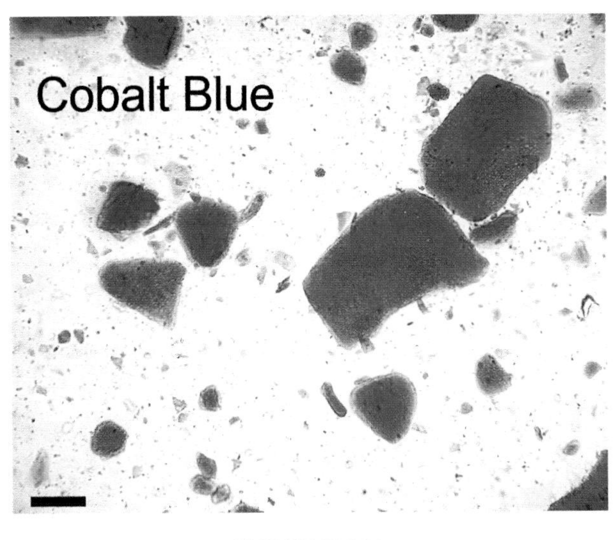

FIGURE D.29A

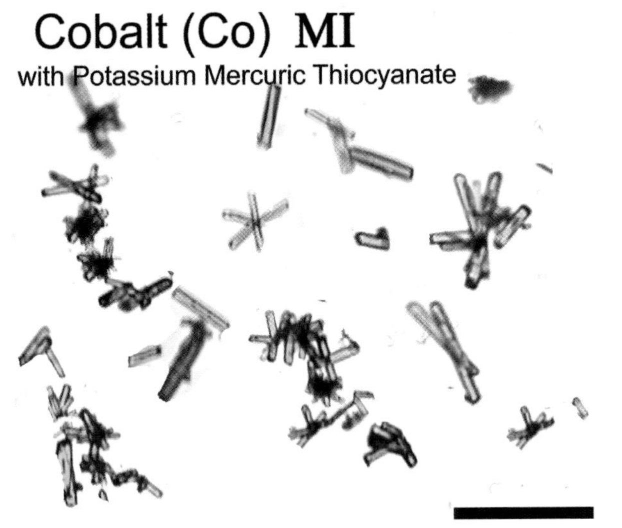

FIGURE D.29B

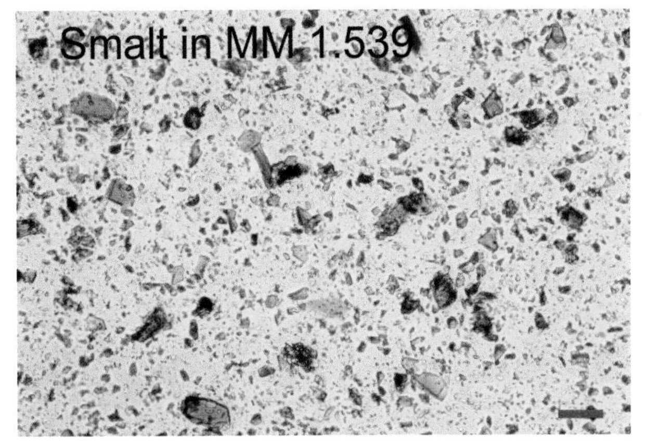

FIGURE D.30

FIGURE D.31A

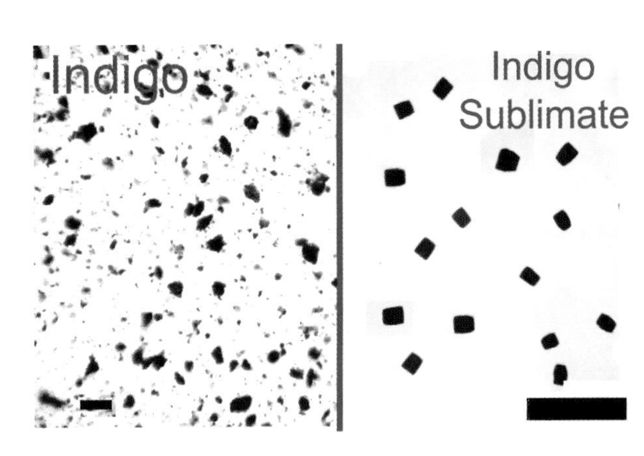

FIGURE D.31B

PIGMENT: GREENS

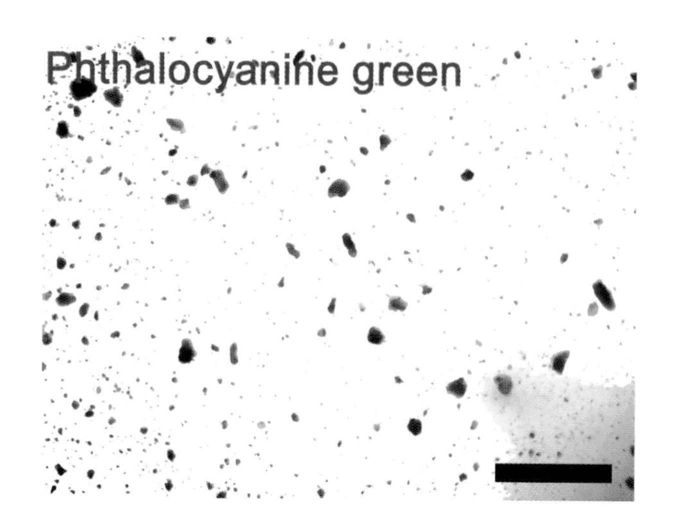

FIGURE D.32A

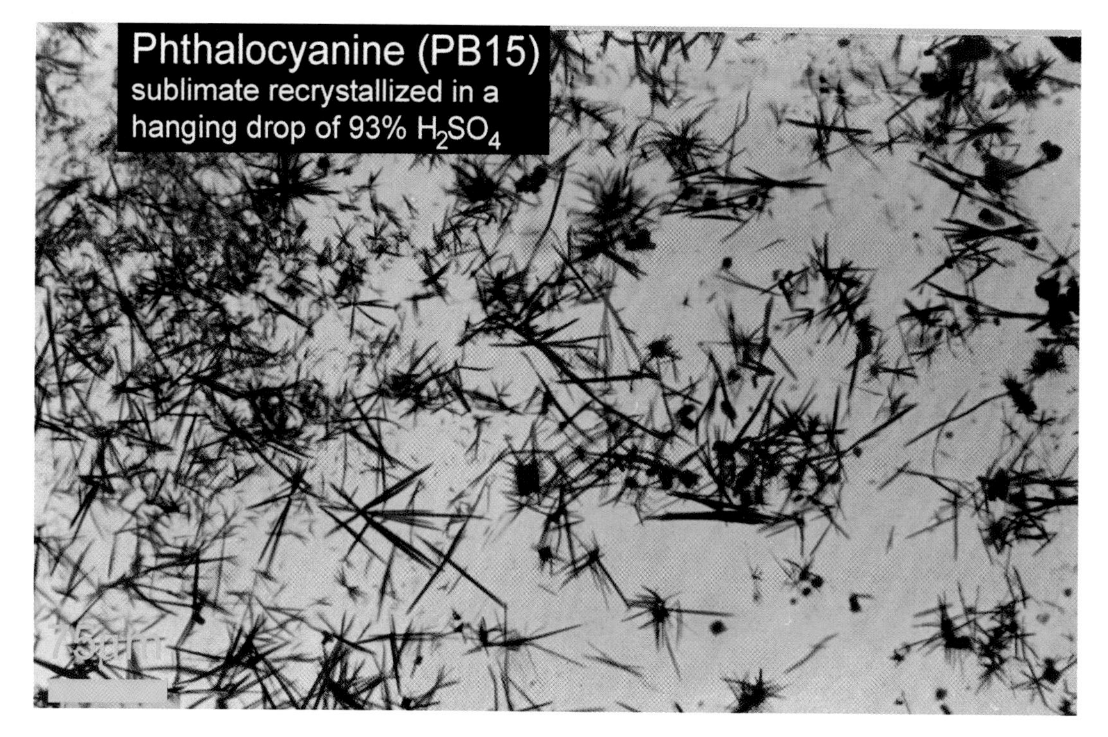

FIGURE D.32B

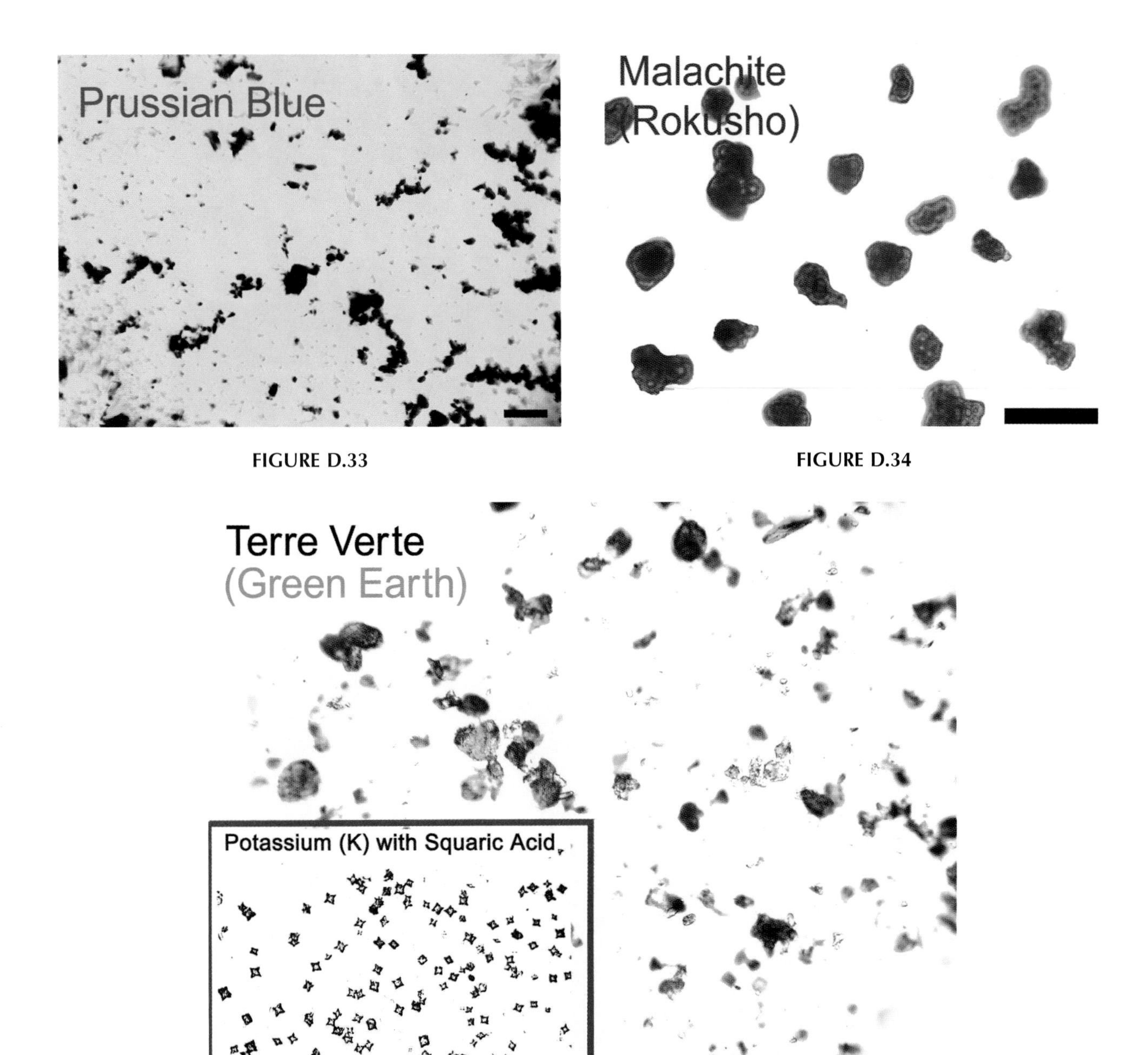

FIGURE D.33

FIGURE D.34

FIGURE D.35

FIGURE D.36A

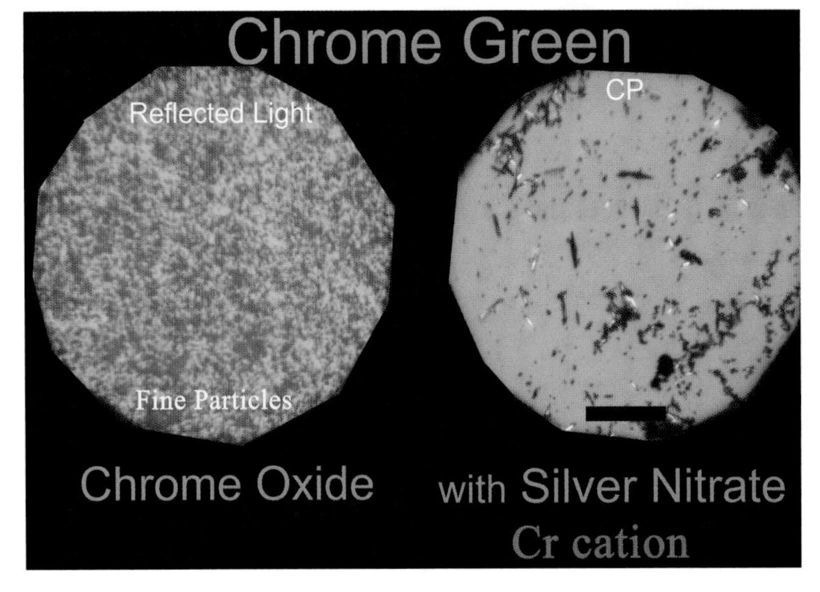

FIGURE D.36B

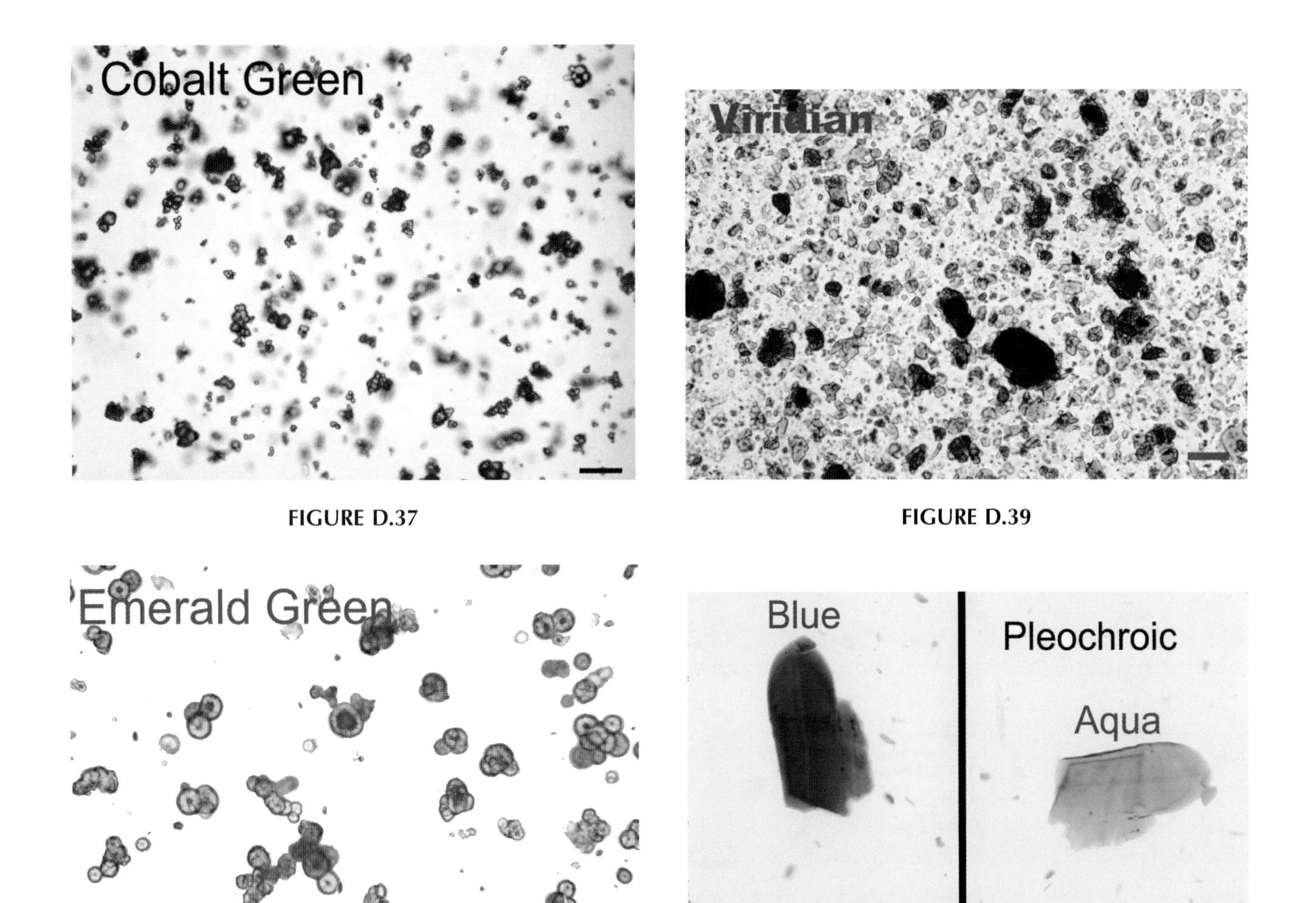

FIGURE D.37

FIGURE D.39

FIGURE D.38

FIGURE D.40A

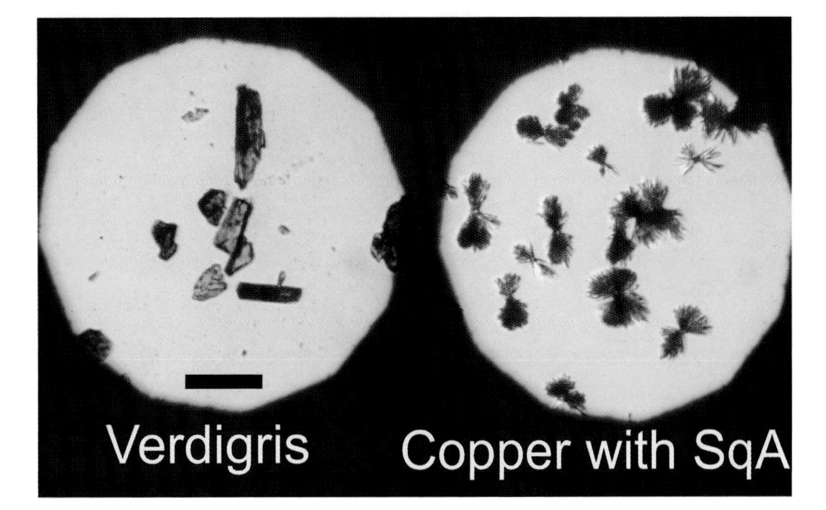

FIGURE D.40B

PIGMENT: BROWNS

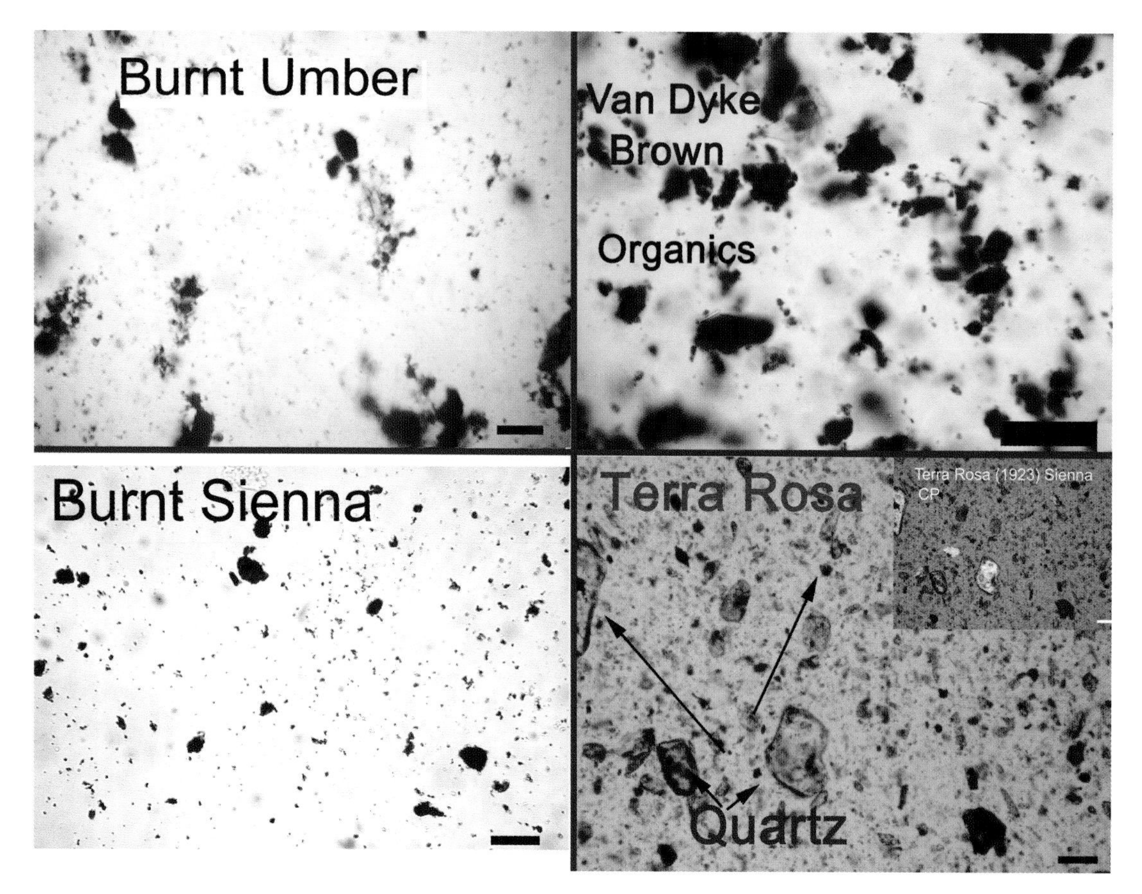

FIGURE D.41

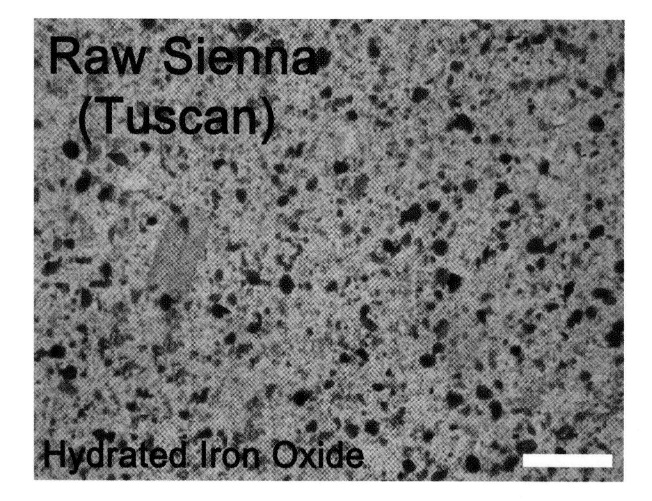

FIGURE D.42

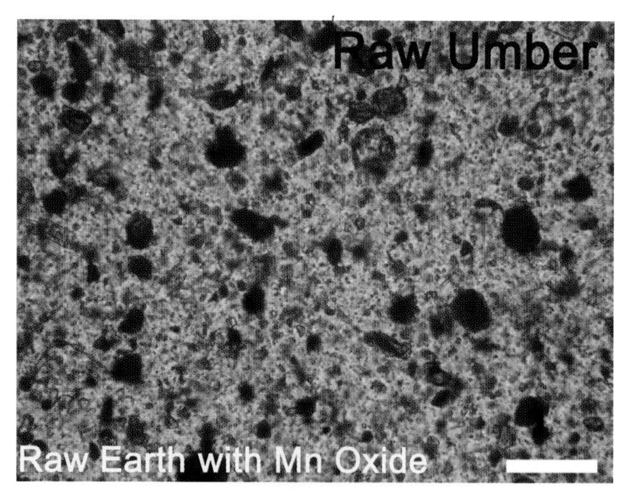

FIGURE D.43

PIGMENT: VIOLETS/PURPLES

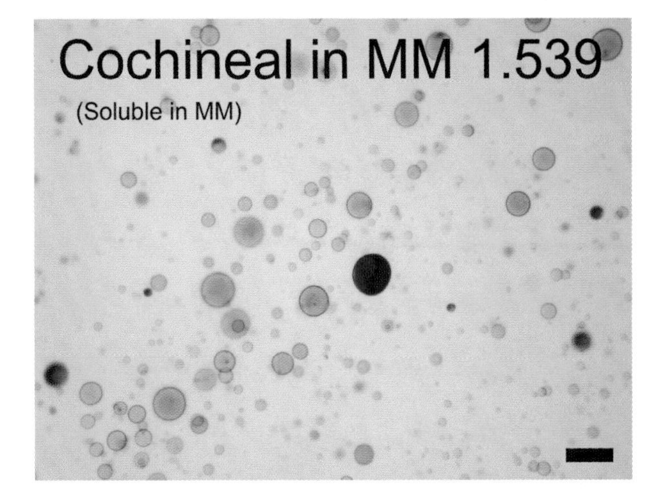

FIGURE D.44

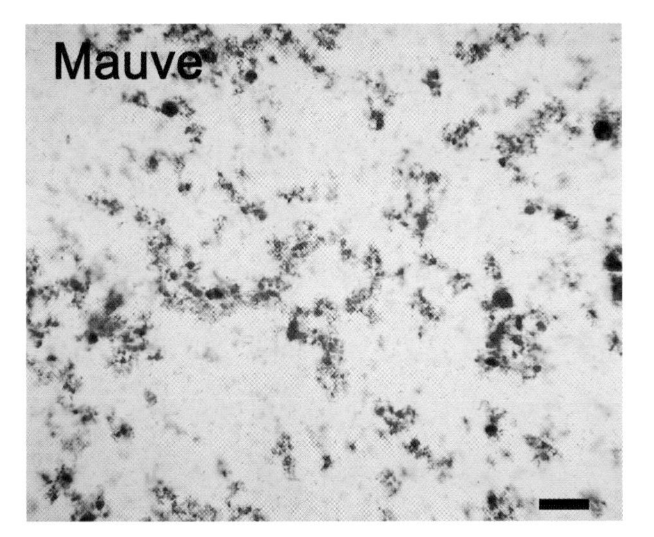

FIGURE D.45

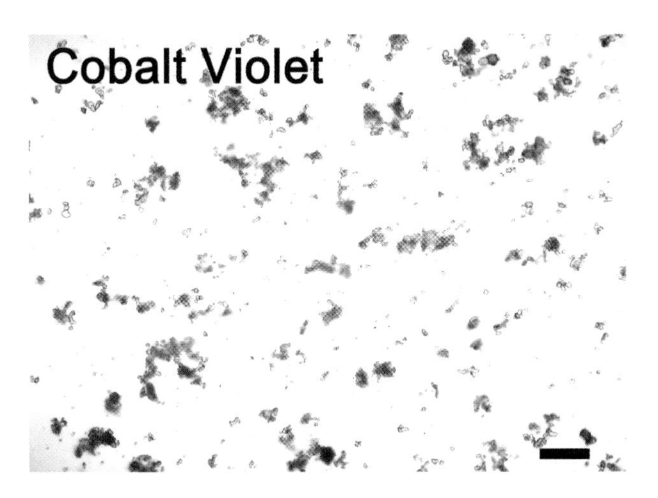

FIGURE D.46

RED

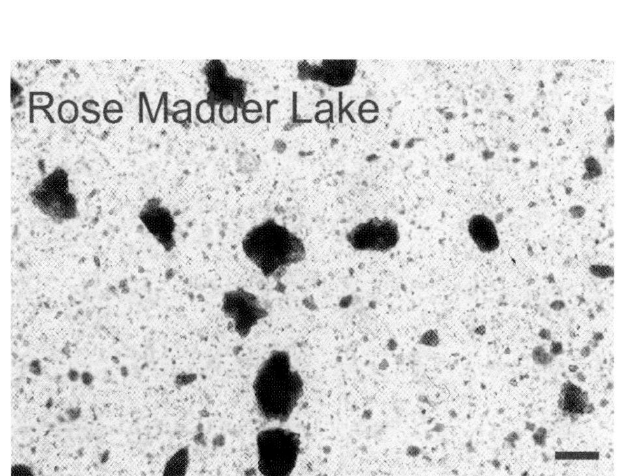

FIGURE D.47

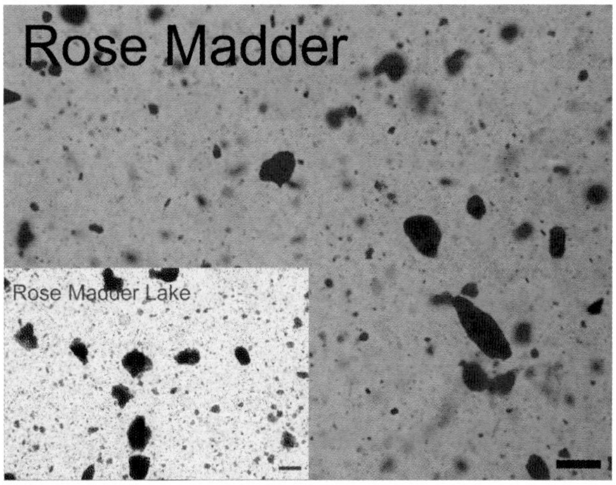

FIGURE D.48A

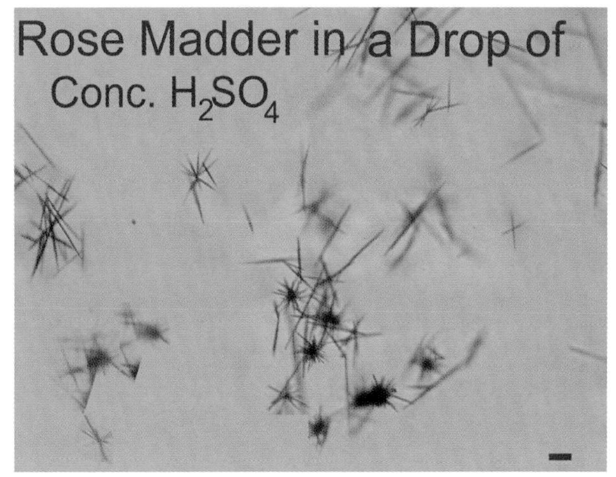

FIGURE D.48B

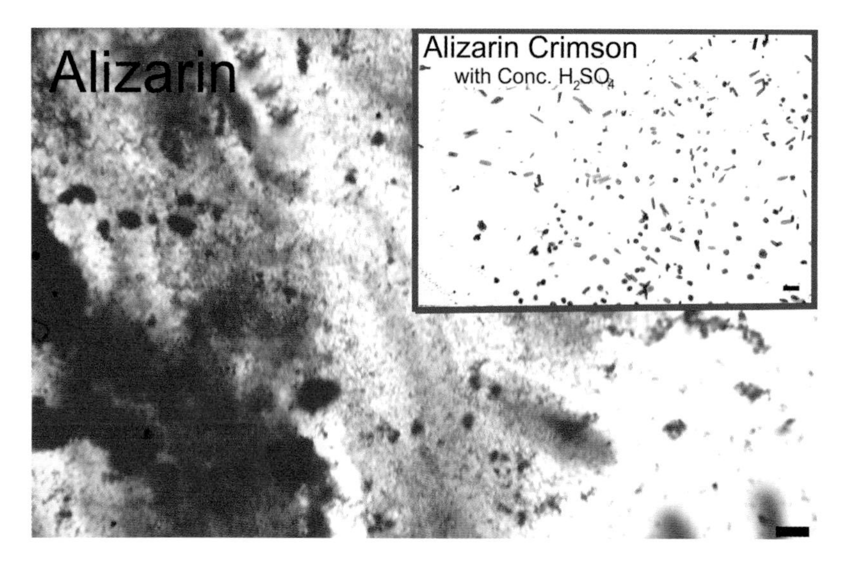

FIGURE D.49

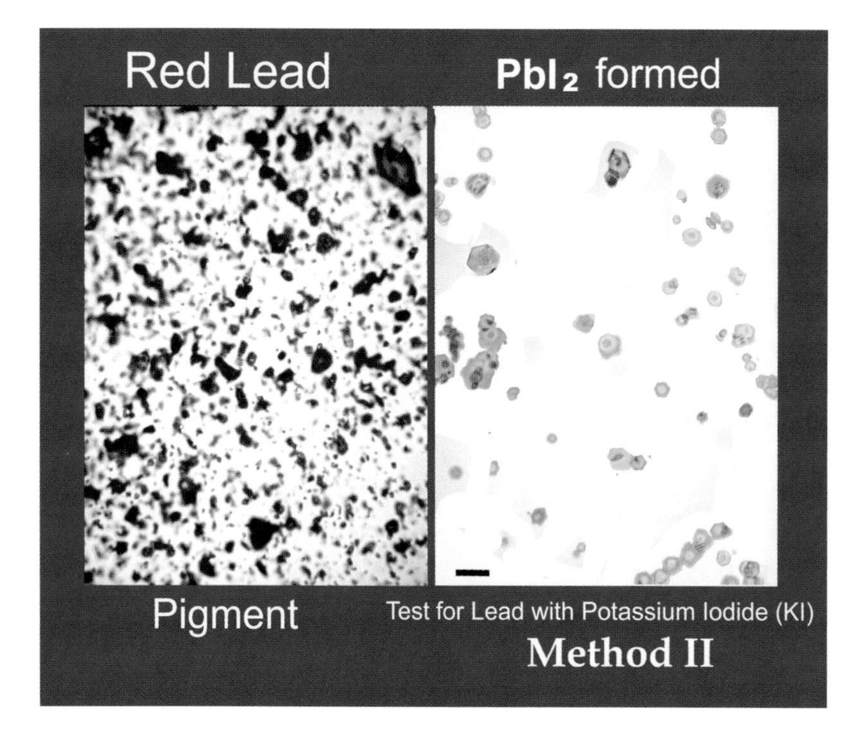

FIGURE D.50

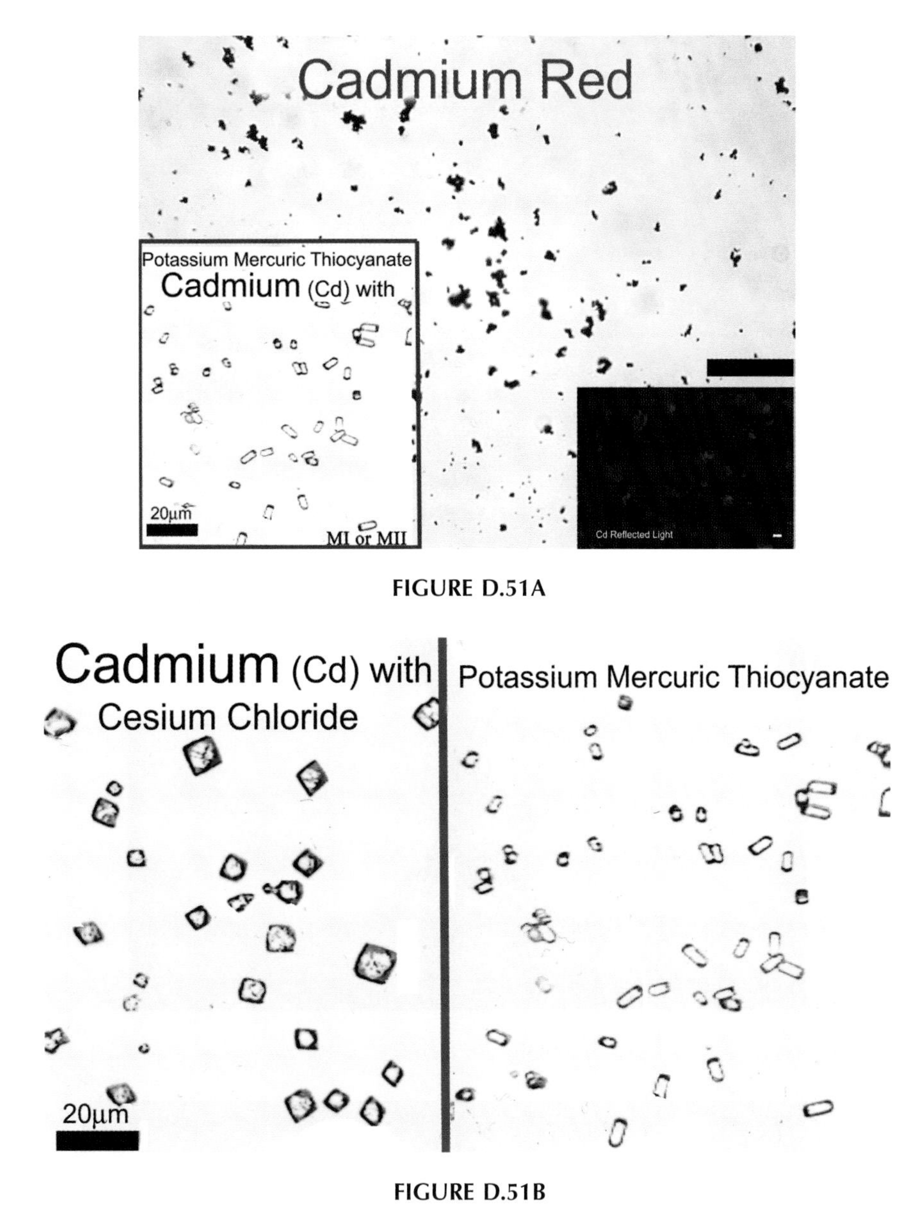

FIGURE D.51A

FIGURE D.51B

FIGURE D.52

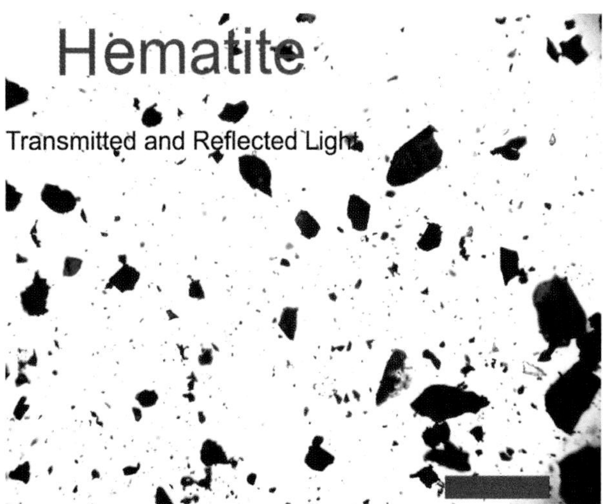

FIGURE D.53

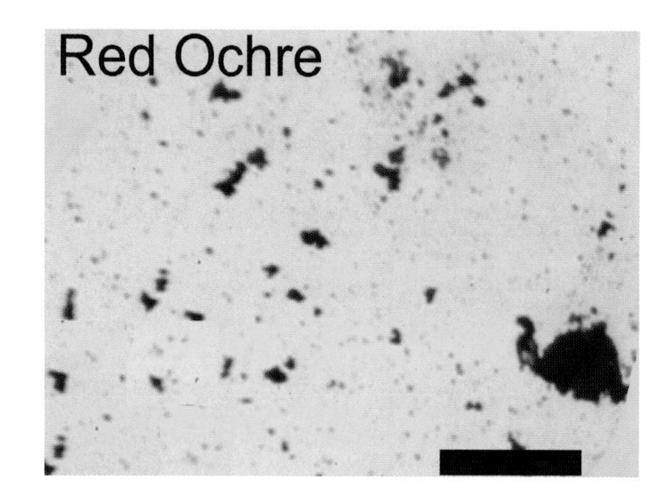

FIGURE D.54A

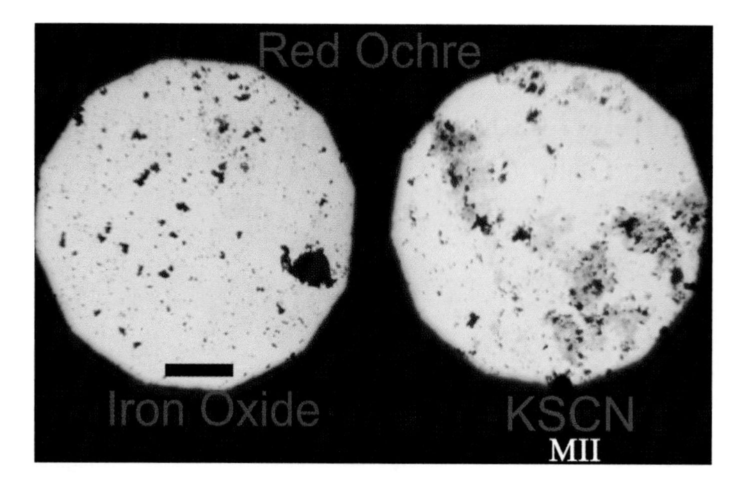

FIGURE D.54B

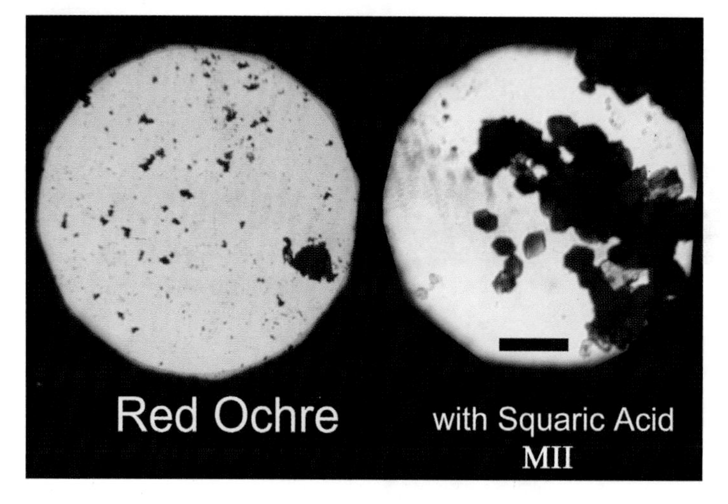

FIGURE D.54C

FIGURE D.55A

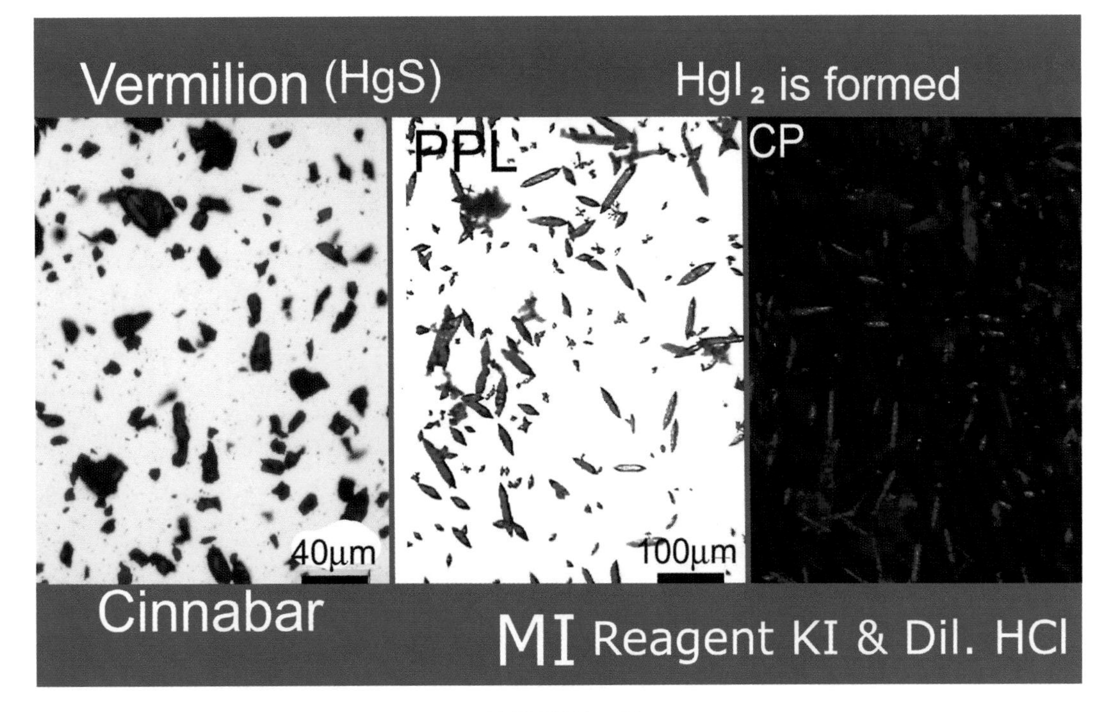

FIGURE D.55B

PIGMENT: YELLOWS

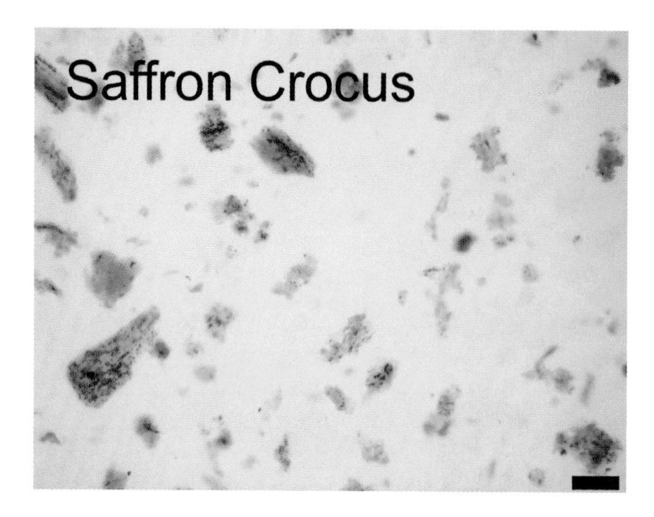

FIGURE D.56

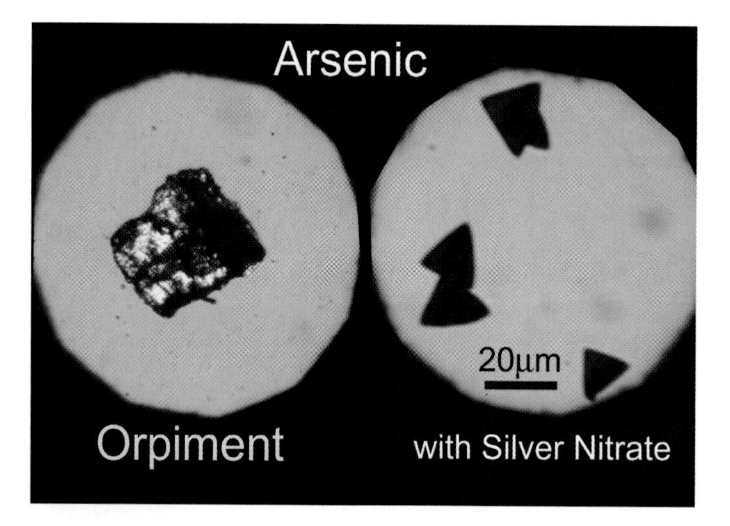

FIGURE D.57

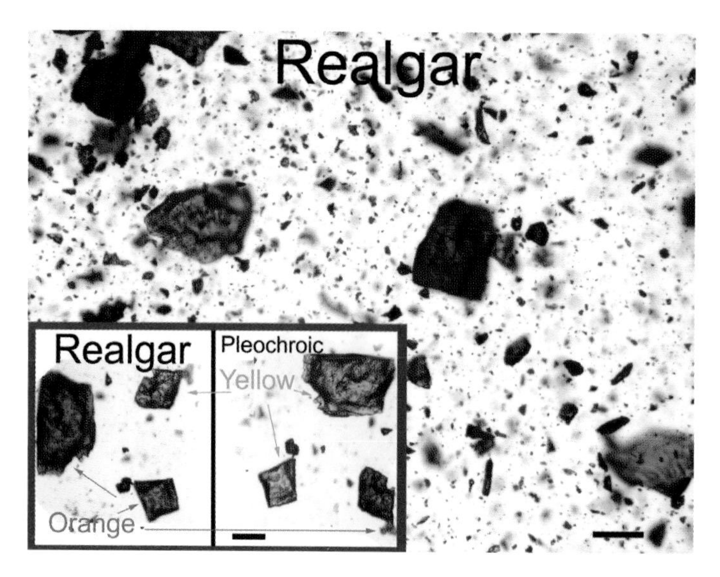

FIGURE D.58

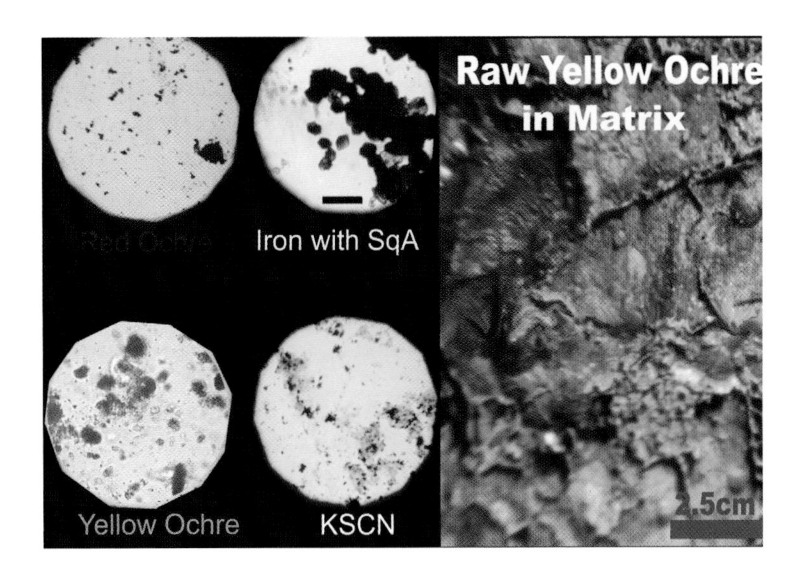

FIGURE D.59

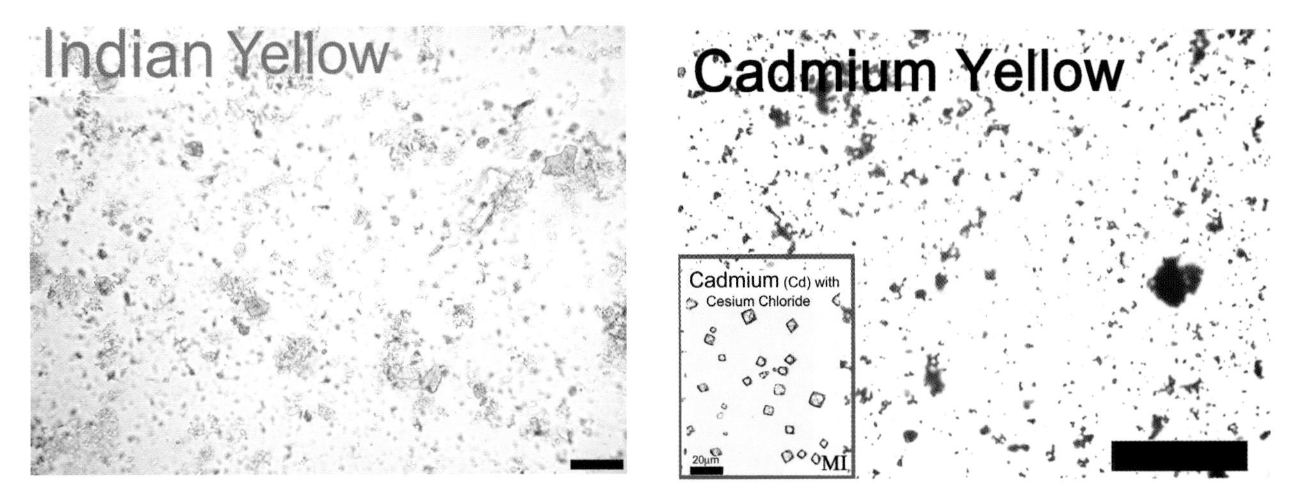

FIGURE D.60

FIGURE D.61

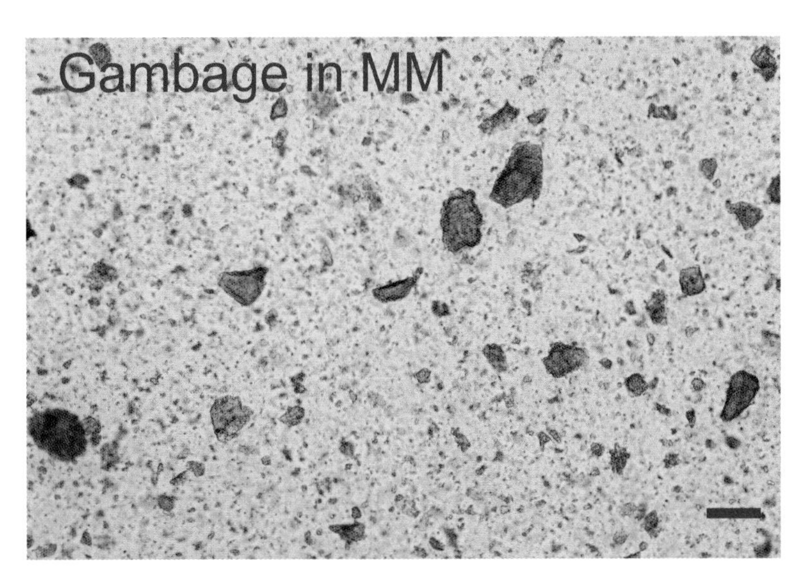

FIGURE D.62

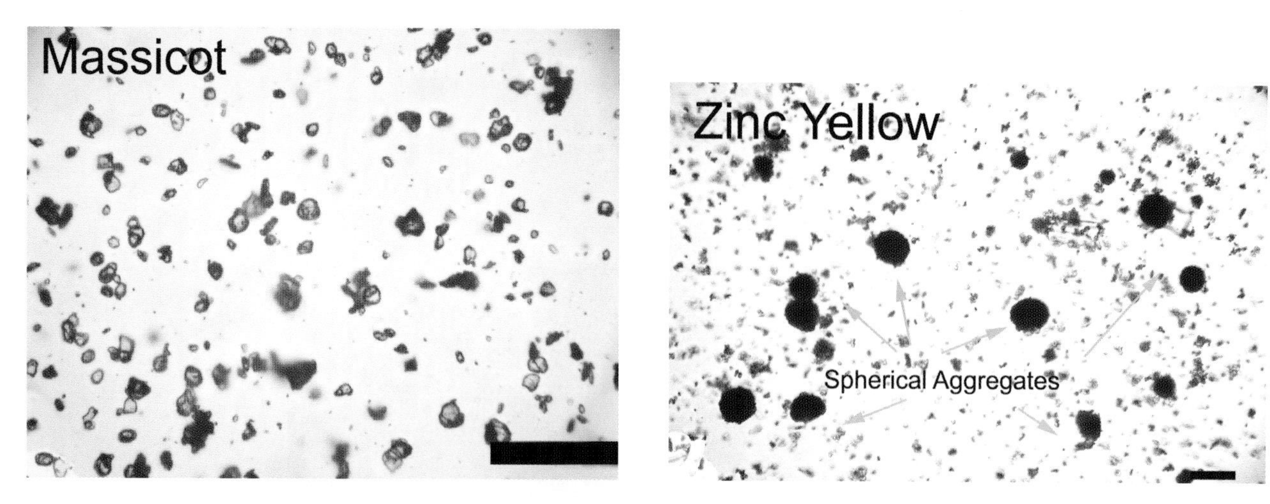

FIGURE D.63A

FIGURE D.64

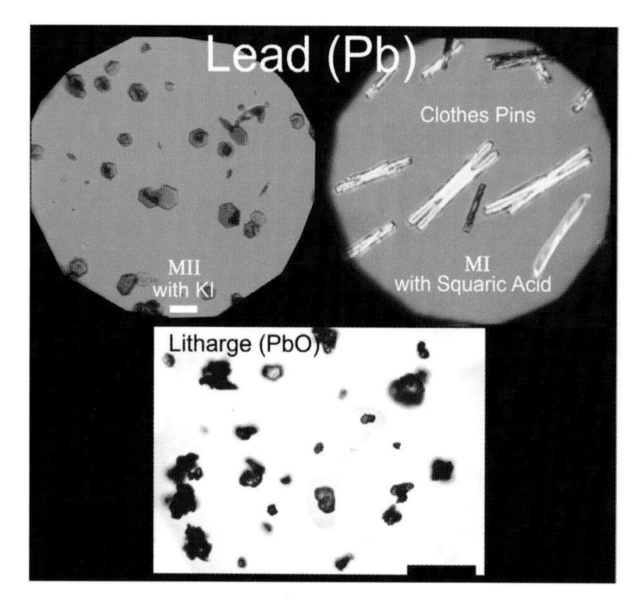

FIGURE D.63B

FIGURE D.65

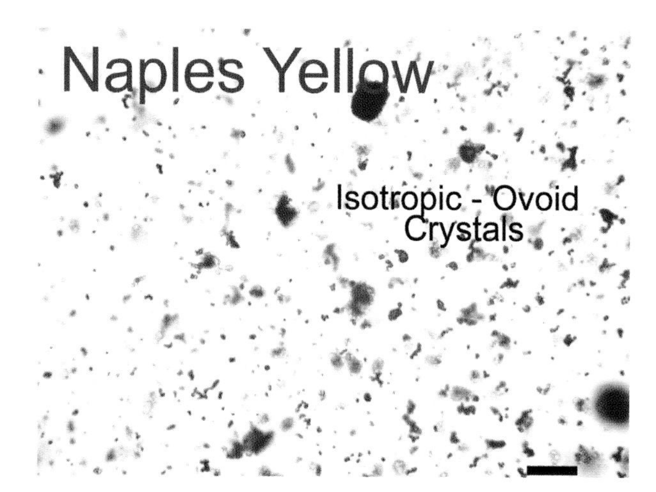

FIGURE D.66

METALS

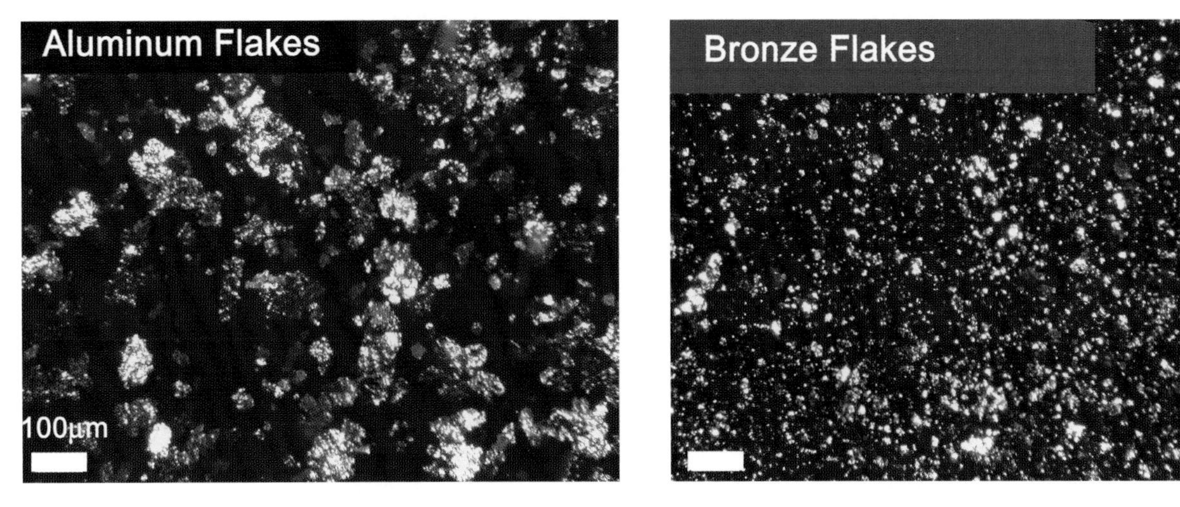

FIGURE D.67

FIGURE D.68

FIGURE D.69B

FIGURE D.69A

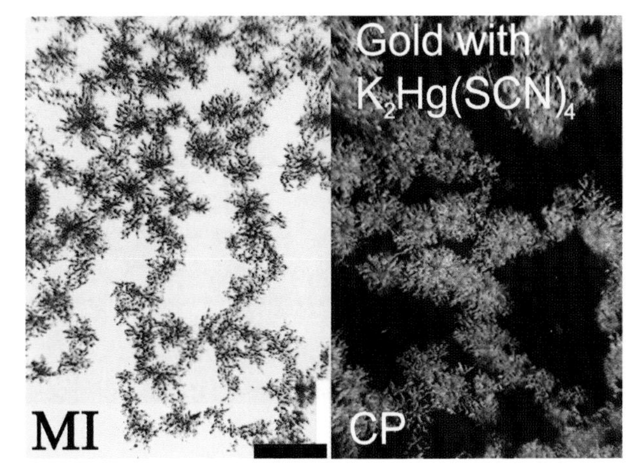

FIGURE D.69C

FIGURE D.70

FIGURE D.71A

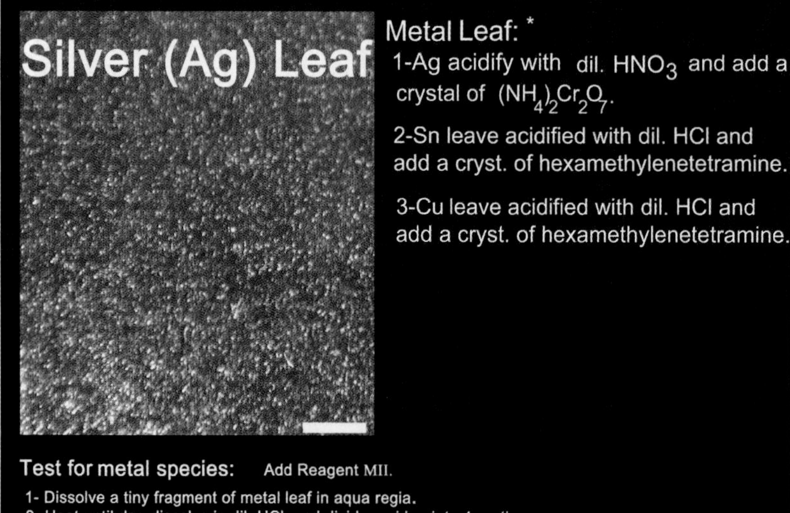

FIGURE D.71B

FIGURE D.72A

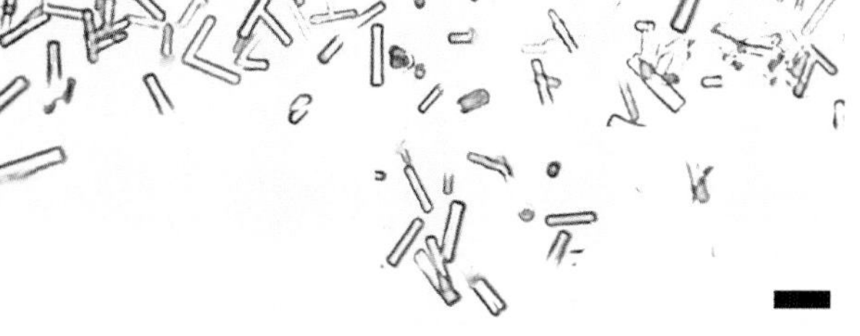

FIGURE D.72B

Index

Numbers

A

B

C

D

E

F

G

H

I

J

K

L

M

N

O

P

Q

R

T

U

V

W

X

Y

Z